CAMBRIDGE TRACTS IN MATHEMATICS

123 Ends of Complexes

Bruce Hughes
Vanderbilt University

Andrew Ranicki
University of Edinburgh

Ends of Complexes

CAMBRIDGE UNIVERSITY PRESS
Cambridge, New York, Melbourne, Madrid, Cape Town, Singapore, São Paulo

Cambridge University Press
The Edinburgh Building, Cambridge CB2 8RU, UK

Published in the United States of America by Cambridge University Press, New York

www.cambridge.org
Information on this title: www.cambridge.org/9780521576253

First published 1996
This digitally printed version 2008

A catalogue record for this publication is available from the British Library

Library of Congress Cataloguing in Publication data

Hughes, Bruce.
End of complexes / Bruce Hughes, Andrew Ranicki.
p. cm. – (Cambridge tracts in mathematics ; 123)
Includes bibliographical references (p. –) and index.
ISBN 0 521 57625 3 (hc)
1. Complexes. I. Ranicki, Andrew, 1948– II. Title.
III. Series.
QA608.H84 1996
514′.223 – dc20 96-4095 CIP

ISBN 978-0-521-57625-3 hardback
ISBN 978-0-521-05519-2 paperback

For Becky and Ida

Contents

Introduction

We take 'complex' to mean both a CW (or simplicial) complex in topology and a chain complex in algebra. An 'end' of a complex is a subcomplex with a particular type of infinite behaviour, involving non-compactness in topology and infinite generation in algebra. The ends of manifolds are of greatest interest; we regard the ends of CW and chain complexes as tools in the investigation of manifolds and related spaces, such as stratified sets. The interplay of the topological properties of the ends of manifolds, the homotopy theoretic properties of the ends of CW complexes and the algebraic properties of the ends of chain complexes has been an important theme in the classification theory of high dimensional manifolds for over 35 years. However, the gaps in the literature mean that there are still some loose ends to wrap up! Our aim in this book is to present a systematic exposition of the various types of ends relevant to manifold classification, closing the gaps as well as obtaining new results. The book is intended to serve both as an account of the existing applications of ends to the topology of high dimensional manifolds and as a foundation for future developments.

We assume familiarity with the basic language of high dimensional manifold theory, and the standard applications of algebraic K- and L-theory to manifolds, but otherwise we have tried to be as self contained as possible.

The algebraic topology of finite CW complexes suffices for the combinatorial topology of compact manifolds. However, in order to understand the difference between the topological and combinatorial properties it is necessary to deal with infinite CW complexes and non-compact manifolds. The classic cases include the Hauptvermutung counterexamples of Milnor [96], the topological invariance of the rational Pontrjagin classes proved by Novikov [103], the topological manifold structure theory of Kirby and Siebenmann [84], and the topological invariance of Whitehead torsion proved by Chapman [22]. The algebraic and geometric topology of non-compact manifolds has been a prominent feature in much of the recent work on the Novikov conjectures – see Ferry, Ranicki and Rosenberg [59] for a survey. (In these applications the non-compact manifolds arise as the universal covers of aspherical compact manifolds, e.g. the Euclidean space $\mathbb{R}^i$ covering the torus $T^i = S^1 \times S^1 \times \ldots \times S^1 = B\mathbb{Z}^i$.) In fact, many current developments in topology, operator theory, differential geometry, hyperbolic geometry, and group theory are concerned with the asymptotic properties of non-compact manifolds and infinite groups – see Gromov [65], Connes [33] and Roe [135] for example.

What is an end of a topological space? Roughly speaking, an end of a non-compact space W is a component of $W \backslash K$ for arbitrarily large compact subspaces $K \subseteq W$. More precisely:

Definition 1. (i) A *neighbourhood of an end* in a non-compact space W is a subspace $U \subset W$ which contains a component of $W \backslash K$ for a non-empty compact subspace $K \subset W$.

(ii) An *end* ϵ of W is an equivalence class of sequences of connected open neighbourhoods $W \supset U_1 \supset U_2 \supset \ldots$ such that

$$\bigcap_{i=1}^{\infty} \mathrm{cl}\,(U_i) = \emptyset$$

subject to the equivalence relation

$$(W \supset U_1 \supset U_2 \supset \ldots) \sim (W \supset V_1 \supset V_2 \supset \ldots)$$

if for each U_i there exists j with $U_i \subseteq V_j$, and for each V_j there exists i with $V_j \subseteq U_i$.

(iii) The *fundamental group* of an end ϵ is the inverse limit

$$\pi_1(\epsilon) = \varprojlim_i \pi_1(U_i) . \qquad \square$$

The theory of ends was initiated by Freudenthal [61] in connection with topological groups. The early applications of the theory concerned the ends of open 3-dimensional manifolds, and the ends of discrete groups (which are the ends of the universal covers of their classifying spaces).

We are especially interested in the ends of manifolds which are 'tame', and in extending the notion of tameness to other types of ends. An end of a manifold is tame if it has a system of neighbourhoods satisfying certain strong restrictions on the fundamental group and chain homotopy type. Any non-compact space W can be compactified by adding a point at infinity, $W^\infty = W \cup \{\infty\}$. A manifold end is 'collared' if it can be compactified by a manifold, i.e. if the point at infinity can be replaced by a closed manifold boundary, allowing the end to be identified with the interior of a compact manifold with boundary. A high dimensional tame manifold end can be collared if and only if an algebraic K-theory obstruction vanishes. The theory of tame ends has found wide application in the surgery classification theory of high dimensional compact manifolds and stratified spaces, and in the related controlled topology and algebraic K- and L-theory.

Example 2. Let K be a connected compact space.

(i) $K \times [0,\infty)$ has one end ϵ, with connected open neighbourhoods

$$U_i \;=\; K \times (i,\infty) \subset K \times [0,\infty)\ ,$$

such that $\pi_1(\epsilon) = \pi_1(K)$.

(ii) $K \times \mathbb{R}$ has two ends ϵ^+, ϵ^-, with connected open neighbourhoods

$$U_i^+ \;=\; K \times (i,\infty)\ \ ,\ \ U_i^- \;=\; K \times (-\infty,-i) \subset K \times \mathbb{R}\ ,$$

such that $\pi_1(\epsilon^\pm) = \pi_1(K)$.

(iii) $K \times \mathbb{R}^2$ has one end ϵ, with connected open neighbourhoods

$$U_i \;=\; K \times \{(x,y) \in \mathbb{R}^2 \,|\, x^2 + y^2 > i^2\}\ ,$$

such that $\pi_1(\epsilon) = \pi_1(K) \times \mathbb{Z}$. □

Example 3. (i) Let W be a space with a proper map $d : W \longrightarrow [0,\infty)$ which is onto, and such that the inverse images $U_t = d^{-1}(t,\infty) \subseteq W$ $(t \geq 1)$ are connected. Then W has one end ϵ with connected open neighbourhoods $W \supset U_1 \supset U_2 \supset \dots$ such that $\mathrm{cl}(U_t) = d^{-1}[t,\infty)$, $\bigcap\limits_{i=0}^{\infty} \mathrm{cl}(U_i) = \emptyset$.

(ii) Let $(W, \partial W)$ be a connected open n-dimensional manifold with connected compact boundary. Then W has one end ϵ if and only if there exists a proper map $d : (W, \partial W) \longrightarrow ([0,\infty),\{0\})$ which is transverse regular at $\mathbb{N} = \{0,1,2,\dots\} \subset [0,\infty)$, with the inverse images

$$(W_i; M_i, M_{i+1}) \;=\; d^{-1}([i,i+1];\{i\},\{i+1\})\ \ (i \in \mathbb{N})$$

connected compact n-dimensional cobordisms such that

$$(W, \partial W) \;=\; (\bigcup_{i=0}^{\infty} W_i, M_0)\ .$$

(iii) Given connected compact n-dimensional cobordisms $(W_i; M_i, M_{i+1})$ $(i \in \mathbb{N})$ there is defined a connected open n-dimensional manifold with compact boundary $(W, \partial W) = (\bigcup\limits_{i=0}^{\infty} W_i, M_0)$. The union of Morse functions $d_i : (W_i; M_i, M_{i+1}) \longrightarrow ([i,i+1];\{i\},\{i+1\})$ defines a proper map $d : (W, \partial W) \longrightarrow ([0,\infty),\{0\})$, and as in (ii) W has one end ϵ. If the inclusions $M_i \longrightarrow W_i$, $M_{i+1} \longrightarrow W_i$ induce isomorphisms in π_1 then

$$\pi_1(M_0) \;=\; \pi_1(W_0) \;=\; \pi_1(M_1) \;=\; \dots \;=\; \pi_1(W) \;=\; \pi_1(\epsilon)\ .$$ □

Definition 4. An end ϵ of an open n-dimensional manifold W can be *collared* if it has a neighbourhood of the type $M\times[0,\infty)\subset W$ for a connected closed $(n-1)$-dimensional manifold M. □

Example 5. (i) An open n-dimensional manifold with one end ϵ is (homeomorphic to) the interior of a closed n-dimensional manifold if and only if ϵ can be collared. More generally, if W is an open n-dimensional manifold with compact boundary ∂W and one end ϵ, then there exists a compact n-dimensional cobordism $(L;\partial W,M)$ with $L\backslash M$ homeomorphic to W rel ∂W if and only if ϵ can be collared.

(ii) If $(V,\partial V)$ is a compact n-dimensional manifold with boundary then for any $x\in V\backslash\partial V$ the complement $W=V\backslash\{x\}$ is an open n-dimensional manifold with a collared end ϵ and $\partial W=\partial V$, with a neighbourhood $M\times[0,\infty)\subset W$ for $M=S^{n-1}$. The one-point compactification of W is $W^\infty=V$. The compactification of W provided by (i) is $L=\mathrm{cl}(V\backslash D^n)$, for any neighbourhood $D^n\subset V\backslash\partial V$ of x, with $(L;\partial W,M)=(W\cup S^{n-1};\partial V,S^{n-1})$. □

Stallings [154] used engulfing to prove that if W is a contractible open n-dimensional PL manifold with one end ϵ such that $\pi_1(\epsilon)=\{1\}$ and $n\geq 5$ then W is PL homeomorphic to $\mathbb{R}^n$ – in particular, the end ϵ can be collared.

Let $(W,\partial W)$ be an open n-dimensional manifold with compact boundary and one end ϵ. Making a proper map $d:(W,\partial W)\longrightarrow([0,\infty),\{0\})$ transverse regular at some $t\in(0,\infty)$ gives a decomposition of $(W,\partial W)$ as

$$(W,\partial W) \;=\; (L;\partial W,M)\cup_M (N,M)$$

with $(L;\partial W,M)=d^{-1}([0,t];\{0\},\{t\})$ a compact n-dimensional cobordism and $N=d^{-1}[t,\infty)$ non-compact. The end ϵ can be collared if and only if N can be chosen such that there exists a homeomorphism $N\cong M\times[0,\infty)$ rel $M=M\times\{0\}$, in which case $L\backslash M\cong L\cup_{M\times\{0\}}M\times[0,\infty)\cong W$ rel ∂W. In terms of Morse theory: it is possible to collar ϵ if and only if $(W,\partial W)$ admits a proper Morse function d with only a finite number of critical points. Browder, Levine and Livesay [14] used codimension 1 surgery on $M\subset W$ to show that if $\pi_1(W)=\pi_1(\epsilon)=\{1\}$ and $n\geq 6$ then ϵ can be collared if and only if the homology groups $H_*(W)$ are finitely generated (with $H_r(W)=0$ for all but finitely many values of r). Siebenmann [140] combined codimension 1 surgery with the finiteness obstruction theory of Wall [163] for finitely dominated spaces, proving that in dimensions ≥ 6 a tame manifold end can be collared if and only if an algebraic K-theory obstruction vanishes.

Definition 6. A space X is *finitely dominated* if there exist a finite CW complex K and maps $f:X\longrightarrow K$, $g:K\longrightarrow X$ with $gf\simeq 1:X\longrightarrow X$. □

Example 7. Any space homotopy equivalent to a finite CW complex is finitely dominated. $\square$

Example 8. A connected CW complex X with $\pi_1(X) = \{1\}$ is finitely dominated if and only if $H_*(X)$ is finitely generated, if and only if X is homotopy equivalent to a finite CW complex. $\square$

For non-simply-connected X the situation is more complicated:

Theorem 9. (Wall [163, 164]) *A connected CW complex X is finitely dominated if and only if $\pi_1(X)$ is finitely presented and the cellular $\mathbb{Z}[\pi_1(X)]$-module chain complex $C(\widetilde{X})$ of the universal cover $\widetilde{X}$ is chain equivalent to a finite f.g. projective $\mathbb{Z}[\pi_1(X)]$-module chain complex P. The reduced projective class of a finitely dominated X*

$$[X] \;=\; [P] \;=\; \sum_{r=0}^{\infty}(-)^r[P_r] \in \widetilde{K}_0(\mathbb{Z}[\pi_1(X)])$$

is the **finiteness obstruction** *of X, such that $[X] = 0$ if and only if X is homotopy equivalent to a finite CW complex.* $\square$

Definition 10. An end ϵ of an open manifold W is *tame* if it admits a sequence $W \supset U_1 \supset U_2 \supset \dots$ of finitely dominated neighbourhoods with

$$\bigcap_{i=1}^{\infty} \mathrm{cl}(U_i) \;=\; \emptyset \;,\quad \pi_1(U_1) \;=\; \pi_1(U_2) \;=\; \dots \;=\; \pi_1(\epsilon)\;. \qquad \square$$

Example 11. If an end ϵ of an open manifold W can be collared then it is tame: if $M \times [0,\infty) \subset W$ is a neighbourhood of ϵ then the open neighbourhoods $W \supset U_1 = M \times (1,\infty) \supset U_2 = M \times (2,\infty) \supset \dots$ satisfy the conditions of Definition 10, with $\mathrm{cl}(U_i) = M \times [i,\infty)$, $\pi_1(\epsilon) = \pi_1(M)$. $\square$

Tameness is a geometric condition which ensures stable (as opposed to wild) behaviour in the topology at infinity of a non-compact space W. The fundamental example is $W = K \times [0,\infty)$ for a compact space K, in which the topology at infinity is that of K.

Theorem 12. (Siebenmann [140]) *A tame end ϵ of an open n-dimensional manifold W has a reduced projective class invariant, the* **end obstruction**

$$[\epsilon] \;=\; \varprojlim_i\, [U_i] \in \widetilde{K}_0(\mathbb{Z}[\pi_1(\epsilon)]) \;=\; \varprojlim_i\, \widetilde{K}_0(\mathbb{Z}[\pi_1(U_i)])$$

such that $[\epsilon] = 0$ if (and for $n \geq 6$ only if) ϵ can be collared. $\square$

Even if a tame manifold end ϵ can be collared, the collarings need not be unique. The various collarings of a tame end ϵ in an open manifold W of dimension ≥ 6 with $[\epsilon] = 0 \in \widetilde{K}_0(\mathbb{Z}[\pi_1(\epsilon)])$ are classified by the Whitehead group $Wh(\pi_1(\epsilon))$: if $M \times [0,\infty)$, $M' \times [0,\infty) \subset W$ are two collar neighbourhoods of ϵ then for sufficiently large $t \geq 0$ there exists an h-cobordism $(N; M, M')$ between $M \times \{0\}$ and $M' \times \{t\} \subset W$, with

$$M \times [0,\infty) \ = \ N \cup_{M' \times \{t\}} M' \times [t,\infty) \subset W \ .$$

By the s-cobordism theorem $(N; M, M')$ is homeomorphic to the product $M \times (I; \{0\}, \{1\})$ if and only if $\tau(M \simeq N) = 0 \in Wh(\pi_1(\epsilon))$. The non-uniqueness of collarings of PL manifold ends was used by Milnor [96] in the construction of homeomorphisms of compact polyhedra which are not homotopic to a PL homeomorphism, disproving the Hauptvermutung for compact polyhedra. The end obstruction theory played an important role in the disproof of the manifold Hauptvermutung by Casson and Sullivan (Ranicki [131]) – the manifold case also requires surgery and L-theory.

Quinn [114, 115, 116] developed a controlled version of the Siebenmann end obstruction theory, and applied it to stratified spaces. (See Ranicki and Yamasaki [132] for a treatment of the controlled finiteness obstruction, and Connolly and Vajiac [34] for an end theorem for stratified spaces.) The tameness condition of Definition 10 for manifold ends was extended by Quinn to stratified spaces, distinguishing *two* tameness conditions for ends of non-compact spaces, involving maps pushing *forward* along the end and in the *reverse* direction. We shall only consider the two-stratum case of a one-point compactification, with the lower stratum the point at infinity. In Chapters 7, 8 we state the definitions of forward and reverse tameness. The original tameness condition of Siebenmann [140] appears in Chapter 8 as *reverse π_1-tameness*, so called since it is a combination of reverse tameness and π_1-stability. In general, forward and reverse tameness are independent of each other, but for π_1-stable manifold ends ϵ with finitely presented $\pi_1(\epsilon)$ the two kinds of tameness are equivalent by a kind of Poincaré duality.

Definition 13. (Quinn [116]) The *end space* $e(W)$ of a space W is the space of proper paths $\omega : [0,\infty) \longrightarrow W$. □

We refer to Appendix B for a brief history of end spaces.

The end space $e(W)$ is a homotopy model for the 'space at infinity' of W, playing a role similar to the ideal boundary in hyperbolic geometry. The topology at infinity of a space W is the inverse system of complements of compact subspaces (i.e. cocompact subspaces or neighbourhoods of infinity) of W, which are the open neighbourhoods of the point ∞ in the one-point compactification $W^\infty = W \cup \{\infty\}$. The homology at infinity $H^\infty_*(W)$ is defined to fit into an exact sequence

$$\dots \longrightarrow H_r^{\infty}(W) \longrightarrow H_r(W) \longrightarrow H_r^{lf}(W) \longrightarrow H_{r-1}^{\infty}(W) \longrightarrow \dots ,$$

and $H_*^{lf}(W) = H_*(W^{\infty}, \{\infty\})$ for reasonable W. The end space $e(W)$ is the 'link of infinity in W^{∞}'. There is a natural passage from the algebraic topology at infinity of W to the algebraic topology of $e(W)$, which is a one-to-one correspondence for forward tame W, with $H_*(e(W)) = H_*^{\infty}(W)$.

If $(W, \partial W)$ is an open n-dimensional manifold with compact boundary and one tame end ϵ the end space $e(W)$ is a finitely dominated $(n-1)$-dimensional Poincaré space with $\pi_1(e(W)) = \pi_1(\epsilon)$, and $(W; \partial W, e(W))$ is a finitely dominated n-dimensional Poincaré cobordism, regarding $e(W)$ as a subspace of W via the evaluation map

$$e(W) \longrightarrow W \ ; \ (\omega : [0, \infty) \longrightarrow W) \longrightarrow \omega(0) .$$

The non-compact spaces of greatest interest to us are the infinite cyclic covers of 'bands':

Definition 14. A *band* (M, c) is a compact space M with a map $c : M \longrightarrow S^1$ such that the infinite cyclic cover $\overline{M} = c^*\mathbb{R}$ of M is finitely dominated, and such that the projection $\overline{M} \longrightarrow M$ induces a bijection of path components $\pi_0(\overline{M}) \cong \pi_0(M)$. □

Example 15. A connected finite CW complex M with a map $c : M \longrightarrow S^1$ inducing an isomorphism $c_* : \pi_1(M) \cong \mathbb{Z}$ defines a band (M, c) (i.e. the infinite cyclic cover $\overline{M} = c^*\mathbb{R}$ is finitely dominated) if and only if the homotopy groups $\pi_*(M) = H_*(\overline{M})$ $(* \geq 2)$ are finitely generated. □

The infinite cyclic cover $\overline{M}$ of a connected manifold band (M, c) has two ends. The projection $c : M \longrightarrow S^1$ lifts to a proper map $\overline{c} : \overline{M} \longrightarrow \mathbb{R}$, such that the inverse images

$$\overline{M}^+ = \overline{c}^{-1}[0, \infty) \ , \ \overline{M}^- = \overline{c}^{-1}(-\infty, 0] \subset \overline{M}$$

are closed neighbourhoods of the two ends. In Chapter 15 we shall prove that the two ends of $\overline{M}$ are tame, with homotopy equivalences

$$e(\overline{M}^+) \simeq e(\overline{M}^-) \simeq \overline{M} .$$

The problem of deciding if an open manifold is the interior of a compact manifold with boundary is closely related to the problem of deciding if a compact manifold M fibres over S^1, i.e. if a map $c : M \longrightarrow S^1$ is homotopic to the projection of a fibre bundle. In the first instance, it is necessary for (M, c) to be a band:

Example 16. Suppose given a fibre bundle $F \longrightarrow M \stackrel{c}{\longrightarrow} S^1$ with F a closed $(n-1)$-dimensional manifold and $M = T(h)$ the mapping torus of a monodromy self homeomorphism $h : F \longrightarrow F$. If h preserves the path components then (M, c) is an n-dimensional manifold band, with the infinite cyclic cover $\overline{M} = F \times \mathbb{R}$ homotopy equivalent to a finite CW complex. □

Stallings [153] used codimension 1 surgery on a surface $c^{-1}(*) \subset M$ to prove that a map $c : M \longrightarrow S^1$ from a compact irreducible 3-dimensional manifold M with $\ker(c_*: \pi_1(M) \longrightarrow \mathbb{Z}) \not\cong \mathbb{Z}_2$ is homotopic to the projection of a fibre bundle if and only if $\ker(c_*)$ is finitely generated, in which case $\ker(c_*) = \pi_1(F)$ is the fundamental group of the fibre F. In particular, the complement of a knot $k : S^1 \subset S^3$

$$(M, \partial M) \;=\; (\mathrm{cl}(S^3 \backslash (k(S^1) \times D^2)), S^1 \times S^1)$$

fibres over S^1 if and only if the commutator subgroup $[\pi, \pi]$ of the fundamental group $\pi = \pi_1(M)$ is finitely generated. Browder and Levine [13] used codimension 1 surgery in higher dimensions to prove that for $n \geq 6$ a compact n-dimensional manifold band (M, c) with $c_* : \pi_1(M) \cong \mathbb{Z}$ fibres. Thus a high-dimensional knot $k : S^{n-2} \subset S^n$ $(n \geq 6)$ with $\pi_1(S^n \backslash k(S^{n-2})) = \mathbb{Z}$ fibres (i.e. the knot complement fibres over S^1) if and only if the higher homotopy groups $\pi_*(S^n \backslash k(S^{n-2}))$ $(* \geq 2)$ are finitely generated. More generally:

Theorem 17. (Farrell [46], Siebenmann [145]) *An n-dimensional manifold band (M, c) has a Whitehead torsion invariant, the* **fibring obstruction**

$$\Phi(M, c) \in Wh(\pi_1(M)) \;,$$

such that $\Phi(M, c) = 0$ if (and for $n \geq 6$ only if) M fibres over S^1, with $c : M \longrightarrow S^1$ homotopic to a fibre bundle projection. □

In the main text we shall actually be dealing with the two fibring obstructions $\Phi^+(M, c)$, $\Phi^-(M, c) \in Wh(\pi_1(M))$ defined for a CW band (M, c). For an n-dimensional manifold band (M, c) the two obstructions determine each other by Poincaré duality

$$\Phi^+(M, c) \;=\; (-)^{n-1}\Phi^-(M, c)^* \in Wh(\pi_1(M)) \;,$$

and in the Introduction we write $\Phi^+(M, c)$ as $\Phi(M, c)$.

Example 18. For any n-dimensional manifold band (M, c) the $(n+1)$-dimensional manifold band $(M \times S^1, d)$ with $d(x, t) = c(x)$ has fibring obstruction

$$\Phi(M \times S^1, d) \;=\; 0 \in Wh(\pi_1(M) \times \mathbb{Z}) \,.$$

For $n \geq 5$ the geometric construction of Theorem 19 below actually gives a canonical fibre bundle

$$F \longrightarrow M \times S^1 \xrightarrow{p} S^1$$

with p homotopic to d. The fibre F is the 'wrapping up' of the tame end $\overline{M}^+$ of $\overline{M}$, a closed n-dimensional manifold such that there are defined homeomorphisms

$$F \times \mathbb{R} \cong \overline{M} \times S^1 \ , \ M \times S^1 \cong T(h)$$

for a monodromy self homeomorphism $h : F \longrightarrow F$. The fibring obstruction $\Phi(M,c) \in Wh(\pi_1(M))$ is the obstruction to splitting off an S^1-factor from $h : F \longrightarrow F$, so that for $n \geq 6$ $\Phi(M,c) = 0$ if and only if up to isotopy

$$h = h_1 \times 1 : F = F_1 \times S^1 \longrightarrow F = F_1 \times S^1$$

with $h_1 : F_1 \longrightarrow F_1$ a self homeomorphism such that $M \cong T(h_1)$. □

Bands are of interest in their own right. For example, the fibring obstruction theory for bands gives a geometric interpretation of the 'fundamental theorem' of algebraic K-theory of Bass [4]

$$Wh(\pi \times \mathbb{Z}) = Wh(\pi) \oplus \widetilde{K}_0(\mathbb{Z}[\pi]) \oplus \widetilde{\mathrm{Nil}}_0(\mathbb{Z}[\pi]) \oplus \widetilde{\mathrm{Nil}}_0(\mathbb{Z}[\pi])$$

– see Ranicki [124] for a recent account. The following uniformization theorem shows that every tame manifold end of dimension ≥ 6 has an open neighbourhood which is the infinite cyclic cover of a manifold band. It was announced by Siebenmann [141], and is proved here in Chapter 17.

Theorem 19. *Let $(W, \partial W)$ be a connected open n-dimensional manifold with compact boundary and one end ϵ, with $n \geq 6$.*

(i) *The end ϵ is tame if and only if it has a neighbourhood $X = \overline{M} \subset W$ which is the finitely dominated infinite cyclic cover of a compact n-dimensional manifold band $\widehat{X} = (M,c)$, the* **wrapping up** *of ϵ, such that*

$$\pi_1(\overline{M}) = \pi_1(\epsilon) \ , \ \pi_1(M) = \pi_1(\epsilon) \times \mathbb{Z} \ , \ e(W) \simeq \overline{M} \ ,$$

$$\Phi(M,c) = [\epsilon] \in \widetilde{K}_0(\mathbb{Z}[\pi_1(\epsilon)]) \subseteq Wh(\pi_1(\epsilon) \times \mathbb{Z}) \ ,$$

and such that the covering translation $\zeta : \overline{M} \longrightarrow \overline{M}$ is isotopic to the identity. The $(n+1)$-dimensional manifold band $(M \times S^1, d)$ with $d(x,t) = c(x)$ fibres over S^1 : the map $d : M \times S^1 \longrightarrow S^1$ is homotopic to the projection of a fibre bundle with fibre M, with a homeomorphism

$$\overline{M} \times S^1 \cong M \times \mathbb{R} \ .$$

Thus $\epsilon \times S^1$ can be collared with boundary M : there exists a compact $(n+1)$-dimensional cobordism $(N; \partial W \times S^1, M)$ with a rel ∂ *homeomorphism*

$$(N \backslash M, \partial W \times S^1) \cong (W, \partial W) \times S^1 \ .$$

(ii) *For tame ϵ the Siebenmann end obstruction of ϵ is the Wall finiteness obstruction of* $\overline{M}^+$

$$[\epsilon] = [\overline{M}^+] \in \widetilde{K}_0(\mathbb{Z}[\pi_1(\epsilon)]) ,$$

with $[\epsilon] = 0$ *if and only if* ϵ *can be collared, in which case there exists a compact* n*-dimensional cobordism* $(K;\partial W, L)$ *with a* rel ∂ *homeomorphism*

$$(K\backslash L, \partial W) \cong (W, \partial W)$$

and a homeomorphism

$$(K;\partial W, L) \times S^1 \cong (N;\partial W \times S^1, M)$$

(N as in (i)), and (M,c) fibres over S^1 with $M \cong L \times S^1$ and $\overline{M} \cong L \times \mathbb{R}$. □

A CW complex X is finitely dominated if and only if $X \times S^1$ is homotopy equivalent to a finite CW complex, by a result of M. Mather [91]. A manifold end ϵ of dimension ≥ 6 is tame if and only if $\epsilon \times S^1$ can be collared – this was already proved by Siebenmann [140], but the wrapping up procedure of Theorem 19 actually gives a canonical collaring of $\epsilon \times S^1$.

In principle, Theorem 19 could be proved using the canonical regular neighbourhood theory of Siebenmann [148] and Siebenmann, Guillou and Hähl [149]. We prefer to give a more elementary approach, using a combination of the geometric, homotopy theoretic and algebraic methods which have been developed in the last 25 years to deal with non-compact spaces. While the wrapping up construction has been a part of the folklore, the new aspect of our approach is that we rely on the end space and the extensively developed theory of *manifold approximate fibrations* rather than ad hoc engulfing methods. An approximate fibration is a map with an approximate lifting property. (Of course, manifold approximate fibration theory relies on engulfing, but we prefer to subsume the details of the engulfing in the theory.) We do not assume previous acquaintance with approximate fibrations and engulfing.

The proof of Theorem 19 occupies most of Parts One and Two (Chapters 1–20). There are three main steps in passing from a tame end ϵ of W to the wrapping up band (M,c) such that the infinite cyclic cover $\overline{M} \subseteq W$ is a neighbourhood of ϵ:

(i) in Chapter 9 we show that tameness conditions on a space W imply that the end space $e(W)$ is finitely dominated and that, near infinity, W looks like the product $e(W) \times [0,\infty)$;
(ii) in Chapter 16 we use (i) to prove that every tame manifold end ϵ of

dimension ≥ 5 has a neighbourhood X which is the total space of a manifold approximate fibration $d : X \longrightarrow \mathbb{R}$;

(iii) in Chapter 17 we show that for every manifold approximate fibration $d : X \longrightarrow \mathbb{R}$ of dimension ≥ 5 there exists a manifold band (M, c) such that $X = \overline{M}$, with a proper homotopy $d \simeq \overline{c} : X \longrightarrow \mathbb{R}$.

The construction in (iii) of the wrapping up (M, c) of (X, d) is by the manifold 'twist glueing' due to Siebenmann [145]. The twist glueing construction of manifold bands is extended to the CW category in Chapters 19 and 20.

In Part Three (Chapters 21–27) we study the algebraic properties of tame ends in the context of chain complexes over a polynomial extension ring and also in bounded algebra. We obtain an abstract version of Theorem 19, giving a chain complex account of wrapping up: manifold wrapping up induces a CW complex wrapping up, which in turn induces a chain complex wrapping up, and similarly for the various types of twist glueing.

In Chapter 15 we introduce the notion of a *ribbon* (X, d), which is a non-compact space X with a proper map $d : X \longrightarrow \mathbb{R}$ with the homotopy theoretic and homological end properties of the infinite cyclic cover $(\overline{W}, \overline{c})$ of a band (W, c). Ribbons are the homotopy analogues of manifold approximate fibration over $\mathbb{R}$. In Chapter 25 we develop the chain complex versions of CW ribbons as well as algebraic versions of tameness.

The study of ends of complexes is particularly relevant to stratified spaces. A *topologically stratified space* is a space X together with a filtration

$$\emptyset = X^{-1} \subseteq X^0 \subseteq X^1 \subseteq \ldots \subseteq X^{n-1} \subseteq X^n = X$$

by closed subspaces such that the *strata* $X^j \backslash X^{j-1}$ are open topological manifolds which satisfy certain tameness conditions and a homotopy link condition. These spaces were first defined by Quinn [116] in order to study purely topological stratified phenomena as opposed to the smoothly stratified spaces of Whitney [170], Thom [161] and J. Mather [90], and the piecewise linear stratified spaces of Akin [1] and Stone [159]. Quinn's paper should be consulted for more precise definitions. Our results only apply directly to the very special case obtained from the one-point compactification $W^\infty = X$ of an open manifold W, regarded as a filtered space by $X^0 = \{\infty\} \subseteq W^\infty = X$. Then X is a topologically stratified space with two strata if and only if W is tame. (The general case requires controlled versions of our results.) Earlier, Siebenmann [147] had studied a class of topologically stratified spaces called *locally conelike stratified spaces.* The one-point compactification of an open manifold W with one end is locally conelike stratified if and only if the end of W can be collared. Hence, Quinn's stratified spaces are much more general than Siebenmann's. The conditions required of topologically stratified spaces by Quinn are designed to imply that strata have neighbourhoods which are homotopy equivalent to mapping cylinders of fibrations, whereas in the classical cases the strata have neighbourhoods which are homeomorphic to mapping cylinders of bundle projections in the appropriate category: fibre bundle projections in the smooth case, block bundle projections in the piecewise linear case. Strata in Siebenmann's locally conelike stratified spaces have neighbourhoods which are locally homeomorphic to mapping cylinders of fibre bundle projections, but not necessarily globally.

A *stratified homotopy equivalence* is a homotopy equivalence in the stratified category (maps must preserve strata, not just the filtration). In the special case of one-point compactifications, stratified homotopy equivalences $(W^\infty,\{\infty\})\longrightarrow(V^\infty,\{\infty\})$ are exactly the proper homotopy equivalences $W\longrightarrow V$. Weinberger [166] has developed a stratified surgery theory which classifies topologically stratified spaces up to stratified homotopy equivalence in the same sense that classical surgery theory classifies manifolds up to homotopy equivalence. Weinberger outlines two separate proofs of his theory. The first proof [166, pp. 182–188] involves stabilizing a stratified space by crossing with high dimensional tori in order to get a nicer stratified space which is amenable to the older stratified surgery theory of Browder and Quinn [15]. The obstruction to codimension i destabilization involves the codimension i lower K-group $K_{1-i}(\mathbb{Z}[\pi])\subseteq Wh(\pi\times\mathbb{Z}^i)$. (Example 18 and Theorem 19 treat the special case $i=1$.) The second proof outlined in [166, Remarks p. 189] uses more directly the existence of appropriate tubular neighbourhoods of strata called *teardrop neighbourhoods.* These neighbourhoods were shown to exist in the case of two strata by Hughes, Taylor, Weinberger and Williams [76] and in general by Hughes [74]. In 16.13 we give a complete proof of the existence of teardrop neighbourhoods in the special case of the topologically stratified space $(W^\infty,\{\infty\})$ determined by an open manifold W with a tame end. The result asserts that W contains an open cocompact subspace $X\subseteq W$ which admits a manifold approximate fibration $X\longrightarrow\mathbb{R}$. In the more rigid smoothly stratified spaces, the tubular neighbourhoods would be given by a genuine fibre bundle projection. The point is that Quinn's definition gives information on the neighbourhoods of strata only up to homotopy. The existence of teardrop neighbourhoods means there is a much stronger geometric structure given in terms of manifold approximate fibrations.

We use the theory of manifold approximate fibrations to perform geometric wrapping up constructions. This is analogous to Weinberger's second approach to stratified surgery, in which teardrop neighbourhoods of strata are used in order to be able to draw on manifold approximate fibration theory rather than stabilization and destabilization. We expect that the general theory of teardrop neighbourhoods will likewise allow generalizations of the wrapping up construction to arbitrary topologically stratified spaces, using the homotopy theoretic and algebraic properties of the ribbons introduced in this book. Such a combination of geometry, homotopy theory and algebra will be necessary to fully understand the algebraic K- and L-theory of stratified spaces.

This book grew out of research begun in 1990–91 when the first-named author was a Fulbright Scholar at the University of Edinburgh. We have received support from the National Science Foundation (U.S.A.), the Science and Engineering Research Council (U.K.), the European Union K-theory Initiative under Science Plan SCI–CT91–0756, the Vanderbilt University Research Council, and the Mathematics Departments of Vanderbilt University and the University of Edinburgh. We have benefited from conversations with Stratos Prassidis and Bruce Williams.

The book was typeset in TEX, with the diagrams created using the LAMS-TEX, PICTEX and XY-pic packages.

Errata (if any) to this book will be posted on the WWW Home Page
http://www.maths.ed.ac.uk/people/aar

Chapter summaries

Part One, *Topology at infinity*, is devoted to the basic theory of the general, geometric and algebraic topology at infinity of non-compact spaces. Various models for the topology at infinity are introduced and compared.

Chapter 1, *End spaces*, begins with the definition of the end space $e(W)$ of a non-compact space W. The set of path components $\pi_0(e(W))$ is shown to be in one-to-one correspondence with the set of ends of W (in the sense of Definition 1 above) for a wide class of spaces.

Chapter 2, *Limits*, reviews the basic constructions of homotopy limits and colimits of spaces, and the related inverse, direct and derived limits of groups and chain complexes. The end space $e(W)$ is shown to be weak homotopy equivalent to the homotopy inverse limit of cocompact subspaces of W and the homotopy inverse limit is compared to the ordinary inverse limit. The 'fundamental group at infinity' $\pi_1^\infty(W)$ of W is defined and compared to $\pi_1(e(W))$.

Chapter 3, *Homology at infinity*, contains an account of locally finite singular homology, which is the homology based on infinite chains. The homology at infinity $H_*^\infty(W)$ of a space W is the difference between ordinary singular homology $H_*(W)$ and locally finite singular homology $H_*^{lf}(W)$.

Chapter 4, *Cellular homology*, reviews locally finite cellular homology, although the technical proof of the equivalence with locally finite singular homology is left to Appendix A.

Chapter 5, *Homology of covers*, concerns ordinary and locally finite singular and cellular homology of the universal cover (and other covers) $\widetilde{W}$ of W. The version of the Whitehead theorem for detecting proper homotopy equivalences of CW complexes is stated.

Chapter 6, *Projective class and torsion*, recalls the Wall finiteness obstruction and Whitehead torsion. A locally finite finiteness obstruction is introduced, which is related to locally finite homology in the same way that the Wall finiteness obstruction is related to ordinary homology, and the difference between the two obstructions is related to homology at infinity.

Chapter 7, *Forward tameness*, concerns a tameness property of ends, which is stated in terms of the ability to push neighbourhoods towards infinity. It is proved that for forward tame W the singular chain complex of the end space $e(W)$ is chain equivalent to the singular chain complex at infinity of W, and that the homotopy groups of $e(W)$ are isomorphic to the inverse limit of the homotopy groups of cocompact subspaces of W. There is a related concept of forward collaring.

Chapter 8, *Reverse tameness*, deals with the other tameness property of ends, which is stated in terms of the ability to pull neighbourhoods in from infinity. It is closely related to finite domination properties of cocompact subspaces of W. There is a related concept of reverse collaring.

Chapter 9, *Homotopy at infinity*, gives an account of proper homotopy theory at infinity. It is shown that the homotopy type of the end space, the two types of tameness, and other end phenomena are invariant under proper homotopy equivalences at infinity. It is also established that in most cases of interest a space W is forward and reverse tame if and only if W is bounded homotopy equivalent at ∞ to $e(W) \times [0, \infty)$, in which case $e(W)$ is finitely dominated.

Chapter 10, *Projective class at infinity*, introduces two finiteness obstructions which the two types of tameness allow to be defined. The finiteness obstruction at infinity of a reverse tame space is an obstruction to reverse collaring. Likewise, the locally finite finiteness obstruction at infinity of a forward tame space is an obstruction to forward collaring. For a space W which is both forward and reverse tame, the end space $e(W)$ is finitely dominated and its Wall finiteness obstruction is the difference of the two finiteness obstructions at infinity. It is also proved that for a manifold end forward and reverse tameness are equivalent under certain fundamental group conditions.

Chapter 11, *Infinite torsion*, contains an account of the infinite simple homotopy theory of Siebenmann for locally finite CW complexes. The infinite Whitehead group of a forward tame CW complex is described algebraically as a relative Whitehead group. The infinite torsion of a proper homotopy equivalence is related to the locally finite finiteness obstruction at infinity. A CW complex W is forward (resp. reverse) tame if and only if $W \times S^1$ is infinite simple homotopy equivalent to a forward (resp. reverse) collared CW complex.

Chapter 12, *Forward tameness is a homotopy pushout*, deals with Quinn's characterization of forward tameness for a σ-compact metric space W in terms of a homotopy property, namely that the one-point compactification W^∞ is the homotopy pushout of the projection $e(W) \longrightarrow W$ and $e(W) \longrightarrow \{\infty\}$, or equivalently that W^∞ is the homotopy cofibre of $e(W) \longrightarrow W$.

Part Two, *Topology over the real line*, concerns spaces W with a proper map $d : W \longrightarrow \mathbb{R}$.

Chapter 13, *Infinite cyclic covers*, proves that a connected infinite cyclic cover $\overline{W}$ of a connected compact ANR W has two ends $\overline{W}^+$, $\overline{W}^-$, and establishes a duality between the two types of tameness: $\overline{W}^+$ is forward tame if and only if $\overline{W}^-$ is reverse tame. A similar duality holds for forward and reverse collared ends.

Chapter 14, *The mapping torus*, works out the end theory of infinite cyclic covers of mapping tori.

Chapter 15, *Geometric ribbons and bands*, presents bands and ribbons. It is proved that $(M, c : M \longrightarrow S^1)$ with M a finite CW complex defines a band (i.e. the infinite cyclic cover $\overline{M} = c^*\mathbb{R}$ of M is finitely dominated) if and only if the ends $\overline{M}^+$, $\overline{M}^-$ are both forward tame, or both reverse tame. The Siebenmann twist glueing construction of a band is formulated for a ribbon $(X, d : X \longrightarrow \mathbb{R})$ and an end-preserving homeomorphism $h : X \longrightarrow X$.

Chapter 16, *Approximate fibrations*, presents the main geometric tool used in the proof of the uniformization Theorem 19 (every tame manifold end of dimension ≥ 5 has a neighbourhood which is the infinite cyclic cover of a manifold band). It is proved that an open manifold W of dimension ≥ 5 is forward and reverse tame if and only if there exists an open cocompact subspace $X \subseteq W$ which admits a manifold approximate fibration $X \longrightarrow \mathbb{R}$.

Chapter 17, *Geometric wrapping up*, uses the twist glueing construction with $h = 1 : X \longrightarrow X$ to prove that the total space X of a manifold approximate fibration $d : X \longrightarrow \mathbb{R}$ is the infinite cyclic cover $X = \overline{M}$ of a manifold band (M, c).

Chapter 18, *Geometric relaxation*, uses the twist glueing construction with $h =$ covering translation $: \overline{M} \longrightarrow \overline{M}$ to pass from a manifold band (M, c) to an h-cobordant manifold band (M', c') such that $c' : M' \longrightarrow S^1$ is a manifold approximate fibration.

Chapter 19, *Homotopy theoretic twist glueing*, and Chapter 20, *Homotopy theoretic wrapping up and relaxation*, extend the geometric constructions for manifolds in Chapters 17 and 18 to CW complex bands and ribbons. Constructions in this generality serve as a bridge to the algebraic theory of Part Three. Moreover, it is shown that any CW ribbon is infinite simple homotopy equivalent to the infinite cyclic cover of a CW band, thereby justifying the concept.

Part Three, *The algebraic theory*, translates most of the geometric, homotopy theoretic and homological constructions of Parts One and Two into an appropriate algebraic context, thereby obtaining several useful algebraic characterizations.

Chapter 21, *Polynomial extensions*, gives background information on chain complexes over polynomial extension rings, motivated by the fact that the cellular chain complex of an infinite cyclic cover of a CW complex is defined over a Laurent polynomial extension.

Chapter 22, *Algebraic bands*, discusses chain complexes over Laurent polynomial extensions which have the algebraic properties of cellular chain complexes of CW complex bands.

Chapter 23, *Algebraic tameness*, develops the algebraic analogues of forward and reverse tameness for chain complexes over polynomial extensions. This yields an algebraic characterization of forward (and reverse) tameness for an end of an infinite cyclic cover of a finite CW complex. End complexes are also defined in this algebraic setting.

Chapter 24, *Relaxation techniques*, contains the algebraic analogues of the constructions of Chapters 18 and 20. When combined with the geometry of Chapter 18 this gives an algebraic characterization of manifold bands which admit approximate fibrations to S^1.

Chapter 25, *Algebraic ribbons*, explores the algebraic analogue of CW ribbons in the context of bounded algebra. The algebra is used to prove that CW ribbons are infinite simple homotopy equivalent to infinite cyclic covers of CW bands.

Chapter 26, *Algebraic twist glueing*, proves that algebraic ribbons are simple chain equivalent to algebraic bands.

Chapter 27, *Wrapping up in algebraic K- and L-theory*, describes the effects of the geometric constructions of Part Two on the level of the algebraic K- and L-groups.

Part Four consists of the three appendices:

Appendix A, *Locally finite homology with local coefficients*, contains a technical treatment of ordinary and locally finite singular and cellular homology theories with local coefficients. This establishes the equivalence of locally finite singular and cellular homology for regular covers of CW complexes.

Appendix B, *A brief history of end spaces*, traces the development of end spaces as homotopy theoretic models for the topology at infinity.

Appendix C, *A brief history of wrapping up*, outlines the history of the wrapping up compactification procedure.

Part One: Topology at infinity

1

End spaces

Throughout the book it is assumed that ANR spaces are locally compact, separable and metric, and that CW complexes are locally finite.

We start with the end space $e(W)$ of a space W, which is a homotopy theoretic model for the behaviour at ∞ of W. The homotopy type of $e(W)$ is determined by the proper homotopy type of W. The set of path components $\pi_0(e(W))$ is related to the number of ends of W, and the fundamental group $\pi_1(e(W))$ is related to the fundamental group at ∞ of W.

Definition 1.1 The *one-point compactification* of a topological space W is the compact topological space

$$W^\infty = W \cup \{\infty\} ,$$

with open sets:

(i) $U \subset W^\infty$ for an open subset $U \subseteq W$,
(ii) $V \cup \{\infty\} \subseteq W^\infty$ for a subset $V \subseteq W$ such that $W \backslash V$ is compact. □

The topology at infinity of W is the topology of W^∞ at ∞.

Definition 1.2 The *end space* $e(W)$ of a space W is the space of paths

$$\omega : ([0,\infty], \{\infty\}) \longrightarrow (W^\infty, \{\infty\})$$

such that $\omega^{-1}(\infty) = \{\infty\}$, with the compact-open topology. □

The end space $e(W)$ is the *homotopy link* $\text{holink}(W^\infty, \{\infty\})$ of $\{\infty\}$ in W^∞ in the sense of Quinn [116]. See 1.8 for the connection with the link in the sense of PL topology, and 12.11 for the general definition of the homotopy link.

We refer to Appendix B for a brief history of end spaces.

An element $\omega \in e(W)$ can also be viewed as a path $\omega : [0,\infty)\longrightarrow W$ such that $\omega(t)$ 'diverges to ∞' as $t\longrightarrow\infty$, meaning that for every compact subspace $K \subset W$ there exists $N > 0$ with $\omega([N,\infty)) \subset W\backslash K$.

Definition 1.3 (i) A map of spaces $f : V\longrightarrow W$ is *proper* if for each compact subspace $K \subseteq W$ the inverse image $f^{-1}(K) \subseteq V$ is compact. This is equivalent to the condition that f extends to a map $f^\infty : V^\infty\longrightarrow W^\infty$ of the one-point compactifications with $f^\infty(\infty) = \infty$.

(ii) A map $f : V\longrightarrow W$ is a *proper homotopy equivalence* if it is a proper map which is a homotopy equivalence in the proper category. □

We refer to Porter [111] for a survey of the applications of proper homotopy theory to ends. The end space $e(W)$ is called the 'Waldhausen boundary' of W in [111, p. 135].

An element $\omega \in e(W)$ is a proper map $\omega : [0,\infty)\longrightarrow W$, which is the same as a path in $\omega^\infty : [0,\infty]\longrightarrow W^\infty$ such that $\omega^\infty[0,\infty) \subseteq W$ and $\omega^\infty(\infty) = \infty$.

Example 1.4 (i) The end space of a compact space W is empty,

$$e(W) = \emptyset ,$$

since $W^\infty = W\cup\{\infty\}$ is disconnected and there are no paths $\omega^\infty : [0,\infty]\longrightarrow W^\infty$ from $\omega^\infty(0) \in W$ to $\omega^\infty(\infty) = \infty \in W^\infty$. The converse is false: the end space of $\mathbb{Z}$ is empty, yet $\mathbb{Z}$ is not compact.

(ii) Let T be a tree, and let $v \in T$ be a base vertex. A *simple edge path* in T is a sequence of adjoining edges $e_1, e_2, e_3, \ldots$ (possibly infinite) without repetition. By the simplicial approximation theorem every proper map $\omega : [0,\infty)\longrightarrow T$ is proper homotopic to an infinite simple edge path starting at v. If T has at most a finite number of vertices of valency > 2 the end space $e(T)$ is homotopy equivalent to the discrete space with one point for each simple edge path of infinite length starting at $v \in T$.

(iii) The end space of $\mathbb{R}^+ = [0,\infty)$ is contractible,

$$e(\mathbb{R}^+) \simeq \{\text{pt.}\} ,$$

corresponding to the unique infinite simple edge path starting at $0 \in \mathbb{R}^+$.

(iv) The end space of $\mathbb{R}$ is such that

$$e(\mathbb{R}) \simeq S^0 = \{+1,-1\} ,$$

corresponding to the two infinite simple edge paths starting at $0 \in \mathbb{R}$. □

In dealing with end spaces $e(W)$, we shall always assume that W is a locally compact Hausdorff space.

Remark 1.5 For any space W the evaluation map

$$p : e(W) \longrightarrow W \; ; \; \omega \longrightarrow \omega(0)$$

fits into a homotopy commutative square

$$\begin{array}{ccc} e(W) & \longrightarrow & \{\infty\} \\ {\scriptstyle p}\downarrow & & \downarrow \\ W & \xrightarrow{\;i\;} & W^{\infty} \end{array}$$

with $i : W \longrightarrow W^{\infty}$ the inclusion. The space W is 'forward tame' if and only if this square is a homotopy pushout rel $\{\infty\}$ – see Chapters 7, 12 for a more detailed discussion. □

Definition 1.6 Let $(K, L \subseteq K)$ be a pair of spaces. The space L is *collared* in K if the inclusion $L = L \times \{0\} \longrightarrow K$ extends to an open embedding $f : L \times [0, \infty) \longrightarrow K$. □

Proposition 1.7 *If (K, L) is a compact pair of spaces such that L is collared in K then the end space of the non-compact space*

$$W = K \backslash L$$

is such that there is defined a homotopy equivalence

$$L \longrightarrow e(W) \; ; \; x \longrightarrow \left(t \longrightarrow f\left(x, \frac{1}{1+t}\right)\right)$$

with $f : L \times [0, \infty) \longrightarrow K$ an open embedding extending the inclusion $L = L \times \{0\} \longrightarrow K$. □

In other words, if W is a non-compact space with a compactification K such that the boundary

$$\partial K = K \backslash W \subset K$$

is a compact subset which is collared in K then there is defined a homotopy equivalence

$$e(W) \simeq \partial K \; .$$

The homotopy theoretic 'space at infinity' $e(W)$ thus has the homotopy type of an actual space at infinity, provided ∂W is collared in the compactification K.

Example 1.8 (i) Let X be a compact polyhedron. For any $x \in X$ there exists a triangulation of X with x as a vertex, with the pair of compact spaces

$$(Y, Z) = (\mathrm{star}(x), \mathrm{link}(x))$$

such that $Y = x * Z$ is the cone on Z, and Z is collared in Y. (See Rourke and Sanderson [139] for the PL theory of stars and links.) The non-compact spaces

$$\begin{aligned} Y \backslash Z &= Z \times [0, \infty)/Z \times \{0\} , \\ W &= X \backslash \{x\} = \mathrm{cl}(X \backslash Y) \cup_{Z \times \{0\}} Z \times [0, \infty) \end{aligned}$$

have one-point compactifications

$$(Y \backslash Z)^{\infty} = Y/Z \ , \ W^{\infty} = X \ \ (\infty = x) \ ,$$

with end spaces such that

$$e(Y \backslash Z) \simeq e(W) \simeq Z \ .$$

The homotopy link of $\{\infty\}$ in W^{∞} is homotopy equivalent to the actual link of x in X.

(ii) Let $(M, \partial M)$ be a compact n-dimensional topological manifold with boundary. The boundary ∂M is collared in M. (In the topological category this was first proved by Brown [16]. See Conelly [31] for a more recent proof.) The interior of M is an open n-dimensional manifold

$$W = \mathrm{int}(M) = M \backslash \partial M$$

with an open embedding $f : \partial M \times [0, \infty) \longrightarrow M$ extending the inclusion $\partial M = \partial M \times \{0\} \longrightarrow M$. The end space of W is such that the map

$$g : \partial M \longrightarrow e(W) \ ; \ x \longrightarrow (t \longrightarrow f\left(x, \frac{1}{t+1}\right))$$

defines a homotopy equivalence, with the adjoint of g

$$\widehat{g} : \partial M \times [0, \infty) \longrightarrow W \ ; \ (x, t) \longrightarrow g(x)(t)$$

homotopic to f.

(iii) In view of (ii) a necessary condition for an open n-dimensional manifold W to be homeomorphic to the interior of a compact n-dimensional manifold with boundary is that the end space $e(W)$ have the homotopy type of a closed $(n-1)$-dimensional manifold. In Chapters 7, 8 we shall be studying geometric tameness conditions on W which ensure that $e(W)$ is at least a finitely dominated $(n-1)$-dimensional geometric Poincaré complex. □

The following result is a useful characterization of continuity for functions into an end space. It is based on elementary facts about the compact-open topology and proper maps.

Proposition 1.9 *For locally compact Hausdorff spaces X, W and a function $f : X \longrightarrow e(W)$, the following are equivalent:*

(i) *f is continuous,*
(ii) *the adjoint $\widehat{f} : X \times [0,\infty) \longrightarrow W; (x,t) \longrightarrow f(x)(t)$ is continuous, and for all compact subspaces $C \subseteq X$, $K \subseteq W$, there exists $N \geq 0$ such that $\widehat{f}(C \times [N,\infty)) \subseteq W \backslash K$,*
(iii) *for every compact subspace $C \subseteq X$, the restriction $\widehat{f}| : C \times [0,\infty) \longrightarrow W$ is a proper map.*

Proof (ii) $\iff$ (iii) is obvious.
(i) $\Longrightarrow$ (iii) If f is continuous, so is the induced function

$$f^* \ : \ X \times [0,\infty] \longrightarrow W^\infty \ ; \ (x,t) \longrightarrow \begin{cases} f(x)(t) & \text{if } t < \infty \ , \\ \infty & \text{if } t = \infty \ . \end{cases}$$

Since $\widehat{f} = f^*|$, $\widehat{f}$ is continuous and $\widehat{f}| : C \times [0,\infty) \longrightarrow W$ is proper.
(iii) $\Longrightarrow$ (i) It suffices to show that the induced function $f^* : X \times [0,\infty] \longrightarrow W^\infty$ is continuous. It is clear that $f^*| : C \times [0,\infty] \longrightarrow W^\infty$ is continuous for each compact subspace $C \subseteq X$. The local compactness of X then implies that f^* is continuous. □

It follows that for a compact Hausdorff space X, a function $f : X \longrightarrow e(W)$ is continuous if and only if the adjoint $\widehat{f} : X \times [0,\infty) \longrightarrow W$ is a proper map. For non-compact X, W a constant map $X \longrightarrow e(W)$ is such that the adjoint $X \times [0,\infty) \longrightarrow W$ is not proper.

Proposition 1.10 *The end space defines a functor $e : W \longrightarrow e(W)$ from the category of topological spaces and proper maps to the category of topological spaces and all maps. A proper map $f : V \longrightarrow W$ induces a map*

$$e(f) \ : \ e(V) \longrightarrow e(W) \ ; \ \omega \longrightarrow f\omega \ ,$$

and a proper homotopy $f \simeq g : V \longrightarrow W$ induces a homotopy

$$e(f) \simeq e(g) \ : \ e(V) \longrightarrow e(W) \ . \qquad \square$$

A subspace $V \subseteq W$ is *cocompact* if the closure of $W \backslash V \subseteq W$ is compact. For a CW complex W a subcomplex $V \subseteq W$ is *cofinite* if it contains all but finitely many cells of W. A cofinite subcomplex is a cocompact subspace.

Definition 1.11 A space W is *σ-compact* if

$$W = \bigcup_{j=1}^{\infty} K_j$$

with each K_j compact and $K_j \subseteq K_{j+1}$. □

In particular, all the ANR's considered by us are σ-compact, since we are assuming that they are locally compact, separable and metric.

It follows from 1.10 that the homotopy type of $e(W)$ is determined by the proper homotopy type of W. A more general result will be established in 9.4 for a metric space W, that the homotopy type of $e(W)$ is determined by the 'proper homotopy type at ∞' of W. The inclusion of a closed cocompact subspace is a special case of a 'proper homotopy equivalence at ∞', and the following result will be used in the proof of 9.4:

Proposition 1.12 *If W is a σ-compact metric space and $u : V \longrightarrow W$ is the inclusion of a closed cocompact subspace then the inclusion of end spaces $e(u) : e(V) \longrightarrow e(W)$ is a homotopy equivalence.*

Proof Since W is a σ-compact metric space, W^∞ and $e(W)$ are metrizable, and so $e(W)$ is paracompact. For each $\omega \in e(W)$ choose a number $t_\omega \in [0, \infty)$ such that

$$\omega([t_\omega, \infty)) \subseteq \text{int}(V) \ .$$

Let $U(\omega)$ be an open neighbourhood of ω in $e(W)$ such that

$$\alpha([t_\omega, \infty)) \subseteq \text{int}(V) \quad (\alpha \in U(\omega)) \ .$$

Let $\{U_i\}$ be a locally finite refinement of the covering $\{U(\omega) \,|\, \omega \in e(W)\}$ of $e(W)$, and let $\{\phi_i\}$ be a partition of unity subordinate to $\{U_i\}$. For each i choose $\omega_i \in e(W)$ such that $U_i \subseteq U(\omega_i)$, and let $t_i = t_{\omega_i}$. For each $\omega \in e(W)$ let

$$m_\omega = \min\{t_i \,|\, \phi_i(\omega) \neq 0\} \ .$$

Note that $\omega([m_\omega, \infty)) \subseteq \text{int}V$ and $\sum\limits_i \phi_i(\omega) t_i \geq m_\omega$. The map

$$\begin{aligned} F \ : e(W) \times I &\longrightarrow e(W) \ ; \\ (\omega, t) &\longrightarrow (s \longrightarrow \omega((1-t)s + (\sum_i \phi_i(\omega)t_i + s)t)) \\ &(\omega \in e(W) \ , \ 0 \leq t \leq 1 \ , \ s \geq 0) \end{aligned}$$

is a deformation of $e(W)$ into $e(V)$ such that $F_t(e(V)) \subseteq e(V)$ for $0 \leq t \leq 1$. □

Example 1.13 (i) The application of 1.12 to the inclusion

$$\{x \in \mathbb{R}^m \mid \|x\| \geq 1\} = S^{m-1} \times [1,\infty) \longrightarrow \mathbb{R}^m \quad (m \geq 1) ,$$

gives a homotopy equivalence

$$e(S^{m-1} \times [1,\infty)) \simeq e(\mathbb{R}^m) .$$

By 1.7 $e(S^{m-1} \times [1,\infty))$ is homotopy equivalent to S^{m-1}, so that

$$e(S^{m-1} \times [1,\infty)) \simeq e(\mathbb{R}^m) \simeq S^{m-1} .$$

(ii) Given a compact space K and an integer $m \geq 1$ let

$$W = K \times \mathbb{R}^m .$$

The one-point compactification $W^\infty = \Sigma^m K^\infty$ is the m-fold reduced suspension of $K^\infty = K \cup \{\text{pt.}\}$, and the end space is such that

$$e(W) \simeq K \times e(\mathbb{R}^m) \simeq K \times S^{m-1} . \qquad \square$$

In dealing with the number of ends of a space W we shall assume the following standing hypothesis for the rest of this chapter: *W is a locally compact, connected, locally connected Hausdorff space* (e.g. a locally finite connected CW complex).

In the literature the end space $e(W)$ has not played as central a role as the 'ends of W' or the 'number of ends of W'. Roughly, an end of W should correspond to a path component of $e(W)$. We now recall these classical notions and their relationship to $\pi_0(e(W))$.

Definition 1.14 (Milnor [100]) An *end* of a space W is a function

$$\epsilon : \{K \mid K \subseteq W \text{ is compact}\} \longrightarrow \{X \mid X \subseteq W\} \; ; \; K \longrightarrow \epsilon(K)$$

such that:

(i) $\epsilon(K)$ is a component of $W \backslash K$ for each K,
(ii) if $K \subseteq L$, then $\epsilon(L) \subseteq \epsilon(K)$.

A *neighbourhood* of ϵ is a connected open subset $U \subseteq W$ such that $U = \epsilon(K)$ for some non-empty compact $K \subseteq W$. $\square$

Remark 1.15 (i) For a σ-compact space W the definition of an end in 1.14 agrees with Definition 1 in the Introduction. A sequence $W \supseteq U_1 \supseteq U_2 \supseteq \dots$ of neighbourhoods of an end (in the sense of Definition 1 of the Introduction) such that $\bigcap\limits_{j=1}^{\infty} \text{cl}(U_j) = \emptyset$ determines an end ϵ of W (in the sense of 1.14) as follows: for a compact subspace $K \subseteq W$ choose j such that $U_j \cap K = \emptyset$ and let $\epsilon(K)$ be the component of $W \backslash K$ which contains U_j. On the other hand, if ϵ is an end of W and $W = \bigcup_{j=1}^{\infty} K_j$ with each K_j compact and $K_j \subseteq K_{j+1}$, then $\epsilon(K_j) = U_j$ defines a sequence of neighbourhoods of an end as above.

(ii) A subspace is *unbounded* if its closure is not compact. Note that if ϵ is an end of W, then $\epsilon(K)$ is unbounded for each compact subspace $K \subseteq W$. (Otherwise, $L = K \cup \mathrm{cl}(\epsilon(K))$ would be a compact subspace of W containing K, so $\epsilon(L) \subseteq \epsilon(K) \subseteq L$, contradicting $\epsilon(L) \subseteq W \backslash L$.) □

Definition 1.16 The *number of ends* of a locally finite CW complex W is the least upper bound of the number (which may be infinite) of infinite components of $W \backslash V$ for finite subcomplexes $V \subset W$. □

Example 1.17 (i) The real line $\mathbb{R}$ has exactly two ends.

(ii) The *dyadic tree* X is the tree embedded in $\mathbb{R}^2$ with each vertex of valency 3, with closure the union of X together with a disjoint Cantor set. The dyadic tree has an uncountable number of ends. See Diestel [37] for more information on ends of graphs. □

An alternative approach to the definition of an end is to focus attention on the number of ends of a space.

Definition 1.18 (Specker [151], Raymond [134]) The space W has *at least k ends* if there exists an open subspace $V \subseteq W$ with compact closure $\mathrm{cl}(V)$ such that $W \backslash \mathrm{cl}(V)$ has at least k unbounded components. The space W has *(exactly) k ends* if W has at least k ends but not at least $k+1$ ends. □

The point set conditions on W imply that if $V \subseteq W$ is an open subspace with compact closure, then $W \backslash \mathrm{cl}(V)$ has at most a finite number of unbounded components (see Hocking and Young [66, Theorem 3–9, p. 111]). If W has exactly k ends then there exists an open subspace $V \subseteq W$ with compact closure so that $W \backslash \mathrm{cl}(V)$ has exactly k unbounded components.

Proposition 1.19 *Let $k \geq 0$ be an integer.*

(i) *If W has at least k ends in the sense of Definition 1.14, then W has at least k ends in the sense of Definition 1.18.*

(ii) *If W is σ-compact and has at least k ends in the sense of Definition 1.18, then W has at least k ends in the sense of Definition 1.14.*

(iii) *For W σ-compact, W has exactly k ends in the sense of Definition 1.14 if and only if W has exactly k ends in the sense of Definition 1.18.*

Proof (i) Let $\epsilon_1, \ldots, \epsilon_k$ be distinct ends of W in the sense of 1.14. For $1 \leq i < j \leq k$, choose a compact subspace $H_{ij} \subseteq W$ such that $\epsilon_i(H_{ij}) \neq \epsilon_j(H_{ij})$.

It follows that $\epsilon_i(H_{ij}) \cap \epsilon_j(H_{ij}) = \emptyset$. Let

$$H = \bigcup_{1 \le i < j \le k} H_{ij} .$$

Since H is compact, there is an open subspace $V \subseteq W$ with compact closure such that $H \subseteq V$. Then $\epsilon_1(\mathrm{cl}(V)), \ldots, \epsilon_k(\mathrm{cl}(V))$ are unbounded components of $W \backslash \mathrm{cl}(V)$. Since $\epsilon_i(\mathrm{cl}(V)) \subseteq \epsilon_i(H_{ij})$, these are in fact k distinct components. Thus, W has at least k ends in the sense of 1.18.

(ii) We may assume that $k \ge 1$, so that W is non-compact. Note that if $K \subseteq W$ is a compact subspace, then $W \backslash K$ has at least one unbounded component. For if $V \subseteq W$ is an open subspace with compact closure such that $K \subseteq V$, then all but finitely many components of $W \backslash K$ are contained in V (see Hocking and Young [66, Theorem 3–9, p. 111]). It follows that one of those finitely many components of $W \backslash K$ must be unbounded.

Next, we shall show that if $K \subseteq W$ is a compact subspace and C is an unbounded component of $W \backslash K$, then there exists an end ϵ of W in the sense of 1.14 such that $\epsilon(K) = C$. For W can be written as

$$W = \bigcup_{j=0}^{\infty} K_j$$

with $K_0 = K$, each K_j compact and $K_j \subseteq K_{j+1}$. Define ϵ as follows. First, let $\epsilon(K) = \epsilon(K_0) = C$. Then, assuming $j \ge 1$ and that $\epsilon(K_{j-1})$ has been defined, define $\epsilon(K_j)$ to be one of the unbounded components of $\mathrm{cl}(\epsilon(K_{j-1})) \backslash K_j$ (which exists by the argument above). Finally, for an arbitrary compact subspace $H \subseteq W$, choose j such that $H \subseteq K_j$, and define $\epsilon(H)$ to be the component of $W \backslash H$ which contains $\epsilon(K_j)$. It is easy to verify that ϵ is an end of W in the sense of 1.14.

Since W has at least k ends in the sense of 1.18, there exists an open subspace $V \subset W$ with compact closure such that $W \backslash \mathrm{cl}(V)$ has at least k unbounded components, say $C_1, \ldots, C_k$. Then there exist ends $\epsilon_1, \ldots, \epsilon_k$ of W in the sense of 1.14 such that $\epsilon_j(\mathrm{cl}(V)) = C_j$ for $j = 1, \ldots, k$.

(iii) Immediate from (i) and (ii). □

If a space W is not assumed to be σ-compact, then we shall assume that an end of W refers to an end in the sense of 1.14 unless otherwise stated. Of course, such an end gives rise to an end in the sense of 1.18.

Proposition 1.20 *A connected space W with exactly k ends can be expressed as*

$$W = K \cup \bigcup_{j=1}^{k} W(j)$$

with $K \subseteq W$ a connected compact subspace, and each $W(j) \subseteq W$ a closed *connected subspace with exactly one end.*

Proof Let $V \subseteq W$ be an open subspace with compact closure such that $W\backslash\text{cl}(V)$ has exactly k unbounded components, say $C_1, C_2, \ldots, C_k$. Let

$$X = W\backslash\bigcup_{j=1}^{k} C_j = \text{cl}(V) \cup \bigcup\{\text{all bounded components of } W\backslash\text{cl}(V)\} .$$

Observe that X is compact. For if $\mathcal{U}$ is a collection of open subsets of W which cover X, extract finitely many $U_1, U_2, \ldots, U_n \in \mathcal{U}$ such that $\text{cl}(V) \subseteq \bigcup_{j=1}^{n} U_j$. Only finitely many of the components of $W\backslash\text{cl}(V)$ are not contained in $\bigcup_{j=1}^{n} U_j$ (see Hocking and Young [66, Theorem 3–9, p. 111]). Let $D_1, D_2, \ldots, D_m$ be the bounded components of $W\backslash\text{cl}(V)$ not contained in $\bigcup_{j=1}^{n} U_j$. Then $\text{cl}(D_j) \subseteq X$ is compact for each $j = 1, 2 \ldots, m$. Thus, there exists a finite subcollection $\mathcal{U}_j$ of $\mathcal{U}$ which covers $\text{cl}(D_j)$. Then

$$\{U_1, U_2, \ldots, U_n\} \cup \mathcal{U}_1 \cup \ldots \cup \mathcal{U}_m$$

is a finite subcollection of $\mathcal{U}$ which covers X.

Now let $K \subseteq W$ be a compact connected subspace containing X (use Dugundji [38, page 254, exercise 2, section 6]) and let $W(j) = \text{cl}(C_j)$ for $j = 1, 2, \ldots, k$.

It only remains to see that each $W(i)$ has one end. Suppose on the contrary that ϵ_1 and ϵ_2 are distinct ends of $W(j)$. These ends induce ends $\widetilde{\epsilon}_1$, $\widetilde{\epsilon}_2$ of W by setting

$$\widetilde{\epsilon}_i(K) = \epsilon_i(K \cap W(j))$$

for $K \subseteq W$ and $i = 1, 2$. This shows that W has at least $k+1$ ends, a contradiction. □

Definition 1.21 The *set of ends* $\mathcal{E}_W$ of a space W is the set of ends of W in the sense of Definition 1.14. □

Proposition 1.22 (i) *The set of path components of the end space* $e(W)$ *is related to the set of ends of a space* W *by the map*

$$\eta_W : \pi_0(e(W)) \longrightarrow \mathcal{E}_W \ ; \ [\omega] \longrightarrow \epsilon_\omega$$

with $\epsilon_\omega(K)$ *the component of* $W\backslash K$ *which contains* $\omega([N, \infty))$, *for any compact* $K \subseteq W$.

(ii) *Given spaces* X, Y, *a closed cocompact subspace* $U \subseteq X$ *and a proper map* $f : U \longrightarrow Y$ *there is induced a map*

$$f_* : \mathcal{E}_X \longrightarrow \mathcal{E}_Y \ ; \ \epsilon \longrightarrow f_*(\epsilon)$$

with $(f_*\epsilon)(K)$ *the component of* $Y\backslash K$ *such that*

$$f(\epsilon(f^{-1}(K)\cup \mathrm{cl}(X\backslash U)))\subseteq (f_*\epsilon)(K)$$

for any compact $K\subseteq Y$. *The induced map* f_* *satisfies the following properties:*

(a) *(Restriction) If* $U'\subseteq X$ *is a closed cocompact subspace with* $U'\subseteq U$, *then* $f_*=(f|U')_*$.
(b) *(Proper homotopy invariance) If* $g: U\longrightarrow Y$ *is a proper map with* f *and* g *properly homotopic, then* $f_*=g_*$.
(c) *(Naturality) Let* $\eta_X:\pi_0(e(X))\longrightarrow\mathcal{E}_X$ *and* $\eta_Y:\pi_0(e(Y))\longrightarrow\mathcal{E}_Y$ *be the maps defined above. Let* $e(f): e(U)\longrightarrow e(Y)$ *be the map induced by* f. *Then the inclusion induced map* $i:\pi_0(e(U))\longrightarrow\pi_0(e(X))$ *is such that* $f_*\circ\eta_X\circ i=\eta_Y\circ e(f)_*:\pi_0(e(U))\longrightarrow\mathcal{E}_Y$.

Proof (i) Immediate from the definitions.

(ii) (a) Let $K\subseteq Y$ be compact, and let

$$A=\epsilon(f|U')^{-1}(K)\cup\mathrm{cl}(X\backslash U')\subseteq X\ ,\quad B=(f|U')(A)\subseteq Y\ .$$

Now

$$X\backslash U\subseteq f^{-1}(K)\cup\mathrm{cl}(X\backslash U')\subseteq (f|U')^{-1}(K)\cup\mathrm{cl}(X\backslash U')\ ,$$

so that

$$A\subseteq\epsilon(f^{-1}(K)\cup\mathrm{cl}(X\backslash U))\subseteq U\backslash f^{-1}(K)$$

and hence

$$B\subseteq f(\epsilon(f^{-1}(K)\cup\mathrm{cl}(X\backslash U)))\subseteq Y\backslash K\ .$$

It follows that the component of $Y\backslash K$ containing B is also the component of $Y\backslash K$ containing $f(\epsilon(f^{-1}(K)\cup\mathrm{cl}(X\backslash U)))$.

(b) Let $h: f\simeq g: U\times I\longrightarrow Y$ be a proper homotopy and let $K\subseteq Y$ be compact. The subspace

$$C=f^{-1}(K)\cup g^{-1}(K)\cup\mathrm{cl}(X\backslash U)\subseteq U\times I$$

is such that $f^{-1}(K)\cup\mathrm{cl}(X\backslash U)\subseteq C$ and $\epsilon(C)\subseteq\epsilon(f^{-1}(K)\cup\mathrm{cl}(X\backslash U))$. Thus $f_*(\epsilon)(K)$ is the component of $Y\backslash K$ which contains $f(\epsilon(C))$. Also, $g_*(\epsilon)(K)$ is the component of $Y\backslash K$ which contains $g(\epsilon(C))$. It suffices to show that $f(\epsilon(C))$ and $g(\epsilon(C))$ are in the same component of $Y\backslash K$. Since $h^{-1}(K)$ is compact and $\epsilon(C)$ is unbounded, there exists $x\in\epsilon(C)$ such that $h(x\times I)\cap K=\emptyset$. It follows that $f(x)\in f(\epsilon(C))$ and $g(x)\in g(\epsilon(C))$ are in the same component of $Y\backslash K$.

(c) This is obvious. □

Remark 1.23 Freudenthal [61] and Raymond [134] defined the *(Freudenthal) end point compactification* W^* of a non-compact space W. In the case that W has exactly k ends $\epsilon_1, \dots, \epsilon_k$, this compactification is essentially by adjoining one point at infinity for each end, with

$$W^* = W \cup \{\epsilon_1, \dots, \epsilon_k\}$$

and the topology on W^* is such that neighbourhoods of ϵ_j are of the form $\epsilon_j(K) \cup \{\epsilon_j\}$ for compact subspaces $K \subseteq W$. (If W has one end this is the one-point compactification, $W^* = W^\infty$.) If W has infinitely many ends, the topology on W^* is more complicated because the ends are no longer isolated. The compactification W^* is characterized in [134] by the properties:

(i) W^* is connected,
(ii) W is open in W^*,
(iii) $W^* \backslash W$ is totally disconnected,
(iv) if $x \in W^* \backslash W$ and U is a connected open neighbourhood of x, then $U \backslash (W^* \backslash W)$ is connected. □

2

Limits

In this chapter we state the basic constructions and properties of homotopy limits and colimits of spaces, and the related direct and inverse systems of groups. In 2.14 we shall show that the end space $e(W)$ of a σ-compact space W has the weak homotopy type of the homotopy limit $\underleftarrow{\operatorname{holim}}_j W_j$ of an inverse system $\{W_j \mid j = 0, 1, 2, \ldots\}$ of closed cocompact subspaces $W_j \subseteq W$ with $\emptyset = \bigcap_{j=0}^{\infty} W_j \subset \ldots \subset W_{j+1} \subset W_j \subset \ldots \subset W_0 = W$, and the homotopy groups $\pi_*(e(W)) = \pi_*(\underleftarrow{\operatorname{holim}}_j W_j)$ fit into short exact sequences

$$0 \longrightarrow \underleftarrow{\lim}_j{}^1 \, \pi_{r+1}(W_j) \longrightarrow \pi_r(\underleftarrow{\operatorname{holim}}_j W_j) \longrightarrow \underleftarrow{\lim}_j \pi_r(W_j) \longrightarrow 0$$

with $\lim^1$ denoting the derived limit (2.11). The 'Mittag–Leffler' and 'stability' conditions for an inverse sequence of groups are recalled (2.20) and related to derived limits. The related geometric condition 'semistability at ∞' for a space W is interpreted in terms of the end space $e(W)$ (2.25).

We refer to Bousfield and Kan [9] for the general theory of homotopy limits and colimits.

Definition 2.1 The *direct limit* of a direct system of sets

$$X_0 \xrightarrow{f_0} X_1 \xrightarrow{f_1} X_2 \longrightarrow \ldots$$

is the quotient of the disjoint union $\coprod_j X_j$

$$\underrightarrow{\lim}_j X_j \;=\; \coprod_{j=0}^{\infty} X_j \Big/ (x_{j+1} = f_j(x_j)) \;. \qquad \square$$

Proposition 2.2 (i) *For any set X and any sequence of subsets $X_0 \subseteq X_1 \subseteq X_2 \subseteq \ldots \subseteq X$ the direct limit of the direct system $\{f_j : X_j \longrightarrow X_{j+1}\}$ defined by the inclusions is the union*

$$\varinjlim_j X_j = \bigcup_{j=0}^{\infty} X_j \subseteq X .$$

(ii) *The direct limit of a direct system of groups $\{f_j : G_j \longrightarrow G_{j+1}\}$ is a group $\varinjlim_j G_j$. For abelian G_j the direct limit is an abelian group, such that up to isomorphism*

$$\varinjlim_j G_j = \operatorname{coker}(1 - \sum_j f_j : \sum_{j=0}^{\infty} G_j \longrightarrow \sum_{j=0}^{\infty} G_j) .$$

(iii) *Homology commutes with direct limits: the direct limit of a direct system of short exact sequences of groups*

$$0 \longrightarrow G_j \longrightarrow H_j \longrightarrow K_j \longrightarrow 0 \quad (j \geq 0)$$

is a short exact sequence of groups

$$0 \longrightarrow \varinjlim_j G_j \longrightarrow \varinjlim_j H_j \longrightarrow \varinjlim_j K_j \longrightarrow 0 . \qquad \square$$

Definition 2.3 The *mapping telescope* or *homotopy direct limit* of a direct system of spaces $X_0 \xrightarrow{f_0} X_1 \xrightarrow{f_1} X_2 \longrightarrow \ldots$ is the identification space

$$\begin{aligned} \mathrm{Tel}(f_j) &= \operatorname*{hocolim}_{\overrightarrow{j}} X_j \\ &= \Big(\coprod_{j=0}^{\infty} X_j \times I\Big) \Big/ ((x_j, 1) = (f_j(x_j), 0)) . \end{aligned}$$

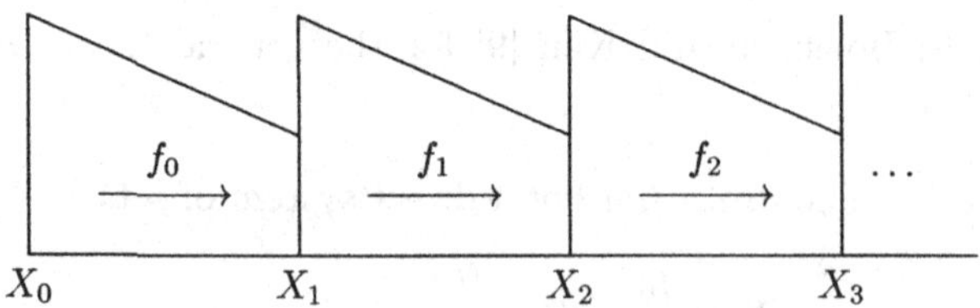

$\square$

Example 2.4 Let X be a CW complex and let

$$X_0 \subseteq X_1 \subseteq \ldots \subseteq X_j \subseteq X_{j+1} \subseteq \ldots \subseteq X$$

be a sequence of subcomplexes. The mapping telescope of the direct system

$\{f_j : X_j \longrightarrow X_{j+1}\}$ defined by the inclusions is a CW complex $\mathrm{Tel}(f_j)$ which is homotopy equivalent to the direct limit $\varinjlim_j X_j = \bigcup_j X_j \subseteq X$. □

Proposition 2.5 *Let $W = \mathrm{Tel}(f_j)$ be the mapping telescope of a direct system of spaces $\{f_j : X_j \longrightarrow X_{j+1}\}$.*

(i) *The homology of W is the direct limit of the induced direct system of homology groups*

$$H_*(W) = H_*(\underrightarrow{\mathrm{hocolim}}_j X_j) = \varinjlim_j H_*(X_j) .$$

(ii) *If each X_j is compact the one-point compactification W^∞ is contractible. Furthermore, the natural projection $p : e(W) \longrightarrow W$ is a homotopy equivalence, so that*

$$e(W) \simeq W .$$

Proof (i) Immediate from 2.2 (iii).

(ii) Define a map

$$g : [0,\infty) \times W \longrightarrow W ; (r,(x,s)) \longrightarrow (f_{j+u-1} \cdots f_{j+1} f_j(x), v)$$
$$(x \in X_j , r+s = u+v , u \in \mathbb{N} , v \in [0,1)) .$$

The map

$$h : W^\infty \times I \longrightarrow W^\infty ; (w,t) \longrightarrow \begin{cases} g\left(\dfrac{t}{1-t}, w\right) & \text{if } w \in W , \\ \infty & \text{if } w = \infty \end{cases}$$

defines a contraction of W^∞

$$h : \mathrm{id} \simeq \{\infty\} : W^\infty \longrightarrow W^\infty ,$$

and the map

$$W \longrightarrow e(W) ; w \longrightarrow (r \longrightarrow g(r,w))$$

defines a homotopy inverse for $p : e(W) \longrightarrow W$. □

Example 2.6 Given a compact space K let

$$W = K \times [0,\infty) = \mathrm{Tel}(f_j)$$

with $f_j = 1 : X_j = K \longrightarrow X_{j+1} = K$. The one-point compactification of W is the cone on K

$$W^\infty = K \times [0,\infty]/K \times \{\infty\} = cK ,$$

and by 1.7 there are defined homotopy equivalences

$$K \longrightarrow W ; x \longrightarrow (x,0) ,$$
$$K \longrightarrow e(W) ; x \longrightarrow (t \longrightarrow (x,t)) .$$

The map $K \longrightarrow e(W)$ sends each $x \in K$ to the ray joining it to the cone point $\infty \in W^\infty = cK$. □

Definition 2.7 The *inverse limit* of an inverse system of sets

$$X_0 \xleftarrow{f_1} X_1 \xleftarrow{f_2} X_2 \longleftarrow \dots$$

is the subset of the product $\prod_j X_j$

$$\varprojlim_j X_j = \{(x_0, x_1, x_2, \dots) \in \prod_{j=0}^{\infty} X_j \,|\, f_j(x_j) = x_{j-1}\} . \qquad \square$$

Proposition 2.8 (i) *For any set X and any sequence of subsets $\dots \subseteq X_2 \subseteq X_1 \subseteq X_0 \subseteq X$ the inverse limit of the inverse system $\{f_j : X_j \longrightarrow X_{j-1}\}$ defined by the inclusions is the intersection*

$$\varprojlim_j X_j = \bigcap_{j=0}^{\infty} X_j \subseteq X .$$

(ii) *The inverse limit of an inverse system of groups $\{G_j \longrightarrow G_{j-1}\}$ is a group $\varprojlim_j G_j$.* □

Definition 2.9 The *homotopy inverse limit* of an inverse system of spaces $\{f_j : X_j \longrightarrow X_{j-1}\}$ is the subspace of the product $\prod_j X_j^I$ of the path spaces X_j^I

$$\underleftarrow{\text{holim}}_j X_j = \{(\omega_0, \omega_1, \dots) \in \prod_{j=0}^{\infty} X_j^I \,|\, f_j(\omega_j(0)) = \omega_{j-1}(1)\} . \qquad \square$$

Example 2.10 (i) Let

$$X \supseteq X_0 \supseteq X_1 \supseteq \dots \supseteq X_j \supseteq X_{j+1} \supseteq \dots$$

be a sequence of subspaces of a space X. The homotopy inverse limit $\underleftarrow{\text{holim}}_j X_j$ of the inverse system $\{f_j : X_j \longrightarrow X_{j-1}\}$ defined by the inclusions is the space of paths $\omega : [0, \infty) \longrightarrow X$ such that $\omega([j, j+1]) \subseteq X_j$ $(j \geq 0)$.

(ii) For any inverse system of spaces $\{f_j : X_j \longrightarrow X_{j-1}\}$ there is defined an inclusion of the inverse limit in the homotopy inverse limit

$$\varprojlim_j X_j \longrightarrow \underleftarrow{\text{holim}}_j X_j \; ; \; (x_0, x_1, \dots) \longrightarrow (\omega_0, \omega_1, \dots) \;\; (\omega_j(I) = x_j) .$$

If each f_j is a fibration this is a homotopy equivalence.

(iii) Every map $f : X \longrightarrow Y$ can be replaced by a fibration: for path-connected Y the path space Y^I is the total space of a fibration $Y^I \longrightarrow Y$; $\omega \longrightarrow \omega(0)$ and is contractible, and the pullback along $f : X \longrightarrow Y$

$$E(f) = f^*(Y^I) = \{(x,\omega) \in X \times Y^I \mid f(x) = \omega(0) \in Y\}$$

is the total space of a fibration $p(f) : E(f) \longrightarrow Y; (x,\omega) \longrightarrow f(x)$ with

$$p(f) : E(f) \simeq X \xrightarrow{f} Y .$$

For any inverse system of spaces $\{f_j : X_j \longrightarrow X_{j-1}\}$ there is thus defined an inverse system $\{g_j : Y_j \longrightarrow Y_{j-1}\}$ with each g_j a fibration and

$$g_j : Y_j \simeq X_j \xrightarrow{f_j} X_{j-1} \simeq Y_{j-1} .$$

The homotopy inverse limit of $\{X_j\}$ is homotopy equivalent to the inverse limit of $\{Y_j\}$:

$$\underset{\overleftarrow{j}}{\operatorname{holim}} X_j \simeq \varprojlim_j Y_j . \qquad \square$$

Definition 2.11 The *derived limit* of an inverse system of groups $\{f_j : G_j \longrightarrow G_{j-1}\}$ is the pointed set

$$\varprojlim_j{}^1 G_j = \prod_{j=0}^{\infty} G_j \Big/ \sim$$

with $(x_j) \sim (y_j)$ if $y_j = z_j x_j f_{j+1}(z_{j+1})^{-1}$ for some $(z_j) \in \prod_j G_j$. $\square$

Proposition 2.12 (i) *The inverse and derived limits of an inverse system of abelian groups* $\{f_j : G_j \longrightarrow G_{j-1}\}$ *are the abelian groups*

$$\varprojlim_j G_j = \ker(1 - \prod_j f_j : \prod_{j=0}^{\infty} G_j \longrightarrow \prod_{j=0}^{\infty} G_j) ,$$

$$\varprojlim_j{}^1 G_j = \operatorname{coker}(1 - \prod_j f_j : \prod_{j=0}^{\infty} G_j \longrightarrow \prod_{j=0}^{\infty} G_j)$$

with an exact sequence

$$0 \longrightarrow \varprojlim_j G_j \longrightarrow \prod_{j=0}^{\infty} G_j \xrightarrow{1-\prod_j f_j} \prod_{j=0}^{\infty} G_j \longrightarrow \varprojlim_j{}^1 G_j \longrightarrow 0 .$$

(ii) *The inverse and derived limits of an inverse system of short exact sequences of abelian groups*

$$0 \longrightarrow G_j \longrightarrow H_j \longrightarrow K_j \longrightarrow 0 \;\; (j \geq 0)$$

are related by a long exact sequence

$$\begin{aligned} 0 \longrightarrow \varprojlim_j G_j \longrightarrow \varprojlim_j H_j \longrightarrow \varprojlim_j K_j \\ \longrightarrow \varprojlim_j{}^1 G_j \longrightarrow \varprojlim_j{}^1 H_j \longrightarrow \varprojlim_j{}^1 K_j \longrightarrow 0 . \end{aligned}$$ □

Proposition 2.13 (Bousfield and Kan [9, p. 254]) *The homotopy groups of a homotopy inverse limit fit into exact sequences*

$$0 \longrightarrow \varprojlim_j{}^1 \pi_{r+1}(X_j) \longrightarrow \pi_r(\underset{\overleftarrow{j}}{\text{holim}}\, X_j) \longrightarrow \varprojlim_j \pi_r(X_j) \longrightarrow 0 .$$ □

A map of spaces $f : X \longrightarrow Y$ is a *weak homotopy equivalence* if it induces a bijection $f_* : \pi_0(X) \longrightarrow \pi_0(Y)$ of the sets of path components, and on each component f induces isomorphisms $f_* : \pi_*(X) \longrightarrow \pi_*(Y)$ of the homotopy groups. A weak homotopy equivalence $f : X \longrightarrow Y$ induces homology isomorphisms $f_* : H_*(X) \longrightarrow H_*(Y)$ by the Hurewicz theorem. If X and Y have the homotopy types of CW complexes then f is a homotopy equivalence, by Whitehead's theorem.

Proposition 2.14 *Let W be a σ-compact space with closed cocompact subspaces $W_j \subseteq W$ for $j = 0, 1, 2, \ldots$ such that*

$$W_{j+1} \subseteq W_j \ , \quad \bigcap_{j=0}^{\infty} W_j \ = \ \emptyset \ ,$$

and write the inclusions as

$$g_j \ : \ W_j \longrightarrow W_{j-1} \quad (j \geq 1) \ .$$

The map

$$f \ : \ \underset{\overleftarrow{j}}{\text{holim}}\, W_j \longrightarrow e(W) \ ; \ (\omega_j) \longrightarrow (t \longrightarrow \omega_j(t-j)) \quad (j \leq t \leq j+1)$$

is a weak homotopy equivalence, so that

$$\pi_*(\underset{\overleftarrow{j}}{\text{holim}}\, W_j) \ = \ \pi_*(e(W))$$

and there are defined exact sequences

$$0 \longrightarrow \varprojlim_j{}^1 \pi_{r+1}(W_j) \longrightarrow \pi_r(e(W)) \longrightarrow \varprojlim_j \pi_r(W_j) \longrightarrow 0 .$$

Proof Let $\alpha : B^n \longrightarrow e(W)$ and $\beta : \partial B^n \longrightarrow \underset{\overleftarrow{j}}{\text{holim}}\, W_j$ be maps such that $f \circ \beta = \alpha | \partial B^n$ where $B^n \subset \mathbb{R}^n$ is the unit ball. It suffices to construct a map $\widetilde{\beta} : B^n \longrightarrow \underleftarrow{\text{holim}}_j W_j$ such that $\widetilde{\beta}|\partial B^n = \beta$ and $f \circ \widetilde{\beta} \simeq \alpha$ rel ∂B^n. Note that an element $\omega \in e(W)$ is in the image of $f : \underleftarrow{\text{holim}}_j W_j \longrightarrow e(W)$ if and only if $\omega([j,\infty)) \subseteq W_j$ for each $j = 0,1,2,\ldots$. Therefore, the adjoint of α is a proper map

$$\widehat{\alpha} \ : \ B^n \times [0,\infty) \longrightarrow W \ ; \ (x,t) \longrightarrow \alpha(x)(t)$$

such that $\widehat{\alpha}(\partial B^n \times [j,\infty)) \subseteq W_j$ for $j = 0,1,2,\ldots$ and the problem is reduced (via 1.9) to showing that $\widehat{\alpha}$ is properly homotopic rel $\partial B^n \times [0,\infty)$ to a map $\widetilde{\alpha} : B^n \times [0,\infty) \longrightarrow W$ such that $\widetilde{\alpha}(B^n \times [j,\infty)) \subseteq W_j$ for $j = 0,1,2,\ldots$. By using a proper homotopy of $\widehat{\alpha}$ rel $\partial B^n \times [0,\infty)$, we may assume $\widehat{\alpha}(x,t) \in W_j$ for $\|x\| \geq \frac{1}{2}$, $t \geq j$, and $j = 0,1,2,\ldots$. Since $\widehat{\alpha}$ is proper, there is a sequence $N_0 < N_1 < N_2 < \ldots$ such that

$$\widehat{\alpha}(B^n \times [N_j,\infty)) \subseteq W_j \quad (j = 0,1,2,\ldots) \ .$$

Thus $\widehat{\alpha}(x,t) \in W_j$ if $t \geq N_j$, or if $t \geq j$ and $\|x\| \geq \frac{1}{2}$ for $j = 0,1,2,\ldots$. Construct a homeomorphism

$$h \ : \ B^n \times [0,\infty) \longrightarrow B^n \times [0,\infty)$$

such that h is isotopic to the identity rel $\partial B^n \times [0,\infty)$ and

$$h(B^n \times [j,\infty)) \subseteq \{(x,t) \in B^n \times [0,\infty) \,|\, t \geq N_j, \text{ or } t \geq j \text{ and } \|x\| \geq \tfrac{1}{2}\} \ .$$

Then $\widetilde{\alpha} = \widehat{\alpha} \circ h$ is properly homotopic to $\widehat{\alpha}$ rel $\partial B^n \times [0,\infty)$ and

$$\widetilde{\alpha}(B^n \times [j,\infty)) \subseteq W_j \quad (j = 0,1,2,\ldots) \ . \qquad \square$$

Remark 2.15 (i) The exact sequences of 2.14 also appear in Brin and Thickstun [10], with $\pi_*(e(W))$ the 'absolute Steenrod homotopy groups' of the ends of W.

(ii) In 7.10 it will be proved that if W is 'forward tame' the $\underleftarrow{\lim}^1$ terms in 2.14 vanish, and that

$$\pi_*(e(W)) \ = \ \underleftarrow{\lim}_j \pi_*(W_j) \ \ , \ \ H_*(e(W)) \ = \ \underleftarrow{\lim}_j H_*(W_j) \ . \qquad \square$$

Definition 2.16 The *mapping cotelescope* of an inverse system of spaces

$$X_0 \xleftarrow{f_1} X_1 \xleftarrow{f_2} X_2 \longleftarrow \ \ldots$$

is the space

$$\mathcal{W}(f_j) \ = \ \Big(\coprod_{k=0}^{\infty} X_k \times I\Big)\Big/(x_k,0) = (f_k(x_k),1) \ .$$

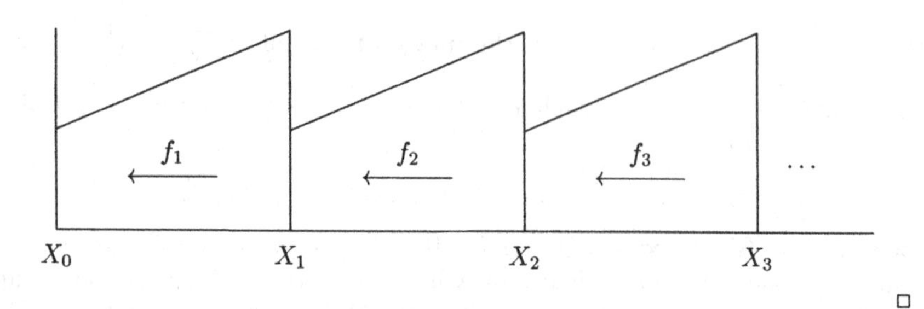

□

Proposition 2.17 *Let $W = \mathcal{W}(f_j)$ be the mapping cotelescope of an inverse system of spaces $\{f_j : X_j \longrightarrow X_{j-1}\}$.*

(i) *The inclusion*

$$X_0 \longrightarrow \mathcal{W}(f_j) \; ; \; x_0 \longrightarrow (x_0, 0)$$

is a homotopy equivalence (but not in general a proper homotopy equivalence).

(ii) *If each X_j is a compact space then the end space of W is weak homotopy equivalent to the homotopy inverse limit*

$$e(W) \simeq \underset{\overleftarrow{j}}{\text{holim}}\, X_j \; .$$

Proof (i) The projection

$$W \longrightarrow X_0 \; ; \; (x_j, t) \longrightarrow f_1 f_2 \dots f_j(x_j)$$

is a homotopy inverse for the inclusion $X_0 \longrightarrow W$.

(ii) Apply 2.14 with

$$W_j = \mathcal{W}(f_i \,|\, i \geq j) \subseteq W = \mathcal{W}(f_j) \; . \qquad \square$$

Example 2.18 Fix an integer $s \geq 2$, and consider the inverse system of fibrations

$$f_j = s : X_j = S^1 \longrightarrow X_{j-1} = S^1 \; ; \; z \longrightarrow z^s \quad (j \geq 1) \; .$$

The end space $e(\mathcal{W}(f_j))$ of the mapping cotelescope $\mathcal{W}(f_j)$ is homotopy equivalent to the *s-adic solenoid*

$$\widehat{S}^1_s = \underset{\overleftarrow{j}}{\lim}\, X_j = \{(x_0, x_1, x_2, \dots) \in \prod_{j=0}^{\infty} S^1 \,|\, x^s_{j+1} = x_j \text{ for } j \geq 0\} \; .$$

See Bourbaki [8, III, 7, Ex. 6, p. 325] for the topological properties of $\widehat{S}^1_s$. See 23.25 and 23.28 for the homological properties of $\widehat{S}^1_s$. □

Proposition 2.19 (i) *The direct limit of a direct system of chain complexes and chain maps*

$$f(j) \;:\; C(j) \longrightarrow C(j+1) \quad (j \geq 0)$$

is a chain complex

$$\varinjlim_j C(j) \;=\; \mathrm{coker}(1 - \sum_j f(j) : \sum_{j=0}^{\infty} C(j) \longrightarrow \sum_{j=0}^{\infty} C(j))$$

such that

$$H_r(\varinjlim_j C(j)) \;=\; \varinjlim_j H_r(C(j)) \;.$$

(ii) *The inverse and derived limits of an inverse system of chain complexes and chain maps*

$$g(j) \;:\; C(j) \longrightarrow C(j-1) \quad (j \geq 1)$$

are the chain complexes

$$\varprojlim_j C(j) \;=\; \ker(1 - \prod_j g(j) : \prod_{j=0}^{\infty} C(j) \longrightarrow \prod_{j=0}^{\infty} C(j)) \;,$$

$$\varprojlim_j{}^1 C(j) \;=\; \mathrm{coker}(1 - \prod_j g(j) : \prod_{j=0}^{\infty} C(j) \longrightarrow \prod_{j=0}^{\infty} C(j)) \;.$$

If $\varprojlim_j{}^1 C(j) = 0$ *there are defined short exact sequences*

$$0 \longrightarrow \varprojlim_j{}^1 H_{r+1}(C(j)) \longrightarrow H_r(\varprojlim_j C(j)) \longrightarrow \varprojlim_j H_r(C(j)) \longrightarrow 0 \;.$$

If $\varprojlim_j C(j) = 0$ *there are defined short exact sequences*

$$0 \longrightarrow \varprojlim_j{}^1 H_r(C(j)) \longrightarrow H_r(\varprojlim_j{}^1 C(j)) \longrightarrow \varprojlim_j H_{r-1}(C(j)) \longrightarrow 0 \;.$$

Proof (i) The direct limit fits into a short exact sequence of chain complexes

$$0 \longrightarrow \sum_{j=1}^{\infty} C(j) \xrightarrow{1-\sum_j f(j)} \sum_{j=1}^{\infty} C(j) \longrightarrow \varinjlim_j C(j) \longrightarrow 0 \;,$$

and the homology exact sequence breaks up into short exact sequences

$$0 \longrightarrow \sum_{j=1}^{\infty} H_*(C(j)) \xrightarrow{1-\sum_j f(j)_*} \sum_{j=1}^{\infty} H_*(C(j)) \longrightarrow H_*(\varinjlim_j C(j)) \longrightarrow 0 \;.$$

(ii) If $\varprojlim_j^1 C(j) = 0$ the inverse limit fits into a short exact sequence of chain complexes

$$0 \longrightarrow \varprojlim_j C(j) \longrightarrow \prod_{j=1}^{\infty} C(j) \xrightarrow{1-\prod_j g(j)} \prod_{j=1}^{\infty} C(j) \longrightarrow 0 ,$$

inducing a long exact sequence of homology groups

$$\dots \longrightarrow \prod_{j=1}^{\infty} H_{r+1}(C(j)) \xrightarrow{1-\prod_j g(j)} \prod_{j=1}^{\infty} H_{r+1}(C(j))$$
$$\longrightarrow H_r(\varprojlim_j C(j)) \longrightarrow \prod_{j=1}^{\infty} H_r(C(j)) \longrightarrow \dots .$$

If $\varprojlim_j C(j) = 0$ the derived limit fits into a short exact sequence of chain complexes

$$0 \longrightarrow \prod_{j=1}^{\infty} C(j) \xrightarrow{1-\prod_j g(j)} \prod_{j=1}^{\infty} C(j) \longrightarrow \varprojlim_j^1 C(j) \longrightarrow 0 ,$$

inducing a long exact sequence of homology groups

$$\dots \longrightarrow \prod_{j=1}^{\infty} H_r(C(j)) \xrightarrow{1-\prod_j g(j)} \prod_{j=1}^{\infty} H_r(C(j))$$
$$\longrightarrow H_r(\varprojlim_j^1 C(j)) \longrightarrow \prod_{j=1}^{\infty} H_{r-1}(C(j)) \longrightarrow \dots .$$

□

Definition 2.20 Let $\{f_j : G_j \longrightarrow G_{j-1} \,|\, j \geq 1\}$ be an inverse system of groups.

(i) The inverse system is *Mittag–Leffler* if there exists $k \geq 1$ such that the morphisms

$$f_j| \; : \; \mathrm{im}(f_{j+1}) \longrightarrow \mathrm{im}(f_j)$$

are onto for all $j \geq k$.

(ii) The inverse system is *stable* if there exists $k \geq 1$ such that the morphisms

$$f_j| \; : \; \mathrm{im}(f_{j+1}) \longrightarrow \mathrm{im}(f_j)$$

are isomorphisms for all $j \geq k$. □

Proposition 2.21 *Let $\{f_j : G_j \longrightarrow G_{j-1} \,|\, j \geq 1\}$ be an inverse system of groups.*

(i) *$\{G_j\}$ has the same inverse and derived limits as $\{\mathrm{im}(f_j) \longrightarrow \mathrm{im}(f_{j-1})\}$:*

$$\varprojlim_j G_j = \varprojlim_j \mathrm{im}(f_j) \ , \ \varprojlim_j{}^1 G_j = \varprojlim_j{}^1 \mathrm{im}(f_j) \ .$$

(ii) *If $\{G_j\}$ is Mittag–Leffler then*

$$\varprojlim_j{}^1 G_j = \varprojlim_j{}^1 \mathrm{im}(f_j) = 0 \ .$$

(iii) *If $\{G_j\}$ is stable then it is Mittag–Leffler, with*

$$\varprojlim_j G_j = \mathrm{im}(f_k) = \mathrm{im}(f_{k+1}) = \dots \ , \ \varprojlim_j{}^1 G_j = 0 \ ,$$

for sufficiently large $k \geq 0$.

(iv) *An inverse system $\{G_j\}$ is stable if and only if there exist a group H and an integer $k \geq 0$ with morphisms*

$$p_j : H \longrightarrow G_j \ , \ q_j : G_j \longrightarrow H \ (j \geq k)$$

such that the diagrams

$$\begin{array}{ccc} H & = & H \\ {\scriptstyle p_j}\downarrow & \overset{q_j}{\nearrow} & \downarrow{\scriptstyle p_{j-1}} \\ G_j & \xrightarrow{f_j} & G_{j-1} \end{array}$$

commute, in which case

$$\varprojlim_j G_j = \mathrm{im}(f_k) = \mathrm{im}(f_{k+1}) = \dots = H \ , \ \varprojlim_j{}^1 G_j = 0$$

and up to isomorphism

$$f_j : G_j = H \times K_j \longrightarrow G_{j-1} = H \times K_{j-1} \ ; \ (x,y) \longrightarrow (x,1) \ \ (j > k)$$

with $K_j = \ker(f_j) = \mathrm{coker}(p_j) = \ker(q_j)$.

Proof (i) By 2.12 (ii) the short exact sequences

$$0 \longrightarrow \ker(f_j) \longrightarrow G_j \longrightarrow \mathrm{im}(f_j) \longrightarrow 0$$

determine a long exact sequence

$$\begin{aligned} 0 \longrightarrow \varprojlim_j \ker(f_j) \longrightarrow \varprojlim_j G_j \longrightarrow \varprojlim_j \mathrm{im}(f_j) \\ \longrightarrow \varprojlim_j{}^1 \ker(f_j) \longrightarrow \varprojlim_j{}^1 G_j \longrightarrow \varprojlim_j{}^1 \mathrm{im}(f_j) \longrightarrow 0 \ . \end{aligned}$$

It follows from $f_j| = 0 : \ker(f_j) \longrightarrow \ker(f_{j-1})$ that

$$\varprojlim_j \ker(f_j) = 0 \ , \ \varprojlim_j{}^1 \ker(f_j) = 0 \ .$$

(ii) By (i) the derived limit of $\{G_j\}$ is the same as the derived limit of $\{\text{im}(f_j)\}$, so it suffices to note that $\varprojlim_j{}^1 G_j = 0$ if each $f_j : G_j \longrightarrow G_{j-1}$ is onto.

(iii) Immediate from (i) and (ii).

(iv) If $\{G_j\}$ is an inverse system which is stable, let

$$\begin{aligned} H &= \text{im}(f_k) = \text{im}(f_{k+1}) = \dots , \\ p_j &= \text{inclusion} : H = \text{im}(f_{j+1}) \longrightarrow G_j , \\ q_j &= \text{projection} : G_j \longrightarrow H = \text{im}(f_j) . \end{aligned}$$

Conversely, given H, k, p_j, q_j note that since $q_j p_j = \text{id} : H \longrightarrow H$ each p_j is one-to-one and each q_j is surjective, and since $f_j = p_{j-1} q_j : G_j \longrightarrow G_{j-1}$ it is possible to identify $H = \text{im}(f_j)$ for $j \geq k$, and hence to apply (iii). □

Remark 2.22 Geoghegan [62] has proved the converse of 2.21 (ii) in the countable case: if $\{G_j\}$ is an inverse system of countable groups such that $\varprojlim_j{}^1 G_j = \{1\}$ then $\{G_j\}$ is Mittag–Leffler. □

As in 1.22 let $\eta_W : \pi_0(e(W)) \longrightarrow \mathcal{E}_W$ be the function which associates an end of a space W to each path component of the end space $e(W)$.

Definition 2.23 (i) A σ-compact space W is *path-connected at* ∞ if every cocompact subspace of W contains a path-connected cocompact subspace of W, or equivalently if there exists a sequence $W \supset W_0 \supset W_1 \supset W_2 \supset \dots$ of path-connected cocompact subspaces with $\bigcap_j \text{cl}(W_j) = \emptyset$.

(ii) A space W is *semistable at* ∞ if any two proper maps $\omega_1, \omega_2 : [0, \infty) \longrightarrow W$ with $\eta_W([\omega_1]) = \eta_W([\omega_2])$ are properly homotopic.

(iii) A space W has *stable* π_1 *at* ∞ if it is path-connected at ∞, and there exists a sequence as in (i) such that the sequence of inclusion induced group morphisms

$$\pi_1(W_0) \xleftarrow{g_1} \pi_1(W_1) \xleftarrow{g_2} \pi_1(W_2) \longleftarrow \cdots$$

(with base points and base paths chosen) is stable, i.e. for some $k \geq 1$ there are induced isomorphisms

$$\text{im}(g_k) \xleftarrow{\cong} \text{im}(g_{k+1}) \xleftarrow{\cong} \dots .$$

(iv) If W has stable π_1 at ∞, define the *fundamental group at* ∞

$$\pi_1^\infty(W) = \varprojlim_j \pi_1(W_j)$$

for some fixed sequence $\{W_j\}$ as above, with

$$\pi_1^\infty(W) = \text{im}(g_k) = \text{im}(g_{k+1}) = \dots . \qquad \square$$

Example 2.24 If $(M, \partial M)$ is a compact manifold with boundary then each component L of ∂M is collared in the complement $W = M \backslash L$, so that W has stable π_1 at infinity with $\pi_1^\infty(W) = \pi_1(L)$. □

Proposition 2.25 (i) *If W is σ-compact and locally path-connected, then $\eta_W : \pi_0(e(W)) \longrightarrow \mathcal{E}_W$ is surjective.*

(ii) *W is semistable at ∞ if and only if η_W is injective.*

(iii) *If W is σ-compact and locally path-connected, then W is semistable at ∞ if and only if η_W is bijective.*

(iv) *If W is σ-compact, locally path-connected and semistable at ∞, then W is path-connected at ∞ if and only if $\pi_0(e(W)) = 0$.*

Proof (i) Write

$$W = \bigcup_{j=0}^{\infty} K_j$$

with each K_j compact and $K_j \subseteq K_{j+1}$. Let ϵ be an end of W (in the sense of 1.14). For each j, choose $x_j \in \epsilon(K_j)$. Then $x_j, x_{j+1} \in \epsilon(K_j)$ and $\epsilon(K_j)$ is a component of an open subset of a locally path-connected space. Hence, $\epsilon(K_j)$ is path-connected, so there is a map

$$\omega_j : [j, j+1] \longrightarrow \epsilon(K_j)$$

with $\omega_j(j) = x_j$ and $\omega_j(j+1) = x_{j+1}$. Then the ω_j's amalgamate to define a proper map $\omega : [0, \infty) \longrightarrow W$ with $\eta_W([\omega]) = \epsilon$.

(ii) Immediate from the Definition 2.23.

(iii) follows from (i) and (ii).

(iv) Since η_W is bijective, we need to show that W is path-connected at ∞ if and only if W has exactly one end. Suppose that W is path-connected at ∞ and that ϵ_1, ϵ_2 are distinct ends of W. Then there is a compact subspace $K \subseteq W$ such that $\epsilon_1(K) \neq \epsilon_2(K)$, and so $\epsilon_1(K) \cap \epsilon_2(K) = \emptyset$. Let X be a path-connected cocompact subspace of $W \backslash K$. Since $\epsilon_1(K), \epsilon_2(K)$ are unbounded

$$\epsilon_1(K) \cap X \neq \emptyset \neq \epsilon_2(K) \cap X.$$

But X must be contained in exactly one of the components of $W \backslash K$, a contradiction.

On the other hand, if ϵ is the only end of W, then for every cocompact subspace $X \subseteq W$, $\epsilon(\mathrm{cl}(W \backslash X))$ must be a path-connected cocompact subspace of W. □

Example 2.26 *Jacob's ladder* can be realized as the subspace $X \subset \mathbb{R}^2$ defined by

$$X = \{(x, y) \mid x = 0, 1,\ y \geq 0\} \cup \{(x, n) \mid 0 \leq x \leq 1,\ n = 1, 2, 3, \ldots\}.$$

Then X has exactly one end, but $\pi_0(e(X))$ is infinite and X is not semistable at ∞.

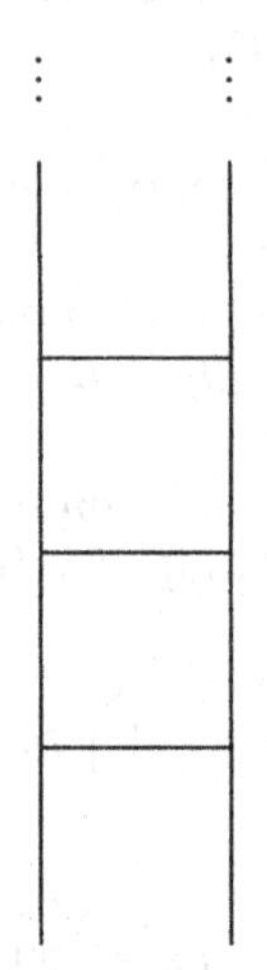

Jacob's ladder (Genesis 28:12)

□

Remark 2.27 (i) In 9.5 below it will be proved that if X and Y are proper homotopy equivalent then X is semistable (resp. has stable π_1) at ∞ if and only if Y is semistable (resp. has stable π_1) at ∞.

(ii) Let $W = \bigcup_j K_j$ be a locally path-connected σ-compact space, which is expressed as a union of compact subspaces $K_j \subseteq K_{j+1}$. The complements $W_j = W \backslash K_j$ are cocompact subsets of W such that

$$W_j \supseteq W_{j+1} \ , \ \bigcap_{j=0}^{\infty} W_j \ = \ \emptyset \ ,$$

so that 2.14 gives an exact sequence

$$0 \longrightarrow \varprojlim_j{}^1 \pi_1(W_j) \longrightarrow \pi_0(e(W)) \longrightarrow \varprojlim_j \pi_0(W_j) \longrightarrow 0 \ ,$$

with $\varprojlim_j \pi_0(W_j) = \mathcal{E}_W$ the number of ends of W. Mihalik [93] proves that W is semistable at ∞ if and only if $\varprojlim_j{}^1 \pi_1(W_j) = 0$, giving another proof of 2.25 (iii). If W has stable π_1 at ∞ the $\varprojlim^1$ term in the exact sequence of 2.14

$$0 \longrightarrow \varprojlim_j{}^1 \pi_2(W_j) \longrightarrow \pi_1(e(W)) \longrightarrow \varprojlim_j \pi_1(W_j) \longrightarrow 0$$

vanishes, and

$$\varprojlim_j \pi_1(W_j) \;=\; \pi_1^\infty(W) \;=\; \pi_1(e(W)) \;.$$

(iii) A well-known conjecture states that if W is a finite connected CW complex, then the universal cover $\widetilde{W}$ is semistable at ∞. This is known to be a property of $\pi_1(W)$ and has been verified in many special cases. See Mihalik [93], Mihalik and Tschantz [94]. □

The homological properties of non-compact spaces are closely related to the localization and completion of rings. Here is a brief account of these constructions, in the special cases of the localization inverting a single element and the completion with respect to a principal ideal.

Definition 2.28 Let A be a ring (associative, with 1) and let $s \in A$ be a central non-zero divisor.

(i) The *localization* of A inverting s is the ring $A[1/s]$ with elements the equivalence classes a/s^j of pairs (a, s^j) $(a \in A,\ j \geq 0)$, subject to the equivalence relation

$$(a, s^j) \sim (b, s^k) \text{ if } as^k \;=\; bs^j \in A \;,$$

and the usual addition and multiplication of fractions. The localization is (up to isomorphism) the direct limit

$$A[1/s] \;=\; \varinjlim(A \xrightarrow{s} A \xrightarrow{s} A \xrightarrow{s} A \longrightarrow \dots) \;,$$

with an actual isomorphism defined by

$$\varinjlim_j A \xrightarrow{\simeq} A[1/s] \;;\; (a_j) \longrightarrow \sum_{j=0}^{\infty} a_j/s^j \;.$$

The ring morphism

$$A \longrightarrow A[1/s] \;;\; a \longrightarrow a/1 \;\;(s^0 = 1)$$

is an injection, with cokernel the derived limit

$$A[1/s]/A \;=\; \varprojlim{}^1(A \xleftarrow{s} A \xleftarrow{s} A \xleftarrow{s} A \longleftarrow \dots) \;.$$

(ii) The *s-adic completion* of A is the inverse limit of the natural projections $A/s^{j+1}A \longrightarrow A/s^jA$, the ring

$$\widehat{A}_s \;=\; \varprojlim(A/sA \longleftarrow A/s^2A \longleftarrow A/s^3A \longleftarrow \dots) \;.$$

The ring morphism

$$A \longrightarrow \widehat{A}_s \;;\; a \longrightarrow (a)$$

has kernel the inverse limit

$$\bigcap_{j=0}^{\infty} s^j A \;=\; \varprojlim(A \xleftarrow{s} A \xleftarrow{s} A \xleftarrow{s} A \longleftarrow \ldots)\,,$$

and cokernel the derived limit

$$\widehat{A}_s/A \;=\; \varprojlim{}^1(A \xleftarrow{s} A \xleftarrow{s} A \xleftarrow{s} A \longleftarrow \ldots) \;=\; A[1/s]/A\,.$$

See 23.20 below for the cartesian square of rings relating localization and completion. □

Example 2.29 The localization inverting

$$s \;=\; z \in A \;=\; \mathbb{Z}[z]$$

is the *Laurent polynomial extension* of $\mathbb{Z}$

$$A[1/s] \;=\; \mathbb{Z}[z, z^{-1}]\,,$$

and the s-adic completion of A is the *formal power series ring* of $\mathbb{Z}$

$$\widehat{A}_s \;=\; \mathbb{Z}[[z]]\,.$$

See Chapter 21 for more on polynomial extension rings. □

Example 2.30 Let $W = W_0 \supset W_1 \supset W_2 \supset \ldots$ be the sequence of subspaces of $\mathbb{R}^{n+2}$ with W_j the union of the line $\{(x, 0, \ldots, 0) \,|\, x \geq j\}$ and a copy of S^{n+1} wedged on at $(k, 0, \ldots, 0)$ for each integer $k \geq j$.

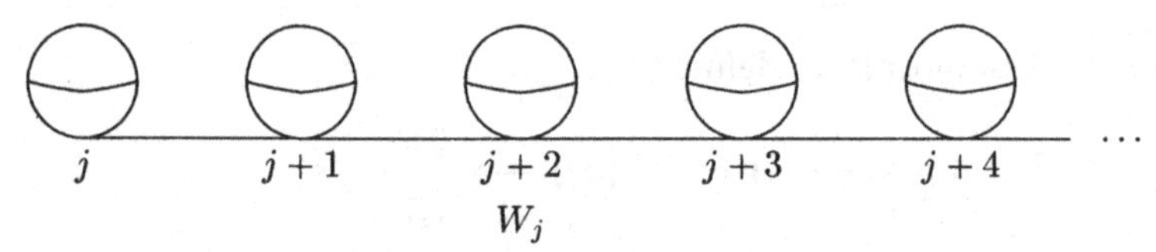

W_j

Use the inclusion $\omega : \mathbb{R}^+ \longrightarrow W$ as the base point $\omega \in e(W)$. For $n = 0$ W is proper homotopy equivalent to Jacob's ladder (2.26). For $n \geq 1$ the homotopy groups of the W_j's in dimensions $n, n+1$ are given by

$$\pi_n(W_j) \;=\; 0 \;\;,\;\; \pi_{n+1}(W_j) \;=\; \mathbb{Z}[z]\,,$$

with the inclusion $W_{j+1} \longrightarrow W_j$ inducing

$$z \;:\; \pi_{n+1}(W_{j+1}) \;=\; \mathbb{Z}[z] \longrightarrow \pi_{n+1}(W_j) \;=\; \mathbb{Z}[z]\,,$$

so that

$$\varprojlim_j \pi_n(W_j) \;=\; 0 \;\;,\;\; \pi_n(e(W)) \;=\; \varprojlim_j{}^1 \pi_{n+1}(W_j) \;=\; \mathbb{Z}[[z]] \neq 0\,.$$

Moreover, for $n \geq 1$ W has stable π_1 at ∞. □

3

Homology at infinity

The homology at infinity $H_*^\infty(W)$ of a space W is the proper homotopy invariant given by the difference between homology $H_*(W)$ and locally finite homology $H_*^{lf}(W)$. The extent to which a space W is non-compact is measured in the first instance by the failure of the natural maps $i : H_*(W) \longrightarrow H_*^{lf}(W)$ to be isomorphisms, or equivalently by the extent to which $H_*^\infty(W)$ is non-zero. The homology groups of the end space $e(W)$ are related to the homology at infinity by morphisms $H_*(e(W)) \longrightarrow H_*^\infty(W)$, which are isomorphisms if W is forward tame (in the sense of Chapter 7).

Locally finite homology is as important in studying non-compact spaces as ordinary homology is important in dealing with compact spaces. Since there is no elementary account of locally finite homology in the literature, we provide one here.

We shall also investigate the connection between the locally finite homology $H_*^{lf}(W)$ of a space W and the reduced homology of the one-point compactification W^∞

$$\widetilde{H}_*(W^\infty) = H_*(W^\infty, \{\infty\}) .$$

In general, these homology groups are not isomorphic – see 3.18 below for an actual example. In 3.16 we identify the singular locally finite chain complex $S^{lf}(W)$ of a σ-compact space W with an inverse limit of singular chain complexes, the singular chain complex at infinity $S^\infty(W)$ with a derived limit of singular chain complexes, and the singular locally finite homology with the homology of an inverse limit of ordinary singular chain complexes involving W^∞. In Chapter 7 we use 3.16 to prove that $H_*^{lf}(W)$ is isomorphic to $H_*(W^\infty, \{\infty\})$ for a forward tame W. 3.16 is used in Appendix A, which relates locally finite singular and cellular homology to each other.

A *singular r-chain* in W is a formal linear combination $\sum\limits_{\alpha} n_\alpha \sigma_\alpha$ with coefficients $n_\alpha \in \mathbb{Z}$ of singular r-simplexes $\sigma_\alpha : \Delta^r \longrightarrow W$. The *singular chain complex* $S(W)$ is the chain complex with $S_r(W)$ the abelian group of singular r-chains, and the usual differentials. The *(singular) homology* of W is defined by

$$H_*(W) = H_*(S(W)) .$$

A map $f : V \longrightarrow W$ induces a chain map $f : S(V) \longrightarrow S(W)$, which in turn induces morphisms in homology $f_* : H_*(V) \longrightarrow H_*(W)$.

Definition 3.1 (i) A *locally finite singular r-chain* in W is a product $\prod\limits_{\alpha} n_\alpha \sigma_\alpha$ of formal multiples by coefficients $n_\alpha \in \mathbb{Z}$ of singular r-simplexes $\sigma_\alpha : \Delta^r \longrightarrow W$ such that for each $x \in W$ there exists an open neighbourhood $U \subseteq W$ of x with $\{\alpha \,|\, U \cap \sigma_\alpha(\Delta^r) \neq \emptyset,\, n_\alpha \neq 0\}$ finite.

(ii) The *locally finite singular chain complex* $S^{lf}(W)$ is the chain complex with $S^{lf}_r(W)$ the abelian group of locally finite singular r-chains, and the usual differentials. The *locally finite homology* of W is defined by

$$H^{lf}_*(W) = H_*(S^{lf}(W)) . \qquad \square$$

A map $f : V \longrightarrow W$ is *closed* if $f(C)$ is a closed subspace of W for every closed subspace $C \subseteq V$.

Proposition 3.2 *A proper closed map $f : V \longrightarrow W$ induces a chain map $f : S^{lf}(V) \longrightarrow S^{lf}(W)$, which in turn induces morphisms in locally finite homology $f_* : H^{lf}_*(V) \longrightarrow H^{lf}_*(W)$.* $\square$

In fact, for W locally compact Hausdorff or metric every proper map $f : V \longrightarrow W$ is closed.

The relative singular and locally finite singular chain complexes of a pair of spaces $(W, V \subseteq W)$

$$S(W,V) = \mathrm{coker}(S(V) \longrightarrow S(W)) ,$$
$$S^{lf}(W,V) = \mathrm{coker}(S^{lf}(V) \longrightarrow S^{lf}(W))$$

fit into the short exact sequences

$$0 \longrightarrow S(V) \longrightarrow S(W) \longrightarrow S(W,V) \longrightarrow 0 ,$$
$$0 \longrightarrow S^{lf}(V) \longrightarrow S^{lf}(W) \longrightarrow S^{lf}(W,V) \longrightarrow 0 .$$

Proposition 3.3 *If W is a space and $V \subseteq W$ is a closed cocompact subspace then*

$$S(W,V) = S^{lf}(W,V) \ , \ H_*(W,V) = H^{lf}_*(W,V) .$$

Proof The inclusion $S(W) \longrightarrow S^{lf}(W)$ induces a chain isomorphism

$$S(W,V) \longrightarrow S^{lf}(W,V) \; ; \; \sum n_\alpha \sigma_\alpha \longrightarrow \sum n_\alpha \sigma_\alpha$$

with inverse

$$S^{lf}(W,V) \longrightarrow S(W,V) \; ; \; \prod n_\alpha \sigma_\alpha \longrightarrow \sum n_\beta \sigma_\beta \; ,$$

where the sum is taken over all the $\beta = \alpha$ with $|\sigma_\beta| \cap \mathrm{cl}(W \backslash V) \neq \emptyset$ and $n_\beta \neq 0$. □

Example 3.4 If K is a compact space

$$S(K) \; = \; S^{lf}(K) \; , \; H_*(K) \; = \; H_*^{lf}(K) \; .$$

This is just the special case $(W,V) = (K, \emptyset)$ of 3.3. □

Definition 3.5 (i) A chain map $f : C \longrightarrow D$ is a *chain equivalence* if there exist a chain map $g : D \longrightarrow C$ and chain homotopies

$$h \; : \; gf \; \simeq \; 1_C \; : \; C \longrightarrow C \quad , \quad k \; : \; fg \; \simeq \; 1_D \; : \; D \longrightarrow D \; .$$

(ii) A chain map $f : C \longrightarrow D$ is a *homology equivalence* if it induces isomorphisms $f_* : H_*(C) \longrightarrow H_*(D)$ in homology. □

Every chain equivalence is a homology equivalence. A chain map $f : C \longrightarrow D$ of projective A-module chain complexes which are bounded below is a chain equivalence if and only if f is a homology equivalence.

The locally finite singular chain groups $S^{lf}(W)_r$ are not in general free $\mathbb{Z}$-modules – see 3.18 for an example. (Note, however, that for any field F the F-modules $S^{lf}(W;F)_r = F \otimes_{\mathbb{Z}} S^{lf}(W)_r$ are free.) In fact, $S^{lf}(W)$ may not even be chain equivalent to a free $\mathbb{Z}$-module chain complex. We shall therefore have to be careful to distinguish between homology and chain equivalences of chain complexes.

Given a chain complex C and any $k \in \mathbb{Z}$ let C_{*+k} denote the k-fold suspension of C, the chain complex defined by

$$d_{C_{*+k}} \; = \; d_C \; : \; (C_{*+k})_r \; = \; C_{r+k} \longrightarrow (C_{*+k})_{r-1} \; = \; C_{r+k-1} \; .$$

The *algebraic mapping cone* $\mathcal{C}(f)$ of a chain map $f : C \longrightarrow D$ is the chain complex defined by

$$d_{\mathcal{C}(f)} \; = \; \begin{pmatrix} d_D & (-)^{r-1} f \\ 0 & d_C \end{pmatrix} \; :$$
$$\mathcal{C}(f)_r \; = \; D_r \oplus C_{r-1} \longrightarrow \mathcal{C}(f)_{r-1} \; = \; D_{r-1} \oplus C_{r-2} \; .$$

The homology groups of $\mathcal{C}(f)$ are the relative homology groups of f:

$$H_*(\mathcal{C}(f)) = H_*(f) .$$

Lemma 3.6 *Let $f : C \longrightarrow D$ be an A-module chain map.*

(i) *The short exact sequence of chain complexes*

$$0 \longrightarrow D \longrightarrow \mathcal{C}(f) \longrightarrow C_{*-1} \longrightarrow 0$$

induces a long exact sequence of homology groups

$$\ldots \longrightarrow H_r(C) \xrightarrow{f_*} H_r(D) \longrightarrow H_r(f) \longrightarrow H_{r-1}(C) \longrightarrow \ldots .$$

(ii) *The chain map $f : C \longrightarrow D$ is a homology equivalence if and only if $H_*(f) = 0$.*

(iii) *The chain map $f : C \longrightarrow D$ is a chain equivalence if and only if $\mathcal{C}(f)$ is chain contractible.*

Proof Standard homological algebra. □

Lemma 3.7 *Let $f : C \longrightarrow D$ be an A-module chain map such that each $f : C_r \longrightarrow D_r$ $(r \in \mathbb{Z})$ is an injection, and let*

$$E = \text{coker}(f : C \longrightarrow D) ,$$

so that there is defined a short exact sequence of A-module chain complexes

$$0 \longrightarrow C \xrightarrow{f} D \xrightarrow{g} E \longrightarrow 0 .$$

(i) *The projection*

$$h : \mathcal{C}(f) \longrightarrow E ; (x, y) \longrightarrow [g(x)]$$

is a homology equivalence.

(ii) *If each $f : C_r \longrightarrow D_r$ is a split injection (e.g. if E is projective) then $h : \mathcal{C}(f) \longrightarrow E$ is a chain equivalence.*

(iii) *If E is projective and bounded below the following conditions are equivalent:*

(a) *f is a chain equivalence,*
(b) *f is a homology equivalence,*
(c) *E is chain contractible,*
(d) *$H_*(E) = 0$.*

Proof (i) Immediate from the homology long exact sequence

$$\ldots \longrightarrow H_r(C) \xrightarrow{f_*} H_r(D) \xrightarrow{g_*} H_r(E) \longrightarrow H_{r-1}(C) \longrightarrow \ldots .$$

(ii) There is no loss of generality in assuming

$$f = \begin{pmatrix} 1 \\ 0 \end{pmatrix} : C_r \longrightarrow D_r = C_r \oplus E_r ,$$

$$g = (0 \quad 1) : D_r = C_r \oplus E_r \longrightarrow E_r ,$$

$$d_D = \begin{pmatrix} d_C & (-)^r j \\ 0 & d_E \end{pmatrix} : D_r = C_r \oplus E_r \longrightarrow D_{r-1} = C_{r-1} \oplus E_{r-1} ,$$

$$h = (0 \quad 1 \quad 0) : \mathcal{C}(f)_r = C_r \oplus E_r \oplus C_{r-1} \longrightarrow E_r$$

for a chain map $j : E \longrightarrow C_{*-1}$. The chain map $k : E \longrightarrow \mathcal{C}(f)$ defined by

$$k = \begin{pmatrix} 0 \\ 1 \\ j \end{pmatrix} : E_r \longrightarrow \mathcal{C}(f)_r = C_r \oplus E_r \oplus C_{r-1}$$

is a chain homotopy inverse for $h : \mathcal{C}(f) \longrightarrow E$.

(iii) (a) $\Longrightarrow$ (b) Trivial.

(b) $\Longleftrightarrow$ (d) Immediate from (i).

(c) $\Longleftrightarrow$ (d) Standard homological algebra.

(c) $\Longrightarrow$ (a) Given a chain contraction

$$\Gamma : 1 \simeq 0 : E \longrightarrow E$$

define a chain homotopy inverse $f^{-1} : D \longrightarrow C$ for f by

$$f^{-1} = (1 \quad (-)^r j \Gamma) : D_r = C_r \oplus E_r \longrightarrow C_r . \qquad \square$$

Definition 3.8 (i) The *singular chain complex at* ∞ of a space W is the algebraic mapping cone (with dimension shift)

$$S^{\infty}(W) = \mathcal{C}(i : S(W) \longrightarrow S^{lf}(W))_{*+1}$$

of the inclusion $i : S(W) \longrightarrow S^{lf}(W)$ defined by regarding singular r-simplexes $\sigma : \Delta^r \longrightarrow W$ as locally finite singular chains.

(ii) The *singular homology at* ∞ of a space W is defined by

$$H^{\infty}_*(W) = H_*(S^{\infty}(W)) . \qquad \square$$

Proposition 3.9 *The various homology groups are related by a long exact sequence*

$$\dots \longrightarrow H^{\infty}_r(W) \longrightarrow H_r(W) \xrightarrow{i} H^{lf}_r(W) \longrightarrow H^{\infty}_{r-1}(W) \longrightarrow \dots . \qquad \square$$

Example 3.10 (i) If K is compact then $S(K) = S^{lf}(K)$, $H_*(K) = H^{lf}_*(K)$ (3.4), so that

$$S^{\infty}(K) \simeq 0 , \quad H^{\infty}_*(K) = 0 .$$

(ii) If $W = K \times [0,\infty)$ for a compact space K then

$$H_*(W) = H^\infty_*(W) = H_*(K) \ , \ H^{lf}_*(W) = 0 \ .$$

(iii) If $W = K \times \mathbb{R}^n$ $(n \geq 1)$ for a compact space K then

$$\begin{aligned} H_*(W) &= H_*(K) \ , \ H^{lf}_*(W) = H_{*-n}(K) \ , \\ H^\infty_*(W) &= H_*(K \times S^{n-1}) = H_*(K) \oplus H_{*-n+1}(K) \ . \end{aligned}$$ □

Remark 3.11 In Chapter 12 we shall investigate the sequence of 3.9 for 'forward tame' W. The locally finite homology of such W is expressed as

$$\begin{aligned} H^{lf}_*(W) &= H_*(W^\infty, \{\infty\}) \\ &= H_*(\varprojlim_K S(W, W\backslash K)) = \varprojlim_K H_*(W, W\backslash K) \end{aligned}$$

with W^∞ the one-point compactification of W, and K running over all the compact subspaces of W. In Chapter 7 we shall prove that for forward tame W

$$H_*(e(W)) = H^\infty_*(W) \ .$$

In Chapter 12 the exact sequence of 3.9

$$\ldots \longrightarrow H^\infty_r(W) \longrightarrow H_r(W) \xrightarrow{i} H^{lf}_r(W) \longrightarrow H^\infty_{r-1}(W) \longrightarrow \ldots$$

is identified with the sequence

$$\ldots \longrightarrow H_r(e(W)) \longrightarrow H_r(W) \xrightarrow{i} H_r(W^\infty, \{\infty\}) \longrightarrow H_{r-1}(e(W)) \longrightarrow \ldots$$

induced by a cofibration sequence

$$e(W) \longrightarrow W \xrightarrow{i} W^\infty \ .$$ □

Just as the ordinary homology groups $H_*(W)$ are homotopy invariant, so the locally finite homology groups $H^{lf}_*(W)$ and the homology groups at ∞ $H^\infty_*(W)$ are proper homotopy invariant. Another essential difference is revealed in the homology of disjoint unions:

Proposition 3.12 (i) *For any collection* $\{X_\lambda \,|\, \lambda \in \Lambda\}$ *of spaces*

$$H_*(\coprod_\lambda X_\lambda) = \sum_\lambda H_*(X_\lambda) \ .$$

(ii) *For any collection* $\{X_\lambda \,|\, \lambda \in \Lambda\}$ *of compact spaces*

$$H^{lf}_*(\coprod_\lambda X_\lambda) = \prod_\lambda H_*(X_\lambda) \ .$$

The natural map

$$i \,:\, H_*(\coprod_\lambda X_\lambda) \;=\; \sum_\lambda H_*(X_\lambda) \longrightarrow H_*^{lf}(\coprod_\lambda X_\lambda) \;=\; \prod_\lambda H_*(X_\lambda)$$

is the inclusion of the direct sum in the direct product, and

$$H_*^\infty(\coprod_\lambda X_\lambda) \;=\; \prod_\lambda H_{*+1}(X_\lambda)\Big/\sum_\lambda H_{*+1}(X_\lambda)\ ,$$

with $H_*^\infty(\coprod_\lambda X_\lambda) = 0$ *if and only if the set* $\{\lambda \in \Lambda \,|\, X_\lambda \neq \emptyset\}$ *is finite.* □

The homology at ∞ $H_*^\infty(W)$ is invariant upon passing to cocompact subspaces of W:

Proposition 3.13 *If* W *is a space and* $V \subseteq W$ *is a closed cocompact subspace the inclusion* $S^\infty(V) \longrightarrow S^\infty(W)$ *is a chain equivalence, and*

$$H_*^\infty(V) \;=\; H_*^\infty(W)\ .$$

Proof For any subspace $V \subseteq W$ there is defined a short exact sequences of chain complexes

$$0 \longrightarrow S^\infty(V) \longrightarrow S^\infty(W) \longrightarrow S^\infty(W,V) \longrightarrow 0$$

with each $S_r^\infty(V) \longrightarrow S_r^\infty(W)$ a split injection and

$$S^\infty(W,V) \;=\; \mathcal{C}(S(W,V) \longrightarrow S^{lf}(W,V))_{*+1}\ .$$

If $V \subseteq W$ is a closed cocompact subspace then $S(W,V) = S^{lf}(W,V)$ by 3.3, so that $S^\infty(W,V)$ is a contractible chain complex by 3.6 (iii), and $S^\infty(V) \longrightarrow S^\infty(W)$ is a chain equivalence by 3.7 (iii). □

Notation 3.14 For any space X let

$$g_k^X \,:\, X \longrightarrow X \times I\ ;\ x \longrightarrow (x,k)\ \ (k=0,1)$$

and let

$$D_X \,:\, g_0^X \simeq g_1^X \,:\, S(X) \longrightarrow S(X \times I)$$

be a natural chain homotopy, with

$$\partial D_X + D_X \partial \;=\; g_0^X - g_1^X \,:\, S_r(X) \longrightarrow S_r(X \times I)\ .$$ □

If one checks any standard source (e.g., Munkres [101, pp. 171–172]) for the acyclic model definition of D_X, one sees that D_X induces a chain homotopy

$$D_X^{lf} \,:\, g_0^X \simeq g_1^X \,:\, S^{lf}(X) \longrightarrow S^{lf}(X \times I)$$

on the locally finite chain level.

We have already mentioned that a proper map between locally compact Hausdorff spaces induces a chain map in locally finite homology. More generally, suppose

$$f = \{f_\beta : X \longrightarrow Y\}$$

is a locally finite family (i.e., the collection of images $\{f_\beta(X)\}$ is locally finite) of proper maps between locally compact Hausdorff spaces, then there is an induced chain map

$$f : S^{lf}(X) \longrightarrow S^{lf}(Y) \; ; \; \prod_\alpha n_\alpha \sigma_\alpha \longrightarrow \prod_{\alpha,\beta} n_\alpha (f_\beta \circ \sigma_\alpha) \, .$$

Proposition 3.15 *For any space W there is defined a natural chain map*

$$\alpha : S(e(W)) \longrightarrow S^\infty(W)$$

such that the homology morphisms induced by the projection

$$p : e(W) \longrightarrow W \; ; \; (\omega : [0,\infty) \longrightarrow W) \longrightarrow \omega(0)$$

factor as

$$p_* : H_*(e(W)) \xrightarrow{\alpha_*} H_*^\infty(W) \longrightarrow H_*(W) \, .$$

Proof In the first instance we prove that for any locally compact space X the inclusion

$$k_X : X \longrightarrow X \times [0,\infty) \; ; \; x \longrightarrow (x,0)$$

is such that there is defined a chain homotopy

$$G_X : k_X \simeq 0 : S^{lf}(X) \longrightarrow S^{lf}(X \times [0,\infty)) \, .$$

The locally finite family of proper closed maps

$$t_X = \{t_k : X \times I \longrightarrow X \times [0,\infty) \, ; \, (x,s) \longrightarrow (x, s+k) \mid k \geq 0\}$$

induce chain maps

$$t_X : S^{lf}(X \times I) \longrightarrow S^{lf}(X \times [0,\infty)) \, .$$

The morphisms

$$G_X : S_r^{lf}(X) \xrightarrow{D_X^{lf}} S_{r+1}^{lf}(X \times I) \xrightarrow{t_X} S_{r+1}^{lf}(X \times [0,\infty))$$

are such that

$$\begin{aligned} \partial^{lf} G_X + G_X \partial^{lf} &= t_X \circ (g_0^X - g_1^X) \\ &= k_X : S_r^{lf}(X) \longrightarrow S_r^{lf}(X \times [0,\infty)) \, , \end{aligned}$$

defining a chain homotopy $G_X : k_X \simeq 0$.

The adjoint of a singular simplex $\sigma : \Delta^r \longrightarrow e(W)$ is a proper map

$$\widehat{\sigma} : \Delta^r \times [0,\infty) \longrightarrow W \; ; \; (x,t) \longrightarrow \sigma(x)(t)$$

such that

$$p\sigma \;=\; \widehat{\sigma}k_{\Delta^r} \;:\; \Delta^r \longrightarrow W \;.$$

The chain map

$$\alpha \;:\; S(e(W)) \longrightarrow S^\infty(W) \;=\; \mathcal{C}(i : S(W) \longrightarrow S^{lf}(W))_{*+1}$$

defined by

$$\begin{aligned}\alpha \;:\; S_r(e(W)) \longrightarrow S^\infty_r(W) \;&=\; S^{lf}_{r+1}(W) \oplus S_r(W) \;;\\ (\sigma : \Delta^r \longrightarrow e(W)) &\longrightarrow (\widehat{\sigma}G_{\Delta^r}(1_{\Delta^r}), p\sigma)\end{aligned}$$

is such that

$$p \;:\; S(e(W)) \xrightarrow{\alpha} S^\infty(W) \longrightarrow S(W) \;. \qquad \square$$

The following result expresses the chain complexes of a σ-compact space W to the chain complexes of an ascending sequence $K_j \subseteq K_{j+1}$ of compact subspaces $K_j \subseteq W$ and also to the chain complexes of a descending subsequence $W_j \supseteq W_{j-1}$ of closed cocompact subspaces $W_j \supseteq W$. In the applications it is convenient to have available both expressions.

Proposition 3.16 *Let W be a σ-compact space, with compact subsets $K_j \subseteq W$ for $j = 0, 1, 2, 3, \ldots$ such that*

$$K_0 \subseteq K_1 \subseteq \ldots \subseteq K_j \subseteq K_{j+1} \subseteq \ldots \subseteq \bigcup_{j=1}^{\infty} K_j \;=\; W \quad , \quad K_j \subseteq \text{int}(K_{j+1}) \;.$$

Write the closed cocompact subsets defined by the closures of the complements as

$$W_j \;=\; \text{cl}(W \backslash K_j) \;,$$

so that

$$W \;\supseteq\; W_0 \supseteq W_1 \supseteq \ldots \supseteq W_j \supseteq W_{j+1} \supseteq \ldots \supseteq \bigcap_{j=1}^{\infty} W_j \;=\; \emptyset \;,$$

$$\ldots \subseteq W\backslash K_j \subseteq W_j \subseteq W\backslash K_{j-1} \subseteq W_{j-1} \subseteq \ldots \;.$$

(i) *The singular locally finite chain complex $S^{lf}(W)$ is the inverse limit*

$$S^{lf}(W) \;=\; \varprojlim_j S(W, W\backslash K_j) \;=\; \varprojlim_j S(W, W_j) \;,$$

and there are defined short exact sequences

$$0 \longrightarrow \varprojlim_j{}^1 H_{r+1}(W, W\backslash K_j) \longrightarrow H^{lf}_r(W) \longrightarrow \varprojlim_j H_r(W, W\backslash K_j) \longrightarrow 0 \;,$$

$$0 \longrightarrow \varprojlim_j{}^1 H_{r+1}(W, W_j) \longrightarrow H^{lf}_r(W) \longrightarrow \varprojlim_j H_r(W, W_j) \longrightarrow 0 \;.$$

(ii) *The singular chain complex at* ∞ $S^\infty(W)$ *is homology equivalent to the derived limit*

$$\mathrm{coker}(i : S(W) \longrightarrow S^{lf}(W))_{*+1} = \varprojlim_j{}^1 S(W\backslash K_j)_{*+1} = \varprojlim_j{}^1 S(W_j)_{*+1} ,$$

so that

$$H^\infty_*(W) = H_{*+1}(\varprojlim_j{}^1 S(W_j))$$

and there are defined short exact sequences

$$0 \longrightarrow \varprojlim_j{}^1 H_{r+1}(W\backslash K_j) \longrightarrow H^\infty_r(W) \longrightarrow \varprojlim_j H_r(W\backslash K_j) \longrightarrow 0 ,$$

$$0 \longrightarrow \varprojlim_j{}^1 H_{r+1}(W_j) \longrightarrow H^\infty_r(W) \longrightarrow \varprojlim_j H_r(W_j) \longrightarrow 0 .$$

(iii) *The composites*

$$\pi_r(e(W)) \xrightarrow{\text{Hurewicz}} H_r(e(W)) \longrightarrow H^\infty_r(W)$$

fit into morphisms of exact sequences

$$\begin{array}{ccccccccc}
0 & \longrightarrow & \varprojlim_j{}^1 \pi_{r+1}(W_j) & \longrightarrow & \pi_r(e(W)) & \longrightarrow & \varprojlim_j \pi_r(W_j) & \longrightarrow & 0 \\
 & & \downarrow & & \downarrow & & \downarrow & & \\
0 & \longrightarrow & \varprojlim_j{}^1 H_{r+1}(W_j) & \longrightarrow & H^\infty_r(W) & \longrightarrow & \varprojlim_j H_r(W_j) & \longrightarrow & 0
\end{array}$$

(iv) *The inclusion* $W \longrightarrow W^\infty$ *induces a chain map*

$$S^{lf}(W) = \varprojlim_j S(W, W\backslash K_j) \longrightarrow \varprojlim_j S(W^\infty, W^\infty\backslash K_j)$$

which is a homology equivalence, inducing isomorphisms in homology

$$H^{lf}_*(W) = H_*(\varprojlim_j S(W, W\backslash K_j)) \xrightarrow{\simeq} H_*(\varprojlim_j S(W^\infty, W^\infty\backslash K_j)) .$$

Proof The inverse systems $\{S(W\backslash K_j)\}$, $\{S(W_j)\}$ have the same inverse and derived limits, since

$$\ldots \subseteq S(W\backslash K_j) \subseteq S(W_j) \subseteq S(W\backslash K_{j-1}) \subseteq S(W_{j-1}) \subseteq \ldots .$$

Similarly for $\{S(W, W\backslash K_j)\}$, $\{S(W, W_j)\}$. Thus in verifying (i) and (ii) it suffices to only consider the expressions in K_j's in detail.

(i) We shall define a chain isomorphism

$$\Psi : S^{lf}(W) \longrightarrow \varprojlim_j S(W, W\backslash K_j) .$$

Suppose given an element $\prod_{\alpha} n_\alpha \sigma_\alpha$ of $S^{lf}(W)$. For every $j = 1, 2, 3, \ldots$ there exist at most finitely many α, say $\alpha_{j_1}, \ldots, \alpha_{j_{n(j)}}$, such that $n_\alpha \neq 0$ and $|\sigma_\alpha| \cap K_j \neq \emptyset$. Here $|\cdot|$ denotes the image of a map. Define

$$\Psi_j \; : \; S^{lf}(W) \longrightarrow S(W, W\backslash K_j) \; ; \; \prod_{\alpha} n_\alpha \sigma_\alpha \longrightarrow \left[\sum_{k=1}^{n(j)} n_{\alpha_{j_k}} \sigma_{\alpha_{j_k}} \right]$$

where $[\cdot]$ denotes the class of an element of $S(W)$ in $S(W, W\backslash K_j)$. Since Ψ_j is the composition

$$S^{lf}(W) \xrightarrow{\Psi_{j+1}} S(W, W\backslash K_{j+1}) \xrightarrow{\text{inc}_*} S(W, W\backslash K_j)$$

we have an induced chain map

$$\Psi \; : \; S^{lf}(W) \longrightarrow \varprojlim_j S(W, W\backslash K_j) \; .$$

To define the inverse of Ψ note that an element of $\varprojlim_j S(W, W\backslash K_j)$ is represented by a sequence of elements $x_j \in S(W)$ with $x_j = x_{j+1} \in S(W, W\backslash K_j)$. It follows that we can assume that the x_j are given as follows: there exist a sequence of singular simplexes σ_k in W and integers n_k for $k = 1, 2, 3, \ldots$, as well as a sequence of integers $1 \leq n(1) \leq n(2) \leq \ldots$ such that $x_j = \sum_{k=1}^{n(j)} n_k \sigma_k$ where $|\sigma_k| \cap K_j \neq \emptyset$ if and only if $k \leq n(j)$. Then Ψ^{-1} is given by

$$\Psi^{-1} \; : \; \varprojlim_j S(W, W\backslash K_j) \longrightarrow S^{lf}(W) \; ; \; \varprojlim_j x_j \longrightarrow \prod_k n_k \sigma_k \; .$$

(ii) The inverse system of short exact sequences of chain complexes

$$0 \longrightarrow S(W\backslash K_j) \longrightarrow S(W) \longrightarrow S(W, W\backslash K_j) \longrightarrow 0 \quad (j \geq 0)$$

is such that

$$\varprojlim_j S(W\backslash K_j) \; = \; S\Big(\bigcap_{j=0}^{\infty} (W\backslash K_j) \Big) \; = \; S(\emptyset) \; = \; 0 \; .$$

By 2.12 (ii) there is defined a short exact sequence

$$0 \longrightarrow \varprojlim_j S(W) \longrightarrow \varprojlim_j S(W, W\backslash K_j) \longrightarrow \varprojlim_j{}^1 S(W\backslash K_j) \longrightarrow 0 \; ,$$

which (by (i)) can be identified with

$$0 \longrightarrow S(W) \xrightarrow{\; i \;} S^{lf}(W) \longrightarrow \text{coker}(i) \longrightarrow 0 \; .$$

(iii) The composite of the chain maps

$$S(\operatorname{holim}_{\overleftarrow{j}} W_j) \longrightarrow S(e(W)) \longrightarrow S^{\infty}(W)$$

given by 2.14 and 3.15 is a chain map such that the induced homology morphisms

$$H_*(\underleftarrow{\mathrm{holim}}_j W_j) \xrightarrow{\simeq} H_*(e(W)) \longrightarrow H_*^\infty(W)$$

are compatible with the inverse and derived limits.

(iv) The inverse system $\{S(W^\infty, W^\infty\backslash K_j)\}$ is such that each of the chain maps

$$S(W^\infty, W^\infty\backslash K_j) \longrightarrow S(W^\infty, W^\infty\backslash K_{j-1})$$

is a surjection, so that

$$\underleftarrow{\lim}_j{}^1 S(W^\infty, W^\infty\backslash K_j) \ = \ 0$$

and by 2.19 (ii) there are defined short exact sequences

$$0 \longrightarrow \underleftarrow{\lim}_j{}^1 H_{r+1}(W^\infty, W^\infty\backslash K_j) \longrightarrow H_r(\underleftarrow{\lim}_j S(W^\infty, W^\infty\backslash K_j)) \longrightarrow \underleftarrow{\lim}_j H_r(W^\infty, W^\infty\backslash K_j) \longrightarrow 0\ .$$

The chain maps

$$S(W, W\backslash K_j) \longrightarrow S(W^\infty, W^\infty\backslash K_j) \ \ (j \geq 0)$$

are chain equivalences by excision. Applying the 5-lemma to the morphism of exact sequences

$$\begin{array}{ccccccccc}
0 \to & \underleftarrow{\lim}_j{}^1 H_{r+1}(W, W\backslash K_j) & \longrightarrow & H_r(\underleftarrow{\lim}_j S(W, W\backslash K_j)) & \longrightarrow & \underleftarrow{\lim}_j H_r(W, W\backslash K_j) & \to 0 \\
& \cong \downarrow & & \downarrow & & \cong \downarrow & \\
0 \to & \underleftarrow{\lim}_j{}^1 H_{r+1}(W^\infty, W^\infty\backslash K_j) & \to & H_r(\underleftarrow{\lim}_j S(W^\infty, W^\infty\backslash K_j)) & \to & \underleftarrow{\lim}_j H_r(W^\infty, W^\infty\backslash K_j) & \to 0
\end{array}$$

gives that $H_*^{lf}(W) \longrightarrow H_*(\underleftarrow{\lim}_j S(W^\infty, W^\infty\backslash K_j))$ are isomorphisms. □

Remark 3.17 (i) The first short exact sequence in 3.16 (i) also occurs in Spanier [150, Thm. 7.3].

(ii) It is not known if the homology equivalence in 3.16 (iv)

$$S^{lf}(W) \longrightarrow \underleftarrow{\lim}_j S(W^\infty, W^\infty\backslash K_j)$$

is a chain equivalence in general. In 7.15 it is shown that it is a chain equivalence if W is a forward tame ANR. □

In general, $H^{\infty}_{-1}(W) \neq 0$ and $H^{lf}_*(W) \neq H_*(W^\infty, \{\infty\})$:

Example 3.18 Let $W = \mathbb{N} = \{0,1,2,\dots\}$, with the discrete topology, and let

$$K_j = \{0,1,2,\dots,j\} \ , \ W_j = \{j+1, j+2, \dots\} \subset W \ (j \geq 0) .$$

The end space is $e(W) = \emptyset$. Now

$$S(K_j) : \dots \longrightarrow \sum_0^j \mathbb{Z} \xrightarrow{0} \sum_0^j \mathbb{Z} \xrightarrow{1} \sum_0^j \mathbb{Z} \xrightarrow{0} \sum_0^j \mathbb{Z} ,$$

$$S(W, W_j) : \dots \longrightarrow \sum_0^j \mathbb{Z} \xrightarrow{0} \sum_0^j \mathbb{Z} \xrightarrow{1} \sum_0^j \mathbb{Z} \xrightarrow{0} \sum_0^j \mathbb{Z} ,$$

$$S(W_j) : \dots \longrightarrow \sum_{j+1}^\infty \mathbb{Z} \xrightarrow{0} \sum_{j+1}^\infty \mathbb{Z} \xrightarrow{1} \sum_{j+1}^\infty \mathbb{Z} \xrightarrow{0} \sum_{j+1}^\infty \mathbb{Z} ,$$

so that

$$S(W^\infty, \{\infty\}) = S(W) = \varinjlim_j S(K_j) :$$

$$\dots \longrightarrow \sum_0^\infty \mathbb{Z} \xrightarrow{0} \sum_0^\infty \mathbb{Z} \xrightarrow{1} \sum_0^\infty \mathbb{Z} \xrightarrow{0} \sum_0^\infty \mathbb{Z} ,$$

$$S^{lf}(W) = \varprojlim_j S(W, W_j) : \dots \longrightarrow \prod_0^\infty \mathbb{Z} \xrightarrow{0} \prod_0^\infty \mathbb{Z} \xrightarrow{1} \prod_0^\infty \mathbb{Z} \xrightarrow{0} \prod_0^\infty \mathbb{Z} ,$$

$$S^\infty(W) \simeq \mathrm{coker}(i)_{*+1} = \varprojlim_j{}^1 S(W_j)_{*+1} :$$

$$\dots \longrightarrow \prod_0^\infty \mathbb{Z}/\sum_0^\infty \mathbb{Z} \xrightarrow{0} \prod_0^\infty \mathbb{Z}/\sum_0^\infty \mathbb{Z} \xrightarrow{1} \prod_0^\infty \mathbb{Z}/\sum_0^\infty \mathbb{Z} \xrightarrow{0} \prod_0^\infty \mathbb{Z}/\sum_0^\infty \mathbb{Z} .$$

Thus

$$H_0(W) = H_0(W^\infty, \{\infty\}) = \sum_0^\infty \mathbb{Z} = \mathbb{Z}[z]$$

$$\neq H^{lf}_0(W) = \prod_0^\infty \mathbb{Z} = \mathbb{Z}[[z]] ,$$

$$H^\infty_{-1}(W) = \prod_0^\infty \mathbb{Z} \Big/ \sum_0^\infty \mathbb{Z} = \mathbb{Z}[[z]]/\mathbb{Z}[z] \neq 0 ,$$

with $H_*(W) = H^{lf}_*(W) = H^\infty_{*-1}(W) = 0$ for $* \neq 0$. □

A *singular r-cochain* in W is a formal product $\prod_\alpha n_\alpha \sigma_\alpha^*$ with coefficients $n_\alpha \in \mathbb{Z}$ of singular r-simplexes $\sigma_\alpha : \Delta^r \longrightarrow W$. The *singular cochain complex*

$$S(W)^* = \mathrm{Hom}_{\mathbb{Z}}(S(W), \mathbb{Z})$$

has $S(W)^r$ the abelian group of singular r-cochains, with *(singular) cohomology* given by

$$H^*(W) = H^*(S(W)) .$$

A *locally finite singular r-cochain* in W is a formal sum $\sum_\alpha n_\alpha \sigma_\alpha^*$. The *locally finite singular cochain complex* $S^{lf}(W)^*$ is the subcomplex of $S(W)^*$ with $S^{lf}(W)^r$ the abelian group of locally finite singular r-cochains. The *locally finite cohomology* of W is defined by

$$H^*_{lf}(W) = H_*(S^{lf}(W)^*) .$$

The *cohomology of W at* ∞ is defined by

$$H^*_\infty(W) = H^*(S^\infty(W)) = H_{*+1}(i^* : S^{lf}(W)^* \longrightarrow S(W)^*) .$$

The cohomology version of 3.9 is given by :

Proposition 3.19 *The various cohomology groups are related by a long exact sequence*

$$\dots \longrightarrow H^r_{lf}(W) \xrightarrow{i} H^r(W) \longrightarrow H^r_\infty(W) \longrightarrow H^{r+1}_{lf}(W) \longrightarrow \dots . \qquad \square$$

Remark 3.20 Epstein [42] identifies the number of ends (1.14) of a locally finite CW complex W with the dimension of the real vector space $H^0_\infty(W;\mathbb{R})$. $\square$

4

Cellular homology

It is well-known that the singular homology groups of a CW complex are isomorphic to the cellular homology groups; it is less well documented (and much harder to prove) that the singular locally finite homology groups of a 'strongly locally finite' CW complex are isomorphic to the cellular locally finite homology groups. This is stated in 4.7, and is proved in Appendix A.

Definition 4.1 The *cellular chain complex* of a CW complex W is the free $\mathbb{Z}$-module chain complex $C(W)$ with chain objects

$$C(W)_r = H_r(W^{(r)}, W^{(r-1)}) = \sum_{I_r} \mathbb{Z} \quad (r \geq 0)$$

the direct sums of $\mathbb{Z}$-modules indexed by the sets I_r of r-cells, and differentials

$$\begin{aligned} d_{C(W)} : C(W)_r &= H_r(W^{(r)}, W^{(r-1)}) \\ &\longrightarrow C(W)_{r-1} = H_{r-1}(W^{(r-1)}, W^{(r-2)}) \end{aligned}$$

the homology boundary maps of the triple $(W^{(r)}, W^{(r-1)}, W^{(r-2)})$. □

Unfortunately, it is not in general possible to regard $C(W)$ as a subcomplex of the singular chain complex $S(W)$. (This is possible in special cases, e.g. if W has the cell structure of a simplicial complex.) We now recall from Wall [164] the precise relationship between $C(W)$ and $S(W)$. The filtration of W by its p-skeleta W^p induces a filtration of $S(W)$ by letting $F^pS(W)$ be the image of $S(W^p) \longrightarrow S(W)$ under the chain map induced by inclusion. Following [164, p. 130] define a free subcomplex $D(W)$ of the singular chain complex $S(W)$ by

$$D_p(W) = \ker(\partial : F^pS_p(W) \longrightarrow F^pS_{p-1}(W)/F^{p-1}S_{p-1}(W)) ,$$

such that $C(W)$ is a quotient complex of $D(W)$.

Proposition 4.2 (Wall [164, Lemma 1]) *The natural chain maps*

$$S(W) \hookleftarrow D(W) \longrightarrow C(W)$$

are chain equivalences of free $\mathbb{Z}$-module chain complexes. In particular, $S(W)$ and $C(W)$ are chain equivalent and

$$H_*(W) = H_*(S(W)) = H_*(C(W)) . \qquad \square$$

This is of course well-known. The locally finite version is somewhat less familiar.

Definition 4.3 The *locally finite cellular chain complex* of a CW complex W is the $\mathbb{Z}$-module chain complex $C^{lf}(W)$ with chain objects

$$C^{lf}(W)_r = H_r^{lf}(W^{(r)}, W^{(r-1)}) = \prod_{I_r} \mathbb{Z} \quad (r \geq 0)$$

the direct products of $\mathbb{Z}$-modules indexed by the sets I_r of r-cells. The differentials

$$\begin{aligned} d_{C^{lf}(W)} : C^{lf}(W)_r &= H_r^{lf}(W^{(r)}, W^{(r-1)}) \\ &\longrightarrow C^{lf}(W)_{r-1} = H_{r-1}^{lf}(W^{(r-1)}, W^{(r-2)}) \end{aligned}$$

are the boundary maps of the triples $(W^{(r)}, W^{(r-1)}, W^{(r-2)})$. $\square$

For an arbitrary locally finite CW complex W, the chain complexes $C^{lf}(W)$ and $S^{lf}(W)$ need not be homology equivalent:

Remark 4.4 For an arbitrary locally finite CW complex W, the chain complexes $C^{lf}(W)$ and $S^{lf}(W)$ need not be homology equivalent. For example, define

$$W = e^0 \cup e^1 \cup e^2 \cup \ldots$$

by attaching each n-cell e^n to $e^0 \cup e^1 \cup \ldots \cup e^{n-1}$ by collapsing all of ∂e^n to a point in the interior of e^{n-1}. By 3.16

$$H_0(S^{lf}(W)) = H_0^{lf}(W) = 0 .$$

Since W has only one cell in each dimension

$$C^{lf}(W) = C(W) : \ldots \xrightarrow{0} \mathbb{Z} \xrightarrow{0} \mathbb{Z} \xrightarrow{0} \mathbb{Z} ,$$

and $H_0(C^{lf}(W)) = H_0(W) = \mathbb{Z}$, so that $C^{lf}(W)$ and $S^{lf}(W)$ are definitely not homology equivalent. $\square$

We now introduce a class of CW complexes for which the locally finite singular chain complex is homology equivalent to the locally finite cellular chain complex.

Definition 4.5 (Farrell, Taylor and Wagoner [51]) A CW complex is *strongly locally finite* if it is the union of a countable, locally finite collection of finite subcomplexes. □

Strongly locally finite CW complexes are ANR's.

It is shown in [51] that every countable locally finite, finite dimensional CW complex is strongly locally finite, and that every countable locally finite simplicial complex is a strongly locally finite CW complex. A proper homotopy extension theorem and a proper cellular approximation theorem for strongly locally finite CW complexes are established in [51].

Lemma 4.6 *If W is a strongly locally finite CW complex and C is a compact subset of W, then there exists a cofinite subcomplex $V \subseteq W$ such that $V \subseteq W\backslash C$.*
Proof Let Ω be a locally finite family of finite subcomplexes which cover W. Note that only finitely many subcomplexes in Ω meet C. Thus

$$V = \bigcap\{K \in \Omega \mid K \cap C = \emptyset\}$$

is a cofinite subcomplex of W such that $V \subseteq W\backslash C$. □

It is pointed out in Farrell, Taylor and Wagoner [51] that the CW complex in Remark 4.4 is not strongly locally finite. W also does not satisfy the conclusion of Lemma 4.6.

Let $D^{lf}(W)$ be the subcomplex of the locally finite singular chain complex $S^{lf}(W)$ defined by

$$D^{lf}_p(W) = \ker(\partial : F^pS^{lf}_p(W) \longrightarrow F^pS^{lf}_{p-1}(W)/F^{p-1}S^{lf}_{p-1}(W)) ,$$

such that $C^{lf}(W)$ is a quotient complex of $D^{lf}(W)$. By analogy with 4.2:

Proposition 4.7 *For a strongly locally finite CW complex W the natural chain maps*

$$S^{lf}(W) \hookleftarrow D^{lf}(W) \longrightarrow C^{lf}(W)$$

are homology equivalences of $\mathbb{Z}$-module chain complexes. In particular, $S^{lf}(W)$ and $C^{lf}(W)$ are homology equivalent and

$$H^{lf}_*(W) = H_*(S^{lf}(W)) = H_*(C^{lf}(W)) .$$

Proof See Proposition A.7 in Appendix A. □

By analogy with the singular chain complex at ∞ $S^\infty(W)$ (3.8):

Definition 4.8 The *cellular chain complex at* ∞ of a CW complex W is defined by

$$C^{\infty}(W) = \mathcal{C}(i : C(W) \longrightarrow C^{lf}(W))_{*+1}$$

with $i : C(W) \longrightarrow C^{lf}(W)$ the inclusion defined by regarding cellular r-simplexes $\sigma : \Delta^r \longrightarrow W$ as locally finite cellular chains. □

Corollary 4.9 *For a strongly locally finite CW complex W the inclusion* $C^{\infty}(W) \longrightarrow S^{\infty}(W)$ *is a homology equivalence, so that*

$$H_*(C^{\infty}(W)) = H_*(S^{\infty}(W)) = H_*^{\infty}(W) . \qquad \square$$

The cohomology version of 4.7 gives:

Proposition 4.10 *The locally finite cohomology groups of a strongly locally finite CW complex W are such that*

$$H^*_{lf}(W) = H_*(S^{lf}(W)^*) = H_*(C^*_{lf}(W)) ,$$

where $C^*_{lf}(W) \subseteq C(W)^*$ *is the locally finite cellular cochain subcomplex defined by*

$$\begin{aligned} C^r_{lf}(W) &= H^r_{lf}(W^{(r)}, W^{(r-1)}) = \sum_{I_r} \mathbb{Z} \\ &\subseteq C^r(W) = H^r(W^{(r)}, W^{(r-1)}) = \prod_{I_r} \mathbb{Z} \end{aligned}$$

with I_r *an indexing set for the r-cells.* □

Proposition 4.11 *Let W be a CW complex with a proper cellular map* $p : W \longrightarrow [0, \infty)$ *such that the inverse images*

$$V_j = p^{-1}[j, j+1] \subseteq W \quad (j \geq 0)$$

are (necessarily finite) subcomplexes, so that W is a strongly locally finite CW complex. Define the subcomplexes

$$U_j = p^{-1}(j) = V_{j-1} \cap V_j ,$$

$$W_j = p^{-1}[j, \infty) = \bigcup_{i=j}^{\infty} V_i ,$$

$$K_j = p^{-1}[0, j] = \bigcup_{i=0}^{j-1} V_i \subseteq W ,$$

so that

$$\begin{aligned} W &= \bigcup_{j=0}^{\infty} V_j = \bigcup_{j=0}^{\infty} W_j = \bigcup_{j=0}^{\infty} K_j , \\ U_0 &= K_0 \subseteq K_1 \subseteq K_2 \subseteq \ldots \subseteq W , \\ W &= W_0 \supseteq W_1 \supseteq W_2 \supseteq \ldots , \bigcap_{j=0}^{\infty} W_j = \emptyset , \end{aligned}$$

and write the inclusions as

$$f_j^+ : U_j \longrightarrow V_j \ , \ f_j^- : U_j \longrightarrow V_{j-1} \ , \ g_j : W_j \longrightarrow W_{j-1} \ .$$

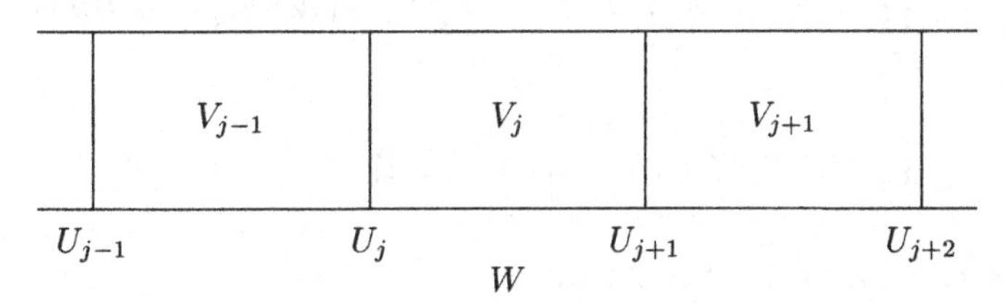

(i) *The cellular chain complex is such that up to homology equivalence*

$$\begin{aligned} C(W) &= \varinjlim_j C(K_j) \\ &\simeq \mathcal{C}(\sum_{j=1}^{\infty}(f_j^+ - f_j^-) : \sum_{j=1}^{\infty} C(U_j) \longrightarrow \sum_{j=0}^{\infty} C(V_j)) \ . \end{aligned}$$

The homology groups are such that

$$H_*(W) = \varinjlim_j H_*(K_j) \ ,$$

and fit into a long exact sequence

$$\begin{aligned} \ldots \longrightarrow \sum_{j=1}^{\infty} H_r(U_j) \xrightarrow{\sum_j (f_j^+ - f_j^-)} \sum_{j=0}^{\infty} H_r(V_j) \longrightarrow H_r(W) \\ \longrightarrow \sum_{j=1}^{\infty} H_{r-1}(U_j) \xrightarrow{\sum_j (f_j^+ - f_j^-)} \sum_{j=0}^{\infty} H_{r-1}(V_j) \longrightarrow \ldots \ . \end{aligned}$$

(ii) *The cellular locally finite chain complex is such that up to homology equivalence*

$$\begin{aligned} C^{lf}(W) &= \varprojlim_j C(W, W_j) = \varprojlim_j C(K_j, U_j) \\ &\simeq \mathcal{C}(\prod_{j=1}^{\infty}(f_j^+ - f_j^-) : \prod_{j=1}^{\infty} C(U_j) \longrightarrow \prod_{j=0}^{\infty} C(V_j)) \ . \end{aligned}$$

The locally finite homology groups fit into a long exact sequence

$$\ldots \longrightarrow \prod_{j=1}^{\infty} H_r(U_j) \xrightarrow{\prod_j (f_j^+ - f_j^-)} \prod_{j=0}^{\infty} H_r(V_j) \longrightarrow H_r^{lf}(W)$$

$$\longrightarrow \prod_{j=1}^{\infty} H_{r-1}(U_j) \xrightarrow{\prod_j (f_j^+ - f_j^-)} \prod_{j=0}^{\infty} H_{r-1}(V_j) \longrightarrow \ldots .$$

The locally finite homology groups also fit into a long exact sequence

$$\ldots \longrightarrow \prod_{j=1}^{\infty} H_{r+1}(W, W_j) \xrightarrow{1 - \prod_j g_j} \prod_{j=1}^{\infty} H_{r+1}(W, W_j) \longrightarrow H_r^{lf}(W)$$

$$\longrightarrow \prod_{j=1}^{\infty} H_r(W, W_j) \xrightarrow{1 - \prod_j g_j} \prod_{j=1}^{\infty} H_r(W, W_j) \longrightarrow \ldots ,$$

which breaks up into short exact sequences

$$0 \longrightarrow \varprojlim_j{}^1 H_{r+1}(W, W_j) \longrightarrow H_r^{lf}(W) \longrightarrow \varprojlim_j H_r(W, W_j) \longrightarrow 0 .$$

(iii) *The cellular cochain complex is such that up to homology equivalence*

$$\begin{aligned} C(W)^* &= \varprojlim_j C(K_j)^* \\ &\simeq \mathcal{C}\Big(\sum_j ((f_j^+)^* - (f_j^-)^*) : \prod_{j=0}^{\infty} C(V_j)^* \longrightarrow \prod_{j=1}^{\infty} C(U_j)^*\Big)_{*-1} . \end{aligned}$$

The cohomology groups fit into a long exact sequence

$$\ldots \longrightarrow \prod_{j=0}^{\infty} H^{r-1}(V_j) \xrightarrow{\prod_j ((f_j^+)^* - (f_j^-)^*)} \prod_{j=1}^{\infty} H^{r-1}(U_j) \longrightarrow H^r(W)$$

$$\longrightarrow \prod_{j=0}^{\infty} H^r(V_j) \xrightarrow{\prod_j ((f_j^+)^* - (f_j^-)^*)} \prod_{j=1}^{\infty} H^r(U_j) \longrightarrow \ldots .$$

The cohomology groups also fit into a long exact sequence

$$\ldots \longrightarrow \prod_{j=1}^{\infty} H^{r-1}(K_j) \xrightarrow{1 - \prod_j f_j^*} \prod_{j=1}^{\infty} H^{r-1}(K_j) \longrightarrow H^r(W)$$

$$\longrightarrow \prod_{j=1}^{\infty} H^r(K_j) \xrightarrow{1 - \prod_j f_j^*} \prod_{j=1}^{\infty} H^r(K_j) \longrightarrow \ldots ,$$

which breaks up into short exact sequences

$$0 \longrightarrow \varprojlim_j{}^1 H^{r-1}(K_j) \longrightarrow H^r(W) \longrightarrow \varprojlim_j H^r(K_j) \longrightarrow 0 .$$

(iv) *The locally finite cellular cochain complex is such that up to homology equivalence*

$$\begin{aligned} C^{lf}(W)^* &= \varinjlim_j C(W,W_j)^* = \varinjlim_j C(K_j,U_j)^* \\ &\simeq \mathcal{C}(\sum_{j=1}^{\infty}((f_j^+)^* - (f_j^-)^*) : \sum_{j=0}^{\infty} C(V_j)^{*-1} \longrightarrow \sum_{j=1}^{\infty} C(U_j)^{*-1}) . \end{aligned}$$

The locally finite cohomology groups of W are such that

$$H^*_{lf}(W) = \varinjlim_j H^*(W,W_j) ,$$

and fit into a long exact sequence

$$\begin{aligned} \dots \longrightarrow \sum_{j=0}^{\infty} H^{r-1}(V_j) &\xrightarrow{\sum_j((f_j^+)^*-(f_j^-)^*)} \sum_{j=1}^{\infty} H^{r-1}(U_j) \longrightarrow H^r_{lf}(W) \\ \longrightarrow \sum_{j=0}^{\infty} H^r(V_j) &\xrightarrow{\sum_j((f_j^+)^*-(f_j^-)^*)} \sum_{j=1}^{\infty} H^r(U_j) \longrightarrow \dots . \end{aligned}$$

(v) *The cellular chain complex at ∞ is such that up to homology equivalence*

$$C^{\infty}(W) \simeq \varprojlim_j{}^1 C(W_j)_{*+1} ,$$

and the homology at ∞

$$H^{\infty}_*(W) = H_{*+1}(\varprojlim_j{}^1 C(W_j))$$

fits into a long exact sequence

$$\begin{aligned} \dots \longrightarrow \prod_{j=1}^{\infty} H_{r+1}(W_j) &\xrightarrow{1-\prod_j g_j} \prod_{j=1}^{\infty} H_{r+1}(W_j) \longrightarrow H^{\infty}_r(W) \\ \longrightarrow \prod_{j=1}^{\infty} H_r(W_j) &\xrightarrow{1-\prod_j g_j} \prod_{j=1}^{\infty} H_r(W_j) \longrightarrow \dots , \end{aligned}$$

which breaks up into short exact sequences

$$0 \longrightarrow \varprojlim_j{}^1 H_{r+1}(W_j) \longrightarrow H^{\infty}_r(W) \longrightarrow \varprojlim_j H_r(W_j) \longrightarrow 0 .$$

The cohomology at ∞ is such that

$$H^*_\infty(W) \;=\; H^*(\varprojlim_j{}^1\, C(W_j)) \;=\; \varinjlim_j H^*(W_j)\ .$$

Proof (i)–(iv) Use the definitions, and the various types of Mayer–Vietoris exact sequence.

(v) Working as in the singular case (3.16) the application of 2.12 (ii) to the inverse system of short exact sequence of chain complexes

$$0 \longrightarrow C(W_j) \longrightarrow C(W) \longrightarrow C(W, W_j) \longrightarrow 0 \quad (j \geq 0)$$

gives a short exact sequence of chain complexes

$$\begin{aligned}0 \longrightarrow \varprojlim_j C(W_j) \longrightarrow \varprojlim_j C(W) \longrightarrow \varprojlim_j C(W, W_j)&\\ \longrightarrow \varprojlim_j{}^1 C(W_j) \longrightarrow \varprojlim_j{}^1 C(W) \longrightarrow 0\ ,&\end{aligned}$$

with

$$\begin{aligned}&\varprojlim_j C(W_j) \;=\; C(\bigcap_j W_j) \;=\; C(\emptyset) \;=\; 0\ ,\quad \varprojlim_j C(W) \;=\; C(W)\ ,\\ &\varprojlim_j C(W, W_j) \;=\; C^{lf}(W)\ ,\quad \varprojlim_j{}^1 C(W) \;=\; 0\ .\end{aligned}$$

□

Remark 4.12 The short exact sequence for $H^r(W)$ of 4.11 (iii) is the prototypical $\varprojlim - \varprojlim^1$ exact sequence of Milnor [97]. □

Example 4.13 If each of the finite CW complexes U_j, V_j $(j \geq 0)$ in 4.11 is path-connected then W is path-connected with

$$\begin{aligned}&H_0(W) \;=\; H^0(W) \;=\; \mathbb{Z}\ ,\quad H_0^{lf}(W) \;=\; H^0_{lf}(W) \;=\; 0\ ,\\ &H^\infty_{-1}(W) \;=\; H^{-1}_\infty(W) \;=\; 0\ .\end{aligned}$$

□

Example 4.14 Let $W = \{(x,0)\,|\,x \geq 0\} \cup \bigcup\limits_{j=0}^{\infty} C_j \subset \mathbb{R}^2$, with C_j the circle centre $(j, 1/4)$ of radius $1/4$ and $p : W \longrightarrow [0,\infty)\,;\,(x,y) \longrightarrow x$.

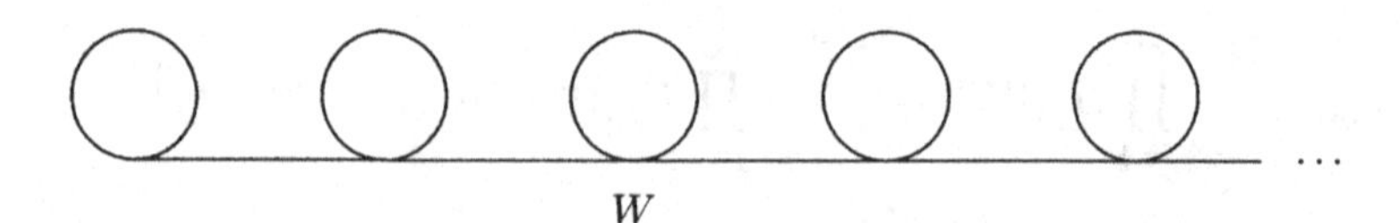

W

Thus W is half an infinite cyclic cover of $S^1 \vee S^1$. Note that W is the special case $n = 0$ of the space in 2.30, and that it is proper homotopy equivalent to Jacob's ladder (2.26). The cellular chain complex of W is given by

$$C(W) \;:\; \ldots \longrightarrow 0 \longrightarrow \mathbb{Z}[z] \oplus \mathbb{Z}[z] \xrightarrow{(1-z\;\;0)} \mathbb{Z}[z]$$

so that

$$H_r(W) \;=\; \begin{cases} \mathbb{Z} & \text{if } r = 0\,, \\ \mathbb{Z}[z] & \text{if } r = 1\,, \\ 0 & \text{otherwise}\,, \end{cases}$$

$$H^r(W) \;=\; \begin{cases} \mathbb{Z} & \text{if } r = 0\,, \\ \mathbb{Z}[[z]] & \text{if } r = 1\,, \\ 0 & \text{otherwise}\,, \end{cases}$$

$$H_r^{lf}(W) \;=\; \begin{cases} \mathbb{Z}[[z]] & \text{if } r = 1\,, \\ 0 & \text{otherwise}\,, \end{cases}$$

$$H^r_{lf}(W) \;=\; \begin{cases} \mathbb{Z}[z] \oplus \mathbb{Z} & \text{if } r = 1\,, \\ 0 & \text{otherwise}\,, \end{cases}$$

$$H_r^{\infty}(W) \;=\; \begin{cases} (\mathbb{Z}[[z]]/\mathbb{Z}[z]) \oplus \mathbb{Z} & \text{if } r = 0\,, \\ 0 & \text{otherwise}\,, \end{cases}$$

$$H^r_{\infty}(W) \;=\; \begin{cases} \mathbb{Z} & \text{if } r = 0\,, \\ \mathbb{Z}[[z]]/\mathbb{Z}[z] & \text{if } r = 1\,, \\ 0 & \text{otherwise}\,. \end{cases}$$

□

Proposition 4.15 *Let $W = \mathrm{Tel}(f_j)$ be the mapping telescope (2.3) of a direct system of cellular maps of finite CW complexes*

$$f_j \;:\; U_j \longrightarrow U_{j+1} \quad (j \geq 0)\,.$$

The various homology and cohomology groups are given by

$$H_*(W) \;=\; H_*^{\infty}(W) \;=\; \varinjlim_j H_*(U_j)\,,$$

$$H^*(W) \;=\; H^*_{\infty}(W)\;,\quad H_*^{lf}(W) \;=\; H^*_{lf}(W) \;=\; 0$$

with short exact sequences

$$0 \longrightarrow \varprojlim_j{}^1 H^{r-1}(U_j) \longrightarrow H^r(W) \longrightarrow \varprojlim_j H^r(U_j) \longrightarrow 0\,.$$

Proof Apply 4.11, with $p : W \longrightarrow [0,\infty)$ such that

$$\begin{aligned}
V_j &= p^{-1}[j,j+1] \\
&= (\text{mapping cylinder of } f_j : U_j \longrightarrow U_{j+1}) \simeq U_{j+1} , \\
W_j &= p^{-1}[j,\infty) = \mathrm{Tel}(f_i \,|\, i \geq j) \simeq W, \\
K_j &= p^{-1}[0,j] \\
&= (\text{mapping cylinder of } f_{j-1}\cdots f_1 f_0 : U_0 \longrightarrow U_j) \simeq U_j , \\
f_j^+ &\simeq f_j : U_j \longrightarrow V_j \simeq U_{j+1} , \ f_j^- \simeq 1 : U_j \longrightarrow V_{j-1} \simeq U_j .
\end{aligned}$$

□

Proposition 4.16 *Let $W = \mathcal{W}(g_j)$ be the mapping cotelescope (2.16) of an inverse system of cellular maps of finite CW complexes*

$$g_j : U_j \longrightarrow U_{j-1} \quad (j \geq 1) .$$

(i) *The homotopy groups of the end space $e(W)$ fit into short exact sequences*

$$0 \longrightarrow \varprojlim_j{}^1 \pi_{r+1}(U_j) \longrightarrow \pi_r(e(W)) \longrightarrow \varprojlim_j \pi_r(U_j) \longrightarrow 0 .$$

(ii) *The various homology and cohomology groups are given by*

$$\begin{aligned}
H_*(W) &= H_*(U_0) \ , \ H^*(W) = H^*(U_0) \ , \\
H^*_\infty(W) &= \varinjlim_j H^*(U_j) \ , \ H^*_{lf}(W) = \varinjlim_j H^*(U_0,U_j)
\end{aligned}$$

with $H^{lf}_*(W)$, $H^\infty_*(W)$ fitting into short exact sequences

$$\begin{aligned}
&0 \longrightarrow \varprojlim_j{}^1 H_{r+1}(U_0,U_j) \longrightarrow H^{lf}_r(W) \longrightarrow \varprojlim_j H_r(U_0,U_j) \longrightarrow 0 , \\
&0 \longrightarrow \varprojlim_j{}^1 H_{r+1}(U_j) \longrightarrow H^\infty_r(W) \longrightarrow \varprojlim_j H_r(U_j) \longrightarrow 0 .
\end{aligned}$$

Proof Apply 2.14 for (i) and 4.11 for (ii), with $p : W \longrightarrow [0,\infty)$ the canonical proper map such that

$$\begin{aligned}
V_j &= p^{-1}[j,j+1] = (\text{mapping cylinder of } g_{j+1} : U_{j+1} \longrightarrow U_j) \simeq U_j , \\
W_j &= p^{-1}[j,\infty) = \mathcal{W}(f_i \,|\, i \geq j) \simeq U_j, \\
K_j &= p^{-1}[0,j] \\
&= (\text{mapping cylinder of } g_1\cdots g_{j-1} g_j : U_j \longrightarrow U_0) \simeq U_0 \simeq W , \\
f_j^+ &\simeq 1 : U_j \longrightarrow V_j \simeq U_j , \ f_j^- \simeq g_j : U_j \longrightarrow V_{j-1} \simeq U_{j-1} .
\end{aligned}$$

□

Example 4.17 Fix an integer $s \geq 2$, and let

$$T = T(s : S^1 \longrightarrow S^1) = (S^1 \times I)/\{(x,0) = (x^s,1) \,|\, x \in S^1\}$$

be the mapping torus of $s : S^1 \longrightarrow S^1; x \longrightarrow x^s$. In (i) and (ii) below we shall apply 4.11, 4.15 and 4.16 to compute the homology at ∞ of the two ends T^+, T^- of the canonical infinite cyclic cover $\overline{T}$ of T, one of which is the mapping telescope of $\{s : S^1 \longrightarrow S^1\}$ regarded as a direct system, and the other is the mapping cotelescope of $\{s : S^1 \longrightarrow S^1\}$ regarded as an inverse system. The general theory of the mapping torus and its canonical infinite cyclic cover is developed in Chapter 14 below.

(i) Let $T^+ = \mathrm{Tel}(s)$ be the mapping telescope of the direct system

$$f_j = s : U_j = S^1 \longrightarrow U_{j+1} = S^1 \quad (j \geq 0) .$$

The chain complexes are such that

$$\begin{aligned} C(T^+) &\simeq \mathcal{C}(1 - sz : \mathbb{Z}[z] \longrightarrow \mathbb{Z}[z])_{*-1} \oplus \mathcal{C}(1 - z : \mathbb{Z}[z] \longrightarrow \mathbb{Z}[z]) , \\ C^{lf}(T^+) &\simeq \mathcal{C}(1 - sz : \mathbb{Z}[[z]] \longrightarrow \mathbb{Z}[[z]])_{*-1} \oplus \mathcal{C}(1 - z : \mathbb{Z}[[z]] \longrightarrow \mathbb{Z}[[z]]) \\ &\simeq 0 , \end{aligned}$$

and the various (co)homology groups are given by

$$H^{lf}_*(T^+) = H^*_{lf}(T^+) = 0 ,$$

$$H_r(T^+) = H^{\infty}_r(T^+) = \varinjlim_j H_r(U_j) = \begin{cases} \mathbb{Z} & \text{if } r = 0 , \\ \mathbb{Z}[1/s] & \text{if } r = 1 , \\ 0 & \text{otherwise} , \end{cases}$$

$$H^r(T^+) = H^r_{\infty}(T^+) = \varprojlim_j H^r(U_j) = \begin{cases} \mathbb{Z} & \text{if } r = 0 , \\ \widehat{\mathbb{Z}}_s & \text{if } r = 1 , \\ 0 & \text{otherwise} , \end{cases}$$

with $\mathbb{Z}[1/s]$ the localization inverting $s \in \mathbb{Z}$ and $\widehat{\mathbb{Z}}_s = \varprojlim_j(\mathbb{Z}/s^j\mathbb{Z})$ the s-adic completion of $\mathbb{Z}$ (2.28).

(ii) Let $T^- = \mathcal{W}(s)$ be the mapping cotelescope of the inverse system

$$g_j = s : U_j = S^1 \longrightarrow U_{j-1} = S^1 \quad (j \geq 1) .$$

The chain complexes are such that

$$\begin{aligned} C(T^-) &\simeq \mathcal{C}(s - z : \mathbb{Z}[z] \longrightarrow \mathbb{Z}[z])_{*-1} \oplus \mathcal{C}(1 - z : \mathbb{Z}[z] \longrightarrow \mathbb{Z}[z]) \\ &\simeq C(S^1) = \mathbb{Z}_{*-1} \oplus \mathbb{Z} , \\ C^{lf}(T^-) &\simeq \mathcal{C}(s - z : \mathbb{Z}[[z]] \longrightarrow \mathbb{Z}[[z]])_{*-1} , \end{aligned}$$

with

$$H_r(T^-) = H^r(T^-) = \begin{cases} \mathbb{Z} & \text{if } r = 0, 1 , \\ 0 & \text{otherwise} , \end{cases}$$

$$H_r^{lf}(T^-) = \begin{cases} \widehat{\mathbb{Z}}_s & \text{if } r = 1 , \\ 0 & \text{otherwise} , \end{cases}$$

$$H_r^{\infty}(T^-) = \begin{cases} \mathbb{Z} \oplus (\widehat{\mathbb{Z}}_s/\mathbb{Z}) & \text{if } r = 0 , \\ 0 & \text{otherwise} , \end{cases}$$

$$H^r_{lf}(T^-) = \begin{cases} \mathbb{Z}[1/s]/\mathbb{Z} & \text{if } r = 2 , \\ 0 & \text{otherwise} , \end{cases}$$

$$H^r_{\infty}(T^-) = \begin{cases} \mathbb{Z} & \text{if } r = 0 , \\ \mathbb{Z}[1/s] & \text{if } r = 1 , \\ 0 & \text{otherwise} . \end{cases}$$

□

The *Steenrod homology groups* $H_*^{st}(X)$ are defined for compact metric spaces X (Steenrod [156]). For a compact subset $X \subseteq S^{n+1}$:

$$\begin{aligned} H_*^{st}(X) &= H^{n+1-*}(S^{n+1}, S^{n+1}\backslash X) \\ &(= H^{n-*}(S^{n+1}\backslash X) \text{ for } * \neq 0, n) , \end{aligned}$$

or equivalently

$$\widetilde{H}_*^{st}(X) = \widetilde{H}^{n-*}(S^{n+1}\backslash X) .$$

The Steenrod homology groups are homotopy invariant, like the singular homology groups, and for finite CW complexes are just the usual singular homology groups. The reduced singular and Steenrod homology groups behave differently on countable infinite one-point unions, with

$$\widetilde{H}_*(\bigvee_{n=0}^{\infty} X_n) = \sum_{n=0}^{\infty} \widetilde{H}_*(X_n) , \quad \widetilde{H}_*^{st}(\bigvee_{n=0}^{\infty} X_n) = \prod_{n=0}^{\infty} \widetilde{H}_*^{st}(X_n) ,$$

assuming that $\varinjlim_n \operatorname{diam}(X_n) = 0$ for Steenrod homology.

More generally, Milnor [95] proved that if $X = \varprojlim_j X_j$ is the inverse limit of an inverse system $X_0 \longleftarrow X_1 \longleftarrow X_2 \longleftarrow \ldots$ of compact metric spaces there are defined exact sequences

$$0 \longrightarrow \varprojlim_j{}^1 H_{r+1}^{st}(X_j) \longrightarrow H_r^{st}(X) \longrightarrow \varprojlim_j H_r^{st}(X_j) \longrightarrow 0 .$$

See Kahn, Kaminker and Schochet [80], Kaminker and Schochet [81]. Also, see Carlsson and Pedersen [21] and Ferry [56] for the applications of Steenrod homology to the Novikov conjecture.

If W is a locally compact separable Hausdorff space (e.g. a countable locally finite CW complex) then W^∞ is compact separable Hausdorff, hence metrizable, and $H_*^{st}(W^\infty)$ is defined.

Proposition 4.18 (Milnor [95]) *The locally finite cellular homology groups of a countable locally finite CW complex W are such that*

$$H_*(C^{lf}(W)) = H_*^{st}(W^\infty, \{\infty\}) ,$$

with H_*^{st} the Steenrod homology groups. □

Remark 4.19 (i) For a strongly locally finite CW complex W the result of 4.18 can also be written as

$$H_*^{lf}(W) = H_*^{st}(W^\infty, \{\infty\}) ,$$

using 4.7 to identify $H_*(C^{lf}(W)) = H_*^{lf}(W)$.

(ii) If W is a countable locally finite CW complex such that $(W^\infty, \{\infty\})$ has the homotopy type of a finite CW pair

$$H_*(C^{lf}(W)) = H_*^{st}(W^\infty, \{\infty\}) = H_*(W^\infty, \{\infty\}) .$$

It will be proved in 7.11 below that for a forward tame countable CW complex W the pair $(W^\infty, \{\infty\})$ has the homotopy type of a finite CW pair, so that

$$H_*(W^\infty, \{\infty\}) = H_*^{st}(W^\infty, \{\infty\}) .$$ □

5

Homology of covers

In dealing with the homological properties of non-compact spaces W and the end spaces $e(W)$ we shall also need to consider the homology of some cover $\widetilde{W}$ of W (usually the universal cover) and the pullback cover $\widetilde{e(W)}$ of $e(W)$. The ordinary Whitehead theorem detects homotopy equivalences by homology isomorphism of the universal covers. The main result of this chapter (5.7) is the analogue of the Whitehead theorem which detects proper homotopy equivalences using locally finite homology isomorphisms of the universal covers.

Let W be a space with a regular cover $\widetilde{W}$, with group of covering translations π. The covering translations

$$g : \widetilde{W} \longrightarrow \widetilde{W} \; ; \; x \longrightarrow gx \quad (g \in \pi)$$

are proper maps inducing isomorphisms

$$g_* : H_*(\widetilde{W}) \xrightarrow{\simeq} H_*(\widetilde{W}) \; , \; g_* : H_*^{lf}(\widetilde{W}) \xrightarrow{\simeq} H_*^{lf}(\widetilde{W})$$

so that the homology groups $H_*(\widetilde{W})$ and the locally finite homology groups $H_*^{lf}(\widetilde{W})$ are $\mathbb{Z}[\pi]$-modules.

Definition 5.1 (i) The *locally π-finite homology $\mathbb{Z}[\pi]$-modules* of $\widetilde{W}$ are

$$H_*^{lf,\pi}(\widetilde{W}) = H_*(S^{lf,\pi}(\widetilde{W})) \; ,$$

with $S^{lf,\pi}(\widetilde{W}) \subseteq S^{lf}(\widetilde{W})$ the subcomplex consisting of the locally finite singular chains in $\widetilde{W}$ which project to locally finite singular chains in W.

(ii) The *singular $\mathbb{Z}[\pi]$-module chain complex at ∞*

$$S^{\infty,\pi}(\widetilde{W}) = \mathcal{C}(i : S(\widetilde{W}) \longrightarrow S^{lf,\pi}(\widetilde{W}))_{*+1} \; ,$$

with i the inclusion. The *locally π-finite homology of W at ∞* is defined by

$$H_*^{\infty,\pi}(\widetilde{W}) = H_*(S^{\infty,\pi}(\widetilde{W})) \; ,$$

to fit into an exact sequence

$$\ldots \longrightarrow H_r^{\infty,\pi}(\widetilde{W}) \longrightarrow H_r(\widetilde{W}) \xrightarrow{i} H_r^{lf,\pi}(\widetilde{W}) \longrightarrow H_{r-1}^{\infty,\pi}(\widetilde{W}) \longrightarrow \ldots .$$

(iii) The π-*cohomology* of $\widetilde{W}$ is

$$H_\pi^*(\widetilde{W}) = H^*(\mathrm{Hom}_{\mathbb{Z}[\pi]}(S(\widetilde{W}),\mathbb{Z}[\pi])) . \qquad \square$$

Example 5.2 (i) In the special case $\widetilde{W} = W$, $\pi = \{1\}$

$$H_*^{lf,\{1\}}(W) = H_*^{lf}(W) , \quad H_*^{\infty,\{1\}}(W) = H_*^{\infty}(W) ,$$
$$H_{\{1\}}^*(W) = H^*(W) .$$

(ii) If W is compact then for any $\widetilde{W}, \pi$

$$H_*^{lf,\pi}(\widetilde{W}) = H_*(\widetilde{W}) , \quad H_*^{\infty,\pi}(\widetilde{W}) = 0 . \qquad \square$$

By analogy with 3.15, 3.16:

Proposition 5.3 *Let $\widetilde{e(W)}$ be the cover of the end space $e(W)$ induced from the cover $\widetilde{W}$ of W by the projection $p : e(W)\longrightarrow W$, and let $\tilde{p} : \widetilde{e(W)}\longrightarrow\widetilde{W}$ be a π-equivariant lift of p.*

(i) *There is defined a natural $\mathbb{Z}[\pi]$-module chain map*

$$\tilde{\alpha} : S(\widetilde{e(W)}) \longrightarrow S^{\infty,\pi}(\widetilde{W})$$

such that

$$\tilde{p}_* : H_*(\widetilde{e(W)}) \xrightarrow{\tilde{\alpha}_*} H_*^{\infty,\pi}(\widetilde{W}) \longrightarrow H_*(\widetilde{W}) .$$

(ii) *If W is a σ-compact space such that*

$$W \supseteq W_0 \supseteq W_1 \supseteq \ldots \supseteq W_j \supseteq W_{j+1} \supseteq \ldots \supseteq \bigcap_{j=1}^{\infty} W_j = \emptyset$$

for closed cocompact subsets $W_j \subset W$ with induced covers $\widetilde{W}_j \subset \widetilde{W}$ there are defined short exact sequences

$$0 \longrightarrow \varprojlim_j{}^1 H_{r+1}(\widetilde{W},\widetilde{W}_j) \longrightarrow H_r^{lf,\pi}(\widetilde{W}) \longrightarrow \varprojlim_j H_r(\widetilde{W},\widetilde{W}_j) \longrightarrow 0 ,$$

$$0 \longrightarrow \varprojlim_j{}^1 H_{r+1}(\widetilde{W}_j) \longrightarrow H_r^{\infty,\pi}(\widetilde{W}) \longrightarrow \varprojlim_j H_r(\widetilde{W}_j) \longrightarrow 0 . \qquad \square$$

A proper map $f : V \longrightarrow W$ between locally compact Hausdorff spaces induces a $\mathbb{Z}[\pi]$-module chain map

$$\widetilde{f} \ : \ S^{lf,\pi}(\widetilde{V}) \longrightarrow S^{lf,\pi}(\widetilde{W}) \ ; \ \prod_{\alpha} n_\alpha \sigma_\alpha \longrightarrow \prod_{\alpha} n_\alpha(\widetilde{f} \circ \sigma_\alpha)$$

for any regular cover $\widetilde{W}$ of W with group of covering translations π, $\widetilde{V} = f^*\widetilde{W}$ the induced cover of V and $\widetilde{f} : \widetilde{V} \longrightarrow \widetilde{W}$ a π-equivariant lift of f.

Remark 5.4 In 7.10 it will be shown that for a 'forward tame' σ-compact metric space W the $\mathbb{Z}[\pi]$-module chain map $\widetilde{\alpha} : S(e(\widetilde{W})) \longrightarrow S^{\infty,\pi}(\widetilde{W})$ of 5.3 (i) is a chain equivalence, inducing isomorphisms

$$\widetilde{\alpha}_* \ : \ H_*(e(\widetilde{W})) \xrightarrow{\simeq} H_*^{\infty,\pi}(\widetilde{W}) \ ,$$

so that there is defined an exact sequence of $\mathbb{Z}[\pi]$-modules

$$\dots \longrightarrow H_r(e(\widetilde{W})) \longrightarrow H_r(\widetilde{W}) \xrightarrow{i} H_r^{lf,\pi}(\widetilde{W}) \longrightarrow H_{r-1}(e(\widetilde{W})) \longrightarrow \dots . \quad \square$$

Let now W be a CW complex with a regular cover $\widetilde{W}$ and with group of covering translations π.

Definition 5.5 (i) The *cellular locally π-finite homology $\mathbb{Z}[\pi]$-modules* of $\widetilde{W}$ are the homology modules $H_*(C^{lf,\pi}(\widetilde{W}))$ of the $\mathbb{Z}[\pi]$-module chain complex $C^{lf,\pi}(\widetilde{W})$ with

$$C_r^{lf,\pi}(\widetilde{W}) \ = \ H_r^{lf,\pi}(\widetilde{W}^{(r)}, \widetilde{W}^{(r-1)}) \ .$$

(ii) The *cellular $\mathbb{Z}[\pi]$-module chain complex of W at ∞* is defined by

$$C^{\infty,\pi}(\widetilde{W}) \ = \ \mathcal{C}(i : C(\widetilde{W}) \longrightarrow C^{lf,\pi}(\widetilde{W}))_{*+1} \ . \quad \square$$

Proposition 5.6 *Let W be a countable strongly locally finite CW complex with a regular cover $\widetilde{W}$ and with group of covering translations π, and for each $r \geq 0$ let I_r be an indexing set for the r-cells of W.*

(i) *The inclusions*

$$C^{lf,\pi}(\widetilde{W}) \longrightarrow S^{lf,\pi}(\widetilde{W}) \ \ , \ \ C^{\infty,\pi}(\widetilde{W}) \longrightarrow S^{\infty,\pi}(\widetilde{W})$$

are homology equivalences.

(ii) *The various homology and cohomology groups of $\widetilde{W}$ are such that*

$$\begin{aligned} H_*(\widetilde{W}) &= H_*(C(\widetilde{W}))\ ,\ H^*(\widetilde{W}) \ =\ H^*(\mathrm{Hom}_{\mathbb{Z}}(C(\widetilde{W}),\mathbb{Z}))\ , \\ H^*_\pi(\widetilde{W}) &= H^*(\mathrm{Hom}_{\mathbb{Z}[\pi]}(C(\widetilde{W}),\mathbb{Z}[\pi]))\ , \\ H^{lf,\pi}_*(\widetilde{W}) &= H_*(C^{lf,\pi}(\widetilde{W}))\ , \\ H^*_{lf,\pi}(\widetilde{W}) &= H^*(C^{lf,\pi}(\widetilde{W})^*)\ , \\ H^{\infty,\pi}_*(\widetilde{W}) &= H_*(C^{\infty,\pi}(\widetilde{W})) \end{aligned}$$

with

$$\begin{aligned} C(\widetilde{W})_r &= H_r(\widetilde{W}^{(r)},\widetilde{W}^{(r-1)}) \ =\ \sum_{I_r}\mathbb{Z}[\pi]\ , \\ \mathrm{Hom}_{\mathbb{Z}}(C(\widetilde{W}),\mathbb{Z})^r &= H^r(\widetilde{W}^{(r)},\widetilde{W}^{(r-1)}) \ =\ \prod_{I_r}\mathbb{Z}[[\pi]]\ , \\ \mathrm{Hom}_{\mathbb{Z}[\pi]}(C(\widetilde{W}),\mathbb{Z}[\pi])^r &= H^r_\pi(\widetilde{W}^{(r)},\widetilde{W}^{(r-1)}) \ =\ \prod_{I_r}\mathbb{Z}[\pi]\ , \\ C^{lf}(\widetilde{W})_r &= H^{lf}_r(\widetilde{W}^{(r)},\widetilde{W}^{(r-1)}) \ =\ \prod_{I_r}\mathbb{Z}[[\pi]]\ , \\ C^{lf}(\widetilde{W})^r &= H^r_{lf}(\widetilde{W}^{(r)},\widetilde{W}^{(r-1)}) \ =\ \sum_{I_r}\mathbb{Z}[\pi]\ , \\ C^{lf,\pi}(\widetilde{W})_r &= H^{lf,\pi}_r(\widetilde{W}^{(r)},\widetilde{W}^{(r-1)}) \ =\ \prod_{I_r}\mathbb{Z}[\pi]\ , \end{aligned}$$

where $\mathbb{Z}[\pi] = \sum\limits_\pi \mathbb{Z}$, $\mathbb{Z}[[\pi]] = \prod\limits_\pi \mathbb{Z}$.

Proof There exists a subcomplex $D^{lf,\pi}(\widetilde{W}) \subset S^{lf,\pi}(\widetilde{W})$ such that the natural $\mathbb{Z}[\pi]$-module chain maps

$$S^{lf,\pi}(\widetilde{W}) \hookleftarrow D^{lf,\pi}(\widetilde{W}) \longrightarrow C^{lf,\pi}(\widetilde{W})$$

are homology equivalences. See Proposition A.7 in Appendix A. □

The Whitehead theorem states that a map of connected CW complexes is a homotopy equivalence if and only if it induces isomorphisms of fundamental groups and the homology groups of the universal covers. Farrell, Taylor and Wagoner [51] established a Whitehead theorem in the proper category; roughly speaking, a homotopy equivalence of locally finite infinite CW complexes is a proper homotopy equivalence if and only if it induces isomorphisms of the fundamental groups at ∞ and of the locally finite cohomology groups. We shall only need the following special case:

Proposition 5.7 *Let W_1, W_2 be connected finite dimensional locally finite infinite CW complexes equipped with proper cellular maps $p_i : W_i \longrightarrow \mathbb{R}$ such that the subcomplexes*

$$W_i^+ = (p_i)^{-1}[0,\infty) \ , \ W_i^- = (p_i)^{-1}(-\infty,0] \subset W_i$$

are connected with the inclusions inducing isomorphisms

$$\pi_1(W_i^\pm) \cong \pi_1^\infty(W_i^\pm) \cong \pi_1(W_i) \ \ (i=1,2) \ .$$

Let $f : W_1 \longrightarrow W_2$ be a proper cellular map with a proper homotopy $p_2 f \simeq p_1 : W_1 \longrightarrow \mathbb{R}$.

(i) *The map f is a proper homotopy equivalence if and only if it induces isomorphisms*

$$f_* : \pi_1(W_1) \xrightarrow{\simeq} \pi_1(W_2) \ , \ \widetilde{f}_* : H_*(\widetilde{W}_1) \xrightarrow{\simeq} H_*(\widetilde{W}_2) \ ,$$
$$\widetilde{f}^* : H^*_{lf}(\widetilde{W}_2) \xrightarrow{\simeq} H^*_{lf}(\widetilde{W}_1)$$

with $\widetilde{W}_1, \widetilde{W}_2$ the universal covers of W_1, W_2.

(ii) *The map f is a proper homotopy equivalence if and only if it induces isomorphisms*

$$f_* : \pi_1(W_1) \xrightarrow{\simeq} \pi_1(W_2) \ , \ \widetilde{f}_* : H_*(\widetilde{W}_1) \xrightarrow{\simeq} H_*(\widetilde{W}_2) \ ,$$
$$\widetilde{f}_* : H_*^{lf,\pi}(\widetilde{W}_1) \xrightarrow{\simeq} H_*^{lf,\pi}(\widetilde{W}_2)$$

with $\pi = \pi_1(W_1) = \pi_1(W_2)$.

Proof (i) A special case of the proper Whitehead theorem of [51].

(ii) Given a $\mathbb{Z}[\pi]$-module which is expressed as a countable direct sum of f.g. free $\mathbb{Z}[\pi]$-modules

$$M = \sum_{j=-\infty}^{\infty} M(j)$$

define the *locally finite $\mathbb{Z}[\pi]$-module*

$$M^{lf} = \prod_{j=-\infty}^{\infty} M(j) \ ,$$

and the *locally finite dual $\mathbb{Z}[\pi]$-module*

$$\begin{aligned}\mathrm{Hom}^{lf}_{\mathbb{Z}[\pi]}(M,\mathbb{Z}[\pi]) &= \mathrm{Hom}_{\mathbb{Z}[\pi]}(M^{lf},\mathbb{Z}[\pi]) \\ &= \sum_{j=-\infty}^{\infty} \mathrm{Hom}_{\mathbb{Z}[\pi]}(M(j),\mathbb{Z}[\pi]) \ .\end{aligned}$$

There are evident identifications

$$\begin{aligned}M(j) &= \mathrm{Hom}_{\mathbb{Z}[\pi]}(\mathrm{Hom}_{\mathbb{Z}[\pi]}(M(j),\mathbb{Z}[\pi]),\mathbb{Z}[\pi]) \ , \\ M^{lf} &= \mathrm{Hom}_{\mathbb{Z}[\pi]}(\mathrm{Hom}^{lf}_{\mathbb{Z}[\pi]}(M,\mathbb{Z}[\pi]),\mathbb{Z}[\pi])\end{aligned}$$

so that

$$\begin{aligned} C^{lf}(\widetilde{W}_i)^* &= \mathrm{Hom}_{\mathbb{Z}[\pi]}(C^{lf,\pi}(\widetilde{W}_i),\mathbb{Z}[\pi]) , \\ C^{lf,\pi}(\widetilde{W}_i) &= \mathrm{Hom}_{\mathbb{Z}[\pi]}(C^{lf}(\widetilde{W}_i)^*,\mathbb{Z}[\pi]) . \end{aligned}$$

Thus the $\mathbb{Z}[\pi]$-module chain map $\widetilde{f}^* : C^{lf}(\widetilde{W}_2)^* \longrightarrow C^{lf}(\widetilde{W}_1)^*$ is a homology equivalence if and only if the $\mathbb{Z}[\pi]$-module chain map $\widetilde{f} : C^{lf,\pi}(\widetilde{W}_1) \longrightarrow C^{lf,\pi}(\widetilde{W}_2)$ is a homology equivalence, and (i) is equivalent to (ii). □

Definition 5.8 (i) An *n-dimensional geometric Poincaré complex* is a finite CW complex W with a fundamental class $[W] \in H_n(W)$ such that the cap product defines $\mathbb{Z}[\pi]$-module isomorphisms

$$[W] \cap - \; : \; H^*_\pi(\widetilde{W}) \xrightarrow{\simeq} H_{n-*}(\widetilde{W}) ,$$

with $\widetilde{W}$ the universal cover of W and $\pi = \pi_1(W)$.

(ii) An *n-dimensional geometric Poincaré pair* is a finite CW pair $(W, \partial W)$ with a fundamental class $[W] \in H_n(W, \partial W)$ such that the cap product defines $\mathbb{Z}[\pi]$-module isomorphisms

$$[W] \cap - \; : \; H^*_\pi(\widetilde{W}, \partial\widetilde{W}) \xrightarrow{\simeq} H_{n-*}(\widetilde{W}) ,$$

with $\widetilde{W}, \pi$ as in (i) and $\partial\widetilde{W} \subset \widetilde{W}$ the induced cover of ∂W. In addition, it is required that ∂W be an $(n-1)$-dimensional geometric Poincaré complex – this is automatic if $\pi_1(\partial W) = \pi_1(W)$.

(iii) An *n-dimensional geometric Poincaré cobordism* $(V; U, U')$ is an n-dimensional geometric Poincaré pair $(V, \partial V)$ such that $\partial V = U \amalg U'$ for disjoint subcomplexes $U, U' \subseteq V$, in which case the cap product defines $\mathbb{Z}[\pi]$-module isomorphisms

$$[V] \cap - \; : \; H^*_\pi(\widetilde{V}, \widetilde{U}) \xrightarrow{\simeq} H_{n-*}(\widetilde{V}, \widetilde{U}') \;\; (\pi \; = \; \pi_1(V)) .$$

(iv) An *open n-dimensional geometric Poincaré pair* is a locally finite CW pair $(W, \partial W)$ with ∂W finite, together with a fundamental class $[W] \in H^{lf}_n(W, \partial W)$ such that the cap product defines $\mathbb{Z}[\pi]$-module isomorphisms

$$[W] \cap - \; : \; H^*_{lf,\pi}(\widetilde{W}, \partial\widetilde{W}) \xrightarrow{\simeq} H_{n-*}(\widetilde{W}) \;\; (\pi \; = \; \pi_1(W)) .$$

In addition, it is required that ∂W be an $(n-1)$-dimensional geometric Poincaré complex – as in (i), this is automatic if $\pi_1(\partial W) = \pi_1(W)$.

(v) A proper map $p : (W, \partial W) \longrightarrow ([0,\infty), \{0\})$ from a locally finite CW pair is *n-dimensional (geometric) Poincaré transverse* if each

$$(V_j; U_j, U_{j+1}) \; = \; p^{-1}([j, j+1]; \{j\}, \{j+1\}) \;\; (j \geq 0)$$

is an n-dimensional geometric Poincaré cobordism. □

Example 5.9 (i) A compact n-dimensional manifold is an n-dimensional geometric Poincaré complex, and similarly for pairs and cobordisms.

(ii) An open n-dimensional manifold with compact boundary $(W, \partial W)$ is an open n-dimensional geometric Poincaré pair with a Poincaré transverse map $p : (W, \partial W) \longrightarrow ([0,\infty), \{0\})$ (cf. 5.11 below).

(iii) If W is an n-dimensional geometric Poincaré complex with a map $c : W \longrightarrow S^1$ then the infinite cyclic cover $\overline{W} = c^*\mathbb{R}$ of W is an open n-dimensional geometric Poincaré complex (= open n-dimensional geometric Poincaré pair of the type $(W, \emptyset)$). □

Proposition 5.10 (i) *The various homology and cohomology groups associated to an open n-dimensional geometric Poincaré pair $(W, \partial W)$ are related by a Poincaré duality isomorphism of exact sequences*

$$\begin{array}{ccccccccc}
\dots \longrightarrow & H^{r-1}_{\infty,\pi}(\widetilde{W}) & \longrightarrow & H^r_{lf,\pi}(\widetilde{W}, \partial\widetilde{W}) & \longrightarrow & H^r_\pi(\widetilde{W}, \partial\widetilde{W}) & \longrightarrow & H^r_{\infty,\pi}(\widetilde{W}) & \longrightarrow \dots \\
& {\scriptstyle [W]\cap -}\big\downarrow{\scriptstyle \cong} & & {\scriptstyle [W]\cap -}\big\downarrow{\scriptstyle \cong} & & {\scriptstyle [W]\cap -}\big\downarrow{\scriptstyle \cong} & & {\scriptstyle [W]\cap -}\big\downarrow{\scriptstyle \cong} & \\
\dots \longrightarrow & H^{\infty,\pi}_{n-r}(\widetilde{W}) & \longrightarrow & H_{n-r}(\widetilde{W}) & \longrightarrow & H^{lf,\pi}_{n-r}(\widetilde{W}) & \longrightarrow & H^{\infty,\pi}_{n-r-1}(\widetilde{W}) & \longrightarrow \dots
\end{array}$$

(ii) *If $(W, \partial W)$ is a locally finite CW pair with an n-dimensional Poincaré transverse map $p : (W, \partial W) \longrightarrow ([0,\infty), \{0\})$ then $(W, \partial W)$ is an open n-dimensional geometric Poincaré pair.*

Proof (i) The dual of the Poincaré duality chain equivalence

$$[W] \cap - \ : \ C^{lf,\pi}(\widetilde{W}, \partial\widetilde{W})^{n-*} \xrightarrow{\simeq} C(\widetilde{W})$$

is the Poincaré duality chain equivalence

$$[W] \cap - \ : \ C(\widetilde{W}, \partial\widetilde{W})^{n-*} \xrightarrow{\simeq} C^{lf,\pi}(\widetilde{W}) \ .$$

The cap product

$$[W] \cap - \ : \ S^{\infty,\pi}(\widetilde{W})^{n-1-*} \longrightarrow S^{\infty,\pi}(\widetilde{W})$$

is a Poincaré duality chain equivalence by the 5-lemma.

(ii) Each

$$(K_j; \partial W, U_j) \ = \ p^{-1}([0,j]; \{0\}, \{j\}) \ = \ \bigcup_{i=0}^{j-1} (V_i; U_i, U_{i+1}) \ \ (j \geq 1)$$

is an n-dimensional geometric Poincaré cobordism, with

$$W_j \ = \ p^{-1}[j, \infty) \ = \ \bigcup_{i=j}^{\infty} V_i \ \subseteq \ W \ .$$

The direct limit of the Poincaré duality isomorphisms

$$[K_j]\cap - \ :\ H^r_\pi(\widetilde{W},\partial\widetilde{W}\amalg\widetilde{W}_j) \ =\ H^r_\pi(\widetilde{K}_j,\partial\widetilde{W}\amalg\widetilde{U}_j) \xrightarrow{\simeq} H_{n-r}(\widetilde{K}_j)$$

is the Poincaré duality isomorphism

$$\begin{aligned}[W]\cap - \ :\ H^r_{lf,\pi}(\widetilde{W},\partial\widetilde{W}) \ &=\ \varinjlim_j H^r_\pi(\widetilde{K}_j,\partial\widetilde{W}\amalg\widetilde{U}_j)\\ &\xrightarrow{\simeq} H_{n-r}(\widetilde{W}) \ =\ \varinjlim_j H_{n-r}(\widetilde{K}_j)\ .\end{aligned}$$

The inverse and derived limits of the Poincaré duality isomorphisms

$$[K_j]\cap - \ :\ H^{n-*}_\pi(\widetilde{K}_j,\partial\widetilde{W}) \xrightarrow{\simeq} H_*(\widetilde{K}_j,\widetilde{U}_j) \ =\ H_*(\widetilde{W},\widetilde{W}_j)$$

determine an isomorphism of short exact sequences

$$\begin{array}{ccccccccc}
0 \longrightarrow & \varprojlim_j^1 H^{r-1}_\pi(\widetilde{K}_j,\partial\widetilde{W}) & \longrightarrow & H^r_\pi(\widetilde{W},\partial\widetilde{W}) & \longrightarrow & \varprojlim_j H^r_\pi(\widetilde{K}_j,\partial\widetilde{W}) & \longrightarrow 0 \\
& \Big\downarrow {\scriptstyle [K_j]\cap -\ \cong} & & \Big\downarrow {\scriptstyle [W]\cap -\ \cong} & & \Big\downarrow {\scriptstyle [K_j]\cap -\ \cong} & \\
0 \longrightarrow & \varprojlim_j^1 H_{n-r+1}(\widetilde{K}_j,\widetilde{U}_j) & \longrightarrow & H^{lf,\pi}_{n-r}(\widetilde{W}) & \longrightarrow & \varprojlim_j H_{n-r}(\widetilde{K}_j,\widetilde{U}_j) & \longrightarrow 0
\end{array}$$

□

Example 5.11 By Morse theory, an (oriented) open n-dimensional manifold with compact boundary $(W,\partial W)$ admits a proper map

$$p\ :\ (W,\partial W) \longrightarrow ([0,\infty),\{0\})$$

which is manifold transverse at $\{0,1,2,\ldots\}\subset[0,\infty)$, with each

$$(V_j;U_j,U_{j+1}) \ =\ p^{-1}([j,j+1];\{j\},\{j+1\}) \quad (j\geq 0)$$

an n-dimensional manifold cobordism. Then p is geometric Poincaré transverse and $(W,\partial W)$ is an open n-dimensional geometric Poincaré pair. □

For any space W satisfying the forward tameness condition of Chapter 7 below the homology and cohomology at ∞ are realized by the end space $e(W)$, with

$$H^\infty_*(W) \ =\ H_*(e(W)) \ ,\ H^*_\infty(W) \ =\ H^*(e(W))\ .$$

By 5.10 the homology and cohomology at ∞ of an open n-dimensional geometric Poincaré pair $(W,\partial W)$ with a Poincaré transverse proper map

$p : (W, \partial W) \longrightarrow ([0, \infty), \{0\})$ are related by $(n-1)$-dimensional Poincaré duality

$$H^{n-1-*}_{\infty}(W) = H^{\infty}_{*}(W) .$$

If $(W, \partial W)$ is forward tame and also reverse tame (Chapter 8) this is realized geometrically, with the end space $e(W)$ a finitely dominated $(n-1)$-dimensional geometric Poincaré space such that

$$H^{n-1-*}(e(W)) = H_{*}(e(W)) .$$

6

Projective class and torsion

This chapter serves two purposes: firstly, a recollection of the fundamental properties of the algebraic theories of the Wall finiteness obstruction and Whitehead torsion, and secondly to introduce the locally finite finiteness obstruction. The relationship between ordinary homology, locally finite homology and homology at ∞ is mirrored in the context of the finiteness obstruction.

Let A be a ring. A *domination* (D, f, g, h) of an A-module chain complex C by an A-module chain complex D is given by chain maps

$$f : C \longrightarrow D \ , \ g : D \longrightarrow C$$

and a chain homotopy $h\colon gf \simeq 1\colon C \longrightarrow C$, so that C is a homotopy direct summand of D. An A-module chain complex C is *chain homotopy finite* if it is chain equivalent to a finite chain complex of f.g. free A-modules

$$F : \ldots \longrightarrow 0 \longrightarrow \ldots \longrightarrow 0 \longrightarrow F_n \longrightarrow F_{n-1} \longrightarrow \ldots \longrightarrow F_0 \ .$$

An A-module chain complex C is *finitely dominated* if it is dominated by a finite chain complex of f.g. free A-modules.

Proposition 6.1 (i) *An A-module chain complex C is dominated by a free A-module chain complex if and only if C is chain equivalent to a free A-module chain complex.*

(ii) *An A-module chain complex C is finitely dominated if and only if it is chain equivalent to a finite f.g. projective A-module chain complex*

$$P : \ldots \longrightarrow 0 \longrightarrow \ldots \longrightarrow 0 \longrightarrow P_n \longrightarrow P_{n-1} \longrightarrow \ldots \longrightarrow P_0 \ .$$

Proof (i) For any A-module chain maps $f : C \longrightarrow D$, $g : D \longrightarrow C$ there is defined in Ranicki [123, Chapter 6] an $A[z, z^{-1}]$-module chain equivalence

$$\mathcal{C}(z - gf : C[z, z^{-1}] \longrightarrow C[z, z^{-1}]) \ \simeq \ \mathcal{C}(z - fg : D[z, z^{-1}] \longrightarrow D[z, z^{-1}]) \ .$$

(This is an abstract version of the mapping torus trick of M. Mather [91].)

If (D,f,g,h) is a domination of C and D is a free A-module chain complex then the left hand side is A-module chain equivalent to C, and the right hand side is a free A-module chain complex. The converse is trivial.

(ii) A finite f.g. projective chain complex is finitely dominated since it is a direct summand of a finite f.g. free chain complex. See Ranicki [120] for the proof of the converse, including the construction from a finite domination (D,f,g,h) of an explicit f.g. projective chain complex P chain equivalent to a finitely dominated C. (In fact, by Lück and Ranicki [87] such P can be constructed from the chain homotopy projection $fg : D \longrightarrow D$.) □

The *projective class* of a finitely dominated A-module chain complex C is defined by

$$[C] = \sum_{r=0}^{\infty}(-)^r[P_r] \in K_0(A)$$

with P any chain equivalent finite f.g. projective A-module chain complex, as usual.

Proposition 6.2 (i) *The reduced projective class of a finitely dominated A-module chain complex C*

$$[C] \in \widetilde{K}_0(A) = \operatorname{coker}(K_0(\mathbb{Z}) \longrightarrow K_0(A))$$

is such that $[C]=0$ if and only if C is chain homotopy finite.

(ii) *If any two chain complexes in a short exact sequence of free A-module chain complexes*

$$0 \longrightarrow C \longrightarrow D \longrightarrow E \longrightarrow 0$$

are finitely dominated then so is the third, with the projective classes related by

$$[C]-[D]+[E] = 0 \in K_0(A)\ .$$

□

A *finite structure* (D,ϕ) on an A-module chain complex C is a finite chain complex D of based f.g. free A-modules together with a chain equivalence $\phi : C \longrightarrow D$. An A-module chain complex C is chain homotopy finite if and only if it admits a finite structure.

The *torsion* of a contractible finite chain complex C of based f.g. free A-modules is

$$\tau(C) = \tau(d+\Gamma : C_{odd} \longrightarrow C_{even}) \in K_1(A)\ ,$$

with $\Gamma : 0 \simeq 1 : C \longrightarrow C$ any chain contraction and

$$C_{even} = C_0 \oplus C_2 \oplus C_4 \dots\ ,\quad C_{odd} = C_1 \oplus C_3 \oplus C_5 \dots\ .$$

The *Whitehead group* of A is defined to be

$$Wh(A) = \begin{cases} Wh(\pi) = K_1(\mathbb{Z}[\pi])/\{\pm\pi\} \text{ if } A = \mathbb{Z}[\pi] \text{ is a group ring,} \\ \widetilde{K}_1(A) = \text{coker}(K_1(\mathbb{Z}) \longrightarrow K_1(A)) = K_1(A)/\{\pm 1\} \text{ otherwise.} \end{cases}$$

The *torsion* of a chain equivalence $f: C \longrightarrow D$ of finite chain complexes of based f.g. free A-modules is the reduced torsion of the algebraic mapping cone

$$\tau(f) = \tau(\mathcal{C}(f)) \in Wh(A) .$$

The chain equivalence is *simple* if $\tau(f) = 0 \in Wh(A)$. A *simple chain homotopy type* on a chain homotopy finite A-module chain complex C is an equivalence class of finite structures $(D, \phi: C \longrightarrow D)$, subject to the equivalence relation

$$(D, \phi) \sim (D', \phi') \text{ if } \tau(\phi'\phi^{-1}: D \longrightarrow D') = 0 \in Wh(A) .$$

Example 6.3 The algebraic mapping cone $\mathcal{C}(1-e: C \longrightarrow C)$ has a canonical simple chain homotopy type, for any self chain map $e: C \longrightarrow C$ of a finitely dominated A-module chain complex C. If $(D, f, g, h: gf \simeq 1_C)$ is any finite domination of C then $(\mathcal{C}(1 - feg: D \longrightarrow D), \phi)$ is a finite structure in the canonical simple chain homotopy type, with

$$\phi = \begin{pmatrix} f & feh \\ 0 & f \end{pmatrix} : \mathcal{C}(1-e)_r = C_r \oplus C_{r-1} \longrightarrow \mathcal{C}(1-feg)_r = D_r \oplus D_{r-1} .$$

(This is another application of the algebraic Mather trick cited in the proof of 6.1 (i).) □

We refer to Milnor [99] and Cohen [30] for accounts of simple homotopy theory, and to Rosenberg [136] for algebraic K-theory. See Ranicki [120, 121, 123] for more detailed accounts of the algebraic theories of finiteness obstruction and torsion.

In the applications to topology $A = \mathbb{Z}[\pi]$ is a group ring. Here are some examples when the algebraic K-groups are known:

Example 6.4 (i) The reduced projective class group of the group ring of the quaternion group $Q(8) = \{\pm 1, \pm i, \pm j, \pm k\}$ is

$$\widetilde{K}_0(\mathbb{Z}[Q(8)]) = \mathbb{Z}_2 ,$$

with generator $[P]$ the projective class of the f.g. projective $\mathbb{Z}[Q(8)]$-module $P = \text{im}(p)$ defined by the image of the projection

$$p = \begin{pmatrix} 1-8N & 21N \\ -3N & 8N \end{pmatrix} : \mathbb{Z}[Q(8)] \oplus \mathbb{Z}[Q(8)] \longrightarrow \mathbb{Z}[Q(8)] \oplus \mathbb{Z}[Q(8)] ,$$

with $N = \sum_{g \in Q(8)} g \in \mathbb{Z}[Q(8)]$ such that $N^2 = 8N$ (cf. Ranicki [129]).

(ii) The Whitehead group of the cyclic group of order 5 $\mathbb{Z}_5 = \{t \mid t^5\}$ is

$$Wh(\mathbb{Z}_5) = \mathbb{Z} ,$$

with generator the torsion $\tau(u)$ of the unit $u = 1 - t + t^2 \in \mathbb{Z}[\mathbb{Z}_5]^\bullet$.

(iii) For many infinite torsion-free groups π with finite classifying space $B\pi$

$$\widetilde{K}_0(\mathbb{Z}[\pi]) = Wh(\pi) = 0 ,$$

by the algebraic K-theory version of the integral Novikov conjecture. In particular, this is the case for the fundamental groups of hyperbolic manifolds (Farrell and Jones [50]). Thus tame ends of high dimensional hyperbolic manifolds have unique collarings. See Chapter D.3 of Benedetti and Petronio [6] for an account of the ends of hyperbolic manifolds, including a geometric proof that certain ends of hyperbolic manifolds can be collared. See also §12.6 of Ratcliffe [133]. □

The *torsion* of a homotopy equivalence $f : K \longrightarrow L$ of finite CW complexes is defined by

$$\tau(f) = \tau(\widetilde{f} : C(\widetilde{K}) \longrightarrow C(\widetilde{L})) \in Wh(\pi_1(L)) .$$

The homotopy equivalence f is *simple* if $\tau(f) = 0$, which is the case if and only if f is homotopic to the composite of a finite sequence of elementary expansions and collapses.

A *finite structure* (Y, ϕ) on a space X is a finite CW complex Y together with a homotopy equivalence $\phi : X \longrightarrow Y$. A topological space is *homotopy finite* if it admits a finite structure, i.e. if it is homotopy equivalent to a finite CW complex. A *simple homotopy type* on a space X is an equivalence class of finite structures (Y, ϕ) on X, subject to the equivalence relation

$$(Y, \phi) \sim (Y', \phi') \text{ if } \tau(\phi'\phi^{-1} : Y \longrightarrow Y') = 0 \in Wh(\pi_1(X)) .$$

The simple homotopy types on a connected CW complex X are in one-to-one correspondence with the simple chain homotopy types (if any) on the cellular $\mathbb{Z}[\pi_1(X)]$-module chain complex $C(\widetilde{X})$ of the universal cover $\widetilde{X}$ of X.

Example 6.5 A compact ANR is homotopy finite (West [168]), and has a canonical simple homotopy type (Chapman [24]). For a finite CW complex this is the simple homotopy type determined by the cellular structure. □

An *h-cobordism* is a manifold cobordism $(W; M, M')$ such that the inclusions $M \longrightarrow W$, $M' \longrightarrow W$ are homotopy equivalences, with *torsion*

$$\tau(W; M, M') = \tau(M \longrightarrow W) \in Wh(\pi_1(W)) .$$

An *s-cobordism* is an h-cobordism $(W; M, M')$ with

$$\tau(W; M, M') = 0 \in Wh(\pi_1(W)) .$$

The *s-cobordism theorem* is given by:

Theorem 6.6 (Barden, Mazur, Stallings) *An n-dimensional h-cobordism $(W; M, M')$ is an s-cobordism if (and for $n \geq 6$ only if) $(W; M, M')$ is homeomorphic* rel M *to* $M \times (I; \{0\}, \{1\})$. □

The original h-cobordism theorem of Smale is the special case $\pi_1(W) = \{1\}$, when every h-cobordism is an s-cobordism, by virtue of $Wh(\{1\}) = 0$. Kervaire [83] is the standard account of the s-cobordism theorem.

A *domination* (Y, f, g, h) of a space X by a space Y is defined by maps $f : X \longrightarrow Y$, $g : Y \longrightarrow X$ and a homotopy $h : gf \simeq 1 : X \longrightarrow X$, so that X is a homotopy direct summand of Y. A topological space is *finitely dominated* if it is dominated by a finite CW complex.

Proposition 6.7 (i) *A topological space X is dominated by a CW complex if and only if it has the homotopy type of a CW complex.*

(ii) *A topological space X is finitely dominated if and only if $X \times S^1$ has the homotopy type of a finite CW complex.*

Proof For any maps $f : X \longrightarrow Y$, $g : Y \longrightarrow X$ M. Mather [91] defines a homotopy equivalence $T(gf) \simeq T(fg)$ of the mapping tori (14.2) – see Chapter 14 below for the definition of the mapping torus.

(i) If (Y, f, g, h) is a domination of a space X by a CW complex Y then $X \times S^1 \simeq T(gf) \simeq T(fg)$ determines a domination of X by the CW complex $T(fg)$. It follows from the homotopy equivalences $X \simeq X \times \mathbb{R} \simeq \overline{T}(fg)$ that X is homotopy equivalent to a CW complex, namely the infinite cyclic cover $\overline{T}(fg)$ of $T(fg)$ (as defined in Chapter 14 below). The converse is trivial.

(ii) As for (i), noting that $T(fg)$ is a finite CW complex for a finite CW complex Y. □

Let $\widetilde{X}$ be a regular cover of a CW complex X, with group of covering translations π. If X is finitely dominated then $C(\widetilde{X})$ is a finitely dominated $\mathbb{Z}[\pi]$-module chain complex, and the *projective class* of X with respect to $\widetilde{X}$ is defined by

$$[X] = [C(\widetilde{X})] \in K_0(\mathbb{Z}[\pi])$$

as usual, with simply-connected component the Euler characteristic

$$\begin{aligned}\chi(X) &= \sum_{r=0}^{\infty}(-)^r \operatorname{rank} C_r(X) \\ &= \sum_{r=0}^{\infty}(-)^r \operatorname{rank} H_r(X)/\text{torsion} \in K_0(\mathbb{Z}) = \mathbb{Z} .\end{aligned}$$

The main results of finiteness obstruction theory are summarized in:

Theorem 6.8 (Wall [163]) (i) *A connected CW complex X is finitely dominated (resp. homotopy finite) if and only if the fundamental group $\pi_1(X)$ is finitely presented and the cellular chain complex $C(\widetilde{X})$ of the universal cover $\widetilde{X}$ is a finitely dominated (resp. chain homotopy finite) $\mathbb{Z}[\pi_1(X)]$-module chain complex.*

(ii) *The reduced projective class of a finitely dominated CW complex X with respect to the universal cover $\widetilde{X}$ is the* finiteness obstruction

$$[X] \in \widetilde{K}_0(\mathbb{Z}[\pi_1(X)]) ,$$

such that $[X] = 0$ if and only if X is homotopy finite.

(iii) *If π is a finitely presented group and P is a f.g. projective $\mathbb{Z}[\pi]$-module there exists a finitely dominated CW complex X with*

$$\pi_1(X) = \pi \ , \ [X] = [P] \in \widetilde{K}_0(\mathbb{Z}[\pi]) .$$

Idea of proof (i) It is clear that if X is finitely dominated (resp. homotopy finite) then $\pi_1(X)$ is finitely presented and $C(\widetilde{X})$ is finitely dominated (resp. chain homotopy finite), so only the converse has to be verified. The original proof in [163] was simplified by Hofer [67] using the algebraic theory of finiteness obstruction of Ranicki [120], as follows. A connected CW complex X with finitely presented $\pi_1(X)$ is homotopy equivalent to a CW complex (also denoted by X) with finite 2-skeleton. If D is a based free $\mathbb{Z}[\pi_1(X)]$-module chain complex with $D_r = C(\widetilde{X})_r$ for $r = 0, 1, 2$ and $f : D \longrightarrow C(\widetilde{X})$ is a chain equivalence which is the identity in dimensions ≤ 2 then the method of attaching cells to kill homotopy classes can be used to realize D by a CW complex Y with a homotopy equivalence $f : Y \longrightarrow X$ inducing $f : C(\widetilde{Y}) = D \longrightarrow C(\widetilde{X})$. Consider first the case when $C(\widetilde{X})$ is chain homotopy finite, so that D can be chosen to be a f.g. free $\mathbb{Z}[\pi_1(X)]$-module chain complex, Y is finite and X is homotopy finite. In the other case $C(\widetilde{X})$ is chain homotopy finitely dominated, and $X \times S^1$ is such that the cellular $\mathbb{Z}[\pi][z, z^{-1}]$-module chain complex of the universal cover $(X \times S^1)^{\sim} = \widetilde{X} \times \mathbb{R}$

$$C(\widetilde{X} \times \mathbb{R}) = C(\widetilde{X}) \otimes_{\mathbb{Z}} C(\mathbb{R})$$

is chain homotopy finite, so that $X \times S^1$ is homotopy finite (by the first case) and X is finitely dominated.

(ii) Immediate from (i) and 6.2.

(iii) Let K be a finite CW complex with $\pi_1(K) = \pi$, and let $p = p^2 : \mathbb{Z}[\pi]^r \longrightarrow \mathbb{Z}[\pi]^r$ be a $\mathbb{Z}[\pi]$-module projection such that $P = \mathrm{im}(p)$. For any $N \geq 2$ the finite CW complex

$$L = (K \times S^1 \vee \bigvee_r S^N) \cup_{-zp+1-p} \bigcup_r D^{N+1}$$

has a finitely dominated infinite cyclic cover $\overline{L}$ with two ends $\overline{L}^+$, $\overline{L}^-$ such that

$$\pi_1(\overline{L}) = \pi_1(\overline{L}^+) = \pi_1(\overline{L}^-) = \pi ,$$
$$[\overline{L}^+] = -[\overline{L}^-] = (-)^N[P] \in \widetilde{K}_0(\mathbb{Z}[\pi]) .$$

(See Chapter 13 for an account of infinite cyclic covers. Note that L is homotopy equivalent to $K \times S^1$, so that $\overline{L}$ is homotopy equivalent to K.) □

Proposition 6.9 *Let W be a connected CW complex.*

(i) *If $\overline{W}$ is a regular cover of W with group of covering translations π such that the classifying map $\pi_1(W) \longrightarrow \pi$ is a split injection and the cellular $\mathbb{Z}[\pi]$-module chain complex $C(\overline{W})$ is finitely dominated, then W is finitely dominated.*

(ii) *If $V \subseteq W$ is a cofinite subcomplex which is finitely dominated then W is also finitely dominated, such that*

$$[W] = u_*[V] \in \widetilde{K}_0(\mathbb{Z}[\pi_1(W)])$$

with $u_* : \pi_1(V) \longrightarrow \pi_1(W)$ the morphism induced by the inclusion $u : V \longrightarrow W$.

(iii) *If W is finitely dominated and $u : V \longrightarrow W$ is the inclusion of a cofinite subcomplex $V \subseteq W$ with $u_* : \pi_1(V) \longrightarrow \pi_1(W)$ a split injection then V is also finitely dominated, with*

$$[V] = r_*[W] \in \widetilde{K}_0(\mathbb{Z}[\pi_1(V)])$$

for any surjection $r : \pi_1(W) \longrightarrow \pi_1(V)$ splitting u_*, and

$$\begin{aligned}[W] &= u_*[V] \in \mathrm{im}(u_* : \widetilde{K}_0(\mathbb{Z}[\pi_1(V)]) \longrightarrow \widetilde{K}_0(\mathbb{Z}[\pi_1(W)])) \\ &= \ker(1 - u_* r_* : \widetilde{K}_0(\mathbb{Z}[\pi_1(W)]) \longrightarrow \widetilde{K}_0(\mathbb{Z}[\pi_1(W)])) .\end{aligned}$$

Proof (i) Since $\pi_1(W) \longrightarrow \pi$ is a split injection the $\mathbb{Z}[\pi_1(W)]$-module chain complex of the universal cover $\widetilde{W}$ of W is induced from the $\mathbb{Z}[\pi]$-module chain complex of the cover $\overline{W}$

$$C(\widetilde{W}) = \mathbb{Z}[\pi_1(W)] \otimes_{\mathbb{Z}[\pi]} C(\overline{W}) .$$

Thus $C(\widetilde{W})$ is a finitely dominated $\mathbb{Z}[\pi_1(W)]$-module chain complex by 6.2 (ii), and W is finitely dominated by 6.8 (i). (Remark: $\overline{W}$ is connected if and only if $\pi_1(W) \longrightarrow \pi$ is a surjection, in which case split injectivity is equivalent to isomorphism and $\overline{W} = \widetilde{W}$.)

(ii) Let $\widetilde{W}$ be the universal cover of W, and let $\widetilde{u} : \widetilde{V} \longrightarrow \widetilde{W}$ be a lift of $u : V \longrightarrow W$ to the induced cover $\widetilde{V}$ of V. The cellular $\mathbb{Z}[\pi_1(W)]$-module chain complex $C(\widetilde{V})$ is finitely dominated, and $C(\widetilde{W}, \widetilde{V})$ is finite f.g. free. It now follows from the short exact sequence

$$0 \longrightarrow C(\widetilde{V}) \xrightarrow{\widetilde{u}} C(\widetilde{W}) \longrightarrow C(\widetilde{W}, \widetilde{V}) \longrightarrow 0$$

and 6.2 (ii) that $C(\widetilde{W})$ is also finitely dominated, so that W is finitely dominated by (i), with

$$[W] = [C(\widetilde{W})] = [C(\widetilde{V})] = u_*[V] \in \widetilde{K}_0(\mathbb{Z}[\pi_1(W)]) .$$

(iii) Let $\overline{W}$ be the cover of W classified by $r_* : \pi_1(W) \longrightarrow \pi_1(V)$, so that the induced cover $\overline{V} \subseteq \overline{W}$ is the universal cover $\overline{V}$ of V. The cellular $\mathbb{Z}[\pi_1(V)]$-module chain complex $C(\overline{W})$ is finitely dominated, and $C(\overline{W}, \overline{V})$ is finite f.g. free. It now follows from the short exact sequence

$$0 \longrightarrow C(\overline{V}) \xrightarrow{\widetilde{u}} C(\overline{W}) \longrightarrow C(\overline{W}, \overline{V}) \longrightarrow 0$$

and 6.2 (ii) that $C(\overline{V})$ is also finitely dominated, so that V is finitely dominated by (i), with

$$[V] = [C(\overline{V})] = [C(\overline{W})] = r_*[W] \in \widetilde{K}_0(\mathbb{Z}[\pi_1(V)]) . \qquad \square$$

Remark 6.10 A cofinite subcomplex $V \subseteq W$ of a finitely dominated CW complex W need not be finitely dominated. A fundamental group condition such as the split injectivity of $g_* : \pi_1(V) \longrightarrow \pi_1(W)$ is necessary in 6.9 (iii) to ensure that V is finitely dominated. Siebenmann [140, 8.8] constructed in every dimension $n \geq 6$ a contractible open n-dimensional manifold W with one end which has stable π_1 at ∞ but which is not reverse π_1-tame (in the terminology of Chapter 8), providing explicit examples of cofinite pairs $(W, V \subset W)$ such that W is finitely dominated but V is not finitely dominated. In 7.19 below it will be shown that a cofinite subcomplex of a forward tame CW complex W with finitely dominated $e(W)$ is finitely dominated. $\square$

Example 6.11 (i) A simply-connected CW complex W is finitely dominated if and only if the homology $H_*(W)$ is finitely generated, in which case W is homotopy finite, and the projective class is the Euler characteristic

$$[W] = \sum_{r=0}^{\infty} (-)^r \operatorname{rank} H_r(W) = \chi(W) \in K_0(\mathbb{Z}) = \mathbb{Z} .$$

(ii) A simply-connected cofinite subcomplex $V \subseteq W$ of a finitely dominated CW complex W is homotopy finite. □

The projective class has the following locally finite analogue:

Definition 6.12 Let W be a space with a regular cover $\widetilde{W}$ such that $S^{lf,\pi}(\widetilde{W})$ is a finitely dominated $\mathbb{Z}[\pi]$-module chain complex, with π the group of covering translations. The *locally finite projective class* of W with respect to $\widetilde{W}$ is

$$[W]^{lf} = [S^{lf,\pi}(\widetilde{W})] \in K_0(\mathbb{Z}[\pi]) .$$ □

The locally finite projective class is a proper homotopy invariant, with simply-connected component the locally finite Euler characteristic:

Example 6.13 The simply-connected component of the locally finite projective class of a CW complex W with finitely generated $H_*^{lf}(W)$ is

$$[W]^{lf} = \sum_{r=0}^{\infty}(-)^r \mathrm{rank}\, H_r^{lf}(W) = \chi^{lf}(W) \in K_0(\mathbb{Z}) = \mathbb{Z} .$$ □

Example 6.14 (i) The projective classes of $\mathbb{R}^+ = [0,\infty)$ and $e(\mathbb{R}^+) \simeq \{\mathrm{pt.}\}$ are given by

$$[\mathbb{R}^+] = [e(\mathbb{R}^+)] = [\mathbb{Z}] \ , \ [\mathbb{R}^+]^{lf} = 0 \in K_0(\mathbb{Z}) .$$

(ii) The projective classes of $\mathbb{R}$ and $e(\mathbb{R}) \simeq S^0$ are given by

$$[\mathbb{R}] = [\mathbb{Z}] \ , \ [\mathbb{R}]^{lf} = -[\mathbb{Z}] \ , \ [e(\mathbb{R})] = [\mathbb{Z} \oplus \mathbb{Z}] \in K_0(\mathbb{Z}) .$$ □

Example 6.15 As in 4.17 let $W = \mathrm{Tel}(s)$ be the mapping telescope of $s : S^1 \longrightarrow S^1$, for some integer $s \geq 2$, an infinite CW complex with cellular $\mathbb{Z}$-module chain complex

$$C(W) : 0 \longrightarrow \mathbb{Z}[z] \xrightarrow{\begin{pmatrix} 1 - sz \\ 0 \end{pmatrix}} \mathbb{Z}[z] \oplus \mathbb{Z}[z] \xrightarrow{(0 \ \ 1 - z)} \mathbb{Z}[z] .$$

The space W is not finitely dominated, since $H_1(W) = \mathbb{Z}[1/s]$ is not finitely generated, so that the projective class $[W] \in K_0(\mathbb{Z})$ is not defined. The locally finite cellular chain complex

$$C^{lf}(W) : 0 \longrightarrow \mathbb{Z}[[z]] \xrightarrow{\begin{pmatrix} 1 - sz \\ 0 \end{pmatrix}} \mathbb{Z}[[z]] \oplus \mathbb{Z}[[z]] \xrightarrow{(0 \ \ 1 - z)} \mathbb{Z}[[z]]$$

is contractible, so that the locally finite projective class is defined, with

$$[W]^{lf} = 0 \in K_0(\mathbb{Z}) = \mathbb{Z} .$$ □

Proposition 6.16 *If W is a CW complex such that the $\mathbb{Z}[\pi]$-module chain complexes $S(\widetilde{W})$, $S^{lf,\pi}(\widetilde{W})$ are finitely dominated then so is $S^{\infty}(\widetilde{W})$, with projective class*

$$[S^{\infty}(\widetilde{W})] = [W] - [W]^{lf} \in K_0(\mathbb{Z}[\pi]) . \qquad \square$$

Here is a preview of the various projective class invariants we shall be associating in Chapter 10 to a CW complex W, subject to various geometric hypotheses on the behaviour of W at ∞. The fundamental group at ∞ $\pi_1^{\infty}(W)$ (2.23) comes equipped with a morphism $\pi_1^{\infty}(W) \longrightarrow \pi_1(W)$. The 'locally finite projective class' of a 'forward tame' W

$$[W]^{lf} = [S^{lf,\pi}(\widetilde{W})] = [C^{lf,\pi}(\widetilde{W})] \in \widetilde{K}_0(\mathbb{Z}[\pi_1(W)])$$

is defined in Chapter 10. If W is 'forward and reverse tame' then W and $e(W)$ are finitely dominated (Chapter 9), and the finiteness obstruction of $e(W)$ has image

$$[e(W)] = [W] - [W]^{lf} \in \widetilde{K}_0(\mathbb{Z}[\pi_1(W)]) .$$

The 'projective class at ∞' of a 'reverse π_1-tame' W

$$[W]_{\infty} \in \widetilde{K}_0(\mathbb{Z}[\pi_1^{\infty}(W)])$$

is defined in Chapter 10, and has image $[W] \in \widetilde{K}_0(\mathbb{Z}[\pi_1(W)])$. The 'locally finite projective class at ∞' of a 'forward tame' W is also defined in Chapter 10,

$$[W]^{lf}_{\infty} = [S^{lf,\pi}(\widetilde{V})] = [C^{lf,\pi}(\widetilde{V})] \in \widetilde{K}_0(\mathbb{Z}[\pi_1^{\infty}(W)]) ,$$

for an appropriate cofinite subcomplex $V \subseteq W$ such that $\pi_1(V) = \pi_1^{\infty}(W)$, with image $[W]^{lf} \in \widetilde{K}_0(\mathbb{Z}[\pi_1(W)])$. The locally finite projective class at ∞ is an obstruction to W being 'forward collared', i.e. to the existence of a cofinite subcomplex $V \subseteq W$ homotopy equivalent to the end space $e(W)$. If W is both forward and reverse tame then $\pi_1(e(W)) = \pi_1^{\infty}(W)$ and the finiteness obstruction of $e(W)$ is given by

$$[e(W)] = [W]_{\infty} - [W]^{lf}_{\infty} \in \widetilde{K}_0(\mathbb{Z}[\pi_1^{\infty}(W)]) .$$

7

Forward tameness

We now recall the definitions of 'forward tameness' and 'forward collaring', and derive various consequences. For a forward tame σ-compact metric space W the homology at ∞ $S^{\infty}(W)$ of Chapter 3 is shown in 7.10 to be just the homology of $e(W)$,

$$H^{\infty}_*(W) \ = \ H_*(e(W)) \ ,$$

so that there is defined an exact sequence

$$\ldots \longrightarrow H_r(e(W)) \longrightarrow H_r(W) \longrightarrow H^{lf}_r(W) \longrightarrow H_{r-1}(e(W)) \longrightarrow \ldots .$$

The locally finite homology of a forward tame ANR W is identified in 7.15 with the reduced homology of the one-point compactification W^{∞}

$$H^{lf}_*(W) \ = \ H_*(W^{\infty}, \{\infty\}) \ .$$

Definition 7.1 (Quinn [116]) Let W be a locally compact Hausdorff space.

(i) The space W is *forward tame* if there exists a closed cocompact subspace $V \subseteq W$ such that the inclusion $V \times \{0\} \longrightarrow W$ extends to a proper map $q : V \times [0, \infty) \longrightarrow W$.

(ii) The space W is *forward collared* if there exists a closed cocompact ANR subspace $V \subseteq W$ such that the identity $V \times \{0\} \longrightarrow V$ extends to a proper map $q : V \times [0, \infty) \longrightarrow V$. □

Forward tameness is a homotopy theoretic version of Siebenmann's compression axiom [148, 149]. Forward tameness will be interpreted as a homotopy pushout property in Chapter 12. In Parts Two and Three we shall be particularly concerned with forward tameness and collaring for the ends of infinite cyclic covers of finite CW complexes. In Chapter 13 we shall give a homotopy theoretic criterion for forward tameness of such an end, and in Chapter 23 we shall give a homological criterion. The 'locally finite projective class at ∞' of a forward tame CW complex constructed in Chapter 10 is an algebraic K-theory obstruction to forward collaring. In Chapter 13 it will be shown that a forward tame end of an infinite cyclic cover of a finite

CW complex is forward collared up to infinite simple homotopy (Chapter 11) if and only if this invariant vanishes.

The proper map q in 7.1 (i) is equivalent to a map of compact spaces

$$\overline{q} : (V \times [0,\infty))^\infty = V^\infty \wedge [0,\infty] \longrightarrow W^\infty$$

such that $(\overline{q})^{-1}(\infty) = \infty$.

Proposition 7.2 (i) *A forward collared ANR space is forward tame.*

(ii) *If a σ-compact metric space W is forward collared and $V \subseteq W$ is the closed cocompact subspace in 7.1 (ii) then there are defined homotopy equivalences*

$$e(W) \simeq e(V) \simeq V .$$

(iii) *A σ-compact metric space W is forward collared if and only if W is forward tame and there exists a closed cocompact ANR subspace $V \subseteq W$ such that the evaluation $p : e(V) \longrightarrow V$ is a homotopy equivalence.*

(iv) *A locally compact Hausdorff space W is forward collared if and only if there exists a closed cocompact ANR subspace $V \subseteq W$ such that the inclusion $V \longrightarrow V \times [0,\infty); x \longrightarrow (x,0)$ is a proper homotopy equivalence.*

Proof (i) Take $V = W$ in 7.1 (i).

(ii) See 1.12 for $e(W) \simeq e(V)$. The adjoint of the proper map $q : V \times [0,\infty) \longrightarrow V$ is a map $\widehat{q} : V \longrightarrow e(V)$ which is a homotopy inverse of the projection $p : e(V) \longrightarrow V$.

(iii) ($\Longleftarrow$) Let $q : V \longrightarrow e(V)$ be a homotopy inverse for p. Thus, there is a homotopy $h : V \times I \longrightarrow V$ such that $h_0 = \mathrm{id}_V$ and $h_1 = pq$. Since W is forward tame, V is also, so there exist a closed cocompact subspace $U \subseteq V$ and a proper map $g : U \times [0,\infty) \longrightarrow V$ extending the inclusion.

Define

$$r : V \times [0,\infty) \longrightarrow V ; (x,t) \longrightarrow \begin{cases} h(x,t) & \text{if } 0 \leq t \leq 1 , \\ q(x)(t-1) & \text{if } 1 < t < \infty . \end{cases}$$

Even though r need not be proper (because neither h nor the adjoint of q need be proper), if $K \subseteq V$ is compact, then $r|_{K\times[0,\infty)}$ is proper. Define

$$G : g \simeq r|_{U\times[0,\infty)} : U \times [0,\infty) \longrightarrow V ;$$
$$(x,t,u) \longrightarrow g(r(x,tu),t(1-u)) .$$

As with r, G need not be proper. However, if $K \subseteq U$ is compact, then $G|K \times [0,\infty) \times I$ is proper.

Choose a closed cocompact subspace $U_0 \subseteq U$ such that

$$\mathrm{cl}(V \backslash U) \cap U_0 = \emptyset$$

and choose a map $\phi : V \longrightarrow I$ such that

$$\phi^{-1}(0) = U_0 , \ \phi^{-1}(1) = \mathrm{cl}(V \backslash U) .$$

Finally, define

$$f : V \times [0,\infty) \longrightarrow V \; ; \; (x,t) \longrightarrow \begin{cases} r(x,t) & \text{if } x \in \text{cl}(V \backslash U) \;, \\ G(x,t,\phi(t)) & \text{if } x \in U \;. \end{cases}$$

Then f is proper and $f_0 = \text{id}_V$.

($\Longrightarrow$) Let $V \subseteq W$ be a closed cocompact ANR subspace for which there is a proper map $f : V \times [0,\infty) \longrightarrow V$ such that $f_0 = \text{id}_V$. Then the adjoint $\widetilde{f} : V \longrightarrow e(V)$ is a homotopy inverse for p because $p\widetilde{f} = \text{id}_V$ and $\widetilde{f}p \simeq \text{id}_{e(V)}$.

(iv) Suppose first that W is forward collared so that there exist a closed cocompact subspace $V \subseteq W$ and a proper map $q : V \times [0,\infty) \longrightarrow V$ extending the identity. To show that q is a proper homotopy inverse for the inclusion $V \longrightarrow V \times [0,\infty)$ define a homotopy

$$\begin{aligned} h \; : \; \text{identity} \; \simeq \; \text{inclusion} \circ q \; : V \times [0,\infty) \longrightarrow V \times [0,\infty) \; ; \\ (x,s,t) \longrightarrow (q(x,st),(1-t)s) \; . \end{aligned}$$

It remains to verify that h is a proper homotopy, for which it suffices to show that for any compact subspace $K \subseteq V$ and $N \geq 0$, $h^{-1}(K \times [0,N])$ is compact. Since q is proper there exist a compact subspace $K' \subseteq V$ and $n' \geq 0$ such that $q^{-1}(K) \subseteq K' \times N'$. It follows that if

$$h(x,s,t) \;=\; (q(x,st),(1-t)s) \in K \times [0,N] \;,$$

then $x \in K', st \leq N', (1-t)s \leq N$. In particular, $s \leq N + N'$ so that

$$h^{-1}(K \times [0,N]) \subseteq K' \times [0, N+N'] \times I \;.$$

Conversely, if $p : V \times [0,\infty) \longrightarrow V$ is a proper map with a proper homotopy $g : \text{identity} \simeq p \circ \text{inclusion} : V \longrightarrow V$, then

$$q \; : \; V \times [0,\infty) \longrightarrow V \; ; \; (x,t) \longrightarrow \begin{cases} g(x,t) & \text{if } 0 \leq t \leq 1 \;, \\ p(x,t-1) & \text{if } t \geq 1 \end{cases}$$

is a proper map extending the identity. □

Example 7.3 (i) Let $(L, K \subseteq L)$ be a pair of compact spaces with K an ANR, and let

$$W \;=\; L \cup_{K \times \{0\}} K \times [0,\infty) \;.$$

Then W is forward collared, with $V = K \times [0,\infty) \subset W$ a closed cocompact ANR subspace such that the identity $V \times \{0\} \longrightarrow V$ extends to the proper map

$$q \; : \; V \times [0,\infty) \longrightarrow V \; ; \; ((x,s),t) \longrightarrow (x, s+t)$$

and

$$W^{\infty} \;=\; L \cup cK \;=\; L \cup_{K \times \{0\}} K \times [0,\infty]/K \times \{\infty\} \;\;,\;\; e(W) \;\simeq\; K \;.$$

(ii) Let $(M, \partial M)$ be a compact manifold with boundary. The boundary is collared (1.8), meaning that the interior

$$W = \mathrm{int}(M) = M \backslash \partial M$$

is homeomorphic to $M \cup \partial M \times [0, \infty)$, and W is forward collared by (i), with

$$W \simeq M \ , \ W^{\infty} = M/\partial M = M \cup c\partial M \ , \ e(W) \simeq \partial M \ .$$

(iii) Let η be a real n-plane vector bundle over a compact space K. The total space $E(\eta)$ is forward collared, with $E(\eta) \simeq K$. The one-point compactification $E(\eta)^{\infty}$ and the end space $e(E(\eta))$ are such that

$$E(\eta)^{\infty} = T(\eta) \ , \ e(E(\eta)) \simeq S(\eta)$$

with $T(\eta)$ the Thom space of η and $S(\eta)$ the $(n-1)$-sphere bundle.

(iv) The special case of (ii) with $(M, \partial M) = (D^n, S^{n-1})$ (or (iii) with $K = \{\text{pt.}\}$) shows that

$$W = \mathrm{int}(M) = \mathbb{R}^n$$

is forward collared, with

$$W^{\infty} = S^n \ , \ e(W) \simeq S^{n-1} \ .$$

(v) The mapping telescope $\mathrm{Tel}(f_j)$ of a direct system of maps $f_j : X_j \longrightarrow X_{j+1}$ (2.3) is forward collared: the projection $e(\mathrm{Tel}(f_j)) \longrightarrow \mathrm{Tel}(f_j)$ is a homotopy equivalence by 2.5, so that 7.2 (iii) applies with $V = W = \mathrm{Tel}(f_j)$. The one-point compactification $\mathrm{Tel}(f_j)^{\infty}$ is contractible.

(vi) If X is a compact subset of the interior of a compact manifold M and X has an I-regular neighbourhood in M in the sense of Siebenmann [148] then a result of Ferry and Pedersen [57, p. 487] can be used to show that $M \backslash X$ is forward tame. In particular, if X is 1-LCC embedded in M, and has the shape of a CW complex (for example, if X has the homotopy type of a CW complex) then $M \backslash X$ is forward tame by [148, p. 56]. □

Remark 7.4 (i) If a space W has finitely many ends, then W is forward tame (resp. forward collared) if and only if each end of W is forward tame (forward collared).

(ii) In 11.14 below we shall show that an ANR space W is forward tame if and only if $W \times S^1$ is infinite simple homotopy equivalent to a forward collared ANR space X. □

Proposition 7.5 *Let W be a forward tame space and let $V \subseteq W$ be a closed cocompact subspace for which the inclusion $u : V \longrightarrow W$ extends to a proper map $q : V \times [0, \infty) \longrightarrow W$. Let $(\widetilde{W}, \widetilde{V})$ be a cover of (W, V) with group of covering translations π.*

(i) *If W is a σ-compact metric space the end space $e(W)$ is dominated by V.*

(ii) *The inclusion of locally π-finite singular chain complexes is chain homotopic to 0,*

$$\widetilde{u} \simeq 0 : S^{lf,\pi}(\widetilde{V}) \longrightarrow S^{lf,\pi}(\widetilde{W}) ,$$

and there is defined a chain equivalence

$$S^{lf,\pi}(\widetilde{W},\widetilde{V}) \simeq S^{lf,\pi}(\widetilde{W}) \oplus S^{lf,\pi}(\widetilde{V})_{*-1} .$$

Thus $S^{lf,\pi}(\widetilde{W})$ is dominated by the chain complex $S^{lf,\pi}(\widetilde{W},\widetilde{V}) = S(\widetilde{W},\widetilde{V})$.

(iii) *If W is an ANR and V is a closed cocompact ANR subspace with the property that (W^{∞},V^{∞}) is a pair of compact ANR's, then $S(\widetilde{W},\widetilde{V})$ is chain equivalent to a f.g. free $\mathbb{Z}[\pi]$-module chain complex and $S^{lf,\pi}(\widetilde{W})$ is finitely dominated.*

Proof (i) The adjoint of q is a map

$$\widehat{q} : V \longrightarrow e(W) ; x \longrightarrow (t \longrightarrow q(x,t))$$

such that $p_W\widehat{q} = u : V \longrightarrow W$. The map

$$F : e(V) \times I \longrightarrow e(W) ; (\omega,t) \longrightarrow (s \longrightarrow \widehat{q}(\omega(ts))((1-t)s))$$

defines a homotopy $F : \widehat{q}p_V \simeq e(u) : e(V) \longrightarrow e(W)$, so there is defined a homotopy commutative diagram

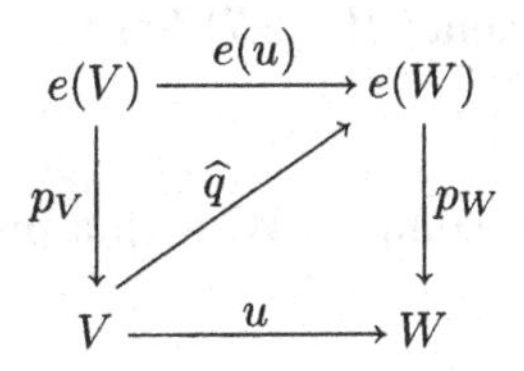

By 1.12 $e(u) : e(V) \longrightarrow e(W)$ is a homotopy equivalence. Use a homotopy inverse to define a map

$$p = p_V e(u)^{-1} : e(W) \longrightarrow V$$

such that

$$\widehat{q}p \simeq 1 : e(W) \longrightarrow e(W) ,$$

so that V dominates $e(W)$.

(ii) For the inclusion $u : V \longrightarrow W$ of any closed cocompact subspace $V \subseteq W$ the short exact sequence of $\mathbb{Z}[\pi]$-module chain complexes

$$0 \longrightarrow S^{lf,\pi}(\widetilde{V}) \xrightarrow{\widetilde{u}} S^{lf,\pi}(\widetilde{W}) \longrightarrow S^{lf,\pi}(\widetilde{W},\widetilde{V}) \longrightarrow 0$$

has $S^{lf,\pi}(\widetilde{W},\widetilde{V}) = S(\widetilde{W},\widetilde{V})$ a free chain complex, so that 3.7 (ii) gives a chain equivalence $\mathcal{C}(\widetilde{u}) \simeq S(\widetilde{W},\widetilde{V})$. Now suppose V, q are as in (i). Let

$$k \,:\, \widetilde{V} \longrightarrow \widetilde{V} \times [0,\infty) \,;\, x \longrightarrow (x,0) \,.$$

The chain homotopy in the proof of 3.15

$$G \,:\, k \,\simeq\, 0 \,:\, S^{lf,\pi}(\widetilde{V}) \longrightarrow S^{lf,\pi}(\widetilde{V} \times [0,\infty))$$

determines a chain homotopy

$$\widehat{q}G \,:\, \widehat{q}k \,=\, \widetilde{u} \,\simeq\, 0 \,:\, S^{lf,\pi}(\widetilde{V}) \longrightarrow S^{lf,\pi}(\widetilde{W}) \,,$$

so that there are defined chain equivalences

$$\mathcal{C}(\widetilde{u}) \,\simeq\, S^{lf,\pi}(\widetilde{W}) \oplus S^{lf,\pi}(\widetilde{V})_{*-1} \,\simeq\, S(\widetilde{W},\widetilde{V}) \,.$$

(iii) The quotient space $W/V = W^\infty/V^\infty$ is a compact ANR and

$$S(\widetilde{W},\widetilde{V}) \,\simeq\, S(\widetilde{W}/\widetilde{V}) \,.$$

Every compact ANR has the homotopy type of a finite CW complex (West [168]), so that $S(\widetilde{W},\widetilde{V})$ is chain equivalent to a f.g. free $\mathbb{Z}[\pi]$-module chain complex. □

Corollary 7.6 *The end space $e(W)$ of a forward tame strongly locally finite CW complex W has the homotopy type of a CW complex.*

Proof Apply 7.5 to a subcomplex $V \subseteq W$ for which there is a proper map $V \times [0,\infty) \longrightarrow W$ extending the inclusion $V \longrightarrow W$. Thus $e(W)$ is dominated by a CW complex V, and hence (by 6.7) has the homotopy type of a CW complex. □

A finitely dominated CW complex W with a proper map $W \longrightarrow \mathbb{R}^+$ need not be forward tame:

Example 7.7 Let W be the subspace of $\mathbb{R}^2$ defined by

$$W \,=\, \{(x,x) \,|\, x \in [0,\infty)\} \cup \{(x,n) \,|\, x \geq n, n = 0,1,2,\ldots\} \,.$$

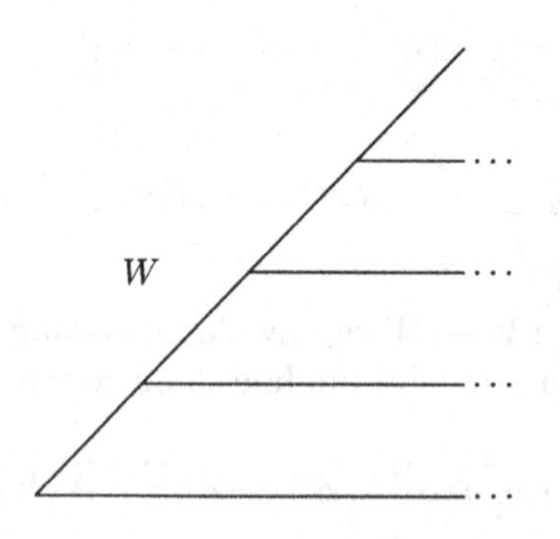

The projection onto the positive real axis is a proper map

$$p : W \longrightarrow [0,\infty) ; (x,y) \longrightarrow x$$

with respect to which W has a bounded CW complex structure. W is contractible, and hence finitely dominated. The end space $e(W)$ is not finitely dominated, since $\pi_0(e(W))$ is (countably) infinite: no two of the proper paths

$$\omega_n : [0,\infty) \longrightarrow W ; t \longrightarrow (n+t,n) \quad (n \geq 0)$$

are proper homotopic. Moreover, since every closed cocompact subspace $V \subseteq W$ has only finitely many path components, $e(W)$ is not dominated by any such V. By 7.5 (i), W is not forward tame.

In fact, W^∞ is homotopy equivalent to the Hawaiian earring, the subspace of the plane consisting of circles of radius $1/n$ and centre $(1/n,0)$ ($n = 1,2,3,\ldots$), which is well-known not to be homotopy equivalent to a CW complex. (This can be proved as follows. By definition, a space X is weakly locally contractible if every point $x \in X$ has a neighbourhood $U \subseteq X$ which is contractible in X. Any space homotopy equivalent to X is then also weakly locally contractible (Dugundji [38, p. 375, Exercise 7]). Every CW complex is (weakly) locally contractible. The Hawaiian earring is not weakly locally contractible at 0.)

□

Remark 7.8 A locally compact space W is *movable at the end* if for each cocompact subspace U of W there exists a cocompact subspace $V \subseteq U$ of W such that for each cocompact subspace Z of W there is a homotopy $f : V \times I \longrightarrow U$ such that $f_0 =$ inclusion : $V \longrightarrow U$ and $f(V \times \{1\}) \subseteq Z$ (see Geoghegan [63]). Movability at the end is a precursor to forward tameness and originated in shape theory with the notion of movability for compacta due to Borsuk [7]. End movability has played a role in the theory of ends of open 3-manifolds (see Brin and Thickstun [11]). Clearly a forward tame space is movable at the end. The converse does not hold as the space W in 7.7 is movable at the end but not forward tame. □

Remark 7.9 (i) The CW complex $W = \{(x,n) \in \mathbb{R}^2 \,|\, x \geq 0, n \in \mathbb{N}\}$ is a forward tame locally finite CW complex with $W \simeq e(W)$ not finitely dominated. (This example appears again in 12.6.)

(ii) If W is forward tame and path-connected at ∞, then W is semistable at ∞ (2.23). □

Proposition 7.10 *Let W be a σ-compact metric space W which is forward tame and path-connected at ∞, and let $W \supseteq W_0 \supseteq W_1 \supseteq \ldots$ be a sequence of closed cocompact subspaces such that $\bigcap\limits_j W_j = \emptyset$.*

(i) *W has stable π_1 at ∞ (2.23), and the fundamental group of the end space $e(W)$ is the fundamental group at ∞ of W,*

$$\pi_1(e(W)) = \pi_1^\infty(W) .$$

(ii) *The homotopy and homology groups of $e(W)$ are such that*

$$\pi_*(e(W)) = \varprojlim_j \pi_*(W_j) \ , \ \varprojlim_j{}^1 \pi_*(W_j) = 0 \ ,$$

$$H_*(e(W)) = H_*^\infty(W) = \varprojlim_j H_*(W_j) \ , \ \varprojlim_j{}^1 H_*(W_j) = 0 \ .$$

(iii) *The chain map $\alpha : S(e(W)) \longrightarrow S^\infty(W)$ of 3.15 is a chain equivalence.*

(iv) *If $\widetilde{W}$ is a regular covering of W with group of covering translations π and $\widetilde{e(W)}$ is the pullback covering of $e(W)$ then*

$$H_*(\widetilde{e(W)}) = H_*^{\infty,\pi}(\widetilde{W}) = \varprojlim_j H_*(\widetilde{W}_j) \ , \ \varprojlim_j{}^1 H_*(\widetilde{W}_j) = 0 \ ,$$

and the $\mathbb{Z}[\pi]$-module chain map of 5.3 is a chain equivalence

$$\widetilde{\alpha} : S(\widetilde{e(W)}) \xrightarrow{\simeq} S^{\infty,\pi}(\widetilde{W}) .$$

Proof (i)+(ii) Let $V \subseteq W$ be a cocompact subspace for which there is a proper map $q : V \times [0,\infty) \longrightarrow W$ with $q(x,0) = x \in W$ $(x \in V)$. Choose path-connected cocompact subspaces of W

$$V = W_0 \supset W_1 \supset W_2 \supset \ldots$$

such that

$$\bigcap_j \mathrm{cl}(W_j) = \emptyset \ , \ q(W_j \times [0,\infty)) \subseteq W_{j-1} \ \ (j \geq 1) .$$

The maps

$$g_j = \text{inclusion} : W_j \longrightarrow W_{j-1} ,$$
$$p_j = p_{W_j} : e(W_j) \longrightarrow W_j \ ; \ \omega \longrightarrow \omega(0) ,$$
$$q_j : W_j \longrightarrow e(W_{j-1}) \ ; \ x \longrightarrow (t \longrightarrow q(x,t))$$

fit into a homotopy commutative diagram

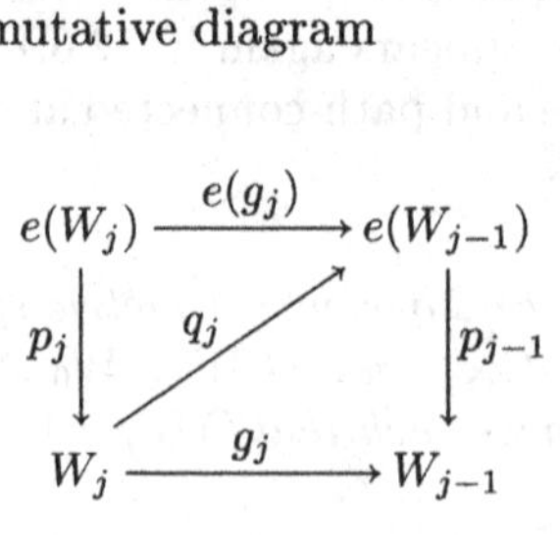

$$\begin{array}{ccc} e(W_j) & \xrightarrow{e(g_j)} & e(W_{j-1}) \\ \big\downarrow p_j & \nearrow q_j & \big\downarrow p_{j-1} \\ W_j & \xrightarrow{g_j} & W_{j-1} \end{array}$$

with $e(g_j) : e(W_j) \longrightarrow e(W_{j-1})$ a homotopy equivalence by 1.12. Apply 2.21 (iv) to the commutative diagrams

$$\begin{array}{ccc} \pi_r(e(W_j)) & \xrightarrow[\cong]{e(g_j)} & \pi_r(e(W_{j-1})) \\ \downarrow p_j & \nearrow q_j & \downarrow p_{j-1} \\ \pi_r(W_j) & \xrightarrow{g_j} & \pi_r(W_{j-1}) \end{array} \qquad \begin{array}{ccc} H_r(e(W_j)) & \xrightarrow[\cong]{e(g_j)} & H_r(e(W_{j-1})) \\ \downarrow p_j & \nearrow q_j & \downarrow p_{j-1} \\ H_r(W_j) & \xrightarrow{g_j} & H_r(W_{j-1}) \end{array}$$

to obtain

$$\pi_r(e(W)) = \varprojlim_j \pi_r(W_j) \ , \ \varprojlim_j{}^1 \pi_r(W_j) = 0 \ ,$$
$$H_r(e(W)) = \varprojlim_j H_r(W_j) \ , \ \varprojlim_j{}^1 H_r(W_j) = 0 \ .$$

By 3.16 and $\varprojlim_j{}^1 H_*(W_j) = 0$

$$H_r^\infty(W) = \varprojlim_j H_r(W_j) \ ,$$

so that the chain map $S(e(W)) \longrightarrow S^\infty(W)$ of 3.15 induces isomorphisms in homology.

(iii) The chain map $\alpha : S(e(W)) \longrightarrow S^\infty(W)$ is a homology equivalence, since by (ii)

$$H_*(e(W)) = \varprojlim_j H_*(W_j) = H_*^\infty(W) \ .$$

The chain complex $S(e(W))$ is free, and $S^\infty(W)$ is chain equivalent to a free chain complex by 7.5 (ii) and 6.1 (i). Any homology equivalence of free chain complexes is a chain equivalence, so that α is a chain equivalence. (In fact, it is possible to define a chain homotopy inverse

$$\alpha^{-1} : S^\infty(W) \xrightarrow{u^{-1}} S^\infty(V) \longrightarrow S(V) \xrightarrow{\widehat{q}} S(e(W))$$

with u^{-1} a chain homotopy inverse to the inclusion $u : S^\infty(V) \longrightarrow S^\infty(W)$, which is a chain equivalence by 3.13.)

(iv) As for (ii), using 5.3. □

Proposition 7.11 (i) *The one-point compactification W^∞ of a forward tame ANR W is an ANR.*

(ii) *The one-point compactification W^∞ of a forward tame ANR W is such that there exists a pointed finite CW complex (X, x_0) with a homotopy equivalence*

$$(W^\infty, \infty) \simeq (X, x_0) \ .$$

Proof (i) Let V be a closed cocompact subset of W for which there is a proper map $q : V \times [0,\infty)\longrightarrow W$ extending the inclusion $q_0 : V\longrightarrow W$. Suppose that W^∞ is a closed subset of some (separable) metric space X. Then W is closed in $X\backslash\{\infty\}$, so there exist a neighbourhood N of W in $X\backslash\{\infty\}$ and a retraction $r : N\longrightarrow W$. Let $U_1 \subseteq X$ be an open subset containing ∞ such that

$$\overline{U}_1 \cap \overline{W\backslash V} = \emptyset .$$

Let $U_2 \subseteq X\backslash\{\infty\}$ be an open subset such that

$$W \subseteq U_2 \subseteq \overline{U}_2 \subseteq N \ , \ (\overline{U}_1\backslash r^{-1}\mathrm{int}(V)) \cap \overline{U}_2 = \{\infty\} .$$

Let $\rho : (\overline{U}_1 \cup \overline{U}_2)\backslash\{\infty\}\longrightarrow[0,\infty]$ be a map such that

$$\rho^{-1}(\infty) = \overline{U}_1\backslash(r^{-1}(\mathrm{int}(V)) \cup \{\infty\}) \ , \ \rho^{-1}(0) = \overline{U}_2 .$$

The map $\widetilde{r} : \overline{U}_1 \cup \overline{U}_2\longrightarrow W^\infty$ defined by

$$\widetilde{r}(x) = \begin{cases} \infty & \text{if } x \in \overline{U}_1\backslash r^{-1}(\mathrm{int}(V)) , \\ q(r(x),\rho(x)) & \text{if } x \in r^{-1}(V) , \\ r(x) & \text{if } x \in \overline{U}_2\backslash r^{-1}(V) \end{cases}$$

is a retraction.

(ii) From (i) we know that W^∞ is a compact ANR. The result now follows from some well-known facts. The Triangulation Theorem of Chapman [23, p. 83] states that every Q-manifold M can be triangulated, i.e. is homeomorphic to $K \times Q$ for a polyhedron K. The ANR Theorem of Edwards ([23, p. 106]) states that if X is an ANR then $X \times Q$ is a Q-manifold. Applied to our context we have that there exists a finite CW complex X such that $W^\infty \times Q \cong X \times Q$ where Q is the Hilbert cube. If $(x_0, q) \in X \times Q$ corresponds to $(\infty, 0)$ under such a homeomorphism, then $(W^\infty, \infty) \simeq (X, x_0)$. West [168] originally proved that compact ANR's have the homotopy type of finite CW complexes. The argument above is a well-known alternative proof of West's theorem (see Chapman [23]). The relevance of this argument is that it shows that pointed compact ANR's have the homotopy type of pointed finite CW complexes. □

Example 7.12 Jacob's ladder X (2.26) is an ANR whose one-point compactification X^∞ is not locally contractible at $\{\infty\}$. Thus X^∞ is not an ANR and 7.11 (i) implies that X is not forward tame. Similarly for the space W of 4.14 (which is proper homotopy equivalent to X – in 9.6 below forward tameness will be shown to be a proper homotopy invariant, in fact an invariant of the 'proper homotopy type at ∞'). □

The one-point compactification of a CW complex does not in general have the homotopy type of a CW complex.

Example 7.13 As in 3.18 let $\mathbb{N} = \{0, 1, 2, \ldots\}$ have the discrete topology, the only topology compatible with CW status, so that

$$H_0(\mathbb{N}) = \sum_0^\infty \mathbb{Z} \ , \ H_0^{lf}(\mathbb{N}) = \prod_0^\infty \mathbb{Z} \ .$$

The countable direct product $\prod_0^\infty \mathbb{Z}$ is an abelian group which is not free, by a result of Baer (Kaplansky [82, p. 48]). The one-point compactification $\mathbb{N}^\infty = \{0, 1, 2, \ldots, \infty\}$ is such that the function

$$\mathbb{N}^\infty \longrightarrow \mathbb{R} \ ; \ n \longrightarrow 1/(n+1) \ , \ \infty \longrightarrow 0$$

is an embedding. $\mathbb{N}^\infty$ does not have the homotopy type of a CW complex, since CW complexes are locally path-connected. Thus $\mathbb{N}$ is not forward tame, by 7.11 (ii). □

Proposition 7.14 *Suppose W is a forward tame ANR written as*

$$W = \bigcup_{j=0}^\infty K_j$$

with K_j compact and $K_j \subseteq K_{j+1}$ for $j = 1, 2, 3, \ldots$. Then the inclusions

$$(W^\infty, \{\infty\}) \longrightarrow (W^\infty, W^\infty \backslash K_j)$$

induce a chain equivalence

$$S(W^\infty, \{\infty\}) \xrightarrow{\simeq} \varprojlim_j S(W^\infty, W^\infty \backslash K_j) \ .$$

Proof There exist a closed cocompact subset $V \subseteq W$ and a proper map $q : V \times [0, \infty) \longrightarrow W$ which extends the inclusion

$$j = q|_{V \times \{0\}} : V \longrightarrow W \ .$$

Extend q to $q_+ : V^\infty \times I \longrightarrow W^\infty$ by insisting that

$$q_+(V^\infty \times \{0\}) = \infty = q_+(\{\infty\} \times [0, \infty]) \ .$$

(Thus q_+ is the composition

$$V^\infty \times [0, \infty] \longrightarrow V^\infty \wedge [0, \infty] \xrightarrow{\overline{q}} W^\infty$$

where $\overline{q}$ is the map mentioned after 7.1.) Without loss of generality assume that K_1 is so large that

$$q((W \backslash K_1) \times [0, \infty)) \subseteq V \ .$$

Since W^∞ is an ANR (7.11), the homotopy extension property implies that q_+ can be extended to a homotopy $\widetilde{q} : W^\infty \times [0, \infty] \longrightarrow W^\infty$ such that:

(i) $\widetilde{q}_0 = \mathrm{id}_{W^\infty}$,
(ii) $\widetilde{q}|_{V\times[0,\infty)} = q$,
(iii) $\widetilde{q}_t(\infty) = \infty$ for each $t \geq 0$,
(iv) $\widetilde{q}_\infty(V^\infty) = \infty$.

Note that for each $j = 1, 2, 3, \ldots$ we have a map of pairs

$$\widetilde{q}_\infty| \ : \ (W^\infty, W^\infty \backslash K_j) \longrightarrow (W^\infty, \{\infty\}) .$$

Hence there is an induced chain map

$$\nu \ : \ \varprojlim_i S(W^\infty, W^\infty \backslash K_j) \longrightarrow S(W^\infty, \{\infty\}) .$$

Let

$$\iota \ : \ S(W^\infty, \{\infty\}) \longrightarrow \varprojlim_j S(W^\infty, W^\infty \backslash K_j)$$

denote the chain map induced by inclusion. We shall show that ι and ν are chain homotopy inverses. Clearly,

$$\nu\iota \ : \ S(W^\infty, \{\infty\}) \longrightarrow S(W^\infty, \{\infty\})$$

is the chain map induced by

$$\widetilde{q}_\infty \ : \ (W^\infty, \{\infty\}) \longrightarrow (W^\infty, \{\infty\}) .$$

Since $\widetilde{q}_\infty$ is homotopic to id_{W^∞} rel $\{\infty\}$, it follows that $\nu\iota$ is chain homotopic to the identity. In order to investigate $\iota\nu$, use the properness of q to define a sequence $i_1 \leq i_2 \leq i_3 \leq \ldots$ of integers such that $i_j \geq j$ and

$$\widetilde{q}((W^\infty \backslash K_{i_j}) \times [0, \infty]) \subseteq W^\infty \backslash K_j$$

for each $j = 1, 2, 3, \ldots$. It follows that if

$$Q \ = \ q_0 \ \simeq \ q_\infty \ : \ S(W^\infty) \longrightarrow S(W^\infty)_{*+1}$$

is the standard chain homotopy then we may consider Q as a chain homotopy

$$Q \ : \ \widetilde{q}_0 \ \simeq \ \widetilde{q}_\infty \ : \ S(W^\infty, W^\infty \backslash K_{i_j}) \longrightarrow S(W^\infty, W^\infty \backslash K_j)_{*+1} .$$

These chain homotopies are compatible with inclusions so that there is a chain homotopy

$$Q \ : \ \widetilde{q}_0 \ \simeq \ \widetilde{q}_\infty \ : \ \varprojlim_j S(W^\infty, W^\infty \backslash K_{i_j}) \longrightarrow \varprojlim_j S(W^\infty, W^\infty \backslash K_j)_{*+1}$$

from $\widetilde{q}_0 = \mathrm{id}$ to $\widetilde{q}_\infty = \iota\nu$. □

Proposition 7.15 (i) *If W is a forward tame ANR then*

$$H_*^{lf}(W) \;=\; H_*(W^\infty, \{\infty\}) \;.$$

(ii) *If W is a forward tame strongly locally finite CW complex then*

$$H_*^{lf}(W) \;=\; H_*(W^\infty, \{\infty\}) \;=\; H_*(C^{lf}(W)) \;.$$

Proof (i) follows from Propositions 3.16 (iv), 7.14.
(ii) follows from (i) and Proposition 4.7 (or Proposition A.7). □

We shall need the following properties of forward tameness in Chapter 9 below.

Lemma 7.16 *If W is a forward tame space with a proper map $p : W \longrightarrow [0,\infty)$, then there exist a closed cocompact subspace $V \subseteq W$ and a proper map $q : V \times [0,\infty) \longrightarrow W$ such that:*

(i) $q| : V = V \times \{0\} \longrightarrow W$ *is the inclusion,*
(ii) $pq(V \times [j,\infty)) \subseteq [j,\infty)$ *for* $j = 0,1,2,\ldots$.

Proof Let $V \subseteq W$ be a closed cocompact subspace for which there exists a proper map $q' : V \times [0,\infty) \longrightarrow W$ extending the inclusion $V \longrightarrow W$. Choose numbers $0 < N_1 < N_2 < \ldots$ such that $pq'(V \times [N_j,\infty)) \subseteq [j,\infty)$ for each $j = 1,2,3,\ldots$. Let $\rho : [0,\infty) \longrightarrow [0,\infty)$ be the PL homeomorphism such that $\rho(j) = N_j$, and which is linear on $[j-1,j]$ for each $j = 1,2,3,\ldots$. The desired map is

$$q \;:\; V \times [0,\infty) \longrightarrow W \;;\; (x,t) \longrightarrow q'(x,\rho(t)) \;.$$ □

Lemma 7.17 *If W is a forward tame metric space with a proper map $p : W \longrightarrow [0,\infty)$, then there is a homotopy $h : e(W) \times I \longrightarrow e(W)$ such that:*

(i) $h_0 = \mathrm{id} : e(W) \longrightarrow e(W)$,
(ii) $(h_1(\omega))[j,\infty) \subseteq p^{-1}[j,\infty)$ *for each* $\omega \in e(W)$ *and* $j = 0,1,2,\ldots$.

Proof By 7.16 there exist a closed cocompact subspace $V \subseteq W$ and a proper map $q : V \times [0,\infty) \longrightarrow W$ extending the inclusion such that $pq(V \times [j,\infty)) \subseteq [j,\infty)$ for $j = 0,1,2,\ldots$. By the proof of 7.5, q is the adjoint of a domination $\widehat{q} : V \longrightarrow e(W)$ with right inverse $s : e(W) \longrightarrow V$. A homotopy $h : \mathrm{id}_{e(W)} \simeq \widehat{q}s$ satisfies the conclusions of the statement. □

Proposition 7.18 *If W is a forward tame metric space with a proper map $p : W \longrightarrow [0,\infty)$ the end space $e(W)$ is homotopy equivalent to the homotopy inverse limit*

$$e(W) \simeq \underset{\overleftarrow{j}}{\text{holim}}\, W_j$$

of the inverse system of inclusions

$$W_{j+1} = p^{-1}[j+1,\infty) \subseteq W_j = p^{-1}[j,\infty) \quad (j \geq 0) .$$

Proof Let $f : \underset{\overleftarrow{j}}{\text{holim}}\, W_j \longrightarrow e(W)$ be the map defined in Proposition 2.14, and define

$$X = \text{im}(f) \subseteq e(W) .$$

Note that f is a homeomorphism onto X, and that by Lemma 7.17 there is a homotopy $h : e(W) \times I \longrightarrow e(W)$ such that $h_0 = \text{id}$ and $h_1(e(W)) \subseteq X$. By examining the explicit formulas used in defining the homotopy, it can be seen that $h_t(X) \subseteq X$ for each $t \in I$, so that f is a homotopy equivalence. □

Proposition 7.19 *Let W be a forward tame metric space such that $e(W)$ is finitely dominated. If U is any cocompact subset of W such that U is an ANR, then U is finitely dominated.*

Proof It suffices to consider the case $U = W$. Let C be a compact space which dominates $e(W)$, say by maps $f : C \longrightarrow e(W)$ and $g : e(W) \longrightarrow C$ with a homotopy $h : fg \simeq \text{id}_{e(W)}$. Let $K = p_W f(C) \subseteq W$, where $p_W : e(W) \longrightarrow W$ is the evaluation. Of course, K is compact. A homotopy $d : Y \times I \longrightarrow W$ such that $d_0(Y) \subseteq K$ and $d_1 = \text{inclusion} : Y \longrightarrow W$ is given by

$$d_s = p_W h_s \widehat{q} : Y \longrightarrow W .$$

Now use the homotopy extension property to define a homotopy $\widetilde{d} : W \times I \longrightarrow W$ such that $\widetilde{d}|_{Y\times I} = d$ and $\widetilde{d}_1 = \text{id}_W$. Then

$$\widetilde{d}_0(W) \subset d_0(Y) \cup \widetilde{d}_0(\overline{W \backslash Y})$$

which is contained in a compact subset of W. □

The end space $e(W)$ of a forward tame CW complex W has the homotopy type of a CW complex (7.6), but it does not have the homotopy type of a particular CW complex, so it is not possible to associate a cellular chain complex to $e(W)$. We shall now show that the cellular chain complex at ∞ $C^\infty(W) = \mathcal{C}(C(W) \longrightarrow C^{lf}(W))_{*+1}$ is an adequate substitute. We shall also show that $C^{lf}(W)$ (and its $\pi_1(W)$-equivariant analogue) is finitely dominated, allowing the locally finite projective classes of W to be defined (in Chapter 10).

Proposition 7.20 *Let W be a forward tame strongly locally finite CW complex, and let $\widetilde{W}$ be a regular cover of W with group of covering translations π. Let $\widetilde{e(W)}$ be the cover of $e(W)$ induced from $\widetilde{W}$ by pullback along the evaluation map $p : e(W) \longrightarrow W$.*

(i) *Let $u : V \longrightarrow W$ be the inclusion of a cofinite subcomplex $V \subseteq W$. The inclusion of the locally finite cellular chain complexes*

$$\widetilde{u}^{lf} \ : \ C^{lf,\pi}(\widetilde{V}) \longrightarrow C^{lf,\pi}(\widetilde{W})$$

is such that there is defined a $\mathbb{Z}[\pi]$-module chain equivalence

$$\mathcal{C}(\widetilde{u}^{lf}) \ \simeq \ C(\widetilde{W},\widetilde{V}) \ .$$

The inclusion of cellular chain complexes at ∞ is a $\mathbb{Z}[\pi]$-module chain equivalence

$$\widetilde{u}^{\infty} \ : \ C^{\infty,\pi}(\widetilde{V}) \xrightarrow{\simeq} C^{\infty,\pi}(\widetilde{W}) \ .$$

(ii) *There exists a cofinite subcomplex $V \subseteq W$ such that $\widetilde{u}^{lf}$ is chain homotopic to 0, and there are defined $\mathbb{Z}[\pi]$-module chain equivalences*

$$\begin{aligned} C(\widetilde{W},\widetilde{V}) \ &\simeq \ C^{lf,\pi}(\widetilde{V})_{*-1} \oplus C^{lf,\pi}(\widetilde{W}) \ , \\ C(\widetilde{V}) \ &\simeq \ C^{lf,\pi}(\widetilde{V}) \oplus C^{\infty,\pi}(\widetilde{W}) \end{aligned}$$

with $C(\widetilde{W},\widetilde{V}) = C^{lf,\pi}(\widetilde{W},\widetilde{V})$ a finite f.g. free $\mathbb{Z}[\pi]$-module chain complex. In particular, $C^{lf,\pi}(\widetilde{W})$ is finitely dominated.

(iii) *If W is finitely dominated and $\pi_1(e(W)) \longrightarrow \pi_1(W)$ is a split injection then $e(W)$ is finitely dominated.*

(iv) *If there exists a cofinite subcomplex $V \subseteq W$ which is finitely dominated with $\pi_1(V) = \pi_1(e(W))$ then $e(V) \simeq e(W)$ is finitely dominated.*

(v) *If W and $e(W)$ are finitely dominated then there exist chain equivalences of finitely dominated $\mathbb{Z}[\pi]$-module chain complexes*

$$C^{lf,\pi}(\widetilde{W}) \ \simeq \ S^{lf,\pi}(\widetilde{W}) \ \ , \ \ S(\widetilde{e(W)}) \ \simeq \ C^{\infty,\pi}(\widetilde{W}) \ \simeq \ S^{\infty,\pi}(\widetilde{W})$$

and the projective classes are such that

$$\begin{aligned} [W]^{lf} \ &= \ [C^{lf,\pi}(\widetilde{W})] \ = \ [S^{lf,\pi}(\widetilde{W})] \ , \\ [e(W)] \ &= \ [C^{\infty,\pi}(\widetilde{W})] \ = \ [S^{\infty,\pi}(\widetilde{W})] \in K_0(\mathbb{Z}[\pi]) \ . \end{aligned}$$

Proof (i) As in the proof of 3.13 we have that $C(\widetilde{W},\widetilde{V}) = C^{lf,\pi}(\widetilde{W},\widetilde{V})$ is a f.g. free $\mathbb{Z}[\pi]$-module chain complex. The short exact sequence of $\mathbb{Z}[\pi]$-module chain complexes

$$0 \longrightarrow C^{lf,\pi}(\widetilde{V}) \xrightarrow{\widetilde{u}^{lf}} C^{lf,\pi}(\widetilde{W}) \longrightarrow C(\widetilde{W},\widetilde{V}) \longrightarrow 0$$

has $C(\widetilde{W},\widetilde{V})$ a finite f.g. free $\mathbb{Z}[\pi]$-module chain complex. By 3.7 (ii) there is a $\mathbb{Z}[\pi]$-module chain equivalence $\mathcal{C}(\widetilde{u}^{lf}) \simeq C(\widetilde{W},\widetilde{V})$.

Next, consider the short exact sequence of $\mathbb{Z}[\pi]$-module chain complexes

$$0 \longrightarrow C^{\infty,\pi}(\widetilde{V}) \xrightarrow{\widetilde{u}^{\infty}} C^{\infty,\pi}(\widetilde{W}) \longrightarrow C^{\infty,\pi}(\widetilde{W},\widetilde{V}) \longrightarrow 0$$

with

$$C^{\infty,\pi}(\widetilde{W},\widetilde{V}) = \mathcal{C}(C(\widetilde{W},\widetilde{V}) \longrightarrow C^{lf}(\widetilde{W},\widetilde{V}))_{*+1}$$

a contractible free $\mathbb{Z}[\pi]$-module chain complex. The inclusion

$$\widetilde{u}^{\infty} : C^{\infty,\pi}(\widetilde{V}) \longrightarrow C^{\infty,\pi}(\widetilde{W})$$

is a $\mathbb{Z}[\pi]$-module chain equivalence by 3.7 (iii).

(ii) Since W is forward tame there exists a cofinite subcomplex $V \subset W$ with an extension of the inclusion $V \times \{0\} \longrightarrow W$ to a proper map $V \times [0,\infty) \longrightarrow W$. Let $\widetilde{V} \subseteq \widetilde{W}$ be the cover of V corresponding to the cover $\widetilde{W}$ of W. The inclusion $\widetilde{u}^{lf} : C^{lf,\pi}(\widetilde{V}) \longrightarrow C^{lf,\pi}(\widetilde{W})$ admits a chain homotopy to 0

$$v : \widetilde{u}^{lf} \simeq 0 : C^{lf,\pi}(\widetilde{V}) \longrightarrow C^{lf,\pi}(\widetilde{W}) ,$$

and

$$\begin{aligned} C(\widetilde{W},\widetilde{V}) &= C^{lf,\pi}(\widetilde{W},\widetilde{V}) \\ &\simeq \mathcal{C}(\widetilde{u}^{lf} : C^{lf,\pi}(\widetilde{V}) \longrightarrow C^{lf,\pi}(\widetilde{W})) \\ &\simeq \mathcal{C}(0 : C^{lf,\pi}(\widetilde{V}) \longrightarrow C^{lf,\pi}(\widetilde{W})) = C^{lf,\pi}(\widetilde{W}) \oplus C^{lf,\pi}(\widetilde{V})_{*-1} . \end{aligned}$$

The chain map

$$p : C^{\infty,\pi}(\widetilde{W}) \simeq C^{\infty,\pi}(\widetilde{V}) \longrightarrow C(\widetilde{V})$$

has a left chain homotopy inverse

$$q : C(\widetilde{V}) \longrightarrow C^{\infty,\pi}(\widetilde{W}) ; x \longrightarrow (u(x), vi(x))$$

with a chain homotopy direct sum system

$$C^{\infty,\pi}(\widetilde{W}) \underset{q}{\overset{p}{\rightleftarrows}} C(\widetilde{V}) \underset{j}{\overset{i}{\rightleftarrows}} C^{lf,\pi}(\widetilde{V}) .$$

(iii) By 7.10 $S(\widetilde{e(W)})$ is $\mathbb{Z}[\pi]$-module chain equivalent to

$$C^{\infty,\pi}(\widetilde{W}) = \mathcal{C}(C(\widetilde{W}) \longrightarrow C^{lf,\pi}(\widetilde{W}))_{*+1} .$$

The $\mathbb{Z}[\pi]$-module chain complex $C(\widetilde{W})$ is finitely dominated by the finite domination of W, and $C^{lf,\pi}(\widetilde{W})$ is finitely dominated by (iii). By 6.9 the finite domination of W follows from the finite domination of the $\mathbb{Z}[\pi]$-module chain complex $S(\widetilde{e(W)})$, with $\pi = \pi_1(W)$ and $\widetilde{e(W)}$ the pullback to $e(W)$ of the universal cover $\widetilde{W}$ of W.

(iv) Apply (iii), with V instead of W, noting that 1.12 gives a homotopy equivalence $e(V) \simeq e(W)$.

(v) The $\mathbb{Z}[\pi]$-module chain map of 5.3 is a chain equivalence

$$\widetilde{\alpha} \; : \; S(e(\widetilde{W})) \xrightarrow{\simeq} S^{\infty,\pi}(\widetilde{W}) \; .$$

By Proposition A.7 in Appendix A there exists a subcomplex

$$D^{lf,\pi} \; = \; D^{lf,\pi}(\widetilde{W}) \subseteq S^{lf,\pi}(\widetilde{W})$$

with homology equivalences

$$D^{lf,\pi} \xrightarrow{\simeq} S^{lf,\pi}(\widetilde{W}) \quad , \quad D^{lf,\pi} \xrightarrow{\simeq} C^{lf,\pi}(\widetilde{W}) \; .$$

The $\mathbb{Z}[\pi]$-module chain complexes $S^{lf,\pi}(\widetilde{W})$, $C^{lf,\pi}(\widetilde{W})$ are finitely dominated by 7.5 and (ii), so that they are chain equivalent to free $\mathbb{Z}[\pi]$-module chain complexes. Let F be a free $\mathbb{Z}[\pi]$-module chain complex with a homology equivalence $F \longrightarrow D^{lf}$. The composites

$$F \longrightarrow D^{lf,\pi} \longrightarrow S^{lf,\pi}(\widetilde{W}) \quad , \quad F \longrightarrow D^{lf,\pi} \longrightarrow C^{lf,\pi}(\widetilde{W})$$

are homology equivalences, and hence chain equivalences. Similarly for the finitely dominated $\mathbb{Z}[\pi]$-module chain complexes $S^{\infty,\pi}(\widetilde{W})$, $C^{\infty,\pi}(\widetilde{W})$, noting that the subcomplexes

$$\begin{aligned} D \; &= \; D^{lf,\pi} \cap S(\widetilde{W}) \; = \; D(\widetilde{W}) \subseteq S(\widetilde{W}) \; , \\ D^{\infty,\pi} \; &= \; \mathcal{C}(D \longrightarrow D^{lf,\pi})_{*+1} \subseteq S^{\infty,\pi}(\widetilde{W}) \end{aligned}$$

are equipped with homology equivalences

$$\begin{aligned} &D \xrightarrow{\simeq} S(\widetilde{W}) \quad , \quad D \xrightarrow{\simeq} C(\widetilde{W}) \; , \\ &D^{\infty,\pi} \xrightarrow{\simeq} S^{\infty,\pi}(\widetilde{W}) \quad , \quad D^{\infty,\pi} \xrightarrow{\simeq} C^{\infty,\pi}(\widetilde{W}) \; . \end{aligned}$$

□

8

Reverse tameness

We now formulate the definition of 'reverse tameness', which is a generalization of the manifold tameness of Siebenmann [140].

Definition 8.1 Let W be a locally compact Hausdorff space.

(i) The space W is *reverse tame* if for every cocompact subspace $U \subseteq W$ there exists a cocompact subspace $V \subseteq W$ with $V \subseteq U$ such that U is dominated by $U \backslash V$, by a homotopy $h : W \times I \longrightarrow W$ such that:

(a) $h_0 = \mathrm{id}_W$,
(b) $h_t|_{(W\backslash U)}$ = inclusion : $W\backslash U \longrightarrow W$ for every $t \in I$,
(c) $h(U \times I) \subseteq U$,
(d) $h_1(W) \subseteq W\backslash V$.

(ii) The space W is *reverse collared* if for every cocompact subspace $U \subseteq W$ there exists a cocompact subspace $V \subseteq U$ such that $U\backslash V$ is a strong deformation retract of U, in which case there exists a homotopy $h : W \times I \longrightarrow W$ as in (i). □

By analogy with 7.2 and 7.3:

Proposition 8.2 *A reverse collared space is reverse tame.* □

In Parts Two and Three we shall be particularly concerned with reverse tameness and collaring for the ends of infinite cyclic covers of finite CW complexes, and with the connections with forward tameness and collarings. In Chapter 13 we shall give a homotopy theoretic criterion for reverse tameness of such an end, and in Chapter 23 we shall give a homological criterion. The 'projective class at ∞' of a reverse tame CW complex constructed in Chapter 10 is an algebraic K-theory obstruction to reverse collaring. In Chapter 13 it will be shown that a reverse tame end of an infinite cyclic cover of a finite CW complex is reverse collared if and only if this invariant vanishes.

Example 8.3 (i) Let $(L, K \subseteq L)$ be a pair of compact spaces, and let

$$W = L \cup_{K \times \{0\}} K \times [0, \infty) .$$

Then W is reverse collared: for each cocompact subspace $U \subset W$ there exists $t > 0$ such that $V = K \times (t, \infty) \subset U$ is a cocompact subspace with $U \backslash V$ a deformation retract of U.

(ii) Let $(M, \partial M)$ be a compact manifold with boundary. The boundary is collared (1.8), so that the interior

$$\mathrm{int}(M) = M \backslash \partial M \cong M \cup_{\partial M \times \{0\}} \partial M \times [0, \infty)$$

is reverse collared by (i). □

Remark 8.4 (i) If a space W has finitely many ends, then W is reverse tame (resp. reverse collared) if and only if each end of W is reverse tame (resp. reverse collared).

(ii) In a reverse tame space W every closed cocompact subspace $U \subseteq W$ is dominated by a compact subspace, the closure of $U \backslash V$ in the terminology of 8.1. In particular, W is dominated by a compact subspace.

(iii) In Chapter 13 below we shall show that a finitely dominated infinite cyclic cover of a finite CW complex is reverse tame (as well as forward tame). □

Proposition 8.5 *For an ANR space W the following are equivalent:*

(i) *W is reverse collared,*

(ii) *there exists a sequence of compact ANR subspaces*

$$K_1 \subseteq K_2 \subseteq K_3 \subseteq \ldots \subseteq W \ , \ W = \bigcup_{j=1}^{\infty} K_j \ ,$$

such that the inclusion $K_j \longrightarrow W$ is a homotopy equivalence for each $j = 1, 2, 3, \ldots$.

Proof (i) $\Longrightarrow$ (ii) Choose a sequence of cocompact subspaces

$$W = V_1 \supseteq V_2 \supseteq V_3 \supseteq \ldots \ , \ \bigcap_{j=1}^{\infty} \mathrm{cl}(V_j) = \emptyset \ ,$$

such that $V_j \backslash V_{j+1}$ is a strong deformation retract of V_j. Let

$$K_j = W \backslash V_{j+1} = V_j \backslash V_{j+1} \cup K_{j-1} .$$

The strong deformation retraction of V_j to $V_j\backslash V_{j+1}$ extends to a strong deformation retraction of W to K_j. Hence, the inclusion $K_j \longrightarrow W$ is a homotopy equivalence. Since K_j is a retract of W, K_j is an ANR.

(ii) $\Longrightarrow$ (i) Since the inclusion $K_j \longrightarrow W$ is a homotopy equivalence and W, K_j are ANR's, K_j is a strong deformation retract of W for each $j = 1, 2, 3, \ldots$. Given a cocompact subspace $U \subseteq W$, there exists $j \geq 1$ such that $W\backslash U \subseteq K_j$. Let $V = W\backslash K_j$. A strong deformation retraction of W to K_j restricts to a strong deformation retraction of U to $U\backslash V$. □

Example 8.6 Jacob's ladder X (2.26) is an ANR with $H_1(X) = \mathbb{Z}[z]$ an infinitely generated f.g. free $\mathbb{Z}$-module. A compact ANR is finitely dominated (and in fact homotopy finite, West [168]), so that its homology consists of f.g. $\mathbb{Z}$-modules. Thus X is not homotopy equivalent to a compact ANR and 8.5 implies that X is not reverse tame. Similarly for the space W of 4.14 (which is proper homotopy equivalent to X – in 9.8 below reverse tameness will be shown to be a proper homotopy invariant for ANR spaces such as X, W). □

Proposition 8.7 *Suppose W is a space with the property that every cocompact subspace $X \subseteq W$ contains an ANR cocompact subspace $Y \subseteq X$ which is closed in W. Then the following conditions are equivalent:*

(i) *W is reverse tame,*
(ii) *every closed ANR cocompact subspace $X \subseteq W$ is finitely dominated,*
(iii) *every cocompact subspace $X \subseteq W$ contains a finitely dominated (ANR) cocompact subspace $Y \subseteq X$ which is closed in W.*

Moreover, if W is also σ-compact, then the above conditions are equivalent to:

(iv) *there exists a sequence of finitely dominated (ANR) closed cocompact subspaces $W = W_0 \supseteq W_1 \supseteq W_2 \supseteq \ldots$ with $\bigcap\limits_j W_j = \emptyset$.*

Proof (i) $\Longrightarrow$ (ii) X is compactly dominated, by 8.4 (ii). Since X is an ANR (and hence homotopy equivalent to a CW complex), it follows that X is finitely dominated.

(ii) $\Longrightarrow$ (iii) This follows immediately from the hypothesis.

(iii) $\Longrightarrow$ (i) Let $U \subseteq W$ be a cocompact subspace. By hypothesis there exists an ANR cocompact subspace $X \subseteq U$ which is closed in W. Now there exists a finitely dominated cocompact subspace $Y \subseteq X$ which is closed in W. We may assume that Y is disjoint from the frontier $\mathrm{Fr}(X)$ of X. Since Y is finitely dominated, there exist a compact subspace $C \subseteq Y$ and a homotopy $g : Y \times I \longrightarrow Y$ such that $g_0 = \mathrm{id}_Y$ and $g_1(Y) \subseteq C$. Extend g to a homotopy $\tilde{g} : (\mathrm{Fr}(X) \cup Y) \times I \longrightarrow X$ by setting $\tilde{g}_t|_{\mathrm{Fr}(X)} = \text{inclusion} : \mathrm{Fr}(X) \longrightarrow X$ for each $t \in I$. Since $\mathrm{Fr}(X) \cup Y$ is a closed subset of the ANR X, the homotopy extension property implies there exists a homotopy $h : X \times I \longrightarrow X$ such that $h_0 = \mathrm{id}_X$ and $h|_{(\mathrm{Fr}(X)\cup Y)\times I} = \tilde{g}$. Now extend h to a homotopy $\tilde{h} : W \times I \longrightarrow W$ by setting $\tilde{h}_t|_{(W\backslash X)} = \text{inclusion} : W\backslash X \longrightarrow W$ for each $t \in I$. Since $\mathrm{cl}(W\backslash Y)$ is compact so is $\tilde{h}_1(\mathrm{cl}(W\backslash Y))$. Thus $V = W\backslash(\tilde{h}_1(\mathrm{cl}(W\backslash Y)) \cup C)$ is cocompact and $V \subseteq U$. Since $\tilde{h}_1(W) \subseteq W\backslash V$ we have shown that W is reverse tame.

Finally, it is clear that if W is σ-compact, then (iii) and (iv) are equivalent. □

Remark 8.8 Lemma 4.6 shows that every strongly locally finite CW complex W satisfies the hypothesis of 8.7 (that W contains arbitrarily small ANR closed cocompact subspaces). It is unclear whether or not every locally finite CW complex has this property, but this would be the case if every locally finite CW complex had a strongly locally finite subdivision. It is apparently unknown if the latter statement is true (cf. Farrell and Wagoner [52, p. 503]). Note that Hilbert cube manifolds satisfy the hypothesis of 8.7. Moreover, all Hilbert cube manifolds are σ-compact (being locally compact, separable and metric) and a countable CW complex is σ-compact. □

Proposition 8.9 *The following conditions on a strongly locally finite CW complex W are equivalent:*

(i) *W is reverse tame,*
(ii) *every cofinite subcomplex $V \subseteq W$ is finitely dominated,*
(iii) *every cocompact subspace $X \subseteq W$ contains a finitely dominated cofinite subcomplex.*

Moreover, if W is also countable, then the above conditions are equivalent to:

(iv) *there exists a sequence of finitely dominated cofinite subcomplexes*

$$W = W_0 \supseteq W_1 \supseteq W_2 \supseteq \ldots$$

with $\bigcap_j W_j = \emptyset$.

Proof (i) $\Longrightarrow$ (ii) This follows from Remark 8.4 (ii).

(ii) $\Longrightarrow$ (iii), (iii) $\Longrightarrow$ (i) These follow from 8.7.
Finally, it is clear that if W is countable, then (iii) and (iv) are equivalent. □

Corollary 8.10 *A reverse tame strongly locally finite CW complex W is finitely dominated.* □

Definition 8.11 A space W is *reverse π_1-tame* if it is reverse tame and each end has stable π_1 at ∞ (2.23). □

Remark 8.12 An open manifold with compact boundary $(W, \partial W)$ and one end is tame in the sense of Siebenmann [140] if:

(i) W is π_1-stable at ∞, so that there exists a sequence $W \supseteq W_0 \supset W_1 \supset W_2 \supset \dots$ of path-connected cocompact subspaces with $\bigcap_j \mathrm{cl}(W_j) = \emptyset$ such that the sequence of inclusion induced group morphisms

$$\pi_1(W_0) \xleftarrow{g_1} \pi_1(W_1) \xleftarrow{g_2} \pi_1(W_2) \longleftarrow \cdots$$

induces isomorphisms

$$\pi_1^\infty(W) = \mathrm{im}(g_1) \xleftarrow{\cong} \mathrm{im}(g_2) \xleftarrow{\cong} \dots, \text{ and}$$

(ii) there is a finitely dominated cocompact subspace $V \subseteq W$ such that $V \subseteq W_1$, $\pi_1(V) = \pi_1^\infty(W)$.

These conditions are equivalent to reverse π_1-tameness. □

Proposition 8.13 *A reverse π_1-tame strongly locally finite countable CW complex W admits a sequence $W \supseteq W_0 \supset W_1 \supset W_2 \supset \dots$ of finitely dominated cofinite subcomplexes, with $\bigcap_j W_j = \emptyset$ and such that the sequence of inclusion induced morphisms*

$$\pi_1(W_0) \xleftarrow{g_1} \pi_1(W_1) \xleftarrow{g_2} \pi_1(W_2) \longleftarrow \dots$$

(with base points and base paths chosen) induces isomorphisms

$$\pi_1^\infty(W) = \mathrm{im}(g_1) \xleftarrow{\cong} \mathrm{im}(g_2) \xleftarrow{\cong} \dots .$$ □

Proposition 8.14 *For a σ-compact metric space W the following conditions are equivalent:*
(i) *W is forward and reverse π_1-tame,*
(ii) *W is forward and reverse tame.*

Proof By 7.10 a forward tame W has stable π_1 at ∞. □

9

Homotopy at infinity

A proper homotopy equivalence at infinity is a proper map which (among other properties) induces a homotopy equivalence of the end spaces. The main result of this chapter is that an ANR is both forward and reverse tame if and only if it is bounded homotopy equivalent at ∞ to a product with $[0, \infty)$.

Definition 9.1 Let X, Y be topological spaces. A *proper homotopy equivalence at* ∞

$$(f, g, U, V) \ : \ X \longrightarrow Y$$

is defined by proper maps

$$f \ : \ U \longrightarrow Y \quad , \quad g \ : \ V \longrightarrow X$$

defined on closed cocompact subspaces $U \subseteq X$, $V \subseteq Y$, such that there exist proper homotopies

$$\begin{aligned} fg| &\simeq \text{inclusion} \ : \ g^{-1}(U) \longrightarrow Y \ , \\ gf| &\simeq \text{inclusion} \ : \ f^{-1}(V) \longrightarrow X \ . \end{aligned}$$

□

We shall usually write (f, g, U, V) as f.

Example 9.2 The inclusion $U \subseteq X$ of a closed cocompact subspace defines a proper homotopy equivalence at ∞. □

Example 9.3 Let $(Y, X \subseteq Y)$ be a pair of metric spaces with a proper homotopy $h_t : Y \longrightarrow Y$ $(t \in I)$ such that:

(i) $h_t(v) = v \in Y$ for all $v \in X$, $t \in I$,
(ii) $h_0(w) = w \in Y$ for all $w \in Y$,

(iii) $h_1(Y) \subseteq X'$ with $X \subseteq X' \subseteq Y$ such that X is a cocompact subspace of X'.

The inclusion $X \longrightarrow Y$ is a proper homotopy equivalence near ∞ in the sense of Siebenmann [144, p. 489]. There is also defined a proper homotopy equivalence at ∞ $(f, g, U, V) : X \longrightarrow Y$ in the sense of 9.1, with

$$\begin{aligned} f &= \text{inclusion} : U = X \longrightarrow Y , \\ g &= h_1| : V = (h_1)^{-1}(X) \longrightarrow X . \end{aligned}$$

□

Proposition 9.4 *A proper homotopy equivalence at* ∞ $(f, g, U, V) : X \longrightarrow Y$ *of* σ*-compact metric spaces* X, Y *induces:*

(i) *a bijection* $f_* : \mathcal{E}_X \longrightarrow \mathcal{E}_Y$ *of the sets of ends if* X, Y *are locally compact, connected, locally connected,*

(ii) *a homotopy equivalence of the end spaces* $e(f) : e(X) \longrightarrow e(Y)$.

Proof (i) Immediate from 1.22 (ii).

(ii) It follows from 1.12 that inclusions induce homotopy equivalences

$$\begin{aligned} i_1 : e(f^{-1}(V)) \longrightarrow e(X) \ , \ i_2 : e(V) \longrightarrow e(Y) \ , \\ i_3 : e(U) \longrightarrow e(X) \ , \ i_4 : e(g^{-1}(U)) \longrightarrow e(Y) \ . \end{aligned}$$

The composite

$$i_2 \, e(f|_{f^{-1}(V)})(i_1)^{-1} : e(X) \longrightarrow e(Y)$$

is a homotopy equivalence with homotopy inverse $(i_4)^{-1} e(g|_{g^{-1}(U)}) i_3$. □

Proposition 9.5 *Suppose* X *and* Y *are* σ*-compact, connected, locally path-connected metric spaces which are proper homotopy equivalent at* ∞.

(i) X *is semistable at* ∞ *if and only if* Y *is semistable at* ∞.

(ii) X *has stable* π_1 *at* ∞ *if and only if* Y *has stable* π_1 *at* ∞.

Proof (i) Propositions 1.22 and 9.4 imply that $\eta_X : \pi_0(e(X)) \longrightarrow \mathcal{E}_X$ is bijective if and only if $\eta_Y : \pi_0(e(Y)) \longrightarrow \mathcal{E}_Y$ is bijective. Now apply Proposition 2.25 (iii).

(ii) Let $(f, g, U, V) : X \longrightarrow Y$ be a proper homotopy equivalence at ∞ with homotopies

$$\begin{aligned} h &: gf| \simeq \text{inclusion} : f^{-1}(V) \longrightarrow X \ , \\ k &: fg| \simeq \text{inclusion} : g^{-1}(U) \longrightarrow Y \ . \end{aligned}$$

The first step is to construct cocompact subspaces

$$X \supseteq f^{-1}(V) = X_0 \supseteq X_1 \supseteq \cdots \ , \ Y \supseteq g^{-1}(U) = Y_0 \supseteq Y_1 \supseteq \cdots$$

such that

$$\bigcap_j \mathrm{cl}(X_j) = \bigcap_j \mathrm{cl}(Y_j) = \emptyset$$

and h and k restrict to homotopies

$$h| : gf| \simeq \text{inclusion} : X_j \longrightarrow X_{j-2} ,$$
$$k| : fg| \simeq \text{inclusion} : Y_j \longrightarrow Y_{j-2}$$

for $j \geq 2$. This is done by an elementary induction argument. Write X and Y as ascending unions of compact subspaces

$$X = \bigcup_{j=0}^{\infty} C_j , \ Y = \bigcup_{j=0}^{\infty} D_j$$

with

$$C_0 = \mathrm{cl}(X \backslash X_0) , \ D_0 = \mathrm{cl}(Y \backslash Y_0) .$$

Assuming that $N \geq 0$ and that X_N, Y_N have been defined, set

$$C'_{N-1} = \mathrm{proj}_X h^{-1}(\mathrm{cl}(X \backslash X_{N-1})) ,$$
$$X_{N+1} = f^{-1}(Y_N) \backslash (C_{N+1} \cup C'_{N-1})$$

and similarly for D'_{N-1}, Y_{N+1}.

Now assume that X has stable π_1 at ∞. In particular, X is path-connected at ∞, and we may assume that each X_j is path-connected for $j \geq 1$. From 2.25 (iv) it follows that $\pi_0(e(X)) = 0$; therefore, $\pi_0(e(Y)) = 0$ by 9.4 (ii). From (i) we know that Y is semistable at ∞. Another application of 2.25 (iv) gives that Y is path-connected at ∞. Therefore, we may also assume that each Y_j is path-connected for $j \geq 1$.

It follows from Siebenmann [140, p. 12] that some subsequence of the X_j satisfies the stability condition of 2.23 (iii). For simplicity, we assume that the original sequence satisfies that condition, i.e. the inclusions induce isomorphisms

$$\mathrm{im}(\pi_1(X_{j+3}) \longrightarrow \pi_1(X_{j+2})) \xrightarrow{\simeq} \mathrm{im}(\pi_1(X_{j+2}) \longrightarrow \pi_1(X_j)) .$$

A diagram chase shows that the inclusions also induce isomorphisms

$$\mathrm{im}(\pi_1(Y_{j+6}) \longrightarrow \pi_1(Y_{j+3})) \xrightarrow{\simeq} \mathrm{im}(\pi_1(Y_{j+3}) \longrightarrow \pi_1(Y_j)) ,$$

so that Y has stable π_1 at ∞. □

Proposition 9.6 *If W, W' are proper homotopy equivalent at ∞ then W is forward tame if and only if W' is forward tame.*

Proof Let $(f, g, U, U') : W \longrightarrow W'$ be a proper homotopy equivalence at ∞. If W is forward tame, then there exist a closed cocompact subspace $V \subseteq W$ and a proper map $q : V \times [0, \infty) \longrightarrow W$ which extends the inclusion

$V \times \{0\} \longrightarrow W$. The inverse image $g^{-1}(V)$ is a closed cocompact subspace of W' and we may assume that $q(V \times [0, \infty)) \subseteq U$, so in particular $V \subseteq U$. Use a proper homotopy

$$h \; : \; fg|_{g^{-1}(U)} \simeq \text{inclusion} \; : \; g^{-1}(U) \longrightarrow W'$$

to define a proper map

$$\begin{aligned} q' &= fq(g \times \mathrm{id}_{[0,\infty)}) \cup h \; : \\ g^{-1}(V \times [-1, \infty)) &= g^{-1}(V \times [0, \infty)) \cup g^{-1}(V \times [-1, 0]) \longrightarrow W' \end{aligned}$$

such that $q'| : g^{-1}(V) \times \{-1\} \longrightarrow W'$ is the inclusion. □

Remark 9.7 The property of being forward collared is not an invariant of the proper homotopy type at ∞ – see 11.7 (ii) below. □

Proposition 9.8 *Suppose W, W' are ANR's which have arbitrarily small closed cocompact subspaces which are ANR's and that W, W' are proper homotopy equivalent at ∞.*

(i) *W is reverse tame if and only if W' is reverse tame.*

(ii) *W is reverse π_1-tame if and only if W' is reverse π_1-tame.*

Proof (i) Let $(f, g, U, U') : W \longrightarrow W'$ be a proper homotopy equivalence at ∞ with a homotopy

$$h \; : \; \text{inclusion} \simeq fg| \; : \; g^{-1}(U) \longrightarrow W' \; .$$

We may assume that $U' \subseteq W'$ is a closed cocompact ANR subspace. To show that W' is reverse tame it suffices to show that U' is finitely dominated (8.7). If W is reverse tame, then there exist a closed cocompact subspace $X \subseteq U$ and a homotopy $k : X \times I \longrightarrow X$ with $k_0 = \mathrm{id}_X$ and $k_1(X) \subseteq C$ for some compact subspace $C \subseteq X$ (8.7). We may assume that $f(X) \subseteq U'$ and $h(g^{-1}(X \times I)) \subseteq U'$ so that $fk_tg(g^{-1}(X)) \subseteq U'$ for each $t \in I$. The homotopy

$$H \; : \; \text{inclusion} \simeq fk_1g| \; : \; g^{-1}(X) \longrightarrow U'$$

defined by

$$H(x,t) \;=\; \begin{cases} h(x, 2t) & \text{if } 0 \leq t \leq \frac{1}{2} \, , \\ fk(g(x), 2t-1) & \text{if } \frac{1}{2} \leq t \leq 1 \end{cases}$$

can be extended (using the homotopy extension property for ANR's) to a homotopy $\widetilde{H} : \mathrm{id}_{U'} \simeq \widetilde{H}_1$ such that

$$\widetilde{H}(U') \subseteq \widetilde{H}_1(\mathrm{cl}(U' \backslash g^{-1}(X))) \cup f(C)$$

which is compact.

(ii) follows from (i) and 9.5 (ii). □

Remark 9.9 Note that 9.8 applies to strongly locally finite CW complexes and Hilbert cube manifolds, since they are ANR spaces which have arbitrarily small closed cocompact subsets which are ANR's (8.8). □

Definition 9.10 Let X, Y be topological spaces with maps

$$p : X \longrightarrow [0,\infty) \ , \ q : Y \longrightarrow [0,\infty) .$$

(i) A *bounded homotopy equivalence at* ∞

$$(f,g,X',Y') : X \longrightarrow Y$$

is defined by maps

$$f : X' \longrightarrow Y \ , \ g : Y' \longrightarrow X$$

defined on subspaces

$$X' = p^{-1}([s,\infty)) \ , \ Y' = q^{-1}([t,\infty))$$

for some $s, t \geq 0$, such that for some $\epsilon > 0$

$$d(p(x), qf(x)) < \epsilon \ , \ d(pg(y), q(y)) < \epsilon \ \ (x \in X', y \in Y')$$

and that there exist homotopies

$$h : fg| \simeq \text{inclusion} : g^{-1}(X') \longrightarrow Y ,$$

$$k : gf| \simeq \text{inclusion} : f^{-1}(Y') \longrightarrow X$$

such that for all $x \in g^{-1}(X')$, $y \in f^{-1}(Y')$

$$\text{diameter } qh(\{x\} \times I) < \epsilon \ , \ \text{diameter } pk(\{y\} \times I) < \epsilon .$$

(ii) An *ϵ-homotopy equivalence at* ∞ is defined as in (i), but with a particular $\epsilon > 0$. □

Note that in 9.10 neither p nor q is required to be proper. Thus, the '∞' referred to is not the ∞ in the one-point compactifications of X and Y, but rather the ∞ in the one-point compactification of $[0,\infty)$.

Proposition 9.11 *Let X, Y be Hausdorff spaces with proper maps $p : X \longrightarrow [0,\infty)$, $q : Y \longrightarrow [0,\infty)$.*

(i) *A bounded homotopy equivalence at ∞ $(f,g,X',Y') : X \longrightarrow Y$ is a proper homotopy equivalence at ∞.*

(ii) *If $(f,g,X',Y') : X \longrightarrow Y$ is a proper homotopy equivalence at ∞ with $X' = p^{-1}([s,\infty))$, $Y' = q^{-1}([t,\infty))$, then there are proper homotopies $p \simeq p'$, $q \simeq q'$ so that (f,g,X',Y') is a bounded homotopy equivalence at ∞ with respect to p', q'.*

Proof (i) Let $\epsilon > 0, h$ and k be as in 9.10 (i). We need to show that f, g, h

and k are proper. If $K \subseteq Y$ is compact, then $q(K) \subseteq [0, N]$ for some $N \geq 0$. Since p is ϵ-close to qf:

$$f^{-1}(K) \subseteq p^{-1}([0, N+\epsilon]) .$$

Thus, $f^{-1}(K)$ is a closed subset of a compact space, hence compact, verifying that f is proper. Likewise,

$$h^{-1}(K) \subseteq q^{-1}([0, N+\epsilon]) \times I ,$$

verifying that h is proper. The proofs that g and k are proper are analogous.

(ii) Let

$$h : fg| \simeq \text{inclusion} : g^{-1}(X') \longrightarrow Y,$$
$$k : gf| \simeq \text{inclusion} : f^{-1}(Y') \longrightarrow X$$

be proper homotopies. Let $n_0 = 0$ and use the properness of p, q, f, g, h, k to inductively choose $n_j \geq j$ for $j = 0, 1, 2, \ldots$ such that:

(a) $fp^{-1}([0, n_{j-1}]) \subseteq q^{-1}([0, n_j])$,
$$gq^{-1}([0, n_{j-1}]) \subseteq p^{-1}([0, n_j]) ,$$
(b) $fp^{-1}([n_j, \infty)) \subseteq q^{-1}([n_{j-1}, \infty))$,
$$gq^{-1}([n_j, \infty)) \subseteq p^{-1}([n_{j-1}, \infty)) ,$$
(c) $qh(q^{-1}([0, n_{j-1}]) \times I) \subseteq q^{-1}([0, n_j])$,
$$pk(p^{-1}([0, n_{j-1}]) \times I) \subseteq p^{-1}([0, n_j]) ,$$
(d) $qh(q^{-1}([n_j, \infty)) \times I) \subseteq q^{-1}([n_{j-1}, \infty))$,
$$pk(p^{-1}([n_j, \infty)) \times I) \subseteq p^{-1}([n_{j-1}, \infty)) .$$

These conditions imply that:

(e) $fp^{-1}([n_j, n_{j+1}]) \subseteq q^{-1}([n_{j-1}, n_{j+2}])$,
$$gq^{-1}([n_j, n_{j+1}]) \subseteq p^{-1}([n_{j-1}, n_{j+2}]) ,$$
(f) $qh(q^{-1}([n_j, n_{j+1}]) \times I) \subseteq q^{-1}([n_{j-1}, n_{j+2}])$,
$$pk(p^{-1}([n_j, n_{j+1}]) \times I) \subseteq p^{-1}([n_{j-1}, n_{j+2}]) .$$

Let $\gamma : [0, \infty) \longrightarrow [0, \infty)$ be the PL homeomorphism such that $\gamma(n_j) = j$ for each $j = 0, 1, 2, \ldots$. Then $p' = \gamma p$ and $q' = \gamma q$ satisfy the requirements. □

Proposition 9.12 *Suppose W is a space with a proper map $p : W \longrightarrow [0, \infty)$ such that W is boundedly homotopy equivalent at ∞ to the projection $q : Y \times [0, \infty) \longrightarrow [0, \infty)$ for some space Y.*

(i) *W is forward tame.*

(ii) *If W is an ANR, then W is reverse tame.*

(iii) *If W is a metric space, then $Y \simeq e(W)$.*

Proof Let

$$\begin{aligned} f &: W' = p^{-1}([s,\infty)) \longrightarrow Y \times [0,\infty) , \\ g &: Y' = Y \times [t,\infty) \longrightarrow W , \\ h &: fg| \simeq \text{inclusion} : g^{-1}(W') \longrightarrow Y , \\ k &: gf| \simeq \text{inclusion} : f^{-1}(Y') \longrightarrow W \end{aligned}$$

be as in Definition 9.10, for some $\epsilon > 0$. Let

$$p_1 : Y \times [0,\infty) \longrightarrow Y \ , \ p_2 : Y \times [0,\infty) \longrightarrow [0,\infty)$$

be the projections.

(i) Since p is proper, it follows that $f^{-1}(Y')$ is cocompact in W and k is a proper homotopy. Then $F : f^{-1}(Y') \times [0,\infty) \longrightarrow W$ defined by

$$F(x,u) = \begin{cases} k(x, 1-u) & \text{if } 0 \le u \le 1 , \\ g(p_1 f(x), u p_2 f(x)) & \text{if } 1 \le u \end{cases}$$

gives the required proper map extending the inclusion.

(ii) Let U be a given open cocompact subset of W. Choose $s_0 \ge \max\{s,t\}$ such that $p^{-1}([s_0,\infty)) \subseteq U$. Let

$$V' = p^{-1}([s_0 + 4\epsilon, \infty))$$

and note that $V' \subseteq W'$ and $f(V') \subseteq Y'$.

Define a homotopy $F : V' \times I \longrightarrow W$ by

$$F(x,u) = \begin{cases} k(x, 1-2u) & \text{if } 0 \le u \le \frac{1}{2} , \\ g(p_1 f(x), (2-2u) p_2 f(x) + (2u-1)(s_0 + 2\epsilon)) & \text{if } \frac{1}{2} \le u \le 1 . \end{cases}$$

Note that $F(V' \times I) \subseteq U$, so we consider F as a homotopy $F : V' \times I \longrightarrow U$ such that

$$F_0 = \text{inclusion} : V' \longrightarrow U \ , \ F_1(V') \subset U \backslash V' .$$

Since W is an ANR so is U. Thus, we can use the homotopy extension property to define a homotopy $\widetilde{F} : U \times I \longrightarrow U$ such that

$$\widetilde{F}|_{V' \times I} = F \ , \ \widetilde{F}_0|_{U \cap p^{-1}([0,s_0])} = \text{inclusion} .$$

Finally, extend via the identity to get a homotopy $\widetilde{F} : W \times I \longrightarrow W$ and let $V = \widetilde{F}_1(W)$. It is easy to see that V is cocompact and that $\widetilde{F}$ gives the required domination of U by $U \backslash V$.

(iii) The map f induces a map

$$f_* : e(W') \longrightarrow Y \ ; \ \omega \longrightarrow p_1 f(\omega(0)) .$$

If t is chosen so large that $g(Y') \subseteq W'$, then g induces the map

$$g_* : Y \longrightarrow e(W') \ ; \ y \longrightarrow (u \longrightarrow g(y, t+u)) \ \ (u \ge 0) .$$

It is clear that h induces a homotopy $f_* g_* \simeq 1_Y$. Also, if W'' is a cocompact subset of W' chosen so that $f(W'') \subset Y'$, then k will induce a homotopy $g_* f_*|_{e(W'')} \simeq 1_{e(W'')}$. Since W is a metric space, Proposition 1.12 implies that the inclusions $W'' \longrightarrow W' \longrightarrow W$ induce homotopy equivalences on end spaces and the result follows. □

Proposition 9.13 *Suppose W is a forward tame metric space with a proper map $p : W \longrightarrow [0,\infty)$. Then there exist a closed cocompact subspace $Y \subseteq W$ and maps*

$$f \; : \; Y \longrightarrow e(W) \times [0,\infty) \quad , \quad g \; : \; e(W) \times [0,\infty) \longrightarrow Y$$

together with homotopies

$$F \; : \; igf \; \simeq \; i \; : \; Y \longrightarrow W \; ,$$
$$G \; : \; fg \; \simeq \; \mathrm{id} \; : \; e(W) \times [0,\infty) \longrightarrow e(W) \times [0,\infty) \; ,$$

with $i : Y \longrightarrow W$ the inclusion such that:

(i) $p = p_2 f : Y \longrightarrow [0,\infty)$ *with* $p_2 : e(W) \times [0,\infty) \longrightarrow [0,\infty)$ *the projection,*
(ii) *for every $N \geq 0$ there exists $M \geq 0$ such that*

$$(pg)^{-1}([0,N]) \subseteq e(W) \times [0,M] \; ,$$

(iii) *$F : Y \times I \longrightarrow W$ is proper,*
(iv) *for every $N \geq 0$ there exists $M \geq 0$ such that*

$$G(e(W) \times [M,\infty) \times I) \subseteq e(W) \times [N,\infty) \; .$$

Proof Let $Y \subseteq W$ be a closed cocompact subspace for which there exists a proper map $q : Y \times [0,\infty) \longrightarrow W$ extending the inclusion such that $pq(Y \times [M,\infty)) \subseteq [M,\infty)$ for $M = 0,1,2,\ldots$ (7.16). For the adjoint $\widehat{q} : Y \longrightarrow e(W)$ choose a closed cocompact subspace $Y' \subseteq Y$ such that $\widehat{q}(Y') \subseteq e(Y)$. Use i also to denote the inclusion $i : Y' \longrightarrow W$. It induces a homotopy equivalence $e(i) : e(Y') \longrightarrow e(W)$, so there is a homotopy inverse $j : e(W) \longrightarrow e(Y')$ with a homotopy

$$k \; : \; e(i)j \; \simeq \; \mathrm{id}_{e(W)} \; : \; e(W) \longrightarrow e(W) \; .$$

By using the explicit construction of 1.12, we may assume that for every $\omega \in e(W)$, $t \in I$, $u \geq 0$, $k(\omega,t)(u) = \omega(s)$ for some $s \geq 0$. Define

$$f \; : \; Y \longrightarrow e(W) \times [0,\infty) \; ; \; x \longrightarrow (\widehat{q}(x), p(x)) \; .$$

Define $g : e(W) \times [0,\infty) \longrightarrow Y$ to be the composition

$$e(W) \times [0,\infty) \xrightarrow{j \times \mathrm{id}} e(Y') \times [0,\infty) \xrightarrow{p_{Y'} \times \mathrm{id}} Y' \times [0,\infty)$$
$$\xrightarrow{\widehat{q} \times \mathrm{id}} e(Y) \times [0,\infty) \xrightarrow{p_Y^+} Y$$

where $p_{Y'}$ and p_Y^+ are the evaluation maps. The homotopy $F : igf \simeq i : Y \longrightarrow W$ is given by

$$F(x,t) = \widehat{q}[k_t\widehat{q}(x)(0)]((1-t)p(x)) .$$

Define

$$\gamma : e(W) \times [0,\infty) \longrightarrow e(W) ; (\omega, t) \longrightarrow \widehat{q}[\widehat{q}(x_\omega)(t)]$$

where $x_\omega = j(\omega)(0) \in Y'$. Define a homotopy

$$G' : e(W) \times [0,\infty) \times I \longrightarrow e(W) \times [0,\infty) ;$$
$$(\omega,t,s) \longrightarrow (\gamma(\omega,t), (1-s)p[\widehat{q}(x_\omega)(t)] + st) .$$

Note that $G'_0 = fg$. We claim there is a homotopy

$$G'' : e(W) \times [0,\infty) \times I \longrightarrow e(W)$$

with $G''_0 = \gamma$ and $G''_1 =$ projection : $e(W) \times [0,\infty) \longrightarrow e(W)$. Contracting $[0,\infty)$ to $\{0\}$ there is defined a homotopy $\gamma \simeq \gamma'$ with

$$\gamma'(\omega,t) = \widehat{q}[\widehat{q}(x_\omega)(0)] = \widehat{q}(x_\omega) = \widehat{q}(p_{Y'}(j(\omega))) .$$

The proof of 7.5 (i) shows that $\widehat{q}p_{Y'} : e(Y') \longrightarrow e(W)$ is homotopic to the inclusion, so $\gamma' \simeq \gamma''$ where $\gamma''(\omega,t) = j(\omega)$. Since $j : e(W) \longrightarrow e(Y')$ was chosen to be a homotopy inverse for the inclusion, the homotopy G'' exists as claimed above. We can now define the homotopy

$$G : e(W) \times [0,\infty) \times I \longrightarrow e(W) \times [0,\infty) ;$$

$$(\omega,t,s) \longrightarrow \begin{cases} G'(\omega,t,2s) & \text{if } 0 \le s \le \frac{1}{2} , \\ (G''(\omega,t,2s-1),t) & \text{if } \frac{1}{2} \le s \le 1 . \end{cases}$$

Finally we verify the four properties of f, g, F, G:

(i) is obvious.

(ii) Since $g(\omega,t) = \widehat{q}(x_\omega)(t) = q(x_\omega,t)$, it follows that $pg(\omega,t) \ge M$ if $t \ge M$ and $M = 0,1,2,\ldots$.

(iii) To verify that F is proper, let $K \subseteq W$ be compact and suppose $F(x,t) \in K$. Since $k_t\widehat{q}(x)(0) = q(x,s)$ for some $s \ge 0$, it follows that

$$F(x,t) = \widehat{q}[q(x,s)]((1-t)p(x)) = q[q(x,s),(1-t)p(x)] .$$

Thus, $[q(x,s),(1-t)p(x)] \in q^{-1}(K)$. Since q is proper, $q^{-1}(K)$ is compact as is $C \subseteq Y$, the projection into Y of $q^{-1}(K)$. Since $q(x,s) \in C$, $(x,s) \in q^{-1}(C)$. Since $q^{-1}(C)$ is compact, so is $C' \subseteq Y$, the projection into Y of $q^{-1}(C)$. Since $\{(1-t)p(x) \,|\, t \in I, x \in C'\} \subseteq [0,\infty)$ is also compact, it follows that $F^{-1}(K) \subseteq Y \times I$ is compact.

(iv) If $t \ge M$ for $M = 0,1,2,\ldots$, and $x \in Y'$, then

$$p(\widehat{q}(x)(t)) = p(q(x,t)) \ge M$$

so

$$(1-s)p[\widehat{q}(x_\omega)(t)] + st \geq M \ .$$

It follows that

$$G'(e(W) \times [M,\infty) \times I) \subseteq e(W) \times [M,\infty) \quad (M = 0,1,2,\ldots) \ . \qquad \square$$

Proposition 9.14 *Suppose W is a forward tame metric space with $e(W)$ finitely dominated and a proper map $p : W \longrightarrow [0,\infty)$. Then for every $\epsilon > 0$, p is properly homotopic to a proper map $p' : W \longrightarrow [0,\infty)$ for which W is ϵ-homotopy equivalent at ∞ to the projection $p_2 : e(W) \times [0,\infty) \longrightarrow [0,\infty)$.*
Proof Let Y, f, g, F, G be as in Proposition 9.13. Since $e(W)$ is finitely dominated, there exist a compact subspace $K \subseteq e(W)$ and a homotopy $D : e(W) \times I \longrightarrow e(W)$ such that $D_0(e(W)) \subseteq K$ and $D_1 = \mathrm{id}_{e(W)}$. Define

$$\begin{aligned} g' &: e(W) \times [0,\infty) \longrightarrow Y \ , \\ F' &: ig'f \simeq i \ : \ Y \longrightarrow W \ , \\ G' &: fg' \simeq \mathrm{id} \ : \ e(W) \times [0,\infty) \longrightarrow e(W) \times [0,\infty) \end{aligned}$$

as follows:

$$\begin{aligned} g' &= g(D_0 \times \mathrm{id}_{[0,\infty)}) \ , \\ F'_s &= \begin{cases} ig(D_{2s} \times \mathrm{id}_{[0,\infty)})f & \text{if } 0 \leq s \leq \frac{1}{2} \ , \\ F_{2s-1} & \text{if } \frac{1}{2} \leq s \leq 1 \ , \end{cases} \\ G'_s &= \begin{cases} G_{2s}(D_0 \times \mathrm{id}_{[0,\infty)}) & \text{if } 0 \leq s \leq \frac{1}{2} \ , \\ D_{2s-1} \times \mathrm{id}_{[0,\infty)} & \text{if } \frac{1}{2} \leq s \leq 1 \ . \end{cases} \end{aligned}$$

It follows from 9.13 (ii), (iii) that F' is proper. From 9.13 (iii) it follows that for every $N \geq 0$ there exists $M \geq 0$ such that

$$G'(e(W) \times [M,\infty) \times I) \subseteq e(W) \times [N,\infty) \ .$$

Let $n_0 = 0$ and choose $n_j \geq j$ inductively such that:

(1) $pF'(p^{-1}([0,n_{j-1}]) \times I) \subseteq [0,n_j]$.
(This uses the properness of p.)
(2) $p_2G'(e(W) \times [0,n_{j-1}] \times I) \subseteq [0,n_j]$.
(This uses the compactness of K.)
(3) $pF'(p^{-1}([n_j,\infty)) \times I) \subseteq [n_{j-1},\infty)$.
(This uses the properness of p and F'.)
(4) $p_2G'(e(W) \times [n_j,\infty) \times I) \subseteq [n_{j-1},\infty)$.
(This uses the property of G' mentioned above.)

Given $\epsilon > 0$ let

$$\gamma \ : \ [0,\infty) \longrightarrow [0,\infty)$$

be the PL homeomorphism such that

$$\gamma(n_j) \ = \ \frac{j\epsilon}{3} \ \ (j = 0,1,2,\dots) \ .$$

Let $p' = \gamma p : W \longrightarrow [0,\infty)$. Then

$$(\mathrm{id}_{e(W)} \times \gamma)f \ : \ Y \longrightarrow e(W) \times [0,\infty)$$

is an ϵ-equivalence with inverse $g'(\mathrm{id}_{e(W)} \times \gamma^{-1})$. □

Proposition 9.15 *Let W be an ANR which has arbitrarily small closed cocompact subsets which are ANR's. The following conditions on W are equivalent:*

(i) *W is both forward and reverse tame,*
(ii) *W is forward tame and the end space $e(W)$ is finitely dominated,*
(iii) *there exist a proper map $W \longrightarrow [0,\infty)$ and a space Y such that W is bounded homotopy equivalent at ∞ to the projection $Y \times [0,\infty) \longrightarrow [0,\infty)$.*

Moreover, if these conditions are satisfied Y is homotopy equivalent to $e(W)$.

Proof (i) implies (ii) by 7.5 (i) and 8.7.
(ii) implies (iii) by 9.14.
(iii) implies (i) by 9.12 (i), (ii).
If these conditions are satisfied $Y \simeq e(W)$ by 9.12 (iii). □

Theorem 9.16 *Let W be an ANR which has arbitrarily small closed cocompact subsets which are ANR's. The following conditions on W are equivalent:*

(i) *W is both forward and reverse tame and $e(W)$ is homotopy equivalent to a finite CW complex,*
(ii) *there exist a proper map $W \longrightarrow [0,\infty)$ and a finite CW complex K so that W is bounded homotopy equivalent at ∞ to the projection $K \times [0,\infty) \longrightarrow [0,\infty)$,*
(iii) *there exists a finite CW complex K so that W is proper homotopy equivalent at ∞ to the projection $K \times [0,\infty) \longrightarrow [0,\infty)$.*

Moreover, if these conditions are satisfied K is homotopy equivalent to $e(W)$.

Proof (ii) and (iii) are equivalent by 9.11. The rest of the proof follows from 9.15. □

Example 9.17 Let

$$X = \{(x,y) \in \mathbb{R}^2 \mid x \geq 0\,,\, y \in \{\tfrac{1}{2}, \tfrac{1}{3}, \tfrac{1}{4}, \ldots, 0\}\}\,.$$

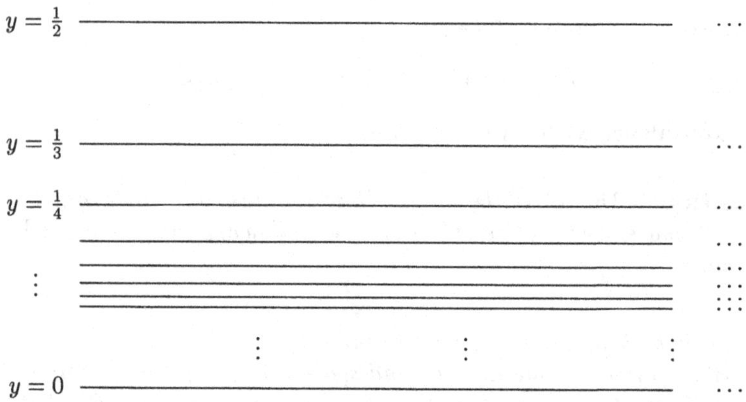

$y = \frac{1}{2}$ ...

$y = \frac{1}{3}$ ...

$y = \frac{1}{4}$ ...

$\vdots$

$y = 0$ ...

X

The metric space X is locally compact, forward tame and reverse tame. However, the end space $e(X)$ has infinitely many components, one for each element of $\{\frac{1}{2}, \frac{1}{3}, \frac{1}{4}, \ldots, 0\}$, so that it is not finitely dominated. Since X is not locally connected, X is not an *ANR* and therefore this does not contradict 9.15. However, this example does contradict Quinn [116, p. 466]. □

Proposition 9.18 *Let W be a forward tame ANR which has arbitrarily small closed cocompact subsets which are ANR's (e.g. a strongly locally finite CW complex or a Hilbert cube manifold). The following conditions on W are equivalent:*

(i) *W is reverse tame,*
(ii) *W is reverse π_1-tame,*
(iii) *the end space $e(W)$ is finitely dominated.*

Proof (i) $\Longrightarrow$ (ii) W has stable π_1 at ∞ by 7.11.
The other implications follow from 9.15. □

10

Projective class at infinity

We associate to a reverse π_1-tame space W the 'projective class at ∞' $[W]_\infty \in \widetilde{K}_0(\mathbb{Z}[\pi_1^\infty(W)])$ (10.1) with image the projective class (= Wall finiteness obstruction) $[W] \in \widetilde{K}_0(\mathbb{Z}[\pi_1(W)])$. The projective class at ∞ is an obstruction to reverse collaring W, which for an open manifold is the end obstruction of Siebenmann [140]. In 10.13 we prove a form of Poincaré duality (originally due to Quinn [116]) that a manifold end is forward tame if and only if it is reverse tame, subject to suitable fundamental group conditions, in which case the locally finite projective class at ∞ is the Poincaré dual of the projective class at ∞ (10.15).

We associate to a forward tame CW complex W the 'locally finite projective class' $[W]^{lf} \in \widetilde{K}_0(\mathbb{Z}[\pi_1(W)])$ (10.4), and a 'locally finite projective class at ∞' $[W]^{lf}_\infty \in \widetilde{K}_0(\mathbb{Z}[\pi_1(e(W))])$ (10.8), such that $[W]^{lf}$ is the image of $[W]^{lf}_\infty$. The locally finite projective class at ∞ is an obstruction to forward collaring W. If W is both forward and reverse tame then $e(W)$ is finitely dominated, with finiteness obstruction

$$[e(W)] \;=\; [W]_\infty - [W]^{lf}_\infty \in \widetilde{K}_0(\mathbb{Z}[\pi])$$

where $\pi = \pi_1(e(W)) = \pi_1^\infty(W)$.

In 10.5 below it will be proved that for an open n-dimensional manifold with boundary $(W, \partial W)$ and a forward and reverse tame end the end space $e(W)$ of W is a finitely dominated $(n-1)$-dimensional Poincaré space with finiteness obstruction

$$\begin{aligned}[e(W)] \;&=\; [W]_\infty + (-)^{n-1}[W]^*_\infty \\ &\in \widetilde{K}_0(\mathbb{Z}[\pi_1(e(W))]) \;=\; \widetilde{K}_0(\mathbb{Z}[\pi_1^\infty(W)])\ ,\end{aligned}$$

and that $(W; \partial W, e(W))$ is an n-dimensional geometric Poincaré cobordism.

Let W be a space with arbitrarily small closed cocompact ANR subspaces, e.g. a strongly locally finite CW complex (4.6). If W has stable π_1 at ∞ there exists a sequence $W_0 \supseteq W_1 \supseteq W_2 \supseteq \ldots$ of closed path-connected cocompact ANR's such that the induced group morphisms

$$\pi_1(W_0) \xleftarrow{g_1} \pi_1(W_1) \xleftarrow{g_2} \pi_1(W_2) \xleftarrow{g_3} \ldots$$

induce isomorphisms between images and

$$\pi_1^\infty(W) = \mathrm{im}(g_1) = \mathrm{im}(g_2) = \ldots .$$

As in 2.21 (iv) each g_j induces a surjection $q_j : \pi_1(W_j) \longrightarrow \mathrm{im}(g_j) = \pi_1^\infty(W)$ which is a left inverse for the injection $p_j : \pi_1^\infty(W) = \mathrm{im}(g_{j+1}) \longrightarrow \pi_1(W_j)$, with

$$\begin{aligned} g_j : \pi_1(W_j) &= \pi_1^\infty(W) \times \ker(q_j) \\ \longrightarrow \pi_1(W_{j-1}) &= \pi_1^\infty(W) \times \ker(q_{j-1}) \; ; \; (x,y) \longrightarrow (x,1) . \end{aligned}$$

If W is σ-compact and reverse tame then each W_j is finitely dominated by 8.7. In particular, if W is a reverse π_1-tame strongly locally finite CW complex there exists such a sequence $W_0 \supseteq W_1 \supseteq W_2 \supseteq \ldots$ of finitely dominated cofinite subcomplexes with each $q_j : \pi_1(W_j) \longrightarrow \pi_1^\infty(W)$ a split surjection and

$$(q_0)_*[W_0] = (q_1)_*[W_1] = (q_2)_*[W_1] = \ldots \in \widetilde{K}_0(\mathbb{Z}[\pi_1^\infty(W)]) .$$

Definition 10.1 The *projective class at* ∞ of a reverse π_1-tame space W with arbitrarily small closed cocompact ANR subspaces is the image of the Wall finiteness obstruction $[V] \in \widetilde{K}_0(\mathbb{Z}[\pi_1(V)])$

$$[W]_\infty = q_*[V] \in \widetilde{K}_0(\mathbb{Z}[\pi_1^\infty(W)])$$

with $V \subseteq W$ any finitely dominated closed cocompact subset such that the natural morphism $p : \pi_1^\infty(W) \longrightarrow \pi_1(V)$ is a split injection (on each component) with a left inverse $q : \pi_1(V) \longrightarrow \pi_1^\infty(W)$. □

The morphism $\widetilde{K}_0(\mathbb{Z}[\pi_1^\infty(W)]) \longrightarrow \widetilde{K}_0(\mathbb{Z}[\pi_1(W)])$ sends the projective class at ∞ $[W]_\infty$ to the finiteness obstruction $[W]$.

Theorem 10.2 (Siebenmann [140]) (i) *Let $(W, \partial W)$ be an open n-dimensional manifold with a compact boundary ∂W and a reverse π_1-tame end. The projective class at ∞*

$$[W]_\infty \in \widetilde{K}_0(\mathbb{Z}[\pi_1^\infty(W)])$$

has image the Wall finiteness obstruction $[W] \in \widetilde{K}_0(\mathbb{Z}[\pi_1(W)])$.

(ii) *The projective class at* ∞ *of* $(W, \partial W)$ *as in (i) is such that* $[W]_\infty = 0$ *if (and for* $n \geq 6$ *only if)* $(W, \partial W)$ *can be collared, i.e. there exists a compact* n*-dimensional cobordism* $(N; \partial W, M)$ *with a homeomorphism* rel ∂W

$$(W, \partial W) \;\cong\; (N \backslash M, \partial W) .$$

If $n \geq 6$ *and* $[W]_\infty = 0$ *any two such cobordisms* $(N; \partial W, M)$, $(N'; \partial W, M')$ *differ by an* h*-cobordism* $(L; M, M')$ *such that*

$$(N'; \partial W, M') \;\cong\; (N; \partial W, M) \cup (L; M, M') .$$

(iii) *For any finitely presented group* π *and any f.g. projective* $\mathbb{Z}[\pi]$*-module* P *there exists* $(W, \partial W)$ *as in (i), with*

$$\pi_1^\infty(W) \;=\; \pi_1(W) \;=\; \pi \;,\; [W]_\infty \;=\; [W] \;=\; [P] \in \widetilde{K}_0(\mathbb{Z}[\pi]) .$$

Idea of proof (i)+(ii) The projective class at ∞ of W is the finiteness obstruction of a cocompact submanifold $V \subseteq W$ with $\pi_1(V) = \pi_1^\infty(W)$

$$[W]_\infty \;=\; [V] \in \widetilde{K}_0(\mathbb{Z}[\pi_1(V)]) \;=\; \widetilde{K}_0(\mathbb{Z}[\pi_1^\infty(W)]) .$$

Let $U = \mathrm{cl}(W \backslash V)$, so that $(U; \partial W, \partial V)$ is a compact cobordism with

$$(W, \partial W) \;=\; (U; \partial W, \partial V) \cup (V, \partial V) .$$

It is possible to choose V such that the inclusion $\partial V \longrightarrow V$ is a homotopy equivalence if (and for $n \geq 6$ only if) W can be collared. The condition $n \geq 6$ occurs here because of the application of the Whitney trick to modify the $(n-1)$-dimensional manifold ∂V by codimension 1 surgeries inside V. The projective class at ∞ $[W]_\infty \in \widetilde{K}_0(\mathbb{Z}[\pi_1^\infty(W)])$ is the obstruction to the construction of such V by handle exchanges on ∂V inside W.

(iii) Let $P = \mathrm{im}(p)$ for a projection $p = p^2 : \mathbb{Z}[\pi]^r \longrightarrow \mathbb{Z}[\pi]^r$. For any $n \geq 5$ there exists a closed $(n-1)$-dimensional manifold M with $\pi_1(M) = \pi$. As in [140, Chapter VIII] construct a 'strange end'

$$(W, \partial W) \;=\; (M \times I \cup \bigcup_0^\infty \text{2-handles} \cup \bigcup_0^\infty \text{3-handles}, M \times \{0\})$$

satisfying the hypothesis of (i), with

$$\pi_1^\infty(W) \;=\; \pi_1(W) \;=\; \pi \;,\; [W]_\infty \;=\; [W] \;=\; [P] .$$

The 2-handles are attached trivially and the 3-handles are attached non-trivially, with

$$C(\widetilde{W}, \partial\widetilde{W}) \;:\; \ldots \longrightarrow 0 \longrightarrow \mathbb{Z}[\pi]^r[z] \xrightarrow{\;d\;} \mathbb{Z}[\pi]^r[z] \longrightarrow 0 \longrightarrow 0$$

such that

$$H_2(\widetilde{W}, \partial\widetilde{W}) \;=\; \mathrm{coker}(d) \;=\; P$$

(e.g. $d = 1 - p \pm zp$ or $1 - zp$). □

Remark 10.3 (i) In 10.5 below it will be shown that if $(W, \partial W)$ is an n-dimensional open manifold as in 10.2 (i) then $(W; \partial W, e(W))$ is a finitely dominated n-dimensional $\mathbb{Z}[\pi_1(W)]$-coefficient geometric Poincaré cobordism. If $n \geq 6$ and $[W]_\infty = 0$ then $(W, \partial W)$ can be collared, and $(W; \partial W, e(W))$ is homotopy equivalent rel ∂W to a compact n-dimensional manifold cobordism $(N; \partial W, M)$ as in 10.2 (ii). Any two such collarings $(N; \partial W, M)$, $(N'; \partial W, M')$ of $(W, \partial W)$ are related by an h-cobordism $(L; M, M')$ with

$$(N'; \partial W, M') = (N; \partial W, M) \cup (L; M, M') ,$$

so that the collarings are classified by the Whitehead group $Wh(\pi_1^\infty(W))$. (See Example 17.3 below for an account of the connections between h-cobordism theory and collarings.)

(ii) In the simply-connected case $\pi_1^\infty(W) = \{1\}$ the projective class at ∞ vanishes, $[W]_\infty = 0 \in \widetilde{K}_0(\mathbb{Z}) = 0$, and 10.2 recovers the result of Browder, Levine and Livesay [14] that for $n \geq 6$ it is possible to collar an open n-dimensional open manifold $(W, \partial W)$ with finitely generated $H_*(W)$ (= reverse π_1-tameness in the simply-connected case).

(iii) The construction of tame manifold ends in 10.2 (iii) can be generalized using bands (15.3), as follows. For $P = \mathrm{im}(p)$, M as in the proof of 10.2 (iii) let $(L; M \times S^1, N)$ be the compact $(n+1)$-dimensional cobordism with

$$L = M \times S^1 \times I \cup \bigcup_r 2\text{-handles} \cup_{1-zp} \bigcup_r 3\text{-handles} .$$

Then $\pi_1(L) = \pi_1(N) = \pi \times \mathbb{Z}$, and the projection $\pi_1(N) \longrightarrow \mathbb{Z}$ is realized by a map $c : N \longrightarrow S^1$ such that (N, c) is an n-dimensional manifold band. The infinite cyclic cover $\overline{N} = c^*\mathbb{R}$ of N has two ends, and

$$(W, \partial W) = (\overline{N}^+, \overline{N}^+ \cap \overline{N}^-)$$

is an open n-dimensional manifold with compact boundary and a reverse π_1-tame end such that $\pi_1^\infty(W) = \pi_1(W) = \pi$ and $[W]_\infty = [W] = [P] \in \widetilde{K}_0(\mathbb{Z}[\pi])$. It will be shown in Proposition 15.9 below that $e(W) \simeq \overline{N}$. □

The locally finite projective class of a forward tame CW complex W is defined as in 6.12, using the finite domination of $C^{lf,\pi}(\widetilde{W})$ given by 7.20:

Definition 10.4 The *locally finite projective class* of a forward tame locally finite CW complex W is the projective class

$$[W]^{lf} = [C^{lf,\pi}(\widetilde{W})] \in K_0(\mathbb{Z}[\pi]) \quad (\pi = \pi_1(W))$$

of the finitely dominated $\mathbb{Z}[\pi]$-module chain complex $C^{lf,\pi}(\widetilde{W})$, with $\widetilde{W}$ the universal cover of W. □

Proposition 10.5 (i) *If W is a locally finite CW complex which is both forward and reverse tame the end space $e(W)$ is finitely dominated, with $\mathbb{Z}[\pi_1(W)]$-coefficient reduced projective class*

$$[e(W)] = [W] - [W]^{lf} \in \widetilde{K}_0(\mathbb{Z}[\pi_1(W)]) .$$

(ii) *If $(W, \partial W)$ is an open n-dimensional geometric Poincaré pair which is both forward and reverse tame then*

$$[W]^{lf} = (-)^n [W]^* \in \widetilde{K}_0(\mathbb{Z}[\pi_1(W)]) .$$

The end space $e(W)$ is a finitely dominated $(n-1)$-dimensional $\mathbb{Z}[\pi_1(W)]$-coefficient Poincaré space and $(W; \partial W, e(W))$ is an n-dimensional $\mathbb{Z}[\pi_1(W)]$-coefficient geometric Poincaré cobordism, with

$$[e(W)] = [W] - [W]^{lf} = [W] + (-)^{n-1}[W]^* \in \widetilde{K}_0(\mathbb{Z}[\pi_1(W)]) .$$

Proof Let $\pi = \pi_1(W)$.

(i) The finite domination of $e(W)$ is given by 7.20. The projective class identity follows from the $\mathbb{Z}[\pi]$-module chain equivalence given by 7.20 (v)

$$S(\widetilde{e(W)}) \simeq C^{\infty,\pi}(\widetilde{W}) = \mathcal{C}(C(\widetilde{W}) \longrightarrow C^{lf,\pi}(\widetilde{W}))_{*+1} .$$

(ii) The universal cover $\widetilde{W}$ of W is such that the $\mathbb{Z}[\pi]$-module chain complexes $C(\widetilde{W})$, $C^{lf,\pi}(\widetilde{W})$ are finitely dominated. Let $[e(W)] \in H_{n-1}(e(W))$ be the image of the fundamental class

$$[W] \in H_n^{lf}(W, \partial W) = H_n(W, \partial W \amalg e(W)) .$$

The commutative diagram

$$\begin{array}{ccc} C^{lf,\pi}(\widetilde{W})^{n-*} & \longrightarrow & C(\widetilde{W})^{n-*} \\ {\scriptstyle [W]\cap -}\downarrow\simeq & & {\scriptstyle [W]\cap -}\downarrow\simeq \\ C(\widetilde{W}, \partial\widetilde{W}) & \longrightarrow & C^{lf,\pi}(\widetilde{W}, \partial\widetilde{W}) \end{array}$$

induces an $(n-1)$-dimensional $\mathbb{Z}[\pi]$-coefficient Poincaré duality chain equivalence

$$\begin{aligned} [e(W)] \cap - \ : S(\widetilde{e(W)})^{n-1-*} &\simeq \mathcal{C}(C(\widetilde{W}) \longrightarrow C^{lf,\pi}(\widetilde{W}))^{n-*} \\ &\xrightarrow{\simeq} S(\widetilde{e(W)}) \simeq \mathcal{C}(C(\widetilde{W}, \partial\widetilde{W}) \longrightarrow C^{lf,\pi}(\widetilde{W}, \partial\widetilde{W}))_{*+1} , \end{aligned}$$

so that $e(W)$ is an $(n-1)$-dimensional $\mathbb{Z}[\pi]$-coefficient Poincaré space. In fact, $(W; \partial W, e(W))$ is an n-dimensional $\mathbb{Z}[\pi]$-coefficient Poincaré cobordism, with a $\mathbb{Z}[\pi]$-coefficient Poincaré duality chain equivalence

$$[W] \cap - \ : \ C^{lf,\pi}(\widetilde{W}, \partial\widetilde{W})^{n-*} \simeq C(\widetilde{W}, \partial\widetilde{W} \amalg \widetilde{e(W)})^{n-*} \xrightarrow{\simeq} C(\widetilde{W}) . \ \square$$

Example 10.6 The polynomial ring $\mathbb{Z}[z]$ consists of the polynomials $\sum\limits_{i=0}^{\infty} a_i z^i$ with only a finite number of the coefficients $a_i \in \mathbb{Z}$ non-zero. Give $\mathbb{R}^+ = [0, \infty)$ the CW structure with one 0-cell at each $n \in \mathbb{N} \subset \mathbb{R}^+$, and 1-cells $[n, n+1]$, so that the cellular chain complex is the $\mathbb{Z}[z]$-module chain complex

$$C(\mathbb{R}^+) \; : \; \mathbb{Z}[z] \xrightarrow{1-z} \mathbb{Z}[z] \; ,$$

with z acting by $\mathbb{R}^+ \longrightarrow \mathbb{R}^+ ; x \longrightarrow x+1$. Let $\mathbb{Z}[[z]]$ be the ring of formal power series $\sum\limits_{i=0}^{\infty} a_i z^i$ $(a_i \in \mathbb{Z})$. The differential in the locally finite cellular chain complex

$$C^{lf}(\mathbb{R}^+) \; : \; \mathbb{Z}[[z]] \xrightarrow{1-z} \mathbb{Z}[[z]]$$

is an isomorphism, with inverse

$$(1-z)^{-1} \; = \; \sum_{k=0}^{\infty} z^k \; : \; \mathbb{Z}[[z]] \longrightarrow \mathbb{Z}[[z]] \; ; \; \sum_{j=0}^{\infty} a_j z^j \longrightarrow \sum_{k=0}^{\infty} (\sum_{j=0}^{k} a_j) z^k \; ,$$

so that $C^{lf}(\mathbb{R}^+)$ is contractible and the locally finite projective class is

$$[\mathbb{R}^+]^{lf} \; = \; [C^{lf}(\mathbb{R}^+)] \; = \; 0 \in K_0(\mathbb{Z}) \; = \; \mathbb{Z} \; .$$

There are defined homology equivalences of $\mathbb{Z}$-module chain complexes

$$C^{\infty}(\mathbb{R}^+) \; \simeq \; C(\mathbb{R}^+) \; \simeq \; \mathbb{Z} \; , \quad C^{lf}(\mathbb{R}^+) \; \simeq \; 0$$

in accordance with $e(\mathbb{R}^+) \simeq \{\text{pt.}\}$. □

Example 10.7 The Laurent polynomial ring $\mathbb{Z}[z, z^{-1}]$ consists of the polynomials $\sum\limits_{i=-\infty}^{\infty} a_i z^i$ with only a finite number of the coefficients $a_i \in \mathbb{Z}$ non-zero. The cellular chain complex of the universal cover $\widetilde{S}^1 = \mathbb{R}$ of the circle S^1 is the $\mathbb{Z}[z, z^{-1}]$-module chain complex

$$C(\mathbb{R}) \; : \; \mathbb{Z}[z, z^{-1}] \xrightarrow{1-z} \mathbb{Z}[z, z^{-1}] \; ,$$

identifying $\mathbb{Z}[\pi_1(S^1)] = \mathbb{Z}[z, z^{-1}]$. The locally finite cellular chain complex of $\mathbb{R}$ is

$$C^{lf}(\mathbb{R}) \; : \; \mathbb{Z}[[z, z^{-1}]] \xrightarrow{1-z} \mathbb{Z}[[z, z^{-1}]] \; ,$$

with $\mathbb{Z}[[z, z^{-1}]] = \mathbb{Z}[[\pi_1(S^1)]]$ the $\mathbb{Z}[z, z^{-1}]$-module of formal Laurent polynomials $\sum\limits_{i=-\infty}^{\infty} a_i z^i$ $(a_i \in \mathbb{Z})$. The $\mathbb{Z}$-module morphism

$$\mathbb{Z} \longrightarrow C^{lf}(\mathbb{R})_1 \; = \; \mathbb{Z}[[z, z^{-1}]] \; ; \; 1 \longrightarrow \sum_{i=-\infty}^{\infty} z^i$$

defines a homology equivalence $\mathbb{Z}\longrightarrow C^{lf}(\mathbb{R})_{*+1}$, and the locally finite projective class of $\mathbb{R}$ is

$$[\mathbb{R}]^{lf} = [C^{lf}(\mathbb{R})] = -[\mathbb{Z}] = -1 \in K_0(\mathbb{Z}) = \mathbb{Z}\ .$$

The $\mathbb{Z}[z,z^{-1}]$-module chain map $i : C(\mathbb{R})\longrightarrow C^{lf}(\mathbb{R})$ defined by inclusion is chain homotopic to 0, with a chain homotopy $h : i \simeq 0$ defined by

$$h\ :\ C(\mathbb{R})_0 = \mathbb{Z}[z,z^{-1}] \longrightarrow C^{lf}(\mathbb{R})_1 = \mathbb{Z}[[z,z^{-1}]]\ ;$$

$$\sum_{j=-\infty}^{\infty} a_j z^j \longrightarrow \sum_{k=-\infty}^{\infty}\Big(\sum_{j=-\infty}^{k} a_j\Big)z^k\ .$$

As $\mathbb{Z}$-module chain complexes $C(\mathbb{R}) \simeq C^{lf}(\mathbb{R})_{*+1} \simeq \mathbb{Z}$, and

$$e(C(\mathbb{R})) = \mathcal{C}(i)_{*+1} \simeq C(\mathbb{R}) \oplus C^{lf}(\mathbb{R})_{*+1} \simeq C(e(\mathbb{R})) \simeq \mathbb{Z}\oplus\mathbb{Z}\ ,$$

in accordance with $e(\mathbb{R}) \simeq S^0$. The locally $\mathbb{Z}$-finite cellular chain complex of $\mathbb{R}$ is

$$C^{lf,\mathbb{Z}}(\mathbb{R}) = C(\mathbb{R})\ :\ \mathbb{Z}[z,z^{-1}] \xrightarrow{1-z} \mathbb{Z}[z,z^{-1}]\ ,$$

so that $i : C(\mathbb{R})\longrightarrow C^{lf,\mathbb{Z}}(\mathbb{R})$ is an isomorphism and

$$C(\widetilde{e(S^1)}) = \mathcal{C}(i : C(\mathbb{R})\longrightarrow C^{lf,\mathbb{Z}}(\mathbb{R}))_{*+1} \simeq 0\ ,$$

in accordance with $e(S^1) = \emptyset$. □

If W is a forward tame strongly locally finite CW complex which is path-connected at ∞ there exists a cofinite subcomplex $V \subseteq W$ such that the inclusion $V\longrightarrow W$ extends to a proper map $q : V\times[0,\infty)\longrightarrow W$. The locally finite $\mathbb{Z}[\pi]$-module chain complex $C^{lf,\pi}(\widetilde{V})$ is finitely dominated by 7.20, with $\pi = \pi_1(e(W))$, and $\widetilde{V}$ the cover of V induced from the universal cover $\widetilde{e(W)}$ of $e(W)$ by the adjoint map $\widehat{q} : V\longrightarrow e(W)$.

Definition 10.8 The *locally finite projective class at* ∞ of a forward tame strongly locally finite CW complex W which is path-connected at ∞ is

$$[W]^{lf}_{\infty} = [C^{lf,\pi}(\widetilde{V})] \in \widetilde{K}_0(\mathbb{Z}[\pi])$$

with $\pi = \pi_1(e(W))$, $V \subseteq W$ any cofinite subcomplex such that the inclusion $V\longrightarrow W$ extends to a proper map $q : V \times [0,\infty)\longrightarrow W$. □

Proposition 10.9 *Let W be a forward tame strongly locally finite CW complex which is path-connected at ∞, and let $\pi = \pi_1(e(W))$.*

(i) *If $V \subseteq W$ is a cofinite subcomplex such that $\pi_1(V) = \pi$ then the locally finite $\mathbb{Z}[\pi]$-module chain complex $C^{lf,\pi}(\widetilde{V})$ is finitely dominated, and*

$$[W]^{lf}_{\infty} = [C^{lf,\pi}(\widetilde{V})] \in \widetilde{K}_0(\mathbb{Z}[\pi])$$

with $\widetilde{V}$ the universal cover of V.

(ii) *If W is both forward and reverse tame then the end space $e(W)$ is finitely dominated, with finiteness obstruction*

$$[e(W)] \;=\; [W]_\infty - [W]^{lf}_\infty \in \widetilde{K}_0(\mathbb{Z}[\pi]) \,.$$

(iii) *If W is forward collared then $[W]^{lf}_\infty = 0 \in \widetilde{K}_0(\mathbb{Z}[\pi])$.*

Proof (i) Let $V' \subseteq V$ be a cofinite subcomplex such that the inclusion $V' \longrightarrow W$ extends to a proper map $q' : V' \times [0,\infty) \longrightarrow W$. Let $\widetilde{V}'$ be the cover of V induced from the universal cover $\widetilde{e(W)}$ of $e(W)$ by the adjoint map $\widehat{q}' : V' \longrightarrow e(W)$. We are assuming that identification $\pi_1(V) = \pi$ is such that

$$i_* \;=\; \widehat{q}'_* \;:\; \pi_1(V') \longrightarrow \pi_1(V) \;=\; \pi_1(e(W)) \;=\; \pi \,,$$

so that there exists a lift of i to an inclusion $\widetilde{i} : \widetilde{V'} \longrightarrow \widetilde{V}$. Now $C^{lf,\pi}(\widetilde{V}')$ is a finitely dominated $\mathbb{Z}[\pi]$-module chain complex such that

$$[W]^{lf}_\infty \;=\; [C^{lf,\pi}(\widetilde{V}')] \in \widetilde{K}_0(\mathbb{Z}[\pi]) \,.$$

The finite domination of $C^{lf,\pi}(\widetilde{V})$ and the identity $[C^{lf,\pi}(\widetilde{V}')] = [C^{lf,\pi}(\widetilde{V})]$ follow from the short exact sequence of $\mathbb{Z}[\pi]$-module chain complexes

$$0 \longrightarrow C^{lf,\pi}(\widetilde{V}') \longrightarrow C^{lf,\pi}(\widetilde{V}) \longrightarrow C^{lf,\pi}(\widetilde{V},\widetilde{V}') \longrightarrow 0$$

with $C^{lf,\pi}(\widetilde{V},\widetilde{V}') = C(\widetilde{V},\widetilde{V}')$ a finite f.g. free $\mathbb{Z}[\pi]$-module chain complex.

(ii) Apply 10.5 (i) to a cofinite subcomplex $V \subseteq W$ such that $\pi_1(V) = \pi_1(e(W))$.

(iii) Let $U \subseteq W$ be a closed cocompact ANR subspace with an extension of the identity $U \times \{0\} \longrightarrow U$ to a proper map $U \times [0,\infty) \longrightarrow U$ and let $V \subseteq U$ be a cofinite subcomplex of W. Let $\pi = \pi_1(e(W))$, and let $\widetilde{U}, \widetilde{V}$ be the covers of U, V induced from the universal cover $\widetilde{e(W)}$ of $e(W)$. By 7.5

$$S(\widetilde{U},\widetilde{V}) \;=\; S^{lf,\pi}(\widetilde{U},\widetilde{V}) \;\simeq\; S^{lf,\pi}(\widetilde{U}) \oplus S^{lf,\pi}(\widetilde{V})_{*-1}$$

with $S(\widetilde{U},\widetilde{V})$ chain homotopy finite. The identity chain map $S^{lf,\pi}(\widetilde{U}) \longrightarrow S^{lf,\pi}(\widetilde{U})$ is chain homotopic to 0, so that there are defined $\mathbb{Z}[\pi]$-module chain equivalences

$$S^{lf,\pi}(\widetilde{U}) \simeq 0 \;,\quad C^{lf,\pi}(\widetilde{V}) \;\simeq\; S^{lf,\pi}(\widetilde{V}) \;\simeq\; S(\widetilde{U},\widetilde{V})_{*+1}$$

and

$$[W]^{lf}_\infty \;=\; [C^{lf,\pi}(\widetilde{V})] \;=\; -[S(\widetilde{U},\widetilde{V})] \;=\; 0 \in \widetilde{K}_0(\mathbb{Z}[\pi]) \,. \qquad \square$$

The locally finite projective class $[W]^{lf}_\infty \in \widetilde{K}_0(\mathbb{Z}[\pi_1(e(W))])$ of a forward tame CW complex W is thus an obstruction to W being forward collared.

For a manifold forward and reverse tameness are Poincaré dual to each other, as was first established by Quinn [116]. We give a proof in 10.13 below. We begin with a simple geometric way to detect forward tameness (10.10). This is combined with the Eventual Hurewicz Theorem (10.11) to give a homological criterion for forward tameness (10.12).

Lemma 10.10 *Let W be a σ-compact space. If for every cocompact subspace $U \subseteq W$ there exists a cocompact subspace $V = V(U) \subseteq U$ such that for every cocompact subspace $X \subseteq V$ there exists a cocompact subspace $Y = Y(U, X) \subseteq X$ so that V deforms in U to X rel Y (i.e. there is a homotopy $h : V \times I \longrightarrow U$ such that $h_0 = \text{inclusion} : V \longrightarrow U$, $h_1(V) \subseteq X$ and $h_t|Y = \text{inclusion} : Y \longrightarrow U$ for each $t \in I$), then W is forward tame.*

Proof Let $W \supseteq W_0 \supseteq W_1 \supseteq \dots$ be closed cocompact subspaces with $\bigcap W_i = \emptyset$. Define closed cocompact subspaces $V_j \subseteq W$, $j = 0, 1, 2, \dots$, inductively as follows. Let $V_0 = V(W_0)$. Assume $j > 0$ and that $V_0 \supseteq V_1 \subseteq \dots \subseteq V_{j-1}$ and $V_i \subseteq W_i$ for $i = 0, 1, \dots, j-1$. Let $V_j = V(W_j \cap V_{j-1})$. For each $j = 0, 1, 2, \dots$ let $X_j = V_{j+1}$ and $Y_j = Y(V_{j-1}, X_j)$. By hypothesis V_j deforms in $W_j \cap V_{j-1}$ to X_j rel Y_j. That is, for each $j = 0, 1, 2, \dots$ there is a homotopy $h^j : V_j \times I \longrightarrow W_j \cap V_{j-1}$ such that $h^j_0 = \text{inclusion} : V_j \longrightarrow W_j \cap V_{j-1}$, $h^j_1(V_j) \subseteq X_j = V_{j+1}$ and $h^j_t|Y_j = \text{inclusion} : Y_j \longrightarrow W_j \cap V_{j-1}$ for all $t \in I$. Define

$$h : V_0 \times [0, \infty) \longrightarrow W \; ; \; (x, t) \longrightarrow h^j(h^{j-1}_1 \circ h^{j-2}_1 \circ \dots \circ h^0_1(x), t - j)$$
$$(x \in V_0, j \leq t \leq j+1) \, .$$

To see that h is a proper homotopy let $K \subseteq W$ be compact. There exists $i > 0$ such that $K \cap V_i = \emptyset$. Thus $h^{-1}(K) = (h|(V_0 \times [0, i+1]))^{-1}(K)$. Since each h_j fixes the cocompact subspace $Y_j \subseteq W$, it follows that each h_j is proper and that $h|(V_0 \times [0, i+1])$ is proper. □

The following Eventual Hurewicz Theorem is a relative version of a result from Ferry [54, p. 570]. A proof of the relative version, in a more general context, can be found in Quinn [114, p. 302].

Lemma 10.11 *For each integer $n > 0$ there exists an integer $k_n > 0$ such that the following holds. Let W be an n-dimensional locally finite CW complex for which there are sequences of cofinite subcomplexes*

$$W \supseteq A_0 \supseteq A_1 \supseteq A_2 \supseteq A_3 \supseteq \dots \supseteq A_{k_n} \, ,$$
$$W \supseteq B_0 \supseteq B_1 \supseteq B_2 \supseteq B_3 \supseteq \dots \supseteq B_{k_n}$$

with $B_j \subseteq A_j$ for each j. Suppose that A_0 has a regular cover $\widetilde{A}_0$ with group of covering translations π and that $\widetilde{A}_j$, $\widetilde{B}_j$ denote the regular covers of A_j, B_j induced by the inclusions $A_j \longrightarrow A_0$ and $B_j \longrightarrow A_0$ for $0 \leq j \leq k_n$.

Suppose the inclusions induce 0 morphisms in π_1

$$0 \; : \; \pi_1(\widetilde{A}_j) \longrightarrow \pi_1(\widetilde{A}_{j-1}) \quad , \quad 0 \; : \; \pi_1(\widetilde{B}_j) \longrightarrow \pi_1(\widetilde{B}_{j-1})$$

and 0 morphisms in homology

$$0 \; : \; H_r(\widetilde{A}_j, \widetilde{B}_j) \longrightarrow H_r(\widetilde{A}_{j-1}, \widetilde{B}_{j-1}) \quad (r \leq n) \; .$$

Then A_{k_n} deforms in A_0 to B_0 rel B_{k_n}, i.e. there is a homotopy $h : A_{k_n} \times I \longrightarrow A_0$ such that $h_0 =$ inclusion : $A_{k_n} \longrightarrow A_0$, $h_1(A_{k_n}) \subseteq B_0$ and $h_t|B_{k_n} =$ inclusion : $B_{k_n} \longrightarrow A_0$ for each $t \in I$. □

Lemma 10.12 *Let W be an n-dimensional locally finite CW complex with stable π_1 at ∞ and let $W \supseteq W_0 \supseteq W_1 \supseteq \ldots$ be a sequence of cofinite subcomplexes such that $\bigcap_j W_j = \emptyset$ and the inclusion induced morphisms*

$$\pi_1(W_0) \xleftarrow{f_0} \pi_1(W_1) \xleftarrow{f_1} \pi_1(W_2) \xleftarrow{f_2} \ldots$$

induce isomorphisms

$$\pi \; = \; \mathrm{im}(f_0) \xleftarrow{\cong} \mathrm{im}(f_1) \xleftarrow{\cong} \mathrm{im}(f_2) \xleftarrow{\cong} \ldots \; .$$

Suppose that for every cocompact subspace $U \subseteq W_0$ there exists a cocompact subspace $V \subseteq U$ such that for every cocompact subspace $X \subseteq V$ there exists a cocompact subspace $Y \subseteq X$ so that the inclusion induces 0 morphisms in homology

$$0 \; : \; H_r(\widetilde{V}, \widetilde{Y}) \longrightarrow H_r(\widetilde{U}, \widetilde{X}) \quad (r \leq n)$$

with $\widetilde{U}, \widetilde{V}, \widetilde{X}, \widetilde{Y}$ the regular covers induced from the universal cover $\widetilde{W}_0 \longrightarrow W_0$ by inclusion. Then W is forward tame.

Proof First note that the inclusions induce 0 morphisms

$$\{0\} \; = \; \pi_1(\widetilde{W}_0) \xleftarrow{0} \pi_1(\widetilde{W}_1) \xleftarrow{0} \pi_1(\widetilde{W}_2) \xleftarrow{0} \ldots \; .$$

In order to verify the hypothesis of 10.10, let $U \subseteq W$ be an arbitrary cocompact subspace and assume that $U \subseteq W_0$. By assumption and induction there exists a sequence of cocompact subspaces $U \supseteq V_0 \supseteq V_1 \supseteq V_2 \supseteq \ldots \supseteq V_{k_n}$ (with k_n given by 10.11) such that for every cocompact subspace $X \subseteq V_{k_n}$ there exists a cocompact subspace $Y \subseteq X$ such that the inclusion induces the 0 morphism in homology

$$0 \; : \; H_r(\widetilde{V}_j, \widetilde{Y}) \longrightarrow H_r(\widetilde{U}, \widetilde{X}) \quad (r \leq n)$$

with the covers induced from $\widetilde{W}_0 \longrightarrow W_0$ by inclusion. Choose integers $i_0 < i_1 < \ldots < i_{k_n}$ such that $W_{i_j} \subseteq V_j$ for each $j = 0, 1, 2, \ldots, k_n$. Then for every cocompact subspace $X \subseteq W_{i_{k_n}}$ there exists a cocompact subspace $Y \subseteq X$ such that the inclusion induces 0 morphisms in homology

$$0 \; : \; H_r(\widetilde{W}_{i_j}, \widetilde{Y}) \longrightarrow H_r(\widetilde{U}, \widetilde{X}) \quad (r \leq n) \; .$$

Let $V = W_{i_{k_n}}$ and let $X \subseteq V$ be an arbitrary cocompact subspace. Again by assumption and induction there exists a sequence of cocompact subspaces

$$X \supseteq Y_0 \supseteq Y_1 \supseteq Y_2 \supseteq \ \dots \ \supseteq Y_{k_n}$$

such that inclusion induces 0 morphisms in homology

$$0 \ : \ H_r(\widetilde{V}_j, \widetilde{Y}_j) \longrightarrow H_r(\widetilde{U}, \widetilde{X}) \ \ (r \leq n) \ .$$

Choose integers $i_{k_n} < l_0 < l_1 < \dots < l_{k_n}$ such that $W_{l_j} \subseteq Y_j$ for each $j = 0, 1, 2, \dots, k_n$. Then inclusion induces 0 morphisms in homology

$$0 \ : \ H_r(\widetilde{V}, \widetilde{W}_{l_j}) \longrightarrow H_r(\widetilde{U}, \widetilde{X}) \ \ (r \leq n) \ .$$

Let $Y = W_{l_{k_n}}$. It follows from 10.11 that V deforms in U to X rel Y so that W is forward tame by 10.10. □

Proposition 10.13 *Let $(W, \partial W)$ be an open n-dimensional manifold with compact boundary. W is forward tame with finitely presented $\pi_1(e(W))$ if and only if W is reverse π_1-tame.*

Proof Suppose first that W is forward tame and $\pi_1(e(W))$ is finitely presented. According to 7.10 (i) W has stable π_1 at ∞ and $\pi_1(e(W)) = \pi_1^\infty(W)$. By 7.5 (i) there exists a closed cocompact subspace $V \subseteq W$ such that the natural morphism

$$\pi \ = \ \pi_1(e(W)) \ = \ \pi_1^\infty(W) \longrightarrow \pi_1(V)$$

is split injective. Since it suffices to prove that V is reverse tame, we assume that $\pi = \pi_1(e(W)) \longrightarrow \pi_1(W)$ is split injective. All covers below are induced from the universal cover $\widetilde{W} \longrightarrow W$. By 9.18 it suffices to prove that $e(W)$ is finitely dominated. Since $e(W)$ has the homotopy type of a CW complex (7.6), it suffices to show that $S(\widetilde{e(W)})$ is a finitely dominated $\mathbb{Z}[\pi]$-module chain complex (6.9 (i)). According to 7.10 (iv) there is defined a $\mathbb{Z}[\pi]$-module chain equivalence

$$S(\widetilde{e(W)}) \ \simeq \ \mathcal{C}(S(\widetilde{W}) \longrightarrow S^{lf,\pi}(\widetilde{W}))_{*+1} \ .$$

Since $S^{lf,\pi}(\widetilde{W})$ is finitely dominated by 7.5 (iii), it suffices to prove that $S(\widetilde{W})$ is a finitely dominated $\mathbb{Z}[\pi]$-module chain complex. The $\mathbb{Z}[\pi]$-module chain complex $S(\widetilde{W})$ is finitely dominated if and only if the n-dual $\mathbb{Z}[\pi]$-module chain complex $S(\widetilde{W})^{n-*} = \mathrm{Hom}_{\mathbb{Z}[\pi]}(S(\widetilde{W}), \mathbb{Z}[\pi])_{n-*}$ is finitely dominated. Now $(W, \partial W)$ is an open n-dimensional Poincaré pair, so there is a $\mathbb{Z}[\pi]$-module chain equivalence $S(\widetilde{W})^{n-*} \simeq S^{lf,\pi}(\widetilde{W}, \partial\widetilde{W})$. Use 7.5 (iii) again to conclude that $S^{lf,\pi}(\widetilde{W})$ is finitely dominated, from which it follows that $S^{lf,\pi}(\widetilde{W}, \partial\widetilde{W})$ is finitely dominated.

Conversely, suppose W is reverse π_1-tame. In order to apply 10.12, let $W \supseteq W_0 \supseteq W_1 \supseteq \dots$ be as in 10.12. Since it suffices to prove that W_0 is forward tame, we assume that $W = W_0$ and all covers below are induced from the universal cover $\widetilde{W}\longrightarrow W$. Since W is reverse tame, for every cocompact subspace $U \subseteq W$ there exists a cocompact subspace $V \subseteq U$ such that U is dominated by $U \setminus V$, by a homotopy $h : W \times I \longrightarrow W$ as in 8.1. For every cocompact subspace $X \subseteq V$ there exists a cocompact subspace $Y \subseteq X$ such that $h((W \setminus X) \times I) \subseteq W \setminus Y$. In particular, the inclusion induced chain map $S(\widetilde{W} \setminus \widetilde{X}, \widetilde{W} \setminus \widetilde{U})\longrightarrow S(\widetilde{W} \setminus \widetilde{Y}, \widetilde{W} \setminus \widetilde{V})$ is chain homotopic to 0. Now Alexander duality gives a commutative diagram

$$\begin{array}{ccc} S(\widetilde{W}\backslash\widetilde{X},\widetilde{W}\backslash\widetilde{U}) & \longrightarrow & S(\widetilde{W}\backslash\widetilde{Y},\widetilde{W}\backslash\widetilde{V}) \\ \simeq\big\downarrow & & \big\downarrow\simeq \\ S(\widetilde{U},\widetilde{X})^{n-*} & \longleftarrow & S(\widetilde{V},\widetilde{Y})^{n-*} \end{array}$$

Since the top horizontal arrow is chain homotopic to 0, so is the bottom arrow. Taking n-duals gives that $S(\widetilde{V},\widetilde{Y})\longrightarrow S(\widetilde{U},\widetilde{X})$ is chain homotopic to 0. Thus for every cocompact subspace $U \subseteq W$ there exists a cocompact subspace $V \subseteq U$ such that for every cocompact subspace $X \subseteq V$ there exists a cocompact subspace $Y \subseteq X$ such that $S(\widetilde{V},\widetilde{Y})\longrightarrow S(\widetilde{U},\widetilde{X})$ is chain homotopic to 0. That W is forward tame now follows from 10.12, since the fundamental group of a finitely dominated space is finitely presented (6.8 (i)). Finally, by 7.5 (i) $e(W)$ is dominated by a closed ANR cocompact subspace $V \subseteq W$. By 8.7 V is finitely dominated. Hence $\pi_1(e(W))$ is finitely presented. □

Remark 10.14 (i) Quinn gave a proof of 10.13 in [116] but without taking into account the fundamental group conditions. However, these conditions are required by Freedman and Quinn [60, p. 214]. The proof of the first half of 10.13 differs from the proof in [116].

(ii) In Chapter 23 we shall show that for a connected finite CW complex X with a connected infinite cyclic cover $\overline{X}$ and $\pi = \pi_1(\overline{X})$ the following conditions are equivalent:

(a) $\overline{X}^+$ is forward tame,
(b) the natural map $e(\overline{X}^+)\longrightarrow\overline{X}$ is a homotopy equivalence,
(c) the locally π-finite cellular $\mathbb{Z}[\pi]$-module chain complex $C^{lf,\pi}(\widetilde{X}^+)$ is finitely dominated, allowing the definition of the locally finite projective class

$$[\overline{X}^+]^{lf} = [C^{lf,\pi}(\widetilde{X}^+)] \in K_0(\mathbb{Z}[\pi]) ,$$

(d) $\overline{X}^-$ is reverse tame,
(e) $\overline{X}^-$ is finitely dominated,
(f) the cellular $\mathbb{Z}[\pi]$-module chain complex $C(\widetilde{X}^-)$ is finitely dominated,

allowing the definition of the projective class

$$[\overline{X}^-] = [C(\widetilde{X}^-)] \in K_0(\mathbb{Z}[\pi]) .$$

If these conditions are satisfied there is defined a $\mathbb{Z}[\pi]$-module chain equivalence

$$C^{lf,\pi}(\widetilde{X}^+) \simeq C(\widetilde{X}^-, \widetilde{X}^+ \cap \widetilde{X}^-)_{*+1}$$

and the projective classes are such that

$$[\overline{X}^+]^{lf} + [\overline{X}^-] = [\overline{X}^+ \cap \overline{X}^-] \in K_0(\mathbb{Z}[\pi]) .$$

(iii) In Chapter 23 we shall use the infinite simple homotopy theory of Chapter 11 to also prove that for $X, \overline{X}, \pi$ satisfying the conditions of (ii) the following conditions are equivalent:

(a) $\overline{X}$ is infinite simple homotopy equivalent to an infinite cyclic cover $\overline{W}$ of a finite CW complex W with $\overline{W}^+$ forward collared,
(b) $\overline{X}$ is infinite simple homotopy equivalent to an infinite cyclic cover $\overline{W}$ of a finite CW complex W with $\overline{W}^-$ reverse collared,
(c) $[\overline{X}^+]^{lf} = 0 \in \widetilde{K}_0(\mathbb{Z}[\pi])$,
(d) $[\overline{X}^-] = 0 \in \widetilde{K}_0(\mathbb{Z}[\pi])$.

(iv) If W is a forward tame strongly locally finite CW complex which is path-connected at ∞ there may not exist a cofinite subcomplex $V \subseteq W$ with $\pi_1(V) = \pi_1(e(W))$ as in 10.9 (i) – the CW complex $\overline{W}^+$ of Example 13.16 is a counterexample. □

Corollary 10.15 *If* $(W, \partial W)$ *is a reverse* π_1*-tame open* n*-dimensional manifold with compact boundary then the locally finite projective class at* ∞ *is the Poincaré dual of the projective class at* ∞ *:*

$$[W]^{lf}_\infty = (-)^n [W]^*_\infty \in \widetilde{K}_0(\mathbb{Z}[\pi_1(e(W))]) .$$

Proof W is forward tame, by 10.13. There exists a cocompact closed neighbourhood $W' \subseteq W$ such that $(W', \partial W')$ is an open n-dimensional manifold with compact boundary and

$$\pi_1(W') = \pi_1(e(W')) = \pi_1(e(W)) \ (= \pi \ , \text{ say}) ,$$
$$[W'] = [W']_\infty = [W]_\infty \ , \ [W']^{lf} = [W']^{lf}_\infty = [W]^{lf}_\infty \in \widetilde{K}_0(\mathbb{Z}[\pi]) .$$

Applying 10.5 (ii) to the forward and reverse tame open n-dimensional geometric Poincaré pair $(W', \partial W')$ we obtain

$$[W]^{lf}_\infty = [W']^{lf} = (-)^n [W']^* = (-)^n [W]^*_\infty \in \widetilde{K}_0(\mathbb{Z}[\pi]) .$$ □

11

Infinite torsion

The 'infinite Whitehead group' $\mathcal{S}(W)$ of Siebenmann [144] is defined geometrically for any locally finite CW complex W, and is denoted here by $Wh^{lf}(W)$. A proper homotopy equivalence $f : V \longrightarrow W$ of locally finite CW complexes has an 'infinite torsion' $\tau^{lf}(f) \in Wh^{lf}(W)$, which for forward tame V, W has image $[W]_\infty - [V]_\infty \in \widetilde{K}_0(\mathbb{Z}[\pi_1^\infty(W)])$. The projective class $[W] \in \widetilde{K}_0(\mathbb{Z}[\pi_1(W)])$ of a finitely dominated space W is a homotopy invariant, whereas the projective class at ∞ $[W]_\infty \in \widetilde{K}_0(\mathbb{Z}[\pi_1^\infty(W)])$ of a reverse π_1-tame space W is only an infinite simple homotopy invariant. In 11.14 and 11.15 we prove that if a strongly locally finite CW complex W is forward (resp. reverse) tame, then $W \times S^1$ is infinite simple homotopy equivalent to a forward (resp. reverse) collared CW complex, analogous to Siebenmann's result that if W is a manifold of dimension $n \geq 5$ with one tame end, then the end of $W \times S^1$ is collared.

For a forward tame W we identify $Wh^{lf}(W)$ with the algebraically defined relative Whitehead group in the exact sequence

$$\begin{aligned} Wh(\pi_1(e(W))) \xrightarrow{p_*} Wh(\pi_1(W)) \longrightarrow Wh^{lf}(W) \\ \longrightarrow \widetilde{K}_0(\mathbb{Z}[\pi_1(e(W))]) \xrightarrow{p_*} \widetilde{K}_0(\mathbb{Z}[\pi_1(W)]) \end{aligned}$$

with $p : e(W) \longrightarrow W$ the projection (11.6). The infinite torsion $\tau^{lf}(f) \in Wh^{lf}(W)$ of a proper homotopy equivalence $f : V \longrightarrow W$ of forward tame CW complexes has image

$$[\tau^{lf}(f)] = [W]^{lf}_\infty - [V]^{lf}_\infty \in \widetilde{K}_0(\mathbb{Z}[\pi_1(e(W))]) .$$

The locally finite projective class $[W]^{lf} \in \widetilde{K}_0(\mathbb{Z}[\pi_1(W)])$ of a forward tame CW complex W is an invariant of proper homotopy type of W, whereas the locally finite projective class at ∞ $[W]^{lf}_\infty \in \widetilde{K}_0(\mathbb{Z}[\pi_1(e(W))])$ is only an infinite simple homotopy invariant.

Definition 11.1 (i) The *infinite Whitehead group* $Wh^{lf}(W)$ of a locally finite CW complex W is the geometrically defined group $\mathcal{S}(W)$ of Siebenmann [144]

$$Wh^{lf}(W) = \mathcal{S}(W) .$$

(ii) The *infinite torsion* of a proper homotopy equivalence $f : V \longrightarrow W$ of locally finite CW complexes is the element defined in [144]

$$\tau^{lf}(f) \in Wh^{lf}(W) ,$$

and f is an *infinite simple homotopy equivalence* if $\tau^{lf}(f) = 0$. □

Remark 11.2 (i) In dealing with infinite torsion, it is necessary to restrict attention to strongly locally finite CW complexes. The need for this restriction is discussed in Chapter 1 of Farrell and Wagoner [52].

(ii) Chapman [22, 23] proved that a proper homotopy equivalence $f : V \longrightarrow W$ of strongly locally finite CW complexes is an infinite simple homotopy equivalence if and only if $f \times \mathrm{id}_Q : V \times Q \longrightarrow W \times Q$ is properly homotopic to a homeomorphism, with Q the Hilbert cube. Hilbert cube manifold theory then allows the extension of proper simple homotopy theory to ANR's: a proper homotopy equivalence $f : V \longrightarrow W$ of ANR's is an *infinite simple homotopy equivalence* if $f \times \mathrm{id}_Q : V \times Q \longrightarrow W \times Q$ is properly homotopic to a homeomorphism. This agrees with the theory of Siebenmann [144] for strongly locally finite CW complexes. □

Proposition 11.3 (Siebenmann [144], Farrell and Wagoner [52]) (i) *A proper homotopy equivalence $f : V \longrightarrow W$ of locally finite CW complexes is a proper simple homotopy equivalence if and only if f is properly homotopic to a finite sequence of proper expansions and collapses.*

(ii) *The infinite Whitehead group of a strongly locally finite CW complex W fits into an exact sequence*

$$Wh^{lf}(W) \longrightarrow \varprojlim_j \widetilde{K}_0(\mathbb{Z}[\pi_1(W_j)]) \longrightarrow \widetilde{K}_0(\mathbb{Z}[\pi_1(W)])$$

for a sequence $W \supset W_1 \supset W_2 \supset \dots$ of cofinite subcomplexes such that $\bigcap\limits_j W_j = \emptyset$. The subgroup

$$Wh^{lf}_b(W) = \ker(Wh^{lf}(W) \longrightarrow \varprojlim_j \widetilde{K}_0(\mathbb{Z}[\pi_1(W_j)])) \subseteq Wh^{lf}(W)$$

fits into an exact sequence

$$\varprojlim_j Wh(\pi_1(W_j)) \longrightarrow Wh(\pi_1(W)) \longrightarrow Wh^{lf}_b(W) \longrightarrow \varprojlim_j{}^1 Wh(\pi_1(W_j)) \longrightarrow 0 .$$

(iii) *The infinite torsion $\tau^{lf}(f) \in Wh^{lf}(W)$ of a proper homotopy equivalence $f : V \longrightarrow W$ of locally finite CW complexes has image*

$$[\tau^{lf}(f)] = [\mathcal{C}(\widetilde{f}_j : C(\widetilde{V}_j) \longrightarrow C(\widetilde{W}_j))] \in \varprojlim_j \widetilde{K}_0(\mathbb{Z}[\pi_1(W_j)]) ,$$

with $\mathcal{C}(\widetilde{f}_j)$ the finitely dominated $\mathbb{Z}[\pi_1(W_j)]$-module chain complex defined by the algebraic mapping cone of the chain map induced by a $\pi_1(W_j)$-equivariant lift $\widetilde{f}_j : \widetilde{V}_j \longrightarrow \widetilde{W}_j$ of $f_j = f| : V_j = f^{-1}(W_j) \longrightarrow W_j$.

(iv) *If $f : V \longrightarrow W$ is a proper homotopy equivalence of reverse π_1-tame CW complexes then it is possible to choose each $W_j \subset W$ and $V_j = f^{-1}(W_j) \subset V$ to be finitely dominated, and*

$$[\tau^{lf}(f)] = [W_j] - [V_j] \in \varprojlim_j \widetilde{K}_0(\mathbb{Z}[\pi_1(W_j)]) ,$$

with image $[W]_\infty - [V]_\infty \in \widetilde{K}_0(\mathbb{Z}[\pi_1^\infty(W)])$.

(v) *(Proper s-cobordism theorem) The infinite torsion of a proper h-cobordism $(W; M, M')$ of open n-dimensional manifolds*

$$\tau^{lf}(W; M, M') = \tau^{lf}(M \longrightarrow W) \in Wh^{lf}(W)$$

is such that $\tau^{lf}(W; M, M') = 0$ if (and for $n \geq 6$ only if) $(W; M, M')$ is homeomorphic rel *M to $M \times (I; \{0\}, \{1\})$.* □

Corollary 11.4 *The projective class $[W] \in \widetilde{K}_0(\mathbb{Z}[\pi_1(W)])$ of a reverse π_1-tame CW complex W is an invariant of the proper homotopy type of W, whereas the projective class at ∞ $[W]_\infty \in \widetilde{K}_0(\mathbb{Z}[\pi_1^\infty(W)])$ is only an invariant of the infinite simple homotopy type.* □

Example 11.5 (i) If K is a finite CW complex with $\pi_1(K) = \pi$ then $e(K \times \mathbb{R}^m) \simeq K \times S^{m-1}$ and

$$Wh^{lf}(K \times \mathbb{R}^m) = \begin{cases} \widetilde{K}_0(\mathbb{Z}[\pi]) & \text{if } m = 1 , \\ \ker(\widetilde{K}_0(\mathbb{Z}[\pi \times \mathbb{Z}]) \longrightarrow \widetilde{K}_0(\mathbb{Z}[\pi])) & \\ = K_{-1}(\mathbb{Z}[\pi]) \oplus \widetilde{\text{Nil}}_{-1}(\mathbb{Z}[\pi]) \oplus \widetilde{\text{Nil}}_{-1}(\mathbb{Z}[\pi]) & \text{if } m = 2 , \\ 0 & \text{if } m \geq 3 \end{cases}$$

as in Siebenmann [144], with K_{-1}, $\widetilde{\text{Nil}}_{-1}$ the lower K- and $\widetilde{\text{Nil}}$-groups of Bass [4] such that

$$K_0(\mathbb{Z}[\pi][z, z^{-1}]) = K_0(\mathbb{Z}[\pi]) \oplus K_{-1}(\mathbb{Z}[\pi]) \oplus \widetilde{\text{Nil}}_{-1}(\mathbb{Z}[\pi]) \oplus \widetilde{\text{Nil}}_{-1}(\mathbb{Z}[\pi]) .$$

(ii) The infinite transfer map

$$Wh(\pi \times \mathbb{Z}) \longrightarrow Wh^{lf}(K \times \mathbb{R}) = \widetilde{K}_0(\mathbb{Z}[\pi]) ;$$

$$\tau(f : L \longrightarrow K \times S^1) \longrightarrow \tau^{lf}(\overline{f} : \overline{L} \longrightarrow K \times \mathbb{R}) = [\overline{L}^+]$$

(with K, π as in (i)) is given algebraically by the split surjection of Bass [4]

$$Wh(\pi \times \mathbb{Z}) \longrightarrow \widetilde{K}_0(\mathbb{Z}[\pi]) \; ;$$
$$\tau(f : M[z,z^{-1}] \longrightarrow M[z,z^{-1}]) \longrightarrow [M^+ \cap f(z^N M^-)] - [M^+ \cap z^N M^-]$$

for any $N \geq 0$ so large that $f(z^N M^+) \subseteq M^+$. See 11.8 below for the application of the calculation $Wh^{lf}(K \times \mathbb{R}) = \widetilde{K}_0(\mathbb{Z}[\pi])$ (in the case $\widetilde{K}_0(\mathbb{Z}[\pi]) \neq 0$) to the construction of an infinite CW complex W which is proper homotopy equivalent to the reverse collared CW complex $K \times \mathbb{R}$, such that W is reverse π_1-tame but not reverse collared. □

A group morphism $f : \pi \longrightarrow \rho$ determines a $(\mathbb{Z}[\rho], \mathbb{Z}[\pi])$-bimodule structure on the group ring $\mathbb{Z}[\rho]$ by

$$\mathbb{Z}[\rho] \times \mathbb{Z}[\rho] \times \mathbb{Z}[\pi] \longrightarrow \mathbb{Z}[\rho] \; ; \; (a, x, b) \longrightarrow axf(b) \; .$$

This is used to define a functor

$$f \; : \; \{\mathbb{Z}[\pi]\text{-modules}\} \longrightarrow \{\mathbb{Z}[\rho]\text{-modules}\} \; ; \; M \longrightarrow \mathbb{Z}[\rho] \otimes_{\mathbb{Z}[\pi]} M$$

inducing morphisms

$$f_* \; : \; \widetilde{K}_0(\mathbb{Z}[\pi]) \longrightarrow \widetilde{K}_0(\mathbb{Z}[\rho]) \quad , \quad f_* \; : \; Wh(\pi) \longrightarrow Wh(\rho) \; .$$

The *relative Whitehead group* of $f : \pi \longrightarrow \rho$ is the abelian group of equivalence classes of triples (P, Q, g) with P a f.g. projective $\mathbb{Z}[\pi]$-module, Q a based f.g. free $\mathbb{Z}[\rho]$-module, and $g : \mathbb{Z}[\rho] \otimes_{\mathbb{Z}[\pi]} P \cong Q$ a $\mathbb{Z}[\rho]$-module isomorphism, subject to the equivalence relation:

$(P, Q, g) \sim (P', Q', g')$ if there exists a $\mathbb{Z}[\pi]$-module isomorphism

$h : P \oplus \mathbb{Z}[\pi]^r \cong P' \oplus \mathbb{Z}[\pi]^{r'}$ with $\tau((g' \oplus 1)(1 \otimes h)(g \oplus 1)^{-1}) = 0 \in Wh(\rho)$.

As usual, addition is by

$$[P_1, Q_1, g_1] + [P_2, Q_2, g_2] \; = \; [P_1 \oplus P_2, Q_1 \oplus Q_2, g_1 \oplus g_2] \in Wh(f) \; .$$

The relative Whitehead group fits into an exact sequence

$$Wh(\pi) \xrightarrow{f_*} Wh(\rho) \longrightarrow Wh(f) \longrightarrow \widetilde{K}_0(\mathbb{Z}[\pi]) \xrightarrow{f_*} \widetilde{K}_0(\mathbb{Z}[\rho])$$

with

$$Wh(f) \longrightarrow \widetilde{K}_0(\mathbb{Z}[\pi]) \; ; \; (P, Q, g) \longrightarrow [P] \; .$$

The involution on the group ring

$$\mathbb{Z}[\pi] \longrightarrow \mathbb{Z}[\pi] \; ; \; a \; = \; \sum_{g \in \pi} n_g g \longrightarrow \bar{a} \; = \; \sum_{g \in \pi} n_g g^{-1}$$

determines an involution

$$\{\text{f.g. projective } \mathbb{Z}[\pi]\text{-modules}\} \longrightarrow \{\text{f.g. projective } \mathbb{Z}[\pi]\text{-modules}\} \; ;$$
$$P \longrightarrow P^* \; = \; \mathrm{Hom}_{\mathbb{Z}[\pi]}(P, \mathbb{Z}[\pi])$$

with

$$\mathbb{Z}[\pi] \times P^* \longrightarrow P^* \; ; \; (a,u) \longrightarrow (x \longrightarrow u(x)\bar{a}) \; .$$

The corresponding duality involutions on the algebraic K-groups

$$\begin{aligned} * \; : \; \widetilde{K}_0(\mathbb{Z}[\pi]) \longrightarrow \widetilde{K}_0(\mathbb{Z}[\pi]) \; ; \; [P] \longrightarrow [P]^* \; &= \; [P^*] \; , \\ * \; : \; Wh(\pi) \longrightarrow Wh(\pi) \; ; \; \tau(\alpha) \longrightarrow \tau(\alpha)^* \; &= \; \tau(\alpha^*) \end{aligned}$$

extend to a duality involution on the relative Whitehead group

$$* \; : \; Wh(f) \longrightarrow Wh(f) \; ; \; (P,Q,g) \longrightarrow (P^*,Q^*,(g^*)^{-1}) \; .$$

See Ranicki [123] for the definition of the relative Whitehead invariant

$$[C,D,\phi] \in Wh(f)$$

of a $\mathbb{Z}[\rho]$-module chain equivalence $\phi : \mathbb{Z}[\rho] \otimes_{\mathbb{Z}[\pi]} C \longrightarrow D$ for a finitely dominated $\mathbb{Z}[\pi]$-module chain complex C and a finite based f.g. free $\mathbb{Z}[\rho]$-module chain complex D, with image $[C] \in \widetilde{K}_0(\mathbb{Z}[\pi])$, and such that

$$[C,D,\phi]^* \; = \; [C^*,D^*,(\phi^*)^{-1}] \in Wh(f) \; .$$

Proposition 11.6 (i) *The infinite Whitehead group of a forward tame locally finite CW complex W is the relative Whitehead group*

$$Wh^{lf}(W) \; = \; Wh(p_* : \pi_1(e(W)) \longrightarrow \pi_1(W))$$

which fits into the exact sequence

$$\begin{aligned} Wh(\pi_1(e(W))) \xrightarrow{p_*} Wh(\pi_1(W)) \longrightarrow Wh^{lf}(W) \\ \longrightarrow \widetilde{K}_0(\mathbb{Z}[\pi_1(e(W))]) \xrightarrow{p_*} \widetilde{K}_0(\mathbb{Z}[\pi_1(W)]) \; . \end{aligned}$$

(ii) *The infinite torsion $\tau^{lf}(f) \in Wh^{lf}(W)$ of a proper homotopy equivalence $f : V \longrightarrow W$ of forward tame locally finite CW complexes has image*

$$\begin{aligned} [\tau^{lf}(f)] \; &= \; [W]^{lf}_\infty - [V]^{lf}_\infty \\ &\in \ker(p_* : \widetilde{K}_0(\mathbb{Z}[\pi_1(e(W))]) \longrightarrow \widetilde{K}_0(\mathbb{Z}[\pi_1(W)])) \; . \end{aligned}$$

(iii) *If $f : V \longrightarrow W$ is a proper homotopy equivalence of strongly locally finite CW complexes which are forward and reverse tame then*

$$\begin{aligned} [e(V)] \; &= \; [V]_\infty - [V]^{lf}_\infty \\ &= \; [W]_\infty - [W]^{lf}_\infty \; = \; [e(W)] \in \widetilde{K}_0(\mathbb{Z}[\pi_1(e(W))]) \; , \end{aligned}$$

and

$$\begin{aligned} [\tau^{lf}(f)] \; &= \; [W]^{lf}_\infty - [V]^{lf}_\infty \; = \; [W]_\infty - [V]_\infty \\ &\in \ker(p_* : \widetilde{K}_0(\mathbb{Z}[\pi_1(e(W))]) \longrightarrow \widetilde{K}_0(\mathbb{Z}[\pi_1(W)])) \; . \end{aligned}$$

Proof (i) As W is forward tame it is possible to choose a π_1-stable sequence $W \supset W_1 \supset W_2 \supset \ldots$ in 11.3. By 2.21 (iv) the stable inverse system $\{\pi_1(W_j)\}$ with inverse limit $\pi_1(e(W))$ induces a stable inverse system $\{Wh(\pi_1(W_j))\}$ with inverse limit $Wh(\pi_1(e(W)))$, and similarly for $\widetilde{K}_0$, so that

$$\begin{aligned}
\varprojlim_j Wh(\pi_1(W_j)) &= Wh(\pi_1(e(W)))\ , \\
\varprojlim_j \widetilde{K}_0(\mathbb{Z}[\pi_1(W_j)]) &= \widetilde{K}_0(\mathbb{Z}[\pi_1(e(W))])\ , \\
\varprojlim_j{}^1\, Wh(\pi_1(W_j)) &= 0\ , \\
Wh_b^{lf}(W) &= \mathrm{coker}(p_* : Wh(\pi_1(e(W)))\longrightarrow Wh(\pi_1(W))) \\
&= \ker(Wh^{lf}(W)\longrightarrow \widetilde{K}_0(\mathbb{Z}[\pi_1(W)]))\ .
\end{aligned}$$

(ii) Let $\widetilde{V}, \widetilde{W}$ be the universal covers of V, W, and let $\widetilde{f} : \widetilde{V}\longrightarrow\widetilde{W}$ be a $\pi_1(W)$-equivariant lift of f. The algebraic mapping cone $\mathcal{C}(\widetilde{f})$ is a contractible $\mathbb{Z}[\pi_1(W)]$-module chain complex. Let $\pi = \pi_1(e(W))$, and let $\overline{e(V)}$, $\overline{e(W)}$ be the universal covers of $e(V)$, $e(W)$. Choose a cofinite subcomplex $W_1 \subseteq W$ such that $\pi_1(W_1) = \pi$, let

$$f_1 \;=\; f| \;:\; V_1 \;=\; f^{-1}(W_1) \longrightarrow W_1$$

and let $\overline{f}_1 : \overline{V}_1\longrightarrow\overline{W}_1$ be a π-equivariant lift of f_1 with $\overline{W}_1$ the universal cover of W_1, and $\overline{V}_1$ the pullback cover of V_1. The homotopy equivalence $e(f_1) : e(V_1)\longrightarrow e(W_1)$ induces a $\mathbb{Z}[\pi]$-module chain equivalence

$$\begin{aligned}
S(\overline{e(V)}) \;\simeq\; C^{\infty,\pi}(\overline{V}_1) \;&=\; \mathcal{C}(C(\overline{V}_1)\longrightarrow C^{lf,\pi}(\overline{V}_1))_{*+1} \\
\longrightarrow S(\overline{e(W)}) \;\simeq\; C^{\infty,\pi}(\overline{W}_1) \;&=\; \mathcal{C}(C(\overline{W}_1)\longrightarrow C^{lf,\pi}(\overline{W}_1))_{*+1}
\end{aligned}$$

so that the inclusion

$$\mathcal{C}(\overline{f}_1 : C(\overline{V}_1)\longrightarrow C(\overline{W}_1)) \longrightarrow \mathcal{C}(\overline{f}_1 : C^{lf,\pi}(\overline{V}_1)\longrightarrow C^{lf,\pi}(\overline{W}_1))$$

is also a $\mathbb{Z}[\pi]$-module chain equivalence. The locally finite cellular $\mathbb{Z}[\pi]$-module chain complexes $C^{lf,\pi}(\overline{V}_1)$, $C^{lf,\pi}(\overline{W}_1)$ are finitely dominated (by 7.20), and hence so is $\mathcal{C}(\overline{f}_1 : C(\overline{V}_1)\longrightarrow C(\overline{W}_1))$. The algebraic mapping cone $\mathcal{C}(\widetilde{f}_1)$ is a finitely dominated $\mathbb{Z}[\pi]$-module chain complex such that there is defined a short exact sequence of $\mathbb{Z}[\pi_1(W)]$-module chain complexes

$$0 \longrightarrow \mathbb{Z}[\pi_1(W)] \otimes_{\mathbb{Z}[\pi]} \mathcal{C}(\overline{f}_1) \longrightarrow \mathcal{C}(\widetilde{f}) \longrightarrow D \longrightarrow 0$$

with

$$D \;=\; \mathcal{C}(\widetilde{f} : C(\widetilde{V}, \widetilde{V}_1)\longrightarrow C(\widetilde{W}, \widetilde{W}_1))$$

a finite based f.g. free $\mathbb{Z}[\pi_1(W)]$-module chain complex. The cellular structure thus determines a finite structure on the chain complex $\mathbb{Z}[\pi_1(W)] \otimes_{\mathbb{Z}[\pi]} \mathcal{C}(\overline{f}_1)$, representing an element $\tau^{lf}(f) \in Wh^{lf}(W)$ with image

$$[\tau^{lf}(f)] = [\mathcal{C}(\overline{f}_1)] = [W]^{lf} - [V]^{lf} \in \widetilde{K}_0(\mathbb{Z}[\pi]) .$$

(iii) The induced map $e(f) : e(V) \longrightarrow e(W)$ is a homotopy equivalence by 9.4. □

Corollary 11.7 (i) *The locally finite projective class* $[W]^{lf} \in \widetilde{K}_0(\mathbb{Z}[\pi_1(W)])$ *of a forward tame CW complex W is an invariant of the proper homotopy type of W, whereas the locally finite projective class at* ∞ $[W]^{lf}_\infty \in \widetilde{K}_0(\mathbb{Z}[\pi_1(e(W))])$ *is only an invariant of the infinite simple homotopy type.*

(ii) *The property of being forward collared is not a proper homotopy invariant.*

Proof (i) If $f : V \longrightarrow W$ is a proper homotopy equivalence then the Wall finiteness obstruction is preserved,

$$[W] = [V] \in \widetilde{K}_0(\mathbb{Z}[\pi_1(W)]) ,$$

so that

$$[W]^{lf} - [V]^{lf} = [W] - [V] = 0 \in \widetilde{K}_0(\mathbb{Z}[\pi_1(W)]) ,$$

and

$$\begin{aligned} [W]_\infty - [V]_\infty &= [W]^{lf}_\infty - [V]^{lf}_\infty \\ &= [\tau^{lf}(f)] \in \ker(p_* : \widetilde{K}_0(\mathbb{Z}[\pi_1(e(W))]) \longrightarrow \widetilde{K}_0(\mathbb{Z}[\pi_1(W)])) \end{aligned}$$

is an invariant of the infinite torsion $\tau^{lf}(f) \in Wh^{lf}(W)$.

(ii) If W is forward collared then $[W]^{lf}_\infty = 0 \in \widetilde{K}_0(\mathbb{Z}[\pi_1(e(W))])$ by 10.9 (iii). For every element $\tau^{lf} \in Wh^{lf}(W)$ there exists a proper homotopy equivalence $f : V \longrightarrow W$ with $\tau^{lf}(f) = \tau^{lf}$, so that V is forward tame (9.6) with $[V]^{lf} = -[\tau^{lf}]$. If $[\tau^{lf}] \neq 0 \in \widetilde{K}_0(\mathbb{Z}[\pi_1(e(W))])$ then V cannot be forward collared. □

Example 11.8 Let K be a connected finite CW complex with a fundamental group $\pi_1(K) = \pi$ such that $\widetilde{K}_0(\mathbb{Z}[\pi]) \neq 0$ (e.g. $\pi = Q(8)$, as in 6.4 (i)). As in the proof of 6.8 (iii) use a $\mathbb{Z}[\pi]$-module projection $p = p^2 : \mathbb{Z}[\pi]^r \longrightarrow \mathbb{Z}[\pi]^r$ with $P = \mathrm{im}(p)$ such that $[P] \neq 0 \in \widetilde{K}_0(\mathbb{Z}[\pi])$ to construct for any $N \geq 2$ a finite CW complex

$$L = (K \times S^1 \vee \bigvee_r S^N) \cup_{-zp+1-p} \bigcup_r D^{N+1}$$

such that the projection defines a homotopy equivalence

$$f : L \xrightarrow{\simeq} K \times S^1$$

with torsion

$$\begin{aligned}\tau(f) &= (-)^N\tau(-zp+1-p : \mathbb{Z}[\pi][z,z^{-1}]^r \longrightarrow \mathbb{Z}[\pi][z,z^{-1}]^r) \\ &= (-)^N\tau(-z : P[z,z^{-1}] \longrightarrow P[z,z^{-1}]) \\ &\neq 0 \in \operatorname{im}(\widetilde{K}_0(\mathbb{Z}[\pi]) \longrightarrow Wh(\pi \times \mathbb{Z})) .\end{aligned}$$

As in 11.5 (ii) the lift of f to the infinite cyclic covers is a proper homotopy equivalence

$$\overline{f} : \overline{L} = f^*(K \times \mathbb{R}) \xrightarrow{\simeq} K \times \mathbb{R}$$

with infinite torsion

$$\begin{aligned}\tau^{lf}(\overline{f}) &= [\overline{L}^+]_\infty = [\overline{L}^+]^{lf}_\infty = [\overline{L}^+] = [\overline{L}^+]^{lf} \\ &= [P] \neq 0 \in Wh^{lf}(K \times \mathbb{R}) = \widetilde{K}_0(\mathbb{Z}[\pi]) .\end{aligned}$$

The infinite CW complex $K \times \mathbb{R}$ (with two ends) is both forward and reverse collared, while the proper homotopy equivalent infinite CW complex $\overline{L}$ is both forward and reverse tame but is neither forward nor reverse collared. □

Proposition 11.9 *Let $(W, \partial W)$ be an open n-dimensional geometric Poincaré pair such that W is both forward and reverse tame.*
(i) *$(W, \partial W)$ has an infinite torsion*

$$\tau^{lf}(W) = (-)^n\tau^{lf}(W)^* \in Wh^{lf}(W) = Wh(p_* : \pi_1(e(W)) \longrightarrow \pi_1(W))$$

with image

$$\begin{aligned}[\tau^{lf}(W)] &= [W]_\infty + (-)^{n-1}([W]^{lf}_\infty)^* \\ &= [W]^{lf}_\infty + (-)^{n-1}[W]^*_\infty \\ &\in \ker(p_* : \widetilde{K}_0(\mathbb{Z}[\pi_1(e(W))]) \longrightarrow \widetilde{K}_0(\mathbb{Z}[\pi_1(W)])) .\end{aligned}$$

(ii) *If there exists a Poincaré transverse map $(W, \partial W) \longrightarrow ([0,\infty), \{0\})$ then*

$$\begin{aligned}\tau^{lf}(W) &\in \operatorname{im}(Wh(\pi_1(W)) \longrightarrow Wh^{lf}(W)) \\ &= \ker(Wh^{lf}(W) \longrightarrow \widetilde{K}_0(\mathbb{Z}[\pi_1(e(W))])) , \\ [W]^{lf}_\infty &= (-)^n[W]^*_\infty \in \widetilde{K}_0(\mathbb{Z}[\pi_1(e(W))]) ,\end{aligned}$$

the end space $e(W)$ is a finitely dominated $(n-1)$-dimensional geometric Poincaré space, and $(W; \partial W, e(W))$ is a finitely dominated n-dimensional geometric Poincaré cobordism with

$$\begin{aligned}[e(W)] &= [W]_\infty - [W]^{lf}_\infty \\ &= [W]_\infty + (-)^{n-1}[W]^*_\infty \in \widetilde{K}_0(\mathbb{Z}[\pi_1(e(W))]) .\end{aligned}$$

Proof (i) Let $\widetilde{W}$ be the universal cover of W. The $\mathbb{Z}[\pi_1(W)]$-module chain complex

$$C = \mathcal{C}([W] \cap - : C^{lf,\pi_1(W)}(\widetilde{W}, \partial\widetilde{W})^{n-*} \longrightarrow C(\widetilde{W}))$$

is contractible. Write $\pi = \pi_1(e(W))$. Let $V \subseteq W$ be a cofinite subcomplex such that $\pi_1(V) = \pi$, and let $\overline{V}$ be the universal cover of V. Let

$$\partial V = V \cap \mathrm{cl}(W \backslash V) ,$$

and let

$$[V] \in H_n^{lf}(V, \partial V) = H_n^{lf}(W, \mathrm{cl}(W \backslash V))$$

be the image of $[W] \in H_n^{lf}(W, \partial W)$. The $\mathbb{Z}[\pi]$-module chain complex

$$D = \mathcal{C}([V] \cap - : C^{lf,\pi}(\overline{V}, \partial\overline{V})^{n-*} \longrightarrow C(\overline{V}))$$

is finitely dominated, and there is defined a short exact sequence of $\mathbb{Z}[\pi_1(W)]$-module chain complexes

$$0 \longrightarrow \mathbb{Z}[\pi_1(W)] \otimes_{\mathbb{Z}[\pi]} D \longrightarrow C \longrightarrow E \longrightarrow 0$$

with E finite based f.g. free. The corresponding finite structure (F, ϕ) on $\mathbb{Z}[\pi_1(W)] \otimes_{\mathbb{Z}[\pi]} D$ determines the infinite torsion of W

$$\tau^{lf}(W) = (D, F, \phi) \in Wh^{lf}(W) = Wh(p_*)$$

with image

$$\begin{aligned}[\tau^{lf}(W)] &= [D] = [V] + (-)^{n-1}([V]^{lf})^* \\ &= [W]_\infty + (-)^{n-1}([W]_\infty^{lf})^* \in \widetilde{K}_0(\mathbb{Z}[\pi]) .\end{aligned}$$

(ii) By Poincaré transversality $(V, \partial V)$ in (i) may be taken to be an open n-dimensional geometric Poincaré pair, i.e. such that $D \simeq 0$. Apply (i) to $(V, \partial V)$, with $\pi_1(V) = \pi_1(e(V)) = \pi$. □

Example 11.10 A forward and reverse tame open n-dimensional manifold $(W, \partial W)$ with a compact boundary ∂W is an open n-dimensional geometric Poincaré pair with a Poincaré transverse map $(W, \partial W) \longrightarrow ([0, \infty), \{0\})$, such that

$$\begin{aligned}&\tau^{lf}(W) = 0 \in Wh^{lf}(W) , \\ &[W]_\infty^{lf} = (-)^n [W]_\infty^* \in \widetilde{K}_0(\mathbb{Z}[\pi_1(e(W))]) \ (10.15) .\end{aligned}$$

In particular, this applies to the open n-dimensional manifold with compact boundary and a reverse π_1-tame end constructed by Siebenmann [140] with prescribed fundamental group at ∞ $\pi_1^\infty(W)$ and prescribed projective class at ∞ $[W]_\infty \in \widetilde{K}_0(\mathbb{Z}[\pi_1^\infty(W)])$ (10.2 (iii)). □

Remark 11.11 (i) Given a locally finite infinite CW complex W let $L_*^{lf,q}(W)$ be the proper surgery obstruction groups of Maumary [92] and Taylor [160], for surgery on proper normal maps of open manifolds up to proper homotopy equivalence for $q = h$ (simple for $q = s$). The proper L-groups are related to the original surgery obstruction groups $L_*^q(\mathbb{Z}[\pi])$ $(q = s, h)$ of Wall [165] and the projective L-groups $L_*^p(\mathbb{Z}[\pi])$ of Novikov [105] and Ranicki [117] by the L-theory analogues of the $\varprojlim$–$\varprojlim^1$ exact sequence for $Wh^{lf}(W)$ of 11.3, as related in Pedersen and Ranicki [109]. For forward tame W it is possible to express $L_*^{lf,q}(W)$ as relative L-groups of $p_* : \mathbb{Z}[\pi_1(e(W))] \longrightarrow \mathbb{Z}[\pi_1(W)]$, by analogy with 11.6, as follows. Let

$$\Pi = \pi_1(e(W)) \ , \quad \pi = \pi_1(W) \ .$$

The $*$-invariant subgroups

$$\begin{aligned} I_0 &= \mathrm{im}(Wh^{lf}(W) \longrightarrow \widetilde{K}_0(\mathbb{Z}[\Pi])) \\ &= \ker(p_* : \widetilde{K}_0(\mathbb{Z}[\Pi]) \longrightarrow \widetilde{K}_0(\mathbb{Z}[\pi])) \subseteq \widetilde{K}_0(\mathbb{Z}[\Pi]) \ , \\ J_1 &= \mathrm{im}(p_* : Wh(\Pi) \longrightarrow Wh(\pi)) \\ &= \ker(Wh(\pi) \longrightarrow Wh^{lf}(W)) \subseteq Wh(\pi) \end{aligned}$$

are such that there is defined a short exact sequence

$$0 \longrightarrow Wh(\pi)/J_1 \longrightarrow Wh^{lf}(W) \longrightarrow I_0 \longrightarrow 0 \ .$$

The groups $L_*^{lf,q}(W)$ for forward tame W fit into the commutative diagram with exact rows and columns

$$\begin{array}{ccccccccc}
 & & \vdots & & \vdots & & \vdots & & \vdots \\
 & & \downarrow & & \downarrow & & \downarrow & & \downarrow \\
\dots & \rightarrow & L_n^h(\mathbb{Z}[\Pi]) & \xrightarrow{p_*} & L_n^{J_1}(\mathbb{Z}[\pi]) & \longrightarrow & L_n^{lf,s}(W) & \longrightarrow & L_{n-1}^h(\mathbb{Z}[\Pi]) \rightarrow \dots \\
 & & \downarrow & & \downarrow & & \downarrow & & \downarrow \\
\dots & \rightarrow & L_n^{I_0}(\mathbb{Z}[\Pi]) & \xrightarrow{p_*} & L_n^h(\mathbb{Z}[\pi]) & \longrightarrow & L_n^{lf,h}(W) & \longrightarrow & L_{n-1}^{I_0}(\mathbb{Z}[\Pi]) \rightarrow \dots \\
 & & \downarrow & & \downarrow & & \downarrow & & \downarrow \\
\dots & \rightarrow & \widehat{H}^n(\mathbb{Z}_2; I_0) & \rightarrow & \widehat{H}^n(\mathbb{Z}_2; Wh(\pi)/J_1) & \rightarrow & \widehat{H}^n(\mathbb{Z}_2; Wh^{lf}(W)) & \rightarrow & \widehat{H}^{n-1}(\mathbb{Z}_2; I_0) \rightarrow \dots \\
 & & \downarrow & & \downarrow & & \downarrow & & \downarrow \\
\dots & \rightarrow & L_{n-1}^h(\mathbb{Z}[\Pi]) & \xrightarrow{p_*} & L_{n-1}^{J_1}(\mathbb{Z}[\pi]) & \longrightarrow & L_{n-1}^{lf,s}(W) & \longrightarrow & L_{n-2}^h(\mathbb{Z}[\Pi]) \rightarrow \dots \\
 & & \downarrow & & \downarrow & & \downarrow & & \downarrow \\
 & & \vdots & & \vdots & & \vdots & & \vdots
\end{array}$$

(ii) Let $W = K \times \mathbb{R}^m$ with K a finite CW complex, so that W is forward collared with $e(K \times \mathbb{R}^m) \simeq K \times S^{m-1}$. The proper surgery obstruction groups of $K \times \mathbb{R}^m$ are given by

$$L_n^{lf,s}(K \times \mathbb{R}^m) = \begin{cases} L_{n-1}^h(\mathbb{Z}[\pi]) & \text{if } m = 1 , \\ L_{n-2}^p(\mathbb{Z}[\pi]) & \text{if } m = 2 , \\ 0 & \text{if } m \geq 3 , \end{cases}$$

$$L_n^{lf,h}(K \times \mathbb{R}^m) = \begin{cases} L_{n-1}^p(\mathbb{Z}[\pi]) & \text{if } m = 1 , \\ L_{n-2}^{\langle -1 \rangle}(\mathbb{Z}[\pi]) & \text{if } m = 2 , \\ 0 & \text{if } m \geq 3 \end{cases}$$

with $\pi = \pi_1(K)$. The lower L-groups $L_*^{\langle -1 \rangle}$ of Ranicki [118] are such that

$$L_n^p(\mathbb{Z}[\pi][z, z^{-1}]) = L_n^p(\mathbb{Z}[\pi]) \oplus L_{n-1}^{\langle -1 \rangle}(\mathbb{Z}[\pi]) ,$$

and there is defined a Rothenberg-type exact sequence

$$\begin{aligned} \ldots \longrightarrow L_n^p(\mathbb{Z}[\pi]) \longrightarrow L_n^{\langle -1 \rangle}(\mathbb{Z}[\pi]) \longrightarrow \widehat{H}^n(\mathbb{Z}_2; K_{-1}(\mathbb{Z}[\pi])) \\ \longrightarrow L_{n-1}^p(\mathbb{Z}[\pi]) \longrightarrow \ldots . \end{aligned}$$

The infinite transfer maps

$$L_n^q(\mathbb{Z}[\pi][z, z^{-1}]) \longrightarrow L_n^{lf,q}(K \times \mathbb{R}) = L_{n-1}^r(\mathbb{Z}[\pi]) ;$$

$$\sigma_*((f, b) : M \longrightarrow X) \longrightarrow \sigma_*^{lf}((\overline{f}, \overline{b}) : \overline{M} \longrightarrow \overline{X}) \quad ((q, r) = (s, h), (h, p))$$

are given algebraically by the projections in the splittings of [118]

$$L_n^q(\mathbb{Z}[\pi][z, z^{-1}]) = L_n^q(\mathbb{Z}[\pi]) \oplus L_{n-1}^r(\mathbb{Z}[\pi]) . \qquad \square$$

We conclude this chapter with the applications of infinite simple homotopy theory to the detection of reverse and forward collaring. In 11.13 we prove that a locally finite CW complex W is infinite simple homotopy equivalent to a reverse collared CW complex if and only if every cofinite subcomplex of W is homotopy equivalent to a finite CW complex. The proof will require the following technical result.

Lemma 11.12 *Let W be a strongly locally finite CW complex with a cofinite subcomplex $V \subseteq W$ homotopy equivalent to a finite CW complex. Let $A \subseteq W$ be a finite subcomplex with $W \backslash V \subseteq A$. Then there exist a cofinite subcomplex $U \subseteq V$ with $A \cap U = \emptyset$, a finite subcomplex $B \subseteq V$ with $W \backslash (A \cup U) \subseteq B$, a finite CW complex B' with $B \cap (A \cup U) = B' \cap (A \cup U)$ such that B, B' are simple homotopy equivalent* rel $B \cap (A \cup U)$, *and a finite subcomplex $K \subseteq W' = A \cup B' \cup U$ with $A \subseteq K \subseteq A \cup B'$ such that the inclusion $K \longrightarrow W'$ is a homotopy equivalence.*

Proof Let $C = A \cap V$ and let L be a finite CW complex homotopy equivalent to V. For some large n, we may assume that L is a subcomplex of

$V \times D^n$ and that inclusion : $L \longrightarrow V \times D^n$ is a homotopy equivalence (i.e. L is a strong deformation retract of $V \times D^n$). Let r : identity $\simeq r_1$: $V \times D^n \longrightarrow V \times D^n$ be a cellular homotopy such that $r_1(V \times D^n) \subseteq L$ and $r_t|L$ = inclusion : $L \longrightarrow V \times D^n$ for each $t \in I$. Let $U \subseteq V$ be a cofinite subcomplex such that

$$r(C \times D^n \times I) \cap (U \times D^n) \ = \ \emptyset \ = \ L \cap (U \times D^n) \ .$$

Let $B \subseteq W$ be a finite subcomplex such that $A \cap B = C$ and $W \backslash (A \cup U) \subseteq B$. The maps

$$g \ : \ C \longrightarrow B \times D^n \ ; \ x \longrightarrow (x, 0) \ ,$$
$$g' \ : \ C \longrightarrow B \times D^n \ ; \ x \longrightarrow r_1(x, 0)$$

are homotopic so that the mapping cylinders $\mathcal{M}(g)$, $\mathcal{M}(g')$ are simple homotopy equivalent rel $C \cup (B \times D^n)$. Let $B' = \mathcal{M}(g')$ with $B' \cap A = C$ (the base of $\mathcal{M}(g)$) and $B' \cap U = (B \cap U) \times \{0\} \subseteq B \times D^n$. Since B is simple homotopy equivalent to $\mathcal{M}(g)$ rel $B \cap (A \cup U)$, it follows that B, B' are simple homotopy equivalent rel $B \cap (A \cup U)$. The map

$$g'' \ : \ C \longrightarrow L \ ; \ x \longrightarrow r_1(x, 0)$$

is such that $\mathcal{M}(g'') \cup_L (B \times D^n) = \mathcal{M}(g')$. Thus, the finite subcomplex $K = A \cup \mathcal{M}(g'')$ of $W' = A \cup B' \cup U$ is such that $A \subseteq K \subseteq A \cup B'$ and the strong deformation retraction of $V \times D^n$ to L induces a strong deformation retraction of W' to K. □

Proposition 11.13 *For a strongly locally finite CW complex W the following conditions are equivalent:*

(i) *W is infinite simple homotopy equivalent to a reverse collared CW complex.*

(ii) *Every cofinite subcomplex of W is infinite simple homotopy equivalent to a reverse collared CW complex.*

(iii) *Every cofinite subcomplex of W is homotopy equivalent to a finite CW complex.*

Proof (i) $\Longrightarrow$ (ii) If W is infinite simple homotopy equivalent to the reverse collared CW complex W' and $V \subseteq W$ is a cofinite subcomplex, then V is infinite simple homotopy equivalent to some cofinite subcomplex $V' \subseteq W'$. Since cofinite subcomplexes of reverse collared CW complexes are reverse collared, V' is reverse collared.

(ii) $\Longrightarrow$ (iii) Reverse collared CW complexes are homotopy equivalent to finite CW complexes.

(iii) $\Longrightarrow$ (i) We shall use 11.12 to construct:

(a) a sequence $A_1 \subseteq A_2 \subseteq A_3 \subseteq \ldots$ of finite subcomplexes of W with $W = \bigcup_{j=1}^{\infty} A_j$,

(b) a sequence $U_1 \supseteq U_2 \supseteq U_3 \supseteq \ldots$ of cofinite subcomplexes of W,

(c) a sequence $B_1, B_2, B_3, \ldots$ of finite subcomplexes of W, and

(d) a sequence $B'_1, B'_2, B'_3, \ldots$ of finite CW complexes

such that for every $i \neq j$ we have:

(e) $A_j \cap U_j = \emptyset$, $W \backslash (A_j \cup U_j) \subseteq B_j$, $B_i \cap B_j = \emptyset = B'_i \cap B'_j$, $B_j \cap (A_j \cup U_j) = B'_j \cap (A_j \cup U_j)$,

(f) B_j and B'_j are simple homotopy equivalent rel $B_j \cap (A_j \cup U_j)$, and

(g) the strongly locally finite CW complex $W_j = A_j \cup B'_j \cup U_j$ contains a finite subcomplex K_j with $A_j \subseteq K_j \subseteq A_j \cup B'_j$ so that the inclusion $K_j \longrightarrow W_j$ is a homotopy equivalence.

The construction is by induction. Let $W = V_1 \supseteq V_2 \supseteq V_3 \supseteq \ldots$ be a sequence of cofinite subcomplexes of W such that $\bigcap_{j=1}^{\infty} V_j = \emptyset$. Let $A_1 = \emptyset$. By 11.12 there exist a cofinite subcomplex $U_1 \subseteq V_1$, a finite subcomplex $B_1 \subseteq W$ with $W \backslash U_1 \subseteq B_1$, a finite CW complex B'_1 with $B_1 \cap U_1 = B'_1 \cap U_1$ such that B_1, B'_1 are simple homotopy equivalent rel $B_1 \cap U_1$, and a finite subcomplex $K_1 \subseteq W_1 = B'_1 \cup U_1$ with $K_1 \subseteq B'_1$ such that the inclusion $K_1 \longrightarrow W_1$ is a homotopy equivalence. Suppose $A_1, \ldots, A_n, U_1, \ldots, U_n, B_1, \ldots, B_n, B'_1, \ldots, B'_n, K_1, \ldots, K_n$ have been constructed with the properties above and $U_j \subseteq V_j$ for $j = 1, \ldots, n$. Let $A_{n+1} \subseteq W$ be a finite subcomplex such that

$$A_n \cup B_n \cup (W \backslash (U_n \cap V_{n+1})) \subseteq A_{n+1} .$$

Let $V \subseteq \mathrm{int}(U_n \cap V_{n+1})$ be a cofinite subcomplex. By 11.12 there exist a cofinite subcomplex $U_{n+1} \subseteq V$ with $A_{n+1} \cap U_{n+1} = \emptyset$, a finite subcomplex $B_{n+1} \subseteq V$ with $W \backslash (A_{n+1} \cup U_{n+1}) \subseteq B_{n+1}$, a finite CW complex B'_{n+1} such that $B_{n+1} \cap (A_{n+1} \cup U_{n+1}) = B'_{n+1} \cap (A_{n+1} \cup U_{n+1})$ such that B_{n+1}, B'_{n+1} are simple homotopy equivalent rel $B_{n+1} \cap (A_{n+1} \cup U_{n+1})$, and a finite subcomplex $K_{n+1} \subseteq W_{n+1} = A_{n+1} \cup B'_{n+1} \cup U_{n+1}$ with $A_{n+1} \subseteq K_{n+1} \subseteq A_{n+1} \cup B'_{n+1}$ such that the inclusion $K_{n+1} \longrightarrow W_{n+1}$ is a homotopy equivalence. Given such a construction, let

$$W' = (W \backslash \bigcup_{j=1}^{\infty} B_j) \cup \bigcup_{j=1}^{\infty} B'_j .$$

Then W and W' are infinite simple homotopy equivalent. The finite subcomplex of W'

$$K'_j = (K_j \backslash \bigcup_{i=1}^{j-1} B_i) \cup \bigcup_{i=1}^{j-1} B'_i$$

is such that the inclusion $K'_j \longrightarrow W'$ is a homotopy equivalence. Since $K'_1 \subseteq K'_2 \subseteq K'_3 \subseteq \ldots \subseteq W'$, $W = \bigcup_{j=1}^{\infty} K'_j$, it follows from 8.5 that W' is reverse collared. □

Proposition 11.14 *The following conditions on an ANR space W are equivalent:*

(i) *W is forward tame,*
(ii) *$W \times S^1$ is infinite simple homotopy equivalent to a forward collared ANR X,*
(iii) *$W \times S^1$ is properly dominated by a forward tame space Z.*

Proof (i) $\Longrightarrow$ (ii) There exist a closed cocompact $V \subseteq W$ and a proper map $q : V \times [0,\infty) \longrightarrow W; (x,t) \longrightarrow q_t(x)$ with $q_0 : V \longrightarrow W$ the inclusion. Inductively select closed cocompact subspaces $V = U_0 \supseteq U_1 \supseteq U_2 \supseteq \ldots$ and non-negative numbers $0 = t_0 \leq t_1 \leq t_2 \leq \ldots$ such that:

(a) $\bigcap_{i=1}^{\infty} U_i = \emptyset$,
(b) $q_t(U_{i+1}) \subseteq U_i$ for each $t \geq 0$ and $i = 1, 2, 3, \ldots$,
(c) $q_t(U_i) \subseteq U_{i+1}$ for each $t \geq t_i$ and $i = 1, 2, 3, \ldots$.

Choose a map $\rho : W \longrightarrow [0,\infty)$ such that $\rho(U_i) \geq t_i$ for each $i = 1, 2, 3, \ldots$, and use it to define a map

$$\zeta : V \longrightarrow W ; x \longrightarrow q(x, \rho(x)) .$$

Note that if $x \in U_i$ then $\rho(x) \geq t_i$ so $\zeta(x) \in U_{i+1}$, and that ζ is proper homotopic to the inclusion $V \longrightarrow W$. The homotopy extension property can be used to extend ζ to a proper map, also denoted ζ, defined on all of W such that ζ is proper homotopic to id_W. It follows that $W \times S^1 = T(\mathrm{id}_W)$ and $X = T(\zeta)$ are infinite simple homotopy equivalent.

To see that $T(\zeta)$ is forward collared, consider

$$A = T(\zeta | U_1) \subseteq T(\zeta) .$$

Then A is closed and cocompact and there exists a proper map

$$g : A \times [0,\infty) \longrightarrow A ; ([x,s],t) \longrightarrow [x, s+t] .$$

(ii) $\Longrightarrow$ (iii) Obvious.

(iii) $\Longrightarrow$ (i) Let $f : W \times S^1 \longrightarrow Z$ be a proper map such that there exist a proper map $g : Z \longrightarrow W \times S^1$ and a proper homotopy $h : \mathrm{id}_{W \times S^1} \simeq gf$ with Z forward tame. Let $A \subseteq Z$ be a closed cocompact subspace for which there exists a proper map $p : A \times [0, \infty) \longrightarrow Z$ extending the inclusion p_0. Choose a closed cocompact subspace $V \subseteq W$ such that $V \times S^1 \subseteq f^{-1}(A)$. Define $q : V \times [0, \infty) \longrightarrow W$ to be the composition

$$\begin{aligned} V \times [0, \infty) \ &= \ V \times \{\text{pt.}\} \times [0, \infty) \subseteq V \times S^1 \times [0, \infty) \\ &\xrightarrow{f| \times \mathrm{id}} A \times [0, \infty) \xrightarrow{p} A \xrightarrow{g|} W \times S^1 \xrightarrow{\text{proj.}} W \ . \end{aligned}$$

Then $q_0 : V \longrightarrow W$ is proper homotopic to the inclusion so that q can be adjusted to get a proper map $q' : V \times [0, \infty) \longrightarrow W$ with q'_0 the inclusion. □

Proposition 11.15 *For a strongly locally finite CW complex W the following conditions are equivalent:*

(i) *W is reverse tame,*
(ii) *$W \times S^1$ is infinite simple homotopy equivalent to a reverse collared CW complex,*
(iii) *$W \times S^1$ is properly dominated by a reverse tame space.*

Proof (i) $\Longrightarrow$ (ii) If $U \subseteq W \times S^1$ is a cofinite subcomplex, then there exists a cofinite subcomplex $V \subseteq W$ with $V \times S^1 \subseteq U$. By 8.9 V is finitely dominated. By 6.7 (ii) $V \times S^1$ is homotopy equivalent to a finite CW complex. It follows that U is homotopy equivalent to a finite CW complex. $W \times S^1$ is infinite simple homotopy equivalent to a reverse collared CW complex by 11.13.

(ii) $\Longrightarrow$ (iii) Obvious.

(iii) $\Longrightarrow$ (i) Let Z be a reverse tame space for which there exist proper maps $f : W \times S^1 \longrightarrow Z$, $g : Z \longrightarrow W \times S^1$ and a proper homotopy h : identity $\simeq gf : W \times S^1 \longrightarrow W \times S^1$. For $V \subseteq W$ a cofinite subcomplex, it suffices to show that V is finitely dominated (8.9). Since V is dominated by $V \times S^1$, we need to show that $V \times S^1$ is finitely dominated. There exists a cocompact subspace $U \subseteq Z$ such that $U \subseteq g^{-1}(V \times S^1)$ and $h(f^{-1}(U) \times I) \subseteq V \times S^1$. Thus $gf| : f^{-1}(U) \longrightarrow V \times S^1$ is homotopic to inclusion : $f^{-1}(U) \longrightarrow V \times S^1$. Since Z is reverse tame, U is dominated by a compact subspace $C \subseteq U$, i.e. there exists a homotopy k : identity $\simeq k_1 : U \longrightarrow U$ with $k_1(U) \subseteq C$. It follows that there is a homotopy of $f^{-1}(U)$ in $V \times S^1$ which deforms $f^{-1}(U)$ into a compact subspace of $V \times S^1$. The homotopy extension property then implies that $V \times S^1$ is finitely dominated. □

12

Forward tameness is a homotopy pushout

Following Quinn [116] we shall now obtain a result of the following type: a σ-compact metric space W is forward tame if and only if the homotopy commutative square

$$\begin{array}{ccc} e(W) & \longrightarrow & \{\infty\} \\ {\scriptstyle p_W}\downarrow & & \downarrow \\ W & \xrightarrow{\ i\ } & W^{\infty} \end{array}$$

is a homotopy pushout, with $i : W \longrightarrow W^{\infty}$ the inclusion in the one-point compactification W^{∞}, $e(W)$ the end space and

$$p_W \ : \ e(W) \longrightarrow W \ ; \ \omega \longrightarrow \omega(0)$$

the evaluation map. The cofibration sequence

$$e(W) \longrightarrow W \xrightarrow{\ i\ } W^{\infty}$$

induces the long exact sequence of homology groups of 3.9

$$\ldots \longrightarrow H_r^{\infty}(W) \longrightarrow H_r(W) \xrightarrow{\ i\ } H_r^{lf}(W) \longrightarrow H_{r-1}(e(W)) \longrightarrow \ldots ,$$

using 7.10 and 7.15 to identify

$$H_*(e(W)) \ = \ H_*^{\infty}(W) \ , \ H_*(W^{\infty}, \infty) \ = \ H_*^{lf}(W) \ .$$

The exact sequence shows that if W is forward tame then

$$H_{-1}^{\infty}(W) \ = \ H_{-1}(e(W)) \ = \ 0 \ .$$

There is a corresponding exact sequence for the cohomology of a forward tame space W

$$\ldots \longrightarrow H_{lf}^r(W) \xrightarrow{\ i\ } H^r(W) \longrightarrow H^r(e(W)) \longrightarrow H_{lf}^{r+1}(W) \longrightarrow \ldots$$

with $H^*(e(W)) = H^*_\infty(W)$. Thus if W is a locally finite forward tame CW complex the number of ends of W (1.14) is the number of path components of the end space $e(W)$.

The standard constructions of mapping cylinders are given by:

Definition 12.1 (i) The *mapping cylinder* of a map $f : X \longrightarrow Y$ is the identification space

$$\mathcal{M}(f) \;=\; (X \times I \coprod Y)/\{(x,1) \sim f(x) \in Y \,|\, x \in X\}\;.$$

(ii) The *double mapping cylinder* of maps $f : X \longrightarrow Y$, $f' : X \longrightarrow Y'$ is the identification space

$$\mathcal{M}(f,f') \;=\; (X \times I \coprod Y \coprod Y')/\!\sim$$

with $\sim$ the equivalence relation on the disjoint union $X \times I \coprod Y \coprod Y'$ generated by

$$(x,0) \sim f(x) \in Y \quad , \quad (x',1) \sim f'(x') \in Y' \quad (x,x' \in X)\;. \qquad \square$$

The double mapping cylinder can also be expressed as a union of single mapping cylinders

$$\mathcal{M}(f,f') \;=\; \mathcal{M}(f) \cup_X \mathcal{M}(f')\;.$$

Proposition 12.2 *The double mapping cylinder $\mathcal{M}(f,f')$ (12.1) of maps $f : X \longrightarrow Y$, $f' : X \longrightarrow Y'$ fits into a homotopy commutative square*

$$\begin{array}{ccc} X & \xrightarrow{\;f\;} & Y \\ {\scriptstyle f'}\big\downarrow & & \big\downarrow \\ Y' & \longrightarrow & \mathcal{M}(f,f') \end{array}$$

which is a homotopy pushout, *with the universal property that for any homotopy commutative square*

$$\begin{array}{ccc} X & \xrightarrow{\;f\;} & Y \\ {\scriptstyle f'}\big\downarrow & & \big\downarrow{\scriptstyle g} \\ Y' & \xrightarrow{\;g'\;} & Z \end{array}$$

there is defined a map uniquely up to homotopy

$$(g,g') \;:\; \mathcal{M}(f,f') \longrightarrow Z$$

which fits into a homotopy commutative square

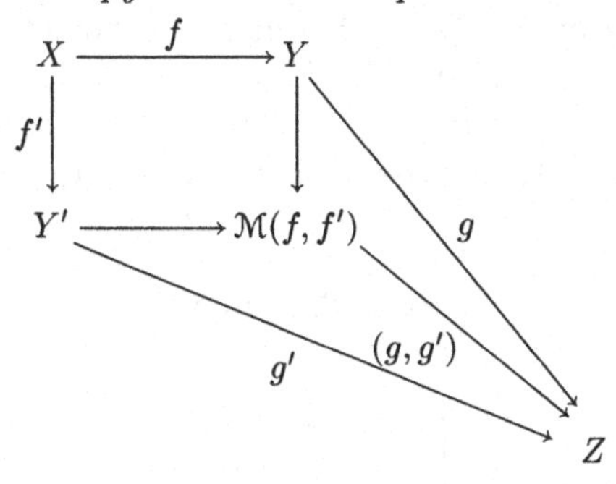

□

We shall also need the following non-standard mapping cylinders, using the 'teardrop topology' of Hughes, Taylor, Weinberger and Williams [76]:

Definition 12.3 (i) The *teardrop mapping cylinder* $\mathrm{cyl}(f)$ of a map $f : X \longrightarrow Y$ is the topological space defined by the set $\mathcal{M}(f)$ with the minimal topology such that:

(a) the inclusion $X \times [0,1) \longrightarrow \mathrm{cyl}(f)$ is an open embedding,
(b) the map

$$c \;:\; \mathcal{M}(f) \longrightarrow Y \times I \;;\; \begin{cases} (x,t) \longrightarrow (f(x),t) & \text{if } (x,t) \in X \times [0,1) \;, \\ y \longrightarrow y & \text{if } y \in Y \end{cases}$$

is continuous.

(ii) The *teardrop double mapping cylinder of* f of maps $f : X \longrightarrow Y$, $f' : X \longrightarrow Y'$ is the topological space defined by the set $\mathcal{M}(f, f')$ with the topology given by $\mathrm{cyl}(f, f')$

$$\mathrm{cyl}(f, f') \;=\; \mathrm{cyl}(f) \cup_X \mathrm{cyl}(f') \;.$$

□

Note that the identity $\mathcal{M}(f) \longrightarrow \mathrm{cyl}(f)$ is continuous, but not in general a homeomorphism. For example, $\mathcal{M}(\mathbb{R} \longrightarrow \{*\})$ is not homeomorphic to $\mathrm{cyl}(\mathbb{R} \longrightarrow \{*\})$.

Proposition 12.4 (i) *For any map* $f : X \longrightarrow Y$ *the identity map* $\mathcal{M}(f) \longrightarrow \mathrm{cyl}(f)$ *is a homotopy equivalence* rel Y.

(ii) *For any maps* $f : X \longrightarrow Y$, $f' : X \longrightarrow Y'$ *the identity map* $\mathcal{M}(f, f') \longrightarrow \mathrm{cyl}(f, f')$ *is a homotopy equivalence* rel $Y \cup Y'$.

Proof (i) In fact, it is easily seen that Y is a strong deformation retract of both $\mathcal{M}(f)$ and $\mathrm{cyl}(f)$.

(ii) The linear map $I \longrightarrow I$ which takes $[0, \frac{1}{3}]$ to $\{0\}$ and $[\frac{2}{3}, 1]$ to $\{1\}$ induces a map $X \times I \longrightarrow X \times I$ which in turn induces a map $\mathrm{cyl}(f, f') \longrightarrow \mathcal{M}(f, f')$ which is the identity on $Y \cup Y'$. This map is a homotopy inverse of the identity $\mathcal{M}(f, f') \longrightarrow \mathrm{cyl}(f, f')$ rel $Y \cup Y'$. □

Proposition 12.5 *Let W be a σ-compact metric space, and let*

$$\begin{array}{ccc} e(W) & \xrightarrow{k} & \{\infty\} \\ {\scriptstyle p_W}\downarrow & & \downarrow \\ W & \xrightarrow{i} & W^\infty \end{array}$$

be the homotopy commutative square given by 1.5.

(i) *W is forward tame if and only if there exists a homotopy equivalence of triads*

$$(W^\infty; W, \{\infty\}) \simeq (\mathrm{cyl}(p_W, k); \mathrm{cyl}(p_W, k)\backslash\{\infty\}, \{\infty\}) .$$

(ii) *If W is forward tame then the induced map*

$$\mathcal{M}(p_W, k) \longrightarrow W^\infty$$

is a homotopy equivalence rel ∞. *In particular, the square is a homotopy pushout.*

Proof (i) For this proof we replace the I coordinates in the double mapping cylinder by $[-\infty, \infty]$ coordinates. That is, write

$$\mathrm{cyl}(p_W, k) = (W \cup e(W) \times [-\infty, \infty] \cup \{\infty\})/\sim$$

with

$$(x, -\infty) \sim p_W(x) \in W \quad , \quad (x, \infty) \sim \infty \quad (x \in e(W)) .$$

Let $\pi : \mathrm{cyl}(p_W, k) \longrightarrow [-\infty, \infty]$ be the natural map so that $\pi^{-1}([0, \infty]) = \mathrm{cyl}(k)$ and $\pi^{-1}([-\infty, 0]) = \mathrm{cyl}(p_W)$. Then $\pi^{-1}([0, \infty)) = \mathrm{cyl}(k)\backslash\{\infty\}$ can be identified with $e(W) \times [0, \infty)$. Note that $\mathrm{cyl}(k)$ has the minimal topology such that $e(W) \times [0, \infty) \longrightarrow \mathrm{cyl}(k)$ is an open embedding and $\pi|\mathrm{cyl}(k) \longrightarrow [0, \infty]$ is continuous.

Suppose now that

$$f : (W^\infty; W, \{\infty\}) \longrightarrow (\mathrm{cyl}(p_W, k); \mathrm{cyl}(p_W, k)\backslash\{\infty\}, \{\infty\})$$

is a homotopy equivalence of triads with inverse g. Let

$$Y = f^{-1}\pi^{-1}([0, \infty))$$

and note that Y is a closed cocompact subspace of W. The proper map

$$[0, \infty) \times [0, \infty) \longrightarrow [0, \infty) \; ; \; (s, t) \longrightarrow s + t$$

induces a map

$$q' : \pi^{-1}([0,\infty)) \times [0,\infty) \longrightarrow \pi^{-1}([0,\infty))$$

which extends the inclusion. Let

$$q = gq'(f|Y \times \mathrm{id}_{[0,\infty)}) : Y \times [0,\infty) \longrightarrow W .$$

Note that $q|Y \times \{0\} = gf|Y$ and, hence, is properly homotopic to the inclusion $Y \longrightarrow W$. Therefore, if q is proper, then q is properly homotopic to a proper map $Y \times [0,\infty) \longrightarrow W$ which extends the inclusion $Y \longrightarrow W$, showing that W is forward tame. The argument will be completed by showing that q is proper. To this end let $K \subseteq W$ be compact. Since $g^{-1}(W \backslash K)$ is a neighbourhood of ∞, there exists $t_1 \geq 0$ with

$$g^{-1}(K) \subseteq \pi^{-1}([-\infty, t_1]) .$$

There also exists $t_2 \geq 0$ with

$$(q')^{-1}\pi^{-1}([0,t_1]) \subseteq \pi^{-1}([0,t_1]) \times [0,t_2] .$$

Thus $q^{-1}(K) \subseteq f^{-1}\pi^{-1}([0,t_1]) \times [0,t_2]$. Observe that $f^{-1}\pi^{-1}([0,t_1])$ is compact because $f^{-1}\pi^{-1}(t_1,\infty]$ is an open neighbourhood of ∞. Thus $q^{-1}(K)$ is a closed subset of a compact subspace, hence compact.

Conversely, assume that W is forward tame. Let $U \subseteq W$ be a closed cocompact subspace for which there is a proper map $q : U \times [0,\infty) \longrightarrow W$ extending the inclusion. Let $p : W \longrightarrow [-\infty,\infty)$ be a proper map such that $p(W \backslash U) = \{-\infty\}$ and let $Y = p^{-1}([0,\infty))$. According to 9.13 there exist maps

$$f : Y \longrightarrow e(W) \times [0,\infty) \ , \ g : e(W) \times [0,\infty) \longrightarrow Y$$

and homotopies

$$F : igf \simeq i : Y \longrightarrow W ,$$
$$G : fg \simeq \text{identity} : e(W) \times [0,\infty) \longrightarrow e(W) \times [0,\infty) ,$$

with $i : Y \longrightarrow W$ the inclusion. The properties of f, g, F, G listed in 9.13 imply that there are continuous extensions

$$f^+ : (Y^\infty, \{\infty\}) \longrightarrow (\mathrm{cyl}(k), \{\infty\}) ,$$
$$g^+ : (\mathrm{cyl}(k), \{\infty\}) \longrightarrow (Y^\infty, \{\infty\}) ,$$
$$F^+ : i^+g^+f^+ \simeq i^+ : Y^\infty \longrightarrow W^\infty ,$$
$$G^+ : f^+g^+ \simeq \text{identity} : \mathrm{cyl}(k) \longrightarrow \mathrm{cyl}(k) ,$$

with $i^+ : Y^\infty \longrightarrow W^\infty$ the inclusion and the homotopies F^+, G^+ both rel $\{\infty\}$. Extend f^+ to $\widetilde{f} : W^\infty \longrightarrow \mathrm{cyl}(p_W, k)$ by

$$\widetilde{f}| : p^{-1}([-\infty,0]) \longrightarrow \mathrm{cyl}(p_W) \ ; \ x \longrightarrow \begin{cases} [\widehat{q}(x), p(x)] & \text{if } x \in p^{-1}(-\infty, 0] , \\ x & \text{if } x \in p^{-1}(-\infty) \end{cases}$$

with $\widehat{q} : U \longrightarrow e(W)$ the adjoint of q. Note from the proof of 9.13 that $g(\omega, 0) = (j\omega)(0)$ for each $\omega \in e(W)$ where $j : e(W) \longrightarrow e(Y')$ is a homotopy inverse of the inclusion $e(i) : e(W) \longrightarrow e(Y')$ for a certain subspace $Y' \subseteq Y$. Let $k : e(i)j \simeq \text{identity}_{e(W)}$. Extend g^+ to $\widetilde{g} : \text{cyl}(p_W, k) \longrightarrow W^\infty$ by

$$\begin{aligned} \widetilde{g}| \ : \ \text{cyl}(p_W) &\longrightarrow W \ ; \\ x &\longrightarrow x \text{ if } x \in W \ , \\ [\omega, t] &\longrightarrow \begin{cases} k(\omega, -t)(0) & \text{if } -1 \leq t \leq 0 \ , \\ \omega(0) & \text{if } -\infty \leq t \leq -1 \end{cases} \quad (\omega \in e(W)) \ . \end{aligned}$$

In order to extend F^+ to a homotopy

$$\widetilde{F} \ : \ \widetilde{g}\widetilde{f} \ \simeq \ \text{id} \ : \ W^\infty \longrightarrow W^\infty \quad \text{rel}\,\{\infty\} \ ,$$

note that $\widetilde{g}\widetilde{f}| : p^{-1}([-\infty, -1]) \longrightarrow W^\infty$ is the inclusion and

$$\widetilde{g}\widetilde{f}| \ : \ p^{-1}([-1, 0]) \longrightarrow W^\infty \ ; \ x \longrightarrow k(\widehat{q}(x), -p(x))(0) \ .$$

Thus

$$\begin{aligned} \widetilde{F}| \ &: p^{-1}([-\infty, 0]) \times I \longrightarrow W \ ; \\ (x, t) &\longrightarrow \begin{cases} k(\widehat{q}(x), t - p(x))(0) & \text{if } t - p(x) \leq 1 \ , \\ x & \text{if } t - p(x) \geq 1 \end{cases} \end{aligned}$$

is a continuous extension of F^+ (one must use the explicit formula for F in 9.13). In order to extend G^+ to a homotopy

$$\widetilde{G} \ : \ \widetilde{f}\widetilde{g} \ \simeq \ \text{id} \ : \ \text{cyl}(p_W, k) \longrightarrow \text{cyl}(p_W, k) \quad \text{rel}\,\{\infty\}$$

note that

$$\begin{aligned} \widetilde{f}\widetilde{g}| \ : W \ &= \ \pi^{-1}(-\infty) \longrightarrow \text{cyl}(p_W, k) \ ; \\ x &\longrightarrow \begin{cases} [\widehat{q}(x), p(x)] \in \text{cyl}(p_W, k) \backslash \{\infty\} & \text{if } x \in p^{-1}(-\infty, \infty) \ , \\ x & \text{if } x \in p^{-1}(-\infty) \ . \end{cases} \end{aligned}$$

Thus $\widetilde{f}\widetilde{g}| : W \longrightarrow \text{cyl}(p_W, k)$ is homotopic to the inclusion. Now use the fact that

$$(W \cup \text{cyl}(k)) \times I \cup (\text{cyl}(p_W, k) \times \{0\})$$

is a strong deformation retract of $\text{cyl}(p_W, k) \times I$ in order to extend G^+.

(ii) By 12.4 (ii) the identity $\mathcal{M}(p_W, k) \longrightarrow \text{cyl}(p_W, k)$ is a homotopy equivalence rel $\{\infty\}$. By (i) there is a homotopy equivalence $\widetilde{g} : \text{cyl}(p_W, k) \longrightarrow W^\infty$ rel $\{\infty\}$. The composition $\mathcal{M}(p_W, k) \longrightarrow W^\infty$ is homotopic rel $\{\infty\}$ to the induced map. □

The following example shows that it is necessary to be careful about the topology on the mapping cylinders in Proposition 12.5.

Example 12.6 Let

$$W_j = \{(x,j) \in \mathbb{R}^2 \mid x \geq 0\} \quad (j = 0,1,2,\dots)\ ,$$

and define

$$W = \coprod_{j=0}^{\infty} W_j \subset \mathbb{R}^2\ .$$

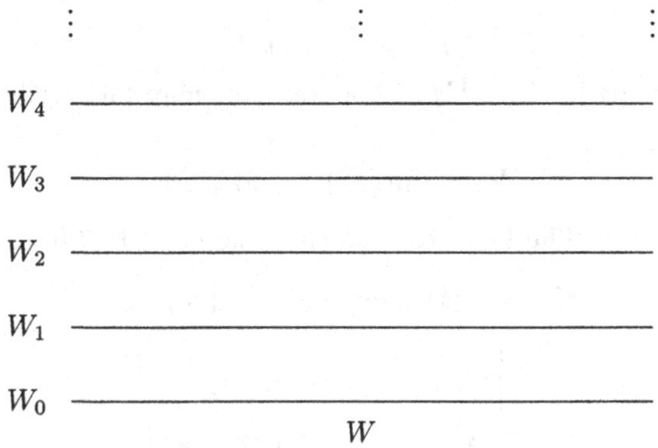

$$\begin{array}{cccc} & \vdots & \vdots & \vdots \\ W_4 & \multicolumn{3}{c}{\rule{6cm}{0.4pt}} \\ W_3 & \multicolumn{3}{c}{\rule{6cm}{0.4pt}} \\ W_2 & \multicolumn{3}{c}{\rule{6cm}{0.4pt}} \\ W_1 & \multicolumn{3}{c}{\rule{6cm}{0.4pt}} \\ W_0 & \multicolumn{3}{c}{\rule{6cm}{0.4pt}} \\ & & W & \end{array}$$

Then W is a forward tame σ-compact metric space. However, the triads $(W^\infty; W, \{\infty\})$ and $(\mathcal{M}(p_W,k); \mathcal{M}(p_W,k)\backslash\{\infty\}, \{\infty\})$ are not homotopy equivalent. For suppose

$$f : (W^\infty; W, \{\infty\}) \longrightarrow (\mathcal{M}(p_W,k); \mathcal{M}(p_W,k)\backslash\{\infty\}, \{\infty\})$$

is a homotopy equivalence. The end space of W is the disjoint union

$$e(W) = \coprod_{j=0}^{\infty} e(W_j)\ .$$

Let $\mathcal{M}_j = \mathcal{M}(p_{W_j}, k|)$ so that $\mathcal{M}(p_W,k) = \bigcup \mathcal{M}_j$. The identification topology insures that no sequence of points from distinct $\mathcal{M}_j$'s converges to $\infty \in \mathcal{M}(p_W,k)$. Let $x_j = (0,j) \in W_j \subseteq W$ so that $x_j \longrightarrow \infty \in W^\infty$. Since f restricts to a homotopy equivalence $W \longrightarrow \mathcal{M}(p_W,k)\backslash\{\infty\}$, for each i there exists a unique $j(i)$ such that $f(W_i) \subseteq \mathcal{M}_{j(i)}$. It follows that the sequence $\{f(x_i)\} \subseteq \mathcal{M}(p_W,k)$ does not converge to $\infty \in \mathcal{M}(p_W,k)$, a contradiction. □

Remark 12.7 The characterization of forward tameness in terms of homotopy pushouts was first obtained by Quinn [116] – the use of the teardrop mapping cylinder in 12.5 corrects certain technical deficiencies in the statement of [116]. □

Example 12.8 (i) As in 7.3 (i) let

$$W = L \cup K \times [0,\infty)$$

for a pair $(L, K \subseteq L)$ of compact spaces, so that W is forward collared, $W \simeq L$, $W^\infty = L \cup cK$ is the mapping cone of the inclusion $K \longrightarrow L$, and $e(W) \simeq K$. The homotopy pushout square of 12.5 is given up to homotopy equivalence by

$$\begin{array}{ccc} K & \longrightarrow & \{\infty\} \\ \big\downarrow & & \big\downarrow \\ L & \longrightarrow & L \cup cK \end{array}$$

(ii) As in 7.3 (ii) let $(M, \partial M)$ be a compact manifold with boundary, so that

$$W \;=\; \mathrm{int}(M) \;=\; M \backslash \partial M$$

is forward collared. The homotopy pushout square of 12.5 is given by

$$\begin{array}{ccc} e(W) \simeq \partial M & \longrightarrow & \{\infty\} \\ \big\downarrow & & \big\downarrow \\ W \simeq M & \longrightarrow & W^\infty = M/\partial M \end{array}$$

(iii) As in 7.3 (iii) let η be a real n-plane vector bundle over a compact space K, so that the total space $W = E(\eta)$ is forward collared. The homotopy pushout square of 12.5 is given by

$$\begin{array}{ccc} e(W) \simeq S(\eta) & \longrightarrow & \{\infty\} \\ \big\downarrow & & \big\downarrow \\ W \simeq K & \longrightarrow & W^\infty = T(\eta) \end{array}$$

□

Example 12.9 As in 1.13 and 3.10 (iii) let $W = K \times \mathbb{R}^n$ $(n \geq 1)$ for a compact space K. Then $W = E(\epsilon^n)$ is the total space of the trivial n-plane bundle ϵ^n over K, so that 7.3 (iii) applies to show that W is forward collared with

$$\begin{aligned} W^\infty &= T(\epsilon^n) = (K \times D^n)/(K \times S^{n-1}) = \Sigma^n K^\infty , \\ e(W) &\simeq S(\epsilon^n) = K \times S^{n-1} . \end{aligned}$$

The homotopy pushout square of 12.5 is given by

$$\begin{array}{ccc} e(W) \simeq K \times S^{n-1} & \longrightarrow & \{\infty\} \\ \big\downarrow & & \big\downarrow \\ W \simeq K \times D^n & \longrightarrow & W^\infty = \Sigma^n K^\infty \end{array}$$

□

Example 12.10 (i) Let X be a locally compact polyhedron. For any $x \in X$ there exists a locally finite triangulation of X with x a vertex, and such that

$$(Y, Z) = (\mathrm{star}(x), \mathrm{link}(x))$$

is a compact polyhedral pair with $Y = x * Z$ a cone on Z (as in 1.8). The open star

$$W = Y \backslash Z$$

is contractible and forward collared with

$$W^\infty = Y/Z \ , \ e(W) \simeq Z \ ,$$

$$\widetilde{H}_*(e(W)) = \widetilde{H}_*(Z) = H^{lf}_{*+1}(W) = H_{*+1}(Y, Z) = H_{*+1}(X, X\backslash\{x\}) \ .$$

The homotopy pushout square of 12.2 gives a homotopy equivalence

$$\Sigma e(W) \simeq W^\infty ,$$

which corresponds to $\Sigma Z \simeq Y/Z$.

(ii) If X is a homology (resp. combinatorial) n-manifold and Y, Z, W are as in (i) then (Y, Z) is homology (resp. homotopy) equivalent to (D^n, S^{n-1}), $e(W)$ is a homology (resp. homotopy) $(n-1)$-sphere and W^∞ is a homology (resp. homotopy) n-sphere.

(iii) If X is a compact polyhedron and x, Z are as in (i) then $V = X\backslash\{x\}$ is forward collared with

$$V^\infty = X \ , \ e(V) \simeq Z \ , \ H^{lf}_*(V) = H_*(X, \{x\}) \ . \qquad \square$$

Remark 12.11 The *homotopy link* of a subspace $Y \subseteq X$ is defined by Quinn [116] to be

$$\mathrm{holink}(X, Y) = \{\omega \in X^I \,|\, \omega[0,1) \in X\backslash Y \ , \ \omega(1) \in Y\} \ .$$

(See Appendix B for historical background on the homotopy link.) The evaluation maps

$$\mathrm{holink}(X, Y) \longrightarrow X\backslash Y \ ; \ \omega \longrightarrow \omega(0) \ ,$$
$$\mathrm{holink}(X, Y) \longrightarrow Y \ ; \ \omega \longrightarrow \omega(1)$$

fit into a homotopy commutative square

$$\begin{array}{ccc} \mathrm{holink}(X,Y) & \longrightarrow & Y \\ \downarrow & & \downarrow \\ X\backslash Y & \longrightarrow & X \end{array}$$

which is a homotopy pushout in many situations of geometric interest. In particular, this is the case if (X, Y) is a CW pair with $X = E(\nu) \cup_{S(\nu)} Z$ for a (D^k, S^{k-1})-fibration $(E(\nu), S(\nu))$ over Y, with $Y \subset E(\nu)$ the zero section and $Z = \mathrm{cl}(X \backslash E(\nu))$, and with homotopy equivalences

$$Z \simeq X \backslash Y \ , \ S(\nu) \simeq \mathrm{holink}(X,Y) \ .$$

Thus if X is a closed n-dimensional manifold and $Y \subseteq X$ is a closed $(n-k)$-dimensional submanifold with a normal topological k-block bundle $\nu = \nu_{Y\subset X} : Y \longrightarrow B\widetilde{TOP}(k)$ then $\mathrm{holink}(X,Y)$ is homotopy equivalent to the total space $S(\nu)$ of the corresponding $(k-1)$-sphere bundle over Y.

We shall be mainly concerned with the situation in which $(X, Y \subseteq X)$ is a pair of spaces such that X is compact, $Y \subseteq X$ is closed and the complement

$$W = X \backslash Y \subseteq X$$

is a dense open subset of X – thus X is a compactification of W and $Y = X \backslash W$ is the 'space at ∞'. Here are some special cases:

(i) For any space W the homotopy link of $(X,Y) = (W^\infty, \{\infty\})$ is the end space of Chapter 1

$$\mathrm{holink}(W^\infty, \{\infty\}) = e(W) \ ,$$

with a homotopy commutative square

$$\begin{array}{ccc} e(W) & \longrightarrow & \{\infty\} \\ \downarrow & & \downarrow \\ W & \longrightarrow & W^\infty \end{array}$$

which is a homotopy pushout if W forward tame (12.5).

(ii) If W is a space with a finite number k of forward tame ends and $(X,Y) = (W^*, \{1,2,\ldots,k\})$ with W^* the Freudenthal compactification of W (1.23). In fact, (i) with W forward tame is just the case $k=1$.

(iii) W is an n-dimensional Hadamard manifold (= a simply-connected complete Riemannian manifold of nonpositive curvature), $Y = \partial X$ is the boundary of a compact n-dimensional manifold with boundary $(X, \partial X)$, such that up to homeomorphism

$$W = \mathbb{R}^n \ , \ X = D^n \ , \ \partial X = Y = X \backslash W = S^{n-1}$$

with $\partial X = S^{n-1}$ the sphere at ∞ – see Ballmann, Gromov and Schroeder [3, pp. 15–22]. In this case W is forward and reverse collared, and there are defined homotopy equivalences

$$\mathrm{holink}(X,Y) \simeq e(W) \simeq Y = S^{n-1} \ , \ W \simeq X = D^n \ . \qquad \square$$

Part Two: Topology over the real line

13

Infinite cyclic covers

The non-compact spaces of greatest interest to us are equipped with a proper map to $\mathbb{R}$. In this chapter we shall be particularly concerned with the infinite cyclic cover $\overline{W}$ of a compact space W classified by a map $c : W \longrightarrow S^1$, which lifts to a proper map $\overline{c} : \overline{W} = c^*\mathbb{R} \longrightarrow \mathbb{R}$. We prove (!) that if W and $\overline{W}$ are connected and W is sufficiently nice (such as a compact ANR) then $\overline{W}$ has two ends with (closed) neighbourhoods

$$\overline{W}^+ = \overline{c}^{-1}[0,\infty) \ , \ \overline{W}^- = \overline{c}^{-1}(-\infty,0] \subset \overline{W}$$

such that

$$\overline{W} = \overline{W}^+ \cup \overline{W}^-$$

with $\overline{W}^+ \cap \overline{W}^- = \overline{c}^{-1}(0)$ compact.

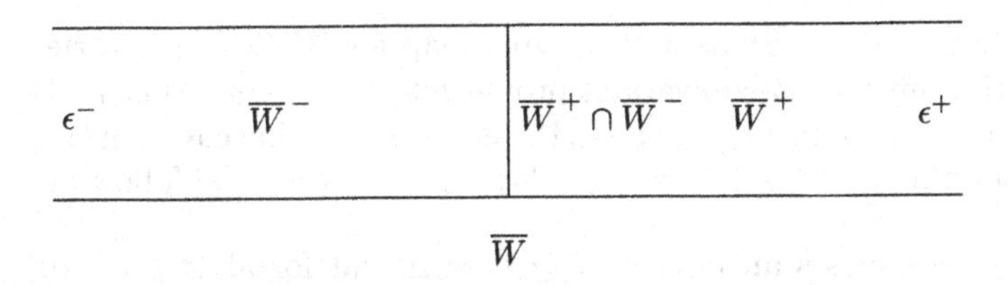

The main result of this chapter is a geometric duality between forward and reverse tameness for the ends of an infinite cyclic cover $\overline{W}$ of a compact ANR W: in 13.13 it is shown that $\overline{W}^+$ is forward tame if and only if $\overline{W}^-$ is reverse tame.

In Chapter 15 we shall study 'bands' (W, c), which are compact spaces W with a map $c : W \longrightarrow S^1$ such that the infinite cyclic cover $\overline{W} = c^*\mathbb{R}$ of W is finitely dominated. It will be shown there that for an ANR band (W, c) the end spaces are such that

$$e(\overline{W}) \simeq \overline{W} \amalg \overline{W} \ , \ e(\overline{W}^+) \simeq e(\overline{W}^-) \simeq \overline{W} .$$

In Chapter 17 it will be shown that a manifold end V which is both forward and reverse tame has an open neighbourhood $\overline{W} \subset V$ which is the infinite cyclic cover of a manifold band (W, c), with $e(V) \simeq \overline{W}$.

Every infinite cyclic cover $\overline{p} : \overline{W} \longrightarrow W$ is induced from $p : \mathbb{R} \longrightarrow S^1$ by a map $c : W \longrightarrow S^1 = I/(0 = 1)$, with

$$\overline{W} = \{(x, t) \in W \times \mathbb{R} \,|\, c(x) = [t] \in S^1\}$$

and generating covering translation

$$\zeta : \overline{W} \longrightarrow \overline{W} \; ; \; (x, t) \longrightarrow (x, t + 1) \, .$$

Proposition 13.1 *The number of path components in an infinite cyclic cover $\overline{W}$ of a path-connected space W is the index of the subgroup* $\mathrm{im}(c_* : \pi_1(W) \longrightarrow \pi_1(S^1))$ *in* $\pi_1(S^1) = \mathbb{Z}$. □

In particular, $\overline{W}$ is path-connected if and only if $c_* : \pi_1(W) \longrightarrow \pi_1(S^1)$ is onto.

The following result is a slight generalization of a result of Hopf [68].

Proposition 13.2 *If $\overline{W}$ is a connected infinite cyclic cover of a compact, path-connected, locally path-connected Hausdorff space W, then $\overline{W}$ has exactly two ends ϵ^+, ϵ^-. Each end is path-connected at ∞, with closed connected neighbourhoods $\overline{W}^+, \overline{W}^- \subset \overline{W}$ such that $\overline{W}^+ \cup \overline{W}^- = \overline{W}$ with $\overline{W}^+ \cap \overline{W}^-$ compact.*

Proof Let $c : W \longrightarrow S^1$ be a classifying map for W (which is surjective on π_1), with a lift to a $\mathbb{Z}$-equivariant proper map $\overline{c} : \overline{W} \longrightarrow \mathbb{R}$. Let $p : \overline{W} \longrightarrow W$ be the covering map. Let $\zeta : \overline{W} \longrightarrow \overline{W}$ be the generating covering translation corresponding to $+1 \in \mathbb{Z} = \pi_1(S^1)$. We need to verify the following:

Claim There exists an integer $N \geq 0$ such that for all $x \in \overline{c}^{-1}(0)$, there exists a path α_x in $\overline{W}$ from x to $\zeta^1 x$ with the image of $\overline{c} \circ \alpha_x$ in $[-N, N] \subset \mathbb{R}$.

Proof For each $y \in c^{-1}(1)$ choose a loop β_y based at y such that $c_*[\beta_y] = +1 \in \pi_1(S^1, 1)$ and choose a path-connected open subspace $U_y \subseteq W$ such that $y \in U_y$ and $c(U_y) \neq S^1$. Choose finitely many $y_1, y_2, \ldots, y_n \in c^{-1}(1)$ such that $\{U_{y_i}\}_{i=1}^n$ covers $c^{-1}(1)$. Let $U_i = U_{y_i}$. Let $x_i = p^{-1}(y_i) \cap \overline{c}^{-1}(0)$ for $i = 1, 2, \ldots, n$. Let α_i be a lift of β_{y_i} from x_i to $\zeta^1 x_i$ for $i = 1, 2, \ldots, n$. For any other $x \in \overline{c}^{-1}(0)$, choose $i(x) \in \{1, 2, \ldots, n\}$ such that $p(x) \in U_{i(x)}$ and choose a path γ_x in $U_{i(x)}$ from $p(x)$ to $y_{i(x)}$. Then let

$$\alpha_x = \tilde{\gamma}_x * \alpha_{i(x)} * \zeta^1 \tilde{\gamma}_x^{-1}$$

where $\tilde{\gamma}_x$ is a lift of γ_x from x to $x_{i(x)}$. □

Returning to the proof of 13.2, define for each $x \in \overline{c}^{-1}(0)$ a proper map

$$\omega_x \; : \; [0,\infty) \longrightarrow \overline{W} \; ; \; t \longrightarrow \zeta^i \alpha_x(t-i) \quad (i \leq t < i+1)$$

where α_x is the path constructed in the claim above. Clearly, ω_x shows that $\overline{W}$ has an end at $+\infty$. A similar construction shows that $\overline{W}$ has an end at $-\infty$, so that $\overline{W}$ has at least two ends.

To see that $\overline{W}$ has exactly two ends suppose that C is an unbounded component of $\overline{W}\backslash\overline{c}^{-1}([-k,k])$ for some $k \in \mathbb{Z}^+$ and assume without loss of generality that $C \subseteq \overline{c}^{-1}(k,\infty)$. Fix $x_0 \in \overline{c}^{-1}(0)$ and let $\omega_0 = \omega_{x_0}$ be as defined in the previous paragraph. We shall show that C is the unbounded component determined by ω_0. The latter component contains $\zeta^n x_0$ for all $n > k$. Choose $x_1 \in \overline{c}^{-1}(0)$ such that for some $m > N+k$, $\zeta^m x_1 \in C$. Then

$$\omega_{x_1}([m,\infty)) \subseteq \overline{W}\backslash\overline{c}^{-1}([-k,k]) \quad , \quad \omega_{x_1}([m,\infty)) \subseteq C \ ,$$

and so $\zeta^p x_1 \in C$ for each $p \geq m$. Choose a path β in $\overline{W}$ from x_0 to $\zeta^n x_1$ for some $n \in \mathbb{Z}$ (which is obtained by lifting a path in W from $p(x_0)$ to $p(x_1)$). Choose $l \in \mathbb{Z}^+$ such that the image of $\zeta^l \beta$ lies in $\overline{c}^{-1}(k,\infty)$. If $L = \max\{l,m\}$, then $\zeta^L \beta$ is a path in $\overline{c}^{-1}(k,\infty)$ from $\zeta^L x_0$ to $\zeta^{L+n} x_1 \in C$.

To see that the end of $\overline{W}$ at $+\infty$ (for example) is path connected at ∞, let $k \in \mathbb{Z}^+$, choose $x_0 \in \overline{c}^{-1}(0)$ and let $\omega_0 = \omega_{x_0} : [0,1) \longrightarrow \overline{W}$ be the proper map constructed above. There exists an integer $n \geq 0$ such that $\omega([n,\infty)) \subseteq \overline{c}^{-1}(k,\infty)$. Let V be the component of $\overline{c}^{-1}(k,\infty)$ which contains $\omega([n,\infty))$. It follows that V is path-connected and is the unique unbounded component of $\overline{c}^{-1}(k,\infty)$. That is, V is a path-connected neighbourhood of the end of $\overline{W}$ at $+\infty$, and $\overline{W}^+ = \text{cl}(V)$ is a connected closed neighbourhood of the end. □

Corollary 13.3 *A connected infinite cyclic cover $\overline{W}$ of a connected finite CW complex W has two ends ϵ^+, ϵ^-.* □

Definition 13.4 A *fundamental domain* $(V;U,\zeta U)$ for an infinite cyclic cover $\overline{W}$ of a space W is a subspace $V \subset \overline{W}$ such that

$$\zeta^{-1}V \cap V \cap \zeta V \;=\; \emptyset \;\; , \;\; \bigcup_{j=-\infty}^{\infty} \zeta^j V \;=\; \overline{W} \; ,$$

with $\zeta : \overline{W} \longrightarrow \overline{W}$ the covering translation and

$$U \;=\; V \cap \zeta^{-1} V \subset \overline{W} \; .$$

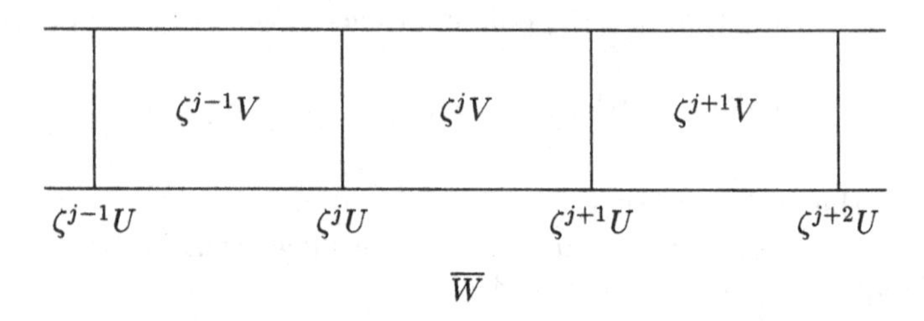

The subspaces defined by

$$\overline{W}^+ = \bigcup_{j=0}^{\infty} \zeta^j V \ , \ \overline{W}^- = \bigcup_{j=-\infty}^{-1} \zeta^j V \subset \overline{W}$$

are then such that

$$U = \overline{W}^+ \cap \overline{W}^- \ , \ V = \overline{W}^+ \cap \zeta \overline{W}^- \ , \ \overline{W} = \overline{W}^+ \cup \overline{W}^- \ .$$

If W is a CW complex and $\zeta : \overline{W} \longrightarrow \overline{W}$ is cellular, then a *CW fundamental domain* is a fundamental domain $(V; U, \zeta U)$ such that $V, U, \zeta U$ are subcomplexes of $\overline{W}$. Similarly, if W is an ANR then an *ANR fundamental domain* is a fundamental domain $(V; U, \zeta U)$ such that $V, U, \zeta U$ are ANR's. □

Example 13.5 Let W be a space with a decomposition of the form

$$W = U \times I \cup V$$

such that

$$(U \times I) \cap V = U \times \{0, 1\}$$

and such that there is given a map

$$c : W \longrightarrow S^1 = I/(0 = 1)$$

with

$$U \times I = c^{-1}[0, 1/2] \ , \ V = c^{-1}[1/2, 1]$$

and

$$c(u, t) = t/2 \ (0 \leq t \leq 1 \ , \ u \in U) \ .$$

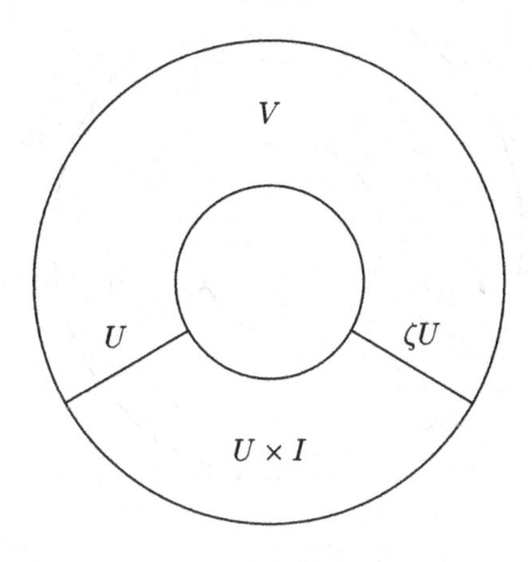

The map $c : W \longrightarrow S^1$ classifies an infinite cyclic cover $\overline{W} = c^*\mathbb{R}$ of W with fundamental domain $(V; U \times \{0\}, U \times \{1\})$. The construction of $\overline{W}$ by cutting W along U and glueing together $\mathbb{Z}$ copies of the fundamental domain generalizes the construction of the canonical infinite cyclic cover of a knot complement by cutting along a Seifert surface. □

Proposition 13.6 (i) *If $(V; U, \zeta U)$ is a fundamental domain for an infinite cyclic cover $\overline{W}$ of a compact space W then U and V are compact.*

(ii) *If a connected infinite cyclic cover $\overline{W}$ of an ANR W admits a fundamental domain $(V; U, \zeta U)$ with U, V connected then the subspaces $\overline{W}^+, \overline{W}^- \subset \overline{W}$ are neighbourhoods of the two ends ϵ^+, ϵ^-.* □

Definition 13.7 The *mapping coequalizer* of maps $f^+, f^- : U \longrightarrow V$ is the identification space

$$\begin{aligned} \mathcal{W}(f^+, f^-) &= U \times I \cup_{f^+ \cup f^-} V \\ &= (U \times I \amalg V)/\{(x,0) = f^-(x), (x,1) = f^+(x) \,|\, x \in U\} . \end{aligned}$$

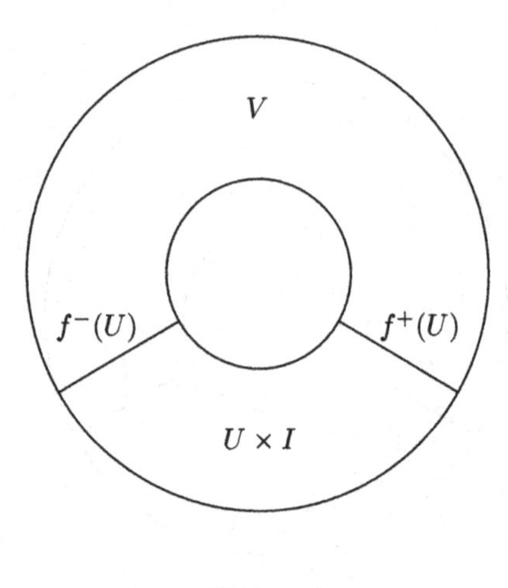

$\mathcal{W}(f^+, f^-)$

Given a commutative diagram of maps

$$\begin{array}{ccccc}
U & \xrightarrow{f^+} & V & \xleftarrow{f^-} & U \\
\big\downarrow{\scriptstyle g} & & \big\downarrow{\scriptstyle h} & & \big\downarrow{\scriptstyle g} \\
U' & \xrightarrow{f'^+} & V' & \xleftarrow{f'^-} & U'
\end{array}$$

let $(g, h) : \mathcal{W}(f^+, f^-) \longrightarrow \mathcal{W}(f'^+, f'^-)$ be the induced map of identification spaces. □

A map $f : U \longrightarrow V$ is the inclusion of a collared subspace (1.6) if it extends to an open embedding $f : L \times [0, \infty) \longrightarrow K$.

Proposition 13.8 *If the maps $f^+, f^- : U \longrightarrow V$ are inclusions of disjoint collared subspaces, with extensions to disjoint embeddings*

$$\overline{f}^+ \,,\, \overline{f}^- \ : \ U \times [0, \infty) \longrightarrow V$$

the mapping coequalizer $\mathcal{W}(f^+, f^-)$ is such that there is defined a homeomorphism

$$\mathcal{W}(f^+, f^-) \ \cong \ V/\{f^+(x) = f^-(x) \,|\, x \in U\} \ ,$$

and $\mathcal{W}(f^+, f^-)$ has a canonical infinite cyclic cover

$$\overline{\mathcal{W}}(f^+, f^-) \ = \ \mathbb{Z} \times V/\{(j, f^-(x)) = (j+1, f^+(x)) \,|\, x \in U, j \in \mathbb{Z}\}$$

with generating covering translation

$$\zeta : \overline{W}(f^+, f^-) \longrightarrow \overline{W}(f^+, f^-) \ ; \ [j, x] \longrightarrow [j+1, x]$$

and fundamental domain $(V; f^+(U), f^-(U))$.

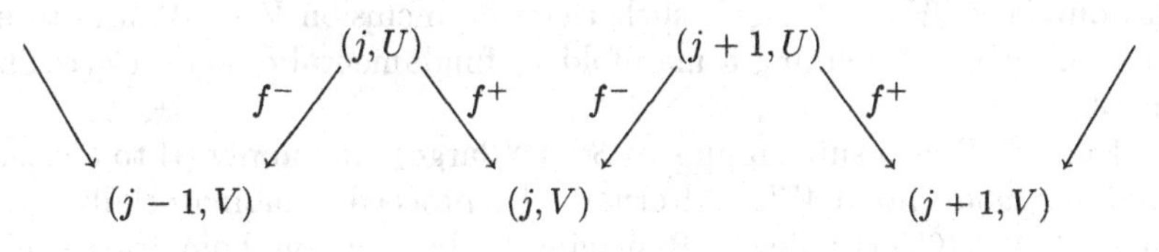

□

Remark 13.9 (i) The hypothesis of 13.8 is satisfied if V is replaced by the mapping cylinder of $f^+ \cup f^- : U \times \{0,1\} \longrightarrow V$.

(ii) If $(V; U, \zeta U)$ is a fundamental domain for an infinite cyclic cover $\overline{W}$ of a space W and the maps

$$f^+ \ : U \longrightarrow V \ ; \ x \longrightarrow x \ ,$$

$$f^- \ : U \longrightarrow V \ ; \ y \longrightarrow \zeta y$$

satisfy the hypothesis of 13.8 then

$$W = W(f^+, f^-) \ , \ \overline{W} = \overline{W}(f^+, f^-) \ .$$ □

Definition 13.10 A π_1*-fundamental domain* $(V; U, \zeta U)$ for an infinite cyclic cover $\overline{W}$ of a space W is a fundamental domain such that the inclusions $U \longrightarrow V$, $\zeta U \longrightarrow V$ induce isomorphisms

$$\pi_1(U) \cong \pi_1(V) \ , \ \pi_1(\zeta U) \cong \pi_1(V)$$

(on each component) in which case the inclusion $V \longrightarrow \overline{W}$ induces an isomorphism $\pi_1(V) \cong \pi_1(\overline{W})$ and $\pi_1(W) = \pi_1(\overline{W}) \times_{\zeta_*} \mathbb{Z}$ with

$$\zeta_* \ : \ \pi_1(\overline{W}) \cong \pi_1(V) \cong \pi_1(U) \cong \pi_1(\zeta U) \cong \pi_1(V) \cong \pi_1(\overline{W}) \ .$$ □

Proposition 13.11 (i) *An infinite cyclic cover* $\overline{W}$ *of a compact manifold* W *admits a manifold* π_1*-fundamental domain* $(V; U, \zeta U)$.

(ii) *For any finite CW complex* W *with an infinite cyclic cover* $\overline{W}$ *there exists a simple homotopy equivalence* $W \simeq W'$ *to a finite CW complex* W' *such that the induced infinite cyclic cover* $\overline{W}'$ *of* W' *admits a CW* π_1*-fundamental domain* $(V'; U', \zeta' U')$.

Proof (i) It is possible to make a classifying map $c : W \longrightarrow S^1$ transverse regular at a point $* \in S^1$, such that the codimension 1 submanifold

$$U = c^{-1}(*) \subset W$$

has an open regular neighbourhood $N(U) = U \times (-1, 1) \subset W$. Surgery on

U inside W ensures that $\pi_1(U) \cong \pi_1(\overline{W})$. The effect of cutting W along U is a codimension 0 submanifold

$$V = W \backslash N(U) \subset W$$

with boundary $\partial V = U \cup \zeta U$, such that the inclusion $V \subset W$ lifts to an embedding $V \subset \overline{W}$ defining a manifold π_1-fundamental domain $(V; U, \zeta U)$ as in 13.5.

(ii) Embed W as a subcomplex of S^N (N large), and apply (i) to a closed regular neighbourhood W'. Alternatively, proceed combinatorially as in Ranicki [124, 8.16], as follows. Replacing W by a simple homotopy equivalent finite CW complex (if necessary), it is possible to choose a finite subcomplex $V \subset \overline{W}$ such that

$$\bigcup_{j=-\infty}^{\infty} \zeta^j V = \overline{W} \ , \quad \pi_1(V \cap \zeta^{-1} V) \cong \pi_1(V) \cong \pi_1(\overline{W})$$

and the maps

$$f^+ : U = V \cap \zeta^{-1} V \longrightarrow V \times I \ ; \ x \longrightarrow (x, 0) \ ,$$
$$f^- : U = V \cap \zeta^{-1} V \longrightarrow V \times I \ ; \ x \longrightarrow (\zeta x, 1)$$

are the inclusions of disjoint subcomplexes inducing π_1-isomorphisms, and such that the mapping coequalizer is a finite CW complex

$$W' = \mathcal{W}(f^+, f^-) \ .$$

The projection

$$W' \longrightarrow \overline{W}/\zeta = W \ ; \ (x, t) \longrightarrow [x]$$

has contractible point inverses, so that it is a simple homotopy equivalence. The induced cyclic cover of W' is the canonical infinite cyclic cover $\overline{W}' = \overline{\mathcal{W}}(f^+, f^-)$ of the mapping coequalizer $\mathcal{W}(f^+, f^-)$ with CW fundamental domain $(V \times I; U \times \{0\}, \zeta U \times \{1\})$. □

Remark 13.12 In Example 13.16 below we shall construct a finite CW complex W with an infinite cyclic cover $\overline{W}$ which does not admit a π_1-fundamental domain, showing the necessity of passing to an infinite simple homotopy equivalent CW complex in 13.11 (ii). □

Proposition 13.13 *Let W be a connected compact ANR with a connected infinite cyclic cover $\overline{W}$ which admits an ANR fundamental domain. The following conditions are equivalent:*

(i) $\overline{W}^+$ *is forward tame,*
(ii) $\overline{W}^-$ *is reverse tame,*

(iii) *there is a compactly supported homotopy*

$$h \,:\, \overline{W}\times I \longrightarrow \overline{W} \;;\; (x,t) \longrightarrow h_t(x)$$

such that

$$h_0 \;=\; \mathrm{id}_{\overline{W}} \;,\; h_1(\overline{W}^+) \subseteq \zeta\overline{W}^+ \;,$$

(iv) *the composite*

$$p \,:\, e(\overline{W}^+) \xrightarrow{p_{\overline{W}^+}} \overline{W}^+ \longrightarrow \overline{W}$$

is a homotopy equivalence.

Proof Let

$$\overline{W} \;=\; \overline{W}^+ \cup \overline{W}^-$$

with $\overline{W}^+ \cap \overline{W}^-$ compact, connected and such that

$$\zeta\overline{W}^+ \subset \overline{W}^+ \;,\; \zeta^{-1}\overline{W}^- \subset \overline{W}^-$$

with $\zeta : \overline{W} \longrightarrow \overline{W}$ a generating covering translation.

(i) $\Longrightarrow$ (iii) There exist a closed cocompact subspace $V \subseteq \overline{W}^+$ and a proper map $q : V \times [0,\infty) \longrightarrow \overline{W}^+$ with $q_0 =$ inclusion : $V \longrightarrow W$. Choose $n, N \geq 1$ so large that

$$\zeta^n\overline{W}^+ \subseteq V \;,\; q_N(V) \subseteq \zeta^{n+1}\overline{W}^+ \;.$$

Set

$$g_t \;=\; \zeta^{-n}\circ q_t \circ \zeta^n \;:\; \overline{W}^+ \longrightarrow \overline{W} \;\; (0 \leq t \leq N) \;,$$

so that

$$g_N\overline{W}^+ \subseteq \zeta\overline{W}^+ \;.$$

Since g is a proper homotopy there exists $k \geq 1$ so large that

$$g_t(\zeta^k\overline{W}^+) \subseteq \overline{W}\backslash\zeta\overline{W}^-$$

for each t. Since $\overline{W}\backslash\zeta\overline{W}^-$ is an ANR, the homotopy extension property can be used to get a homotopy

$$\widetilde{g}_t \;:\; \overline{W}^+ \longrightarrow \overline{W} \;\; (0 \leq t \leq N)$$

such that $\widetilde{g}$ is compactly supported and

$$\widetilde{g}_t|_{\overline{W}^+\backslash\zeta^k\overline{W}^+} \;=\; g_t| \;,\; \widetilde{g}_t(\zeta^k\overline{W}^+) \subseteq \overline{W}\backslash\zeta\overline{W}^- \;.$$

Then we still have

$$\widetilde{g}_N(\overline{W}^+) \subseteq \zeta\overline{W}^+ \;.$$

Finally, use the homotopy extension property again to extend $\widetilde{g}$ to a compactly supported homotopy defined on all of $\overline{W}$ and reparametrize $[0, N]$ to get the required homotopy h.

(iii) $\Longrightarrow$ (i) We have to find a closed cocompact subspace $V \subseteq \overline{W}^+$ such that the inclusion $V \times \{0\} \longrightarrow \overline{W}^+$ extends to a proper map

$$q : V \times [0, \infty) \longrightarrow \overline{W}^+ .$$

Choose $n \geq 1$ so large that $h(\overline{W}^+ \times I) \subseteq \zeta^{-n}\overline{W}^+$, let $V = \zeta^n\overline{W}^+$ and define a map $q : V \times [0, \infty) \longrightarrow \overline{W}^+ ; (x, t) \longrightarrow q_t(x)$ as follows. First set

$$q_t = \zeta^n \circ h_t \circ \zeta^{-n} : V \longrightarrow \overline{W}^+ \quad (0 \leq t \leq 1) .$$

Then assuming $k \geq 1$ and $q|_{V\times[0,k]}$ has been defined, set

$$q_t = \zeta^{n+k} \circ h_{t-k} \circ \zeta^{-n-k} \circ q_k : V \longrightarrow \overline{W}^+ \quad (k \leq t \leq k+1) .$$

Of course

$$q_0 = \text{inclusion} : V \longrightarrow \overline{W}^+ ,$$

so it remains to verify that q is proper. One can verify inductively that

$$q_{k+1}(V) \subseteq \zeta^{n+k+1}\overline{W}^+ \ , \ q_t(V) \subseteq \zeta^k\overline{W}^+ \quad (k \leq t \leq k+1) .$$

From this it follows that if $C \subseteq \overline{W}^+$ is compact, then there exists $N \geq 1$ so large that $q_t(V) \cap C = \emptyset$ for $t \geq N$. Thus $q^{-1}(C) \subseteq V \times [0, N]$. Since $q|_{V\times[0,N]}$ is clearly proper, $q^{-1}(C)$ is compact as required.

(ii) $\Longrightarrow$ (iii) By reverse tameness there exist a cocompact $V \subseteq \overline{W}^-$ and a homotopy $g : \overline{W}^- \times I \longrightarrow \overline{W}^-$ such that

$$g_0 = \text{id}_{\overline{W}^-} \ , \ g_1(\overline{W}^-) \subseteq \overline{W}^- \backslash V$$

and

$$g_t| = \text{inclusion} : \overline{W}^- \cap \overline{W}^+ \longrightarrow \overline{W}^-$$

for each t. It follows that g extends via the identity to a homotopy $\widetilde{g} : \overline{W} \times I \longrightarrow \overline{W}$.

Choose $n \geq 1$ so large that

$$\zeta^{-n+1}\overline{W}^- \subseteq V .$$

Define $h' : \overline{W}^+ \times I \longrightarrow \overline{W} ; (x, t) \longrightarrow h'_t(x)$ by

$$h'_t = \zeta^n \circ \widetilde{g}_t \circ \zeta^{-n} : \overline{W}^+ \longrightarrow \overline{W} .$$

Then

$$h'_1(\overline{W}^+) = \zeta^n \circ \widetilde{g}_1 \circ \zeta^{-n}\overline{W}^+ \subseteq \zeta^n(\overline{W}\backslash V) \subseteq \zeta\overline{W}^+ .$$

Since $\widetilde{g}$ is the identity on $\zeta^n\overline{W}^+$, it follows that h' is compactly supported.

Finally, use the homotopy extension property to extend h' to a compactly supported homotopy $h : \overline{W} \times I \longrightarrow \overline{W}$.

(iii) $\Longrightarrow$ (ii) For each $k = 1, 2, 3, \ldots$ define a homotopy

$$h^k : \overline{W} \times I \longrightarrow \overline{W} \; ; \; (x,t) \longrightarrow h^k_t(x)$$

by

$$h^k_t = \zeta^{-k} \circ h_t \circ \zeta^k : \overline{W} \longrightarrow \overline{W} \quad (0 \leq t \leq 1) \, .$$

The formulas which follow are designed to make sense of the infinite right concatenation

$$h^1 * h^2 * h^3 * \ldots \, .$$

For $k = 1, 2, 3, \ldots$ let

$$X_k = \mathrm{cl}(\zeta^{-k}\overline{W}^+ \backslash \zeta^{-k+1}\overline{W}^+)$$

and define $\rho : \overline{W}^- \longrightarrow [1, \infty)$ to be such that $\rho^{-1}([k, k+1]) = X_k$. For $x \in X_k$ and $t \in I$ define

$$g_t(x) = \begin{cases} h^{k+1}\Big(x, t\Big(\dfrac{\rho(x)}{\rho(x) - k}\Big)\Big) & \text{if } 0 \leq t \leq \dfrac{\rho(x) - k}{\rho(x)} \, , \\ h^k(x, t\rho(x) - \rho(x) + k) & \text{if } \dfrac{\rho(x) - k}{\rho(x)} \leq t \leq \dfrac{\rho(x) - k + 1}{\rho(x)} \, , \\ \vdots & \\ h^1(x, t\rho(x) - \rho(x) + 1) & \text{if } \dfrac{\rho(x) - k + (k-1)}{\rho(x)} \leq t \leq 1 \, . \end{cases}$$

This defines a homotopy

$$g : \overline{W}^- \times I \longrightarrow \overline{W} \; ; \; (x,t) \longrightarrow g_t(x)$$

such that

$$g_0 = \text{inclusion} : \overline{W}^- \longrightarrow \overline{W} \; , \; g_1(\overline{W}^-) \subseteq \overline{W}^+ \, ,$$

and there exists $n \geq 1$ so large that

$$g_t|_{\zeta^n \overline{W}^+} = \text{inclusion} : \zeta^n \overline{W}^+ \longrightarrow \overline{W} \quad (t \in I) \, .$$

If $U \subseteq \overline{W}^-$ is cocompact, choose $N \geq 1$ so large that

$$\overline{W}^- \backslash U \subseteq \zeta^{-N+n}\overline{W}^+ \, .$$

Then $\zeta^{-N} \circ g_t \circ \zeta^N$ is a homotopy establishing the reverse tameness of $\overline{W}^-$.

(i) $\Longrightarrow$ (iv) By (iii) and forward tameness there exist a closed cocompact subset $V \subseteq \overline{W}^+$ with $\zeta \overline{W}^+ \subseteq V$ and a proper map $q : V \times [0, \infty) \longrightarrow \overline{W}^+$

such that q_1 = inclusion and a homotopy $h : \overline{W} \times I \longrightarrow \overline{W}$ with $h_0 = \mathrm{id}$, $h_1(\overline{W}) \subseteq \zeta\overline{W}^+$ and $h_t|V$ = inclusion for each $t \in I$. Define

$$g : \overline{W} \xrightarrow{h_1} V \xrightarrow{\widehat{q}} e(\overline{W}^+) ,$$

where $\widehat{q}$ is the adjoint of q. We claim that g is a homotopy inverse of p. To see this, first note that

$$pg = h_1 \simeq 1 : \overline{W} \longrightarrow \overline{W} .$$

The existence of a homotopy

$$gp \simeq 1 : e(\overline{W}^+) \longrightarrow e(\overline{W}^+)$$

follows from the homotopy commutativity of the diagram

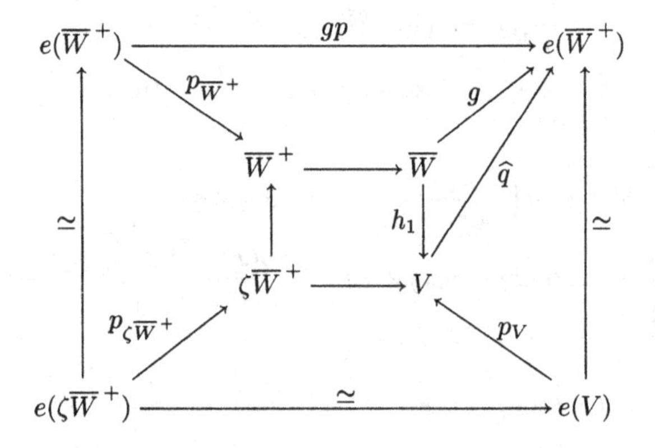

in which the unnamed arrows are inclusions.

(iv) $\Longrightarrow$ (iii) Suppose that $p : e(\overline{W}^+) \longrightarrow \overline{W}$ is a homotopy equivalence with homotopy inverse $g : \overline{W} \longrightarrow e(\overline{W}^+)$ and homotopy $F : \overline{W} \times I \longrightarrow \overline{W}$ such that $F_0 = \mathrm{id}_{\overline{W}}$ and $F_1 = pg$. Elements in $e(\overline{W}^+)$ are considered as proper maps $[0, \infty) \longrightarrow \overline{W}^+$. Define

$$G : \overline{W} \times [0, \infty) \longrightarrow \overline{W} ; (x, t) \longrightarrow \begin{cases} F(x, t) & \text{if } 0 \leq t \leq 1 , \\ g(x)(t-1) & \text{if } 1 \leq t < \infty . \end{cases}$$

Let

$$D = \overline{W}^+ \cap \zeta(\overline{W}^-)$$

(with $\zeta(\overline{W}^+) \subseteq \overline{W}^+$). There exists an integer $k \geq 0$ such that $F(\zeta(D) \times I) \subseteq \zeta^{-k+1}(\overline{W}^+)$. Let

$$A = \bigcup_{i=0}^{k+1} \zeta^i(D) = \overline{W}^+ \cap \zeta^{k+2}(\overline{W}^-) ,$$

and define

$$\widehat{G} : A \times [0,\infty) \longrightarrow \overline{W} \; ; \; (x,t) \longrightarrow \zeta^k G(\zeta^{-k}(x),t) \, .$$

Note that $\widehat{G}_0 =$ inclusion and $\widehat{G}(\zeta^{k+1}(D) \times [0,\infty)) \subseteq \zeta\overline{W}^+$.
There exists an integer $N \geq 0$ such that $\widehat{G}(A \times \{N\}) \subseteq \zeta(\overline{W}^+)$. Since $\zeta(\overline{W}^+)$ is an ANR we can extend $\widehat{G}|A \times [0,N]$ to a homotopy $H : \overline{W}^+ \times [0,N] \longrightarrow \overline{W}$ such that:

(a) $H|A \times [0,N] = \widehat{G}|$,
(b) $H_0 =$ inclusion,
(c) $H(\zeta^{k+1}(\overline{W}^+) \times [0,N]) \subseteq \zeta(\overline{W}^+)$,
(d) $H_t|\zeta^{k+2}(\overline{W}^+) =$ inclusion for each $t \in [0,N]$.

Now use the fact that $\overline{W}$ is an ANR to extend H to a compactly supported homotopy $h : \overline{W} \times [0,N] \longrightarrow \overline{W}$. Since $h|\overline{W}^+ \times [0,N] = H$ it follows that $h_N(\overline{W}^+) \subseteq \zeta(\overline{W}^+)$. After reparametrizing $[0,N]$ by I, we have a homotopy satisfying (iii). □

Proposition 13.14 *The following conditions on a connected infinite cyclic cover $\overline{W}$ of a compact ANR W are equivalent:*

(i) *$\overline{W}^+$ is forward collared,*
(ii) *$\overline{W}^-$ is reverse collared,*
(iii) *there exist a closed cocompact ANR subspace $Y \subseteq \overline{W}^+$ and an integer $M \geq 1$ such that $\zeta^\ell Y$ is a strong deformation retract of Y for every integer $\ell \geq M$.*

Proof (i) $\Longrightarrow$ (iii) According to 7.2 (iii) there exists a closed cocompact ANR subspace $Y \subseteq W$ such that the evaluation map $e(Y) \longrightarrow Y$ is a homotopy equivalence. Choose a positive integer M so large that $\zeta^\ell Y \subseteq Y$ for every $\ell \geq M$. Then there is a commutative diagram

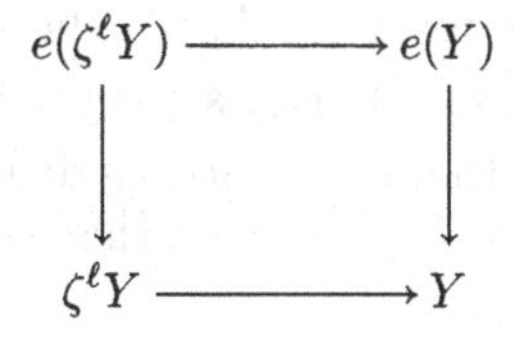

where the vertical arrows are evaluation maps and the horizontal ones are inclusions. The evaluation map $e(Y) \longrightarrow Y$ is a homotopy equivalence, and hence so is $e(\zeta^\ell Y) \longrightarrow \zeta^\ell Y$. The inclusion $e(\zeta^\ell Y) \longrightarrow e(Y)$ is a homotopy equivalence by 1.12. Thus, the inclusion $\zeta^\ell Y \longrightarrow Y$ is also a homotopy equivalence. Since Y is an ANR it follows that $\zeta^\ell Y$ is a strong deformation retract of Y.

(iii) $\Longrightarrow$ (i) Assume M is so large that $\zeta^\ell Y \subseteq Y$ for every $\ell \geq M$ and there exists a homotopy $h : Y \times I \longrightarrow Y$ such that $h_0 = \mathrm{id}_Y$, $h_t|\zeta^M Y$ =inclusion for each $t \in I$, and $h_1(Y) = \zeta^M Y$. Define a proper map

$$f \ : \ Y \times [0, \infty) \longrightarrow Y \ ; \ (y, t) \longrightarrow f_t(y)$$

by induction as follows. Let $f_t = h_t$ for $0 \leq t \leq 1$. Assume $k \geq 1$ and $f|Y \times [0, k]$ has been defined. For $k \leq t \leq k+1$, set

$$f_t \ = \ \zeta^{kM} \circ h_{t-k} \circ \zeta^{-kM} \circ f_k \ : \ Y \longrightarrow Y \ .$$

Note that $f_k(Y) \subseteq \zeta^{kM} Y$.

(i) $\Longrightarrow$ (ii) Let $U \subseteq \overline{W}^-$ be a cocompact subspace. By the proof above we may assume that condition (iii) holds. Thus, there exists a closed cocompact ANR subspace $Y \subseteq \overline{W}^+$ and an integer $M \geq 1$ such that $\zeta^\ell Y$ is a strong deformation retract of Y for each integer $\ell \geq M$. Choose a positive integer k so large that $\overline{W}^- \backslash U \subseteq \zeta^{-k} Y$. Since $\overline{W}^+$ is forward tame, 13.13 implies that $\overline{W}^-$ is reverse tame so that there exist a cocompact subspace $V \subseteq \overline{W}^-$ with $V \subseteq \overline{W}^- \backslash \zeta^{-k} Y$ and a homotopy $h : \overline{W}^- \times I \longrightarrow \overline{W}^-$ such that:

(1) $h_0 = \mathrm{id}_{\overline{W}^-}$,
(2) $h_t|\overline{W} \cap \zeta^{-k} Y$ = inclusion for each $t \in I$,
(3) $h_t(\overline{W}^- \backslash \zeta^{-k} Y) \subseteq \overline{W}^- \backslash \zeta^{-k} Y$ for each $t \in I$,
(4) $h_1(\overline{W}^-) \subseteq \overline{W}^- \backslash V$.

Choose an integer $N > k$ so large that $\overline{W}^- \backslash \zeta^{-N}(Y) \subseteq V$ and $N - k \geq M$. It follows that $\zeta^{-k} Y$ is a strong deformation retract of $\zeta^{-N}(Y)$, so let $g : \zeta^{-N}(Y) \times I \longrightarrow \zeta^{-N}(Y)$ be a homotopy such that $g_0 = \mathrm{id}$, $g_t|\zeta^{-k} Y =$ inclusion for each $t \in I$ and $g_1(\zeta^{-N}(Y)) \subseteq \zeta^{-k} Y$. Then the concatenation of g and h shows that $U \backslash (\overline{W}^- \backslash \zeta^{-k} Y)$ is a strong deformation retract of U and, hence, $\overline{W}^-$ is reverse collared.

(ii) $\Longrightarrow$ (iii) Since $\overline{W}^-$ is reverse collared there exists a cocompact subspace $V \subseteq \overline{W}^-$ such that $\overline{W}^- \backslash V$ is a strong deformation retract of $\overline{W}^-$. Let $X = (\overline{W}^- \backslash V) \cup \overline{W}^+$ so that X is a strong deformation retract of $\overline{W}$ and there is a homotopy $g : \overline{W} \times I \longrightarrow \overline{W}$ such that $g_0 = \mathrm{id}_{\overline{W}}$, $g_t|X =$ inclusion for each $t \in I$ and $g_1(\overline{W}) \subseteq X$.

Choose an integer $M \geq 1$ so large that if $\ell \geq M$ then $\zeta^\ell X \subseteq X$ and $g_1 \zeta^\ell X \subseteq \zeta X$. It will to suffice to show that if $\ell \geq M$ then X is a strong deformation retract of $\zeta^{-\ell} X$. Define a homotopy $h : \zeta^{-\ell} X \times I \longrightarrow \overline{W}$ by

$h(x,t) = h_t(x)$, with

$$h_t = \zeta^{-\ell} \circ g_1 \circ \zeta^{\ell} \circ g_t : \zeta^{-\ell}X \longrightarrow \overline{W} \ (t \in I) .$$

It can be verified that $h_0 =$ inclusion, $h_t(\zeta^{-\ell}X) \subseteq \zeta^{\ell}X$ for each $t \in I$, and $h_1(\zeta^{-\ell}X) \subseteq X$. □

We now specialize to the case of an infinite cyclic cover $\overline{W}$ of a finite CW complex W.

Proposition 13.15 *Let W be a connected finite CW complex, and let $\overline{W}$ be a connected infinite cyclic cover of W which admits a π_1-fundamental domain. Let $\widetilde{W}$ be the universal cover of W, and let $\pi = \pi_1(\overline{W})$, so that $\overline{W} = \widetilde{W}/\pi$.*

(i) *The following conditions are equivalent:*

(a) *$\overline{W}^+$ is forward tame,*
(b) *$\overline{W}^-$ is reverse tame,*
(c) *the projection $e(\overline{W}^+) \longrightarrow \overline{W}$ is a homotopy equivalence,*
(d) *$\overline{W}^-$ is finitely dominated.*

(ii) *If the conditions of (i) are satisfied and $\overline{W} = \overline{W}^+ \cup \overline{W}^-$ then*

$$\begin{aligned} C(\widetilde{W}^+) &\simeq C(\widetilde{W}) \oplus C^{lf,\pi}(\widetilde{W}^+) , \\ C(\widetilde{W}^+ \cap \widetilde{W}^-) &\simeq C(\widetilde{W}^-) \oplus C^{lf,\pi}(\widetilde{W}^+) , \\ [\overline{W}^+]^{lf} &= -[\overline{W}^-] \in \widetilde{K}_0(\mathbb{Z}[\pi]) , \end{aligned}$$

with $\widetilde{W}^\pm \subset \widetilde{W}$ the cover of $\overline{W}^\pm \subset \overline{W}$ induced from $\widetilde{W}$ regarded as a cover of $\overline{W}$.

(iii) *The following conditions are equivalent:*

(a) *$\overline{W}$ is infinite simple homotopy equivalent to an infinite cyclic cover $\overline{X}$ of a finite CW complex X with $\overline{X}^+$ forward collared,*
(b) *$\overline{W}$ is infinite simple homotopy equivalent to an infinite cyclic cover $\overline{X}$ of a finite CW complex X with $\overline{X}^-$ reverse collared,*
(c) *$\overline{W}^+$ is forward tame and $[\overline{W}^+]^{lf} = 0 \in \widetilde{K}_0(\mathbb{Z}[\pi])$,*
(d) *$\overline{W}^-$ is reverse tame and $[\overline{W}^-] = 0 \in \widetilde{K}_0(\mathbb{Z}[\pi])$,*
(e) *$\overline{W}^-$ is homotopy equivalent to a finite CW complex.*

Proof (i) (a) $\Longleftrightarrow$ (b) $\Longleftrightarrow$ (c) by 13.13.

(b) $\Longrightarrow$ (d) By 8.7.

(d) $\Longrightarrow$ (b) By 8.7, since

$$\overline{W}^- \supset \zeta^{-1}\overline{W}^- \supset \zeta^{-2}\overline{W}^- \supset \dots$$

is a sequence of cofinite subcomplexes isomorphic to $\overline{W}^-$ with $\bigcap_j \zeta^{-j}\overline{W}^- = \emptyset$.

(ii) If the conditions of (i) are satisfied then by 7.20 (v) there is defined a chain equivalence of finitely dominated $\mathbb{Z}[\pi]$-module chain complexes

$$S(\widetilde{e(\overline{W}^+)}) \simeq C^{\infty,\pi}(\widetilde{W}^+) .$$

Also, the homotopy equivalence $e(\overline{W}^+) \simeq \overline{W}$ induces a chain equivalence

$$S(\widetilde{e(\overline{W}^+)}) \simeq S(\widetilde{W}) ,$$

so that the composite chain map

$$f : C^{\infty,\pi}(\widetilde{W}^+) = \mathcal{C}(C(\widetilde{W}^+) \longrightarrow C^{lf,\pi}(\widetilde{W}^+))_{*+1} \longrightarrow C(\widetilde{W}^+) \longrightarrow C(\widetilde{W})$$

is a chain equivalence. The algebraic mapping cone of the chain map

$$g : C(\widetilde{W}^+) \longrightarrow C(\widetilde{W}) \oplus C^{lf,\pi}(\widetilde{W}^+)$$

defined by inclusion on each component is also the algebraic mapping cone of f, $\mathcal{C}(g) = \mathcal{C}(f)$, so that g is also a chain equivalence. The inclusion

$$\mathcal{C}(C(\widetilde{W}^+ \cap \widetilde{W}^-) \longrightarrow C(\widetilde{W}^-)) \longrightarrow \mathcal{C}(C(\widetilde{W}^+) \longrightarrow C(\widetilde{W}))$$

is a chain equivalence (excision), so that the chain map

$$C(\widetilde{W}^+ \cap \widetilde{W}^-) \longrightarrow C(\widetilde{W}^-) \oplus C^{lf,\pi}(\widetilde{W})$$

defined by inclusion on each component is also a chain equivalence.

(iii) (a) $\Longleftrightarrow$ (b) by 13.14.

(c) $\Longleftrightarrow$ (d) by (ii).

(a) $\Longrightarrow$ (c) $\overline{W}^+$ is forward tame by 7.2 (i) and 9.6, and $[\overline{W}^+]^{lf} = [\overline{W}^+]^{lf}_{\infty} = 0$ by 13.13 (iv), 11.7 (i) and 10.9 (iii).

(d) $\Longrightarrow$ (e) The reduced projective class $[\overline{W}^-] \in \widetilde{K}_0(\mathbb{Z}[\pi])$ is the finiteness obstruction of $\overline{W}^-$.

(e) $\Longrightarrow$ (d) $\overline{W}^-$ is finitely dominated, so that $\overline{W}^-$ is reverse tame by (i). Moreover, $[\overline{W}^-] = 0$ since $\overline{W}^-$ is homotopy equivalent to a finite CW complex.

(e) $\Longrightarrow$ (b) Let $\overline{W}^-$ be homotopy equivalent to a finite CW complex K. By replacing W by $W \times D^n$ for some large n, we may assume that K is a subcomplex of $\overline{W}^-$ and K is a strong deformation retract of $\overline{W}^-$. Let

$$r : \mathrm{id}_{\overline{W}^-} \simeq r_1 : \overline{W}^- \longrightarrow \overline{W}^-$$

be a cellular homotopy with $r_t|K = \text{inclusion} : K \longrightarrow \overline{W}^-$ for each $t \in I$ and $r_1(\overline{W}^-) = K$. Let $\zeta : \overline{W} \longrightarrow \overline{W}$ be the $(+1)$-generating covering translation and $(V; U, \zeta U)$ a fundamental domain for $\overline{W}$. There is an integer $N \leq -1$ such that

$$r(\zeta^{-1}V \times I) \subseteq K \cup \bigcup_{j=N}^{-1} \zeta^j V = Z .$$

In particular, $K \subseteq Z$ and the maps

$$g = \text{inclusion} : \zeta^{-1}V \longrightarrow Z ,$$
$$g' = r_1| : \zeta^{-1}V \longrightarrow Z$$

are homotopic. Hence $\mathcal{M}(g)$, $\mathcal{M}(g')$ are simple homotopy equivalent rel $\zeta^{-1}V \cup Z$. Let

$$\overline{X} = (\mathcal{M}(g') \times \mathbb{Z})/\sim$$

where $\sim$ identifies $\zeta^{-1}V \times \{i\}$ in the base of $\mathcal{M}(g') \times \{i\}$ with $\zeta^N V \times \{i+1\}$ in the top of $\mathcal{M}(g') \times \{i+1\}$ for each $i \in \mathbb{Z}$: $(x, 0, i) \sim (\zeta^{N+1}x, i+1)$ for $(x, 0) \in \zeta^{-1}V \times \{0\} \subseteq \mathcal{M}(g')$ and $\zeta^{N+1}x \in \zeta^N V \subseteq Z \subseteq \mathcal{M}(g')$. Similarly let

$$\overline{Y} = (\mathcal{M}(g) \times \mathbb{Z})/\sim$$

where $\sim$ identifies $\zeta^{-1}V \times \{i\}$ in the base of $\mathcal{M}(g) \times \{i\}$ with $\zeta^N V \times \{i+1\}$ in the top of $\mathcal{M}(g) \times \{i+1\}$. Clearly $\overline{Y}$ and $\overline{W}$ are infinite simple homotopy equivalent. The simple homotopy equivalence of mapping cylinders $\mathcal{M}(g')$, $\mathcal{M}(g)$ induces an infinite simple homotopy equivalence of $\overline{X}$ and $\overline{Y}$. Define a generating covering translation

$$\zeta' : \overline{X} \longrightarrow \overline{X} ; (x, i) \longrightarrow (x, i+1)$$

so that $X = \overline{X}/\zeta'$ is a finite CW complex. Let

$$V' = (\mathcal{M}(g') \bigcup_{\zeta^{-1}V} Z) \times \{0\}$$

so that V' contains two natural copies of Z, firstly

$$U' = Z \times \{1\} \times \{0\} \subseteq \mathcal{M}(g') \times \{0\}$$

which is the top of $\mathcal{M}(g')$, and secondly $\zeta' U'$ which intersects $\mathcal{M}(g') \times \{0\}$ in the bottom. Then $(V'; U', \zeta' U')$ is a fundamental domain for $\overline{X}$. It remains to show that $\overline{X}^-$ is reverse collared. For each $k \leq 0$, $(\zeta')^k(\overline{X}^-)$ is infinite simple homotopy equivalent to $\zeta^{|N|k}\overline{W}^-$ rel $(\zeta')^k(U') = \zeta^{|N|k}Z$. Since $\zeta^{|N|k}\overline{W}^-$ strong deformation retracts to $\zeta^{|N|k}K \subseteq \zeta^{|N|k}Z$, it follows that $\overline{X}^-$ strong deformation retracts to $\bigcup_{j=0}^{k+1} (\zeta')^j V' \cup (\zeta')^k(\mathcal{M}(r_1 : \zeta^{-1}V \longrightarrow K))$.

Thus $\overline{X}^-$ is reverse collared. □

The following example is a CW complex with bad local properties at infinity, such as a nasty two-point compactification:

Example 13.16 Let $S^k_1 \vee S^k_2$ be the wedge of two k-spheres with $\{p_0\} = S^k_1 \cap S^k_2$. Let D^{k+1} be a $(k+1)$-disk with $S^k_2 = \partial D^{k+1}$ so that $S^k_1 \vee S^k_2 \subseteq S^k_1 \vee D^{k+1}$. Represent D^{k+1} by $[-1,1]^{k+1}$ where $p_0 = (-1,0,0,\ldots,0) \in D^{k+1}$. Define the k-cell

$$Y = \{(x_1, x_2, \ldots, x_{k+1}) \in D^{k+1} \,|\, x_1 \le 0\,,\, x_{k+1} = 0\}$$

and let $q : Y \longrightarrow S^k_1$ be a surjective map such that $q(p_0) = p_0$. Define also

$$X = (S^k_1 \vee D^{k+1})/{\sim}$$

where $\sim$ is generated by $y \sim q(y)$ for each $y \in Y$. Thus X is a CW complex with subcomplex $S^k_1 \vee S^k_2$. Note that:

(i) S^k_1 is not contractible in X,
(ii) S^k_2 is contractible in X,
(iii) S^k_2 is not contractible in $X \backslash \{x\}$ for any $x \in S^k_1$,
(iv) X is simple homotopy equivalent to $S^k_1 \vee D^{k+1}$ rel $S^k_1 \vee S^k_2$.

Let

$$\overline{W} = (X \times \mathbb{Z})/{\sim}$$

where $\sim$ homeomorphically identifies $S^k_1 \times \{n+1\}$ with $S^k_2 \times \{n\}$, $n \in \mathbb{Z}$. Then

$$\zeta : \overline{W} \longrightarrow \overline{W}\,;\; [x,n] \longrightarrow [x, n+2]$$

is a covering translation and

$$W = \overline{W}/\zeta$$

is a finite CW complex. The simple homotopy equivalence $X \simeq S^k_1 \vee D^{k+1}$ induces an infinite simple homotopy equivalence

$$\begin{aligned}\overline{W} &\simeq \ldots \vee D^{k+1} \vee D^{k+1} \vee \ldots \\ &= \{(x_1, x_2, \ldots, x_{k+1}) \in \mathbb{R}^{k+1} \,|\, (x_1 - i)^2 + x_2^2 + \ldots + x_{k+1}^2 \le 1/4 \\ &\qquad\qquad \text{for some } i \in \mathbb{Z}\}\,.\end{aligned}$$

Thus $\overline{W}$ is infinite simple homotopy equivalent to $\mathbb{R}$. In fact, W is simple homotopy equivalent to S^1. There is a fundamental domain $(V; U, \zeta U)$ for $\zeta : \overline{W} \longrightarrow \overline{W}$ with

$$\begin{aligned}V &= X \times \{0\} \cup D^{k+1} \times \{1\} \simeq S^k\,, \\ U &= S^k_1 \times \{0\} = S^k_2 \times \{-1\} \cong S^k\,, \\ \zeta U &= S^k_1 \times \{2\} = S^k_2 \times \{1\} \cong S^k\,.\end{aligned}$$

The inclusion $U \longrightarrow V$ is a homotopy equivalence, but ζU is contractible in V. There are infinite simple homotopy equivalences

$$\overline{W} \simeq \mathbb{R} \ , \ \overline{W}^+ \simeq S^k \vee [0,\infty) \ , \ \overline{W}^- \simeq [0,\infty) \ .$$

For $k = 1$ the one-point compactification C of $\overline{W}^+$ is the compact metric space described in Hocking and Young [66, page 350 and Figure 8-9(b)]. The space C is locally simply-connected at the point at infinity, but that point does not have arbitrarily small simply-connected open neighbourhoods in C. For $k > 1$

$$\pi_1(U) = \pi_1(V) = \pi_1(\overline{W}^+) = \pi_1(\overline{W}^-) = \pi_1(\overline{W}) = \{1\}$$

so that the conventions of this chapter are satisfied.

Claim The infinite cyclic cover $\overline{W}$ is such that:

(i) $\overline{W}$ is both forward and reverse tame,
(ii) the end spaces $e(\overline{W}^+)$, $e(\overline{W}^-)$ are contractible,
(iii) $\overline{W}^+$ is not forward collared,
(iv) $\overline{W}^-$ is not reverse collared,
(v) $\overline{W}^+$ is reverse collared,
(vi) $\overline{W}^-$ is forward collared,
(vii) $[\overline{W}^+]^{lf} = [\overline{W}^-] = 0 \in \widetilde{K}_0(\mathbb{Z}) = \{0\}$.

Proof (i) follows from 9.6 and 9.8 since $\overline{W}$ is proper homotopy equivalent to $\mathbb{R}$.

(ii) follows from 9.4 (ii).

(iii) First observe that if $b \in S^k_1 \subseteq X$ and $b \neq p_0$, then $X\backslash\{b\} \simeq S^k \vee S^k$ with one of the k-spheres given by S^k_2. It follows that if $b \in S^k_1 \times \{n_0\} \subseteq \overline{W}$ and $b \notin \{p_0\} \times \mathbb{Z}$, then $\overline{W}\backslash\{b\} \simeq S^k \vee S^k$ with one of the k-spheres given by $S^k_1 \times \{n_0+1\} = S^k_2 \times \{n_0\}$. Now for any closed cocompact subspace $Z \subseteq \overline{W}^+$, let $n_0 \in \{-1,0,1,\dots\}$ be the largest integer such that $S^k_1 \times \{n_0\} = S^k_2 \times \{n_0 - 1\} \not\subseteq Z$ so that $S^k_1 \times \{n_0+1\} = S^k_2 \times \{n_0\} \subseteq Z$. Choose $b \in S^k_1 \times \{n_0\} \subseteq \overline{W}$ with $b \notin \{p_0\} \times \mathbb{Z}$ and $b \notin Z$. It follows from the observation above that $S^k_1 \times \{n_0+1\} = S^k_2 \times \{n_0\}$ is not contractible in $\overline{W}\backslash\{b\}$ and also not contractible in Z. Thus $\pi_k(Z)$ is not trivial so that $\overline{W}^+$ is not forward collared by (ii) and 7.2 (ii).

(iv) follows from (iii) and 13.14.

(v) and (vi) follow from 13.14 since $\zeta^{-1}\overline{W}^-$ is a strong deformation retract of $\overline{W}^-$.

(vii) is obvious. □

Remark 13.17 Example 13.16 shows that neither the property of being forward collared nor that of being reverse collared is an invariant of infinite simple homotopy type (cf. 9.7, 11.7 (ii), 11.8). Moreover, this example shows the necessity of passing to an infinite simple homotopy equivalent CW complex in 13.15 (iii). The CW complex P obtained from $\overline{W}^+$ by filling in $U \cong S^k$ with a $(k+1)$-cell is forward collared, showing that a cofinite subcomplex of a forward collared CW complex need not be forward collared. □

The following result will be used in Chapters 14, 19:

Proposition 13.18 *For any cellular maps of CW complexes*

$$\begin{aligned} i^+ &: K \longrightarrow X \ , \quad i^- : K \longrightarrow Y \ , \\ j^+ &: L \longrightarrow Y \ , \quad j^- : L \longrightarrow X \end{aligned}$$

the CW complexes $\mathcal{W}(f(i,j)^+, f(i,j)^-), \mathcal{W}(g(i,j)^+, g(i,j)^-)$ determined by the maps

$$\begin{aligned} f(i,j)^+ &: K \xrightarrow{i^+} X \longrightarrow X \cup_L Y \ , \\ f(i,j)^- &: K \xrightarrow{i^-} Y \longrightarrow X \cup_L Y \ , \\ g(i,j)^+ &: L \xrightarrow{j^+} Y \longrightarrow X \cup_K Y \ , \\ g(i,j)^- &: L \xrightarrow{j^-} X \longrightarrow X \cup_K Y \end{aligned}$$

are related by a canonical homotopy equivalence

$$\mathcal{W}(f(i,j)^+, f(i,j)^-) \simeq \mathcal{W}(g(i,j)^+, g(i,j)^-) \ ,$$

which is simple if K, L, X, Y are finite CW complexes.

Proof Replacing X, Y by mapping cylinders (if necessary) it may be assumed that i^+, i^-, j^+, j^- are embeddings of subcomplexes such that

$$i^+(K) \cap j^-(L) = \emptyset \subset X \ , \quad i^-(K) \cap j^+(L) = \emptyset \subset Y \ .$$

The CW complex

$$Z = (X \amalg Y)/\{i^+(K) = i^-(K), j^+(L) = j^-(L)\}$$

can be cut open along either K or L, so that both $\mathcal{W}(f(i,j)^+, f(i,j)^-)$ and $\mathcal{W}(g(i,j)^+, g(i,j)^-)$ are homotopy equivalent to Z.

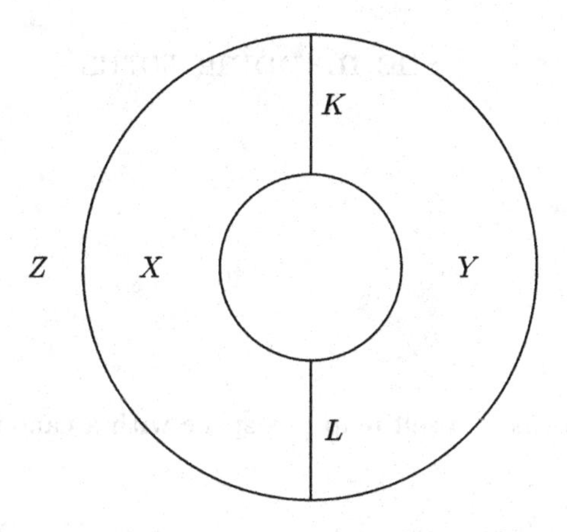
K
Z
X
Y
L

□

14

The mapping torus

The mapping torus of a self map is a space with a canonical infinite cyclic cover.

Definition 14.1 The *mapping torus* of a map $h : X \longrightarrow X$ is the identification space

$$T(h) = X \times I/\{(x,1) = (h(x),0) \mid x \in X\} . \qquad \square$$

In the terminology of 13.7 the mapping torus is the mapping coequalizer

$$T(h) = \mathcal{W}(1 : X \longrightarrow X, h : X \longrightarrow X) .$$

The canonical infinite cyclic cover $\overline{T}(h)$ of $T(h)$ is given by 13.8, as described in detail in 14.6 below.

A homotopy $h \simeq h' : X \longrightarrow X$ determines a homotopy equivalence

$$T(h) \simeq T(h') .$$

Proposition 14.2 (M. Mather [91], Ferry [55], Ranicki [123]) *For any maps $f : X \longrightarrow Y$, $g : Y \longrightarrow X$ the map*

$$T(gf : X \longrightarrow X) \longrightarrow T(fg : Y \longrightarrow Y) \; ; \; (x,t) \longrightarrow (f(x),t) .$$

is a homotopy equivalence. If X, Y are finite CW complexes it is a simple homotopy equivalence. $\square$

Definition 14.3 Let X be a finitely dominated space, and let $h : X \longrightarrow X$ be a map. The *canonical simple homotopy type* on the mapping torus $T(h)$ is represented by $(T(fhg : Y \longrightarrow Y), e)$ for any domination of X

$$(Y, f : X \longrightarrow Y, g : Y \longrightarrow X, gf \simeq 1 : X \longrightarrow X)$$

by a finite CW complex Y, with

$$e : T(h) \simeq T(hgf) \simeq T(fhg) = \text{finite } CW \text{ complex} . \qquad \square$$

Remark 14.4 (i) For a finite CW complex X the canonical simple homotopy type on $T(h: X \longrightarrow X)$ is just the simple homotopy type given by the finite CW structure of $T(h)$.

(ii) For maps $f: X \longrightarrow Y$, $g: Y \longrightarrow X$ of finitely dominated CW complexes the homotopy equivalence $T(gf) \simeq T(fg)$ of 14.2 is simple,

$$\tau(T(gf) \simeq T(fg)) = 0 \in Wh(\pi_1(T(gf))) ,$$

giving $T(gf)$, $T(fg)$ the canonical simple homotopy types. This is a special case of the simple homotopy equivalence

$$\mathcal{W}(f(i,j)^+, f(i,j)^-) \simeq_s \mathcal{W}(g(i,j)^+, g(i,j)^-)$$

of 13.18, with

$$\begin{aligned} i^+ &= 1 : K = X \longrightarrow X \ , \ i^- = f : K = X \longrightarrow Y \ , \\ j^+ &= 1 : L = Y \longrightarrow Y \ , \ j^- = g : L = Y \longrightarrow X \ . \end{aligned}$$

□

Example 14.5 For any finitely dominated space X

$$X \times S^1 = T(1 : X \longrightarrow X)$$

is homotopy finite (Mather [91]), with a canonical simple homotopy type. □

The canonical infinite cyclic cover $\overline{T}(h)$ of $T(h)$ (13.8) has the following properties:

Proposition 14.6 (i) *A mapping torus $T(h: X \longrightarrow X)$ has a canonical projection*

$$c : T(h) \longrightarrow S^1 = I/(0=1) \ ; \ (x,t) \longrightarrow t$$

which classifies the canonical infinite cyclic cover

$$\overline{T}(h) = c^*\mathbb{R} = \Big(\coprod_{j=-\infty}^{\infty} (X \times I \times \{j\}) \Big) \Big/ (x,1,j) = (h(x),0,j+1)$$

with generating covering translation

$$\zeta : \overline{T}(h) \longrightarrow \overline{T}(h) \ ; \ (x,t,j) \longrightarrow (x,t,j+1)$$

and fundamental domain

$$(\mathcal{M}(h: X \longrightarrow X); X \times \{0\}, X \times \{1\}) .$$

(ii) *If X is connected with fundamental group $\pi_1(X) = \pi$ and $h: X \longrightarrow X$ induces $h_* = \alpha : \pi \longrightarrow \pi$ then $T(h)$ is connected with fundamental group*

$$\pi_1(T(h)) = \pi *_\alpha \mathbb{Z} = \pi * \{z\}/\{\alpha x = zxz^{-1} \,|\, x \in \pi\} .$$

The canonical infinite cyclic cover $\overline{T}(h)$ *of* $T(h)$ *is classified by the projection*

$$\pi *_\alpha \mathbb{Z} \longrightarrow \mathbb{Z} \; ; \; \pi \longrightarrow 1 \; , \; z \longrightarrow z$$

and

$$\pi_1(\overline{T}(h)) \;=\; \langle \mathrm{im}(\pi) \rangle \;=\; \ker(\pi *_\alpha \mathbb{Z} \longrightarrow \mathbb{Z})$$

is the normal subgroup generated by $\mathrm{im}(\pi)$ *in* $\pi *_\alpha \mathbb{Z}$*. The natural map* $\pi \longrightarrow \pi *_\alpha \mathbb{Z}$ *is an injection if and only if* $\alpha : \pi \longrightarrow \pi$ *is an injection.*

(iii) *If* X *is a finite connected CW complex and* $h : X \longrightarrow X$ *is a cellular map then* $T(h)$ *is a finite CW complex, such that the infinite cyclic cover* $\overline{T}(h)$ *has exactly two ends (13.2).*

(iv) *If* $h : X \longrightarrow X$ *is a homotopy equivalence then*

$$X \longrightarrow T(h) \longrightarrow S^1$$

is a homotopy fibration, and the inclusion $X \longrightarrow \overline{T}(h); x \longrightarrow (x, 0, 0)$ *is a homotopy equivalence such that*

$$\zeta_{\overline{T}(h)} \;\simeq\; h^{-1} \;:\; \overline{T}(h) \;\simeq\; X \longrightarrow \overline{T}(h) \;\simeq\; X \; .$$

(v) *If* $h : X \longrightarrow X$ *is a homeomorphism then* $X \longrightarrow T(h) \longrightarrow S^1$ *is a fibre bundle, with a homeomorphism*

$$T(h) \longrightarrow X \times_{\mathbb{Z}} \mathbb{R} \; ; \; [x, s, j] \longrightarrow [h^{-j}(x), s + j] \; ,$$

where the action of $\mathbb{Z}$ *on* $X \times \mathbb{R}$ *is by*

$$\mathbb{Z} \times (X \times \mathbb{R}) \longrightarrow (X \times \mathbb{R}) \; ; \; (i, (x, j)) \longrightarrow (h^{-i}(x), i + j) \; .$$

(vi) *If* $p : \overline{W} \longrightarrow W$ *is the covering projection of an infinite cyclic cover with generating covering translation* $\zeta : \overline{W} \longrightarrow \overline{W}$ *there is defined a homotopy equivalence*

$$T(\zeta) \;=\; \overline{W} \times_{\mathbb{Z}} \mathbb{R} \longrightarrow W \;=\; \overline{W}/\mathbb{Z} \; ; \; (x, t) \longrightarrow p(x) \; ,$$

where the action of $\mathbb{Z}$ *on* $\overline{W} \times \mathbb{R}$ *is by*

$$\mathbb{Z} \times (\overline{W} \times \mathbb{R}) \longrightarrow (\overline{W} \times \mathbb{R}) \; ; \; (i, (x, j)) \longrightarrow (\zeta^i(x), i + j) \; .$$

(vii) *The subspaces of* $\overline{T}(h)$ *defined by*

$$\overline{T}^+(h) \;=\; \mathrm{Tel}(h) \;=\; \Big(\coprod_{j=0}^{\infty} X \times I \times \{j\}\Big) \Big/ (x, 1, j) = (h(x), 0, j + 1) \; ,$$

$$\overline{T}^-(h) \;=\; \Big(\coprod_{j=-\infty}^{-1} X \times I \times \{i\}\Big) \Big/ (x, 1, j) = (h(x), 0, j + 1)$$

are such that the inclusions

$$\overline{T}^+(h) \longrightarrow \overline{T}^+(h) \cup \overline{T}^-(h) \;=\; \overline{T}(h) \; ,$$

$$\overline{T}^+(h) \cap \overline{T}^-(h) \;=\; X \longrightarrow \overline{T}^-(h)$$

are homotopy equivalences (although not in general proper homotopy equivalences), and

$$\overline{T}^+(h) \simeq \underrightarrow{\text{hocolim}}\,(X \xrightarrow{h} X \xrightarrow{h} X \longrightarrow \dots) .$$

If X is compact then

$$e(\overline{T}^-(h)) \simeq \underleftarrow{\text{holim}}\,(X \xleftarrow{h} X \xleftarrow{h} X \longleftarrow \dots) ,$$

the one-point compactification $\overline{T}^+(h)^\infty$ is contractible, $\overline{T}^+(h)$ is forward collared with locally finite projective class

$$[\overline{T}^+(h)]^{lf} = 0 \in K_0(\mathbb{Z}[\pi_1(\overline{T}^+(h))]) ,$$

and the evaluation map $p_{\overline{T}^+(h)} : e(\overline{T}^+(h)) \longrightarrow \overline{T}^+(h)$ is a homotopy equivalence.

(viii) *If X is a finitely dominated CW complex then $\overline{T}^-(h)$ is finitely dominated for any map $h : X \longrightarrow X$, with projective class*

$$[\overline{T}^-(h)] = [X] \in K_0(\mathbb{Z}[\pi_1(X)]) .$$

If X is finite then $\overline{T}^-(h)$ is reverse collared, with $[\overline{T}^-(h)] = [X] = 0$. □

Remark 14.7 It will follow from 23.22 that the following conditions are equivalent for the mapping torus $T(h)$ of a map $h : X \longrightarrow X$ of a finite CW complex X with $h_* = 1 : \pi_1(X) \longrightarrow \pi_1(X)$:

(a) $\overline{T}^+(h)$ is reverse tame,
(b) $\overline{T}^-(h)$ is forward tame,
(c) $\overline{T}(h)$ is finitely dominated,

in which case

$$e(\overline{T}^+(h)) \simeq e(\overline{T}^-(h)) \simeq \overline{T}^+(h) \simeq \overline{T}(h) ,$$
$$\pi_1^\infty(\overline{T}^+(h)) = \pi_1^\infty(\overline{T}^-(h)) = \pi_1(\overline{T}(h)) = \pi_1(X)$$

and the projective classes are such that

$$[e(\overline{T}^+(h))] = [e(\overline{T}^-(h))] = [\overline{T}^+(h)] = [\overline{T}(h)] ,$$
$$[\overline{T}^-(h)]^{lf} = [X] - [\overline{T}(h)] \in K_0(\mathbb{Z}[\pi_1(X)]) .$$ □

In general, the infinite cyclic cover $\overline{T}(h)$ of the mapping torus $T(h)$ of a map $h : X \longrightarrow X$ of a finite CW complex X is not finitely dominated, $\overline{T}^+(h)$

is not reverse tame, and $\overline{T}^{-}(h)$ is not forward tame – see 23.25 below for an explicit example.

Proposition 14.8 (i) *Let $\zeta : X \longrightarrow X$ be a self map which fits into a commutative square*

$$\begin{array}{ccc} X & \xrightarrow{\zeta} & X \\ {\scriptstyle q}\downarrow & & \downarrow{\scriptstyle q} \\ \mathbb{R} & \longrightarrow & \mathbb{R} \end{array}$$

with $q : X \longrightarrow \mathbb{R}$ some map and $\mathbb{R} \longrightarrow \mathbb{R}; s \longrightarrow s+1$. The mapping torus $T(\zeta)$ is such that there is defined a homeomorphism

$$T(\zeta) \cong X/\zeta \times \mathbb{R}$$

with X/ζ the quotient of X by the equivalence relation $\sim$ generated by $x \sim \zeta(x)$ $(x \in X)$.

(ii) *Let $\overline{W} = c^{*}\mathbb{R}$ be the infinite cyclic cover $\overline{W}$ of a space W classified by a map $c : W \longrightarrow S^1$. The mapping torus of a generating covering translation $\zeta : \overline{W} \longrightarrow \overline{W}$ is such that there is defined a homeomorphism $T(\zeta) \cong W \times \mathbb{R}$.*

Proof (i) Let $p : X \longrightarrow X/\zeta$ be the projection, and define a homeomorphism

$$T(\zeta) \longrightarrow X/\zeta \times \mathbb{R} \; ; \; (x,s) \longrightarrow (p(x), q(x)+s) \; .$$

(ii) Apply (i) with

$$\begin{aligned} X &= \overline{W} = \{(w,s) \in W \times \mathbb{R} \,|\, c(w) = [s] \in S^1 = \mathbb{R}/\mathbb{Z}\} \; , \\ \zeta &: X \longrightarrow X \; ; \; (w,s) \longrightarrow (w, s+1) \; , \\ q &: X \longrightarrow \mathbb{R} \; ; \; (w,s) \longrightarrow s \; . \end{aligned}$$

□

15

Geometric ribbons and bands

A band is a compact space W with a finitely dominated infinite cyclic cover $\overline{W}$. The main result (15.10) of this chapter is that an infinite cyclic cover $\overline{W}$ of a finite CW complex W is finitely dominated if and only if $\overline{W}$ is both forward and reverse tame, in which case the end space $e(\overline{W})$ is homotopy equivalent to the disjoint union of two copies of $\overline{W}$.

Ribbons are non-compact spaces with the homotopy theoretic and homological end properties of the infinite cyclic covers of bands. In Chapters 15–20 we shall use engulfing and homotopy theory to prove that ribbons are in fact proper homotopy equivalent to the infinite cyclic covers of bands.

Definition 15.1 A *ribbon* (X,d) is a non-compact space X with a proper map $d : X \longrightarrow \mathbb{R}$ such that the subspaces

$$X^+ = d^{-1}[0,\infty) \ , \ X^- = d^{-1}(-\infty,0] \subset X$$

satisfy :

(i) the inclusions

$$X^+ \cap X^- \longrightarrow X^+ \ , \ X^+ \cap X^- \longrightarrow X^- \ , \ X^+ \longrightarrow X \ , \ X^- \longrightarrow X$$

induce bijections between the path components and induce isomorphisms

$$\pi_1(X^+ \cap X^-) \cong \pi_1(X^\pm) \cong \pi_1^\infty(X^\pm) \cong \pi_1(X)$$

on each component,

(ii) the composites

$$e(X^+) \longrightarrow X^+ \longrightarrow X \ , \ e(X^-) \longrightarrow X^- \longrightarrow X$$

are homotopy equivalences,

(iii) inclusions (on each component) induce $\mathbb{Z}[\pi]$-module isomorphisms

$$H_*(\widetilde{X}^+) \cong H_*^{lf,\pi}(\widetilde{X}^+) \oplus H_*(\widetilde{X}) \ , \ H_*(\widetilde{X}^-) \cong H_*^{lf,\pi}(\widetilde{X}^-) \oplus H_*(\widetilde{X})$$

with $\pi = \pi_1(X)$ and $\widetilde{X}$, $\widetilde{X}^+$, $\widetilde{X}^-$ the universal covers of X, X^+, X^-.

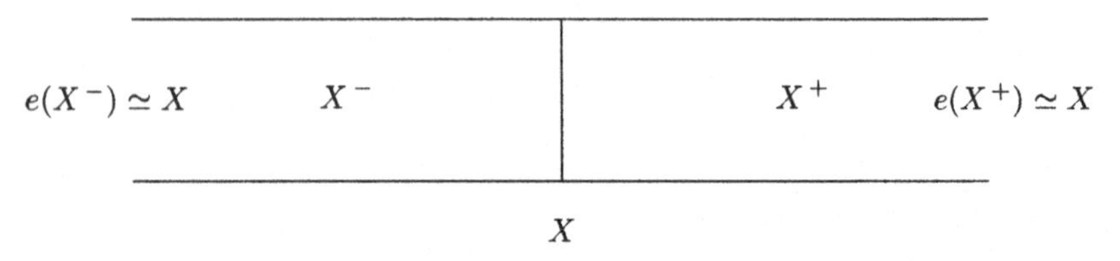

□

Definition 15.2 (i) A *CW ribbon* is a ribbon (X, d) such that X is a CW complex, and the subspaces $X^+, X^-, X^+ \cap X^-$ are subcomplexes. Similarly for an *ANR ribbon.*

(ii) An *n-dimensional manifold ribbon* is a ribbon (X, d) such that X is an n-dimensional manifold and $d : X \longrightarrow \mathbb{R}$ is transverse regular at $0 \in \mathbb{R}$, so that

$$X^+ = d^{-1}[0, \infty) \ , \ X^- = d^{-1}(-\infty, 0] \subset X$$

are codimension 0 submanifolds with

$$\partial X^+ = \partial X^- = X^+ \cap X^- = d^{-1}(0) \subset X$$

a compact $(n-1)$-dimensional submanifold.

(iii) An *n-dimensional geometric Poincaré ribbon* (X, d) is a CW ribbon which is an open n-dimensional geometric Poincaré complex, i.e. equipped with a locally finite homology class $[X] \in H_n^{lf}(X)$ such that the cap products define $\mathbb{Z}[\pi]$-module isomorphisms

$$[X] \cap - : H_\pi^{n-*}(\widetilde{X}) \xrightarrow{\simeq} H_*^{lf,\pi}(\widetilde{X}) \ (\pi = \pi_1(X)) \ . \qquad \square$$

In particular, manifold ribbons are geometric Poincaré ribbons.

It is not required in 15.2 (iii) that $d : X \longrightarrow \mathbb{R}$ be geometric Poincaré transverse at $0 \in \mathbb{R}$, so that $X^+ \cap X^-$ need not be an $(n-1)$-dimensional geometric Poincaré complex (as is the case in 15.2 (ii)).

Manifolds with tame ends arise in the obstruction theory of Farrell [46, 47] and Siebenmann [145] for fibring manifolds over S^1, as finitely dominated infinite cyclic covers of compact manifolds. These are particular cases of 'bands':

Definition 15.3 (Siebenmann [143]) A *band* (W, c) is a compact space W with a map $c : W \longrightarrow S^1$ such that the pullback infinite cyclic cover $\overline{W} = c^*\mathbb{R}$ of W is finitely dominated, and such that the covering projection $\overline{W} \longrightarrow W$ determines a bijection between the path components of $\overline{W}$ and those of W. □

Similarly for *ANR band*, *CW band*, *geometric Poincaré band* and *manifold band.*

In Chapter 15 we shall be mainly concerned with bands $(W, c : W \longrightarrow S^1)$, particularly ones for which the infinite cyclic cover $(\overline{W}, \overline{c} : \overline{W} \longrightarrow \mathbb{R})$ is a ribbon. In 15.9 below it will be shown that if (W, c) is a ANR band with a π_1-fundamental domain (e.g. a manifold band) then $(\overline{W}, \overline{c})$ is an ANR ribbon. However, in general the infinite cyclic cover $(\overline{W}, \overline{c})$ of a band (W, c) is *not* a ribbon, since the π_1-conditions of 15.1 may fail:

Example 15.4 The infinite cyclic cover $(\overline{W}, \overline{c})$ of the CW band (W, c) constructed in Example 13.16 for $k = 1$ is not a ribbon, since $\pi_1(\overline{W}) = \{1\}$ and $\pi_1(Z) \neq \{1\}$ for any closed cocompact $Z \subseteq \overline{W}^+$. □

In Chapter 19 we shall develop the 'wrapping up' construction of bands from ribbons, tying the two ends of a ribbon (X, d) to obtain a band $(W, c) = (\widehat{X}, \widehat{d})$ with an infinite simple proper homotopy equivalence $(\overline{W}, \overline{c}) \simeq (X, d)$. In particular, the infinite cyclic cover of the band $(S^1, 1)$ is the ribbon $(\overline{S}^1, \overline{1}) = (\mathbb{R}, 1)$, and the wrapping up of the ribbon $(\mathbb{R}, 1)$ is the band $(\widehat{\mathbb{R}}, \widehat{1}) = (S^1, 1)$.

Wrapping up will be used in Chapters 17–20 to prove that:

(i) if W is a finite CW complex with a map $c : W \longrightarrow S^1$ then (W, c) is a band (i.e. the infinite cyclic cover $\overline{W} = c^*\mathbb{R}$ is finitely dominated) with a π_1-fundamental domain for the infinite cyclic cover $\overline{W} = c^*\mathbb{R}$ of W if and only if $(\overline{W}, \overline{c} : \overline{W} \longrightarrow \mathbb{R})$ is a ribbon,

(ii) if X is an infinite CW complex with a proper map $d : X \longrightarrow \mathbb{R}$ then (X, d) is infinite simple homotopy equivalent to a CW ribbon if and only if (X, d) is infinite simple homotopy equivalent to the infinite cyclic cover $(\overline{W}, \overline{c})$ of a CW band (W, c), if and only if $d : X \longrightarrow \mathbb{R}$ is proper homotopic to a bounded fibration, in which case X^+, X^- are both forward and reverse tame, and $H^{lf,\pi}_*(\widetilde{X}) = H_{*-1}(\widetilde{X})$,

(iii) if (X, d) is an n-dimensional geometric Poincaré ribbon then X is a finitely dominated $(n-1)$-dimensional geometric Poincaré complex,

(iv) if X is an open manifold of dimension $n \geq 5$ with a proper map $d : X \longrightarrow \mathbb{R}$ then (X, d) is a ribbon if and only if (X, d) is homeomorphic to the infinite cyclic cover $(\overline{W}, \overline{c})$ of an n-dimensional manifold band (W, c), if and only if $d : X \longrightarrow \mathbb{R}$ is proper homotopic to an approximate fibration. (Some of these results can be proved directly.)

Remark 15.5 (i) By 13.1 the condition on path components in 15.3 is equivalent to the surjectivity of $c_* : \pi_1(W_i) \longrightarrow \pi_1(S^1)$ for each path component W_i of W.

(ii) For the mapping torus $T(f)$ of a map $f : K \longrightarrow K$ of a compact space K the condition on path components in 15.3 is equivalent to f preserving the path components of K.

(iii) In dealing with CW (resp. ANR) bands (W, c) we shall always assume that the infinite cyclic cover $\overline{W}$ admits a compact CW (resp. ANR) fundamental domain. □

Example 15.6 (i) Let $f_1, f_2 : S^0 \longrightarrow S^0$ be the maps defined by

$$\begin{aligned} f_1 &= \text{identity} : S^0 \longrightarrow S^0 \; ; \; \pm 1 \longrightarrow \pm 1 \, , \\ f_2 &= \text{flip} : S^0 \longrightarrow S^0 \; ; \; \pm 1 \longrightarrow \mp 1 \, . \end{aligned}$$

The mapping torus $T(f_1)$ is a band, with

$$c_1 = \text{proj.} : T(f_1) = S^0 \times S^1 \longrightarrow S^1 \quad , \quad \overline{T}(f_1) = S^0 \times \mathbb{R} \, ,$$

since $\overline{p}_1 : \overline{T}(f_1) \longrightarrow T(f_1)$ induces a bijection between the path components. The mapping torus $T(f_2)$ is not a band, with

$$c_2 = 2 : T(f_2) = S^1 \longrightarrow S^1 \quad , \quad \overline{T}(f_2) = S^0 \times \mathbb{R} \, ,$$

since $\overline{p}_2 : \overline{T}(f_2) \longrightarrow T(f_2)$ does not induce a bijection between the path components.

(ii) If K is compact and $f : K \longrightarrow K$ is a homotopy equivalence then $\overline{T}(f) \simeq K$ is homotopy finite, and $T(f)$ is a band if and only if f preserves path components.

(iii) The mapping torus $T(h)$ of a homeomorphism $h : F \longrightarrow F$ of a compact $(n-1)$-dimensional manifold F which preserves path components is an n-dimensional manifold band $T(h)$ with

$$\zeta_{\overline{T}(h)} : \overline{T}(h) = F \times \mathbb{R} \longrightarrow \overline{T}(h) = F \times \mathbb{R} \; ; \; (x, t) \longrightarrow (h(x), t+1) \, .$$

(iv) A compact n-dimensional manifold band (M, c) is a fibre bundle over S^1 if and only if M is homeomorphic to the mapping torus $T(h)$ of a homeomorphism $h : F \longrightarrow F$ of a compact $(n-1)$-dimensional manifold F which preserves path components, such that c is homotopic to $M \cong T(h) \longrightarrow S^1$, so that the homeomorphism $M \cong T(h)$ lifts to a $\mathbb{Z}$-equivariant homeomorphism $\overline{M} \cong \overline{T}(h) = F \times \mathbb{R}$. □

Remark 15.7 (i) If (W, c) is a connected ANR band, then $\overline{W} = c^*\mathbb{R}$ has exactly two ends $\overline{W}^+, \overline{W}^-$ which will be shown in 15.9 to be both forward and reverse tame, with

$$e(\overline{W}^+) \simeq e(\overline{W}^-) \simeq \overline{W} .$$

The one-point compactification $\overline{W}^\infty$ is homotopy equivalent to the reduced suspension of $\overline{W}_+ = \overline{W} \cup \{\text{pt.}\}$

$$\Sigma(\overline{W}_+) = (\overline{W} \times I)/(\overline{W} \times \{0,1\}) ,$$

so that for a CW band (W, c)

$$H_*^{lf,\pi}(\widetilde{W}) = H_{*-1}(\widetilde{W})$$

with $\widetilde{W}$ any regular cover of $\overline{W}$ with group of covering translations π.

(ii) Let W be a connected finite CW complex with a map $c : W \longrightarrow S^1$ such that the infinite cyclic cover $\overline{W} = c^*\mathbb{R}$ of W is connected. By 13.13 $\overline{W}^+$ is forward tame if and only if $\overline{W}^-$ is reverse tame, with

$$\pi_1(\overline{W}^+) = \pi_1(\overline{W}^-) = \pi_1(\overline{W}) = \pi .$$

In 15.10 it will be proved that (W, c) is a band if and only if $\overline{W}^+$ is both forward and reverse tame. In Chapter 23 it will be proved that $\overline{W}^+$ is reverse (resp. forward) tame if and only if the cellular $\mathbb{Z}[\pi]$-module chain complex $C(\widetilde{W}^+)$ (resp. the π-locally finite cellular $\mathbb{Z}[\pi]$-module chain complex $C^{lf,\pi}(\widetilde{W}^+)$) is finitely dominated, with $\widetilde{W}^+$ the universal cover of W^+. It follows that if W is a finite n-dimensional geometric Poincaré complex the end $\overline{W}^+$ is forward tame if and only if $\overline{W}^+$ is reverse tame, if and only if (W, c) is a band, in which case $\overline{W}$ is a finitely dominated $(n-1)$-dimensional geometric Poincaré complex and cap product with the fundamental class

$$[\overline{W}] \in H_n^{lf}(\overline{W}) = H_{n-1}(\overline{W})$$

defines Poincaré duality isomorphisms

$$[\overline{W}] \cap - : H^{n-*}(\overline{W}) \xrightarrow{\simeq} H_*^{lf}(\overline{W}) = H_{*-1}(\overline{W}) .$$

(iii) Let $(V, \partial V)$ be an open n-dimensional manifold with a compact boundary ∂V and one end which is both reverse and forward tame, so that $(V; \partial V, e(V))$ is a finitely dominated n-dimensional geometric Poincaré cobordism. In Chapter 17 it will be shown that for $n \geq 5$ there exists an open neighbourhood of the end $\overline{W} \subset V$ which is the infinite cyclic cover of a compact n-dimensional manifold band (W, c) (the 'wrapping up' of V) such that there exists a compact $(n+1)$-dimensional cobordism $(M; \partial V \times S^1, W)$ with homotopy equivalences

$$(V; \partial V, e(V)) \times S^1 \simeq (M; \partial V \times S^1, W) \ , \ e(V) \simeq \overline{W}$$

and homeomorphisms

$$(M\backslash W, \partial V \times S^1) \cong (V, \partial V) \times S^1 \ , \ W \times \mathbb{R} \cong \overline{W} \times S^1 \ . \qquad \square$$

Recall from Chapter 13 that a fundamental domain $(V; U, \zeta U)$ for the infinite cyclic cover $\overline{W} = c^*\mathbb{R}$ of a band $(W, c : W \longrightarrow S^1)$ is 'π_1-fundamental' if the inclusions induce isomorphisms $\pi_1(U) \cong \pi_1(V)$, $\pi_1(\zeta U) \cong \pi_1(V)$.

Definition 15.8 A π_1-*band* is a band (W, c) which admits a π_1-fundamental domain. □

Recall from 13.11, 13.12 that there exist CW bands which are not π_1-bands, but that every CW band is simple homotopy equivalent to a π_1-band.

Proposition 15.9 *Let (W, c) be a compact ANR band.*

(i) *The spaces $\overline{W}^+, \overline{W}^-$ are forward and reverse π_1-tame.*

(ii) *The end spaces are such that there are defined homotopy equivalences*

$$e(\overline{W}) \simeq \overline{W} \amalg \overline{W} \ , \ e(\overline{W}^+) \simeq \overline{W} \simeq e(\overline{W}^-) \ .$$

(iii) *If (W, c) is a π_1-band*

$$\begin{aligned}
\pi_1^\infty(\overline{W}^+) &= \pi_1^\infty(\overline{W}^-) = \pi_1(\overline{W}) = \pi \ , \\
H_*(\widetilde{W}^+) &= H_*^{lf,\pi}(\widetilde{W}^+) \oplus H_*(\widetilde{W}) \ , \\
H_*(\widetilde{W}^-) &= H_*^{lf,\pi}(\widetilde{W}^-) \oplus H_*(\widetilde{W}) \ , \\
H_*(\widetilde{W}^+ \cap \widetilde{W}^-) &= H_*^{lf,\pi}(\widetilde{W}^+) \oplus H_*^{lf,\pi}(\widetilde{W}^-) \oplus H_*(\widetilde{W}) \ .
\end{aligned}$$

In particular, $(\overline{W}, \overline{c})$ is an ANR ribbon.

(iv) *If W is a CW π_1-band*

$$\begin{aligned}
&[\overline{W}^+]^{lf}_\infty = [\overline{W}^+]^{lf} \ , \ [\overline{W}^-]^{lf}_\infty = [\overline{W}^-]^{lf} \ , \\
&[\overline{W}] = [\overline{W}^+] + [\overline{W}^-] - [\overline{W}^+ \cap \overline{W}^-] \ , \\
&[\overline{W}^+]^{lf} = [\overline{W}^+] - [\overline{W}] = [\overline{W}^+ \cap \overline{W}^-] - [\overline{W}^-] \ , \\
&[\overline{W}^-]^{lf} = [\overline{W}^-] - [\overline{W}] = [\overline{W}^+ \cap \overline{W}^-] - [\overline{W}^+] \in K_0(\mathbb{Z}[\pi]) \ .
\end{aligned}$$

Proof (i) The finite domination of $\overline{W}$ implies that there exists a homotopy

$$f : \overline{W} \times I \longrightarrow \overline{W} \ ; \ (x, t) \longrightarrow f_t(x)$$

such that

$$f_0 = 1 : \overline{W} \longrightarrow \overline{W} \ , \ f_1(\overline{W}) \subseteq \overline{c}^{-1}[-k, k]$$

for some integer $k \geq 0$. Use the homotopy extension property to construct a homotopy

$$F : \overline{W} \times I \longrightarrow \overline{W} \; ; \; (x,t) \longrightarrow F_t(x)$$

such that

$$\begin{aligned} F_0 &= 1 : \overline{W} \longrightarrow \overline{W} \; , \; F_1(\overline{W}) \subseteq \overline{c}^{-1}[-k,\infty) \; , \\ F_t| &= \text{inclusion} : \overline{c}^{-1}[k+1,\infty) \longrightarrow \overline{W} \; \; (t \in I) \; . \end{aligned}$$

The generating covering translation $\zeta : \overline{W} \longrightarrow \overline{W}$ can be used to construct for each integer $m \geq 0$ a homotopy

$$h(m) : \overline{W} \times I \longrightarrow \overline{W} \; ; \; (x,t) \longrightarrow \zeta^{k+m} F_t \zeta^{-(k+m)}(x)$$

such that

$$\begin{aligned} h(m)_0 &= 1 : \overline{W} \longrightarrow \overline{W} \; , \; h(m)_1(\overline{W}) \subseteq \overline{c}^{-1}[m,\infty) \; , \\ h(m)_t| &= \text{inclusion} : \overline{c}^{-1}[2k+m+1,\infty) \longrightarrow \overline{W} \; \; (t \in I) \; . \end{aligned}$$

Define a sequence of positive integers $m_1, m_2, \dots$ by setting

$$m_1 = 1 \; , \; m_i = 2k + m_{i-1} + 1 \; \; (i > 1) \; .$$

Then the homotopies $h(m_i)$ for $i = 1, 2, \dots$ can be concatenated to define a proper homotopy $\overline{W}^+ \times [0,1) \longrightarrow \overline{W}$ extending the inclusion. Reversing the role of the two ends of $\overline{W}$ similarly gives a proper homotopy $\overline{W}^- \times [0,1) \longrightarrow \overline{W}$ extending the inclusion. Let $N \geq 0$ be an integer so large that $\zeta^N \overline{W}^+$ is disjoint from $\overline{W}^-$. The subspace

$$V = \zeta^N \overline{W}^+ \cup \overline{W}^- \subset \overline{W}$$

is cocompact, and such that there exists a proper homotopy $V \times [0,1) \longrightarrow \overline{W}$ extending the inclusion, so that $\overline{W}$ is forward tame, and hence $\overline{W}^+, \overline{W}^-$ are forward tame. The spaces $\overline{W}^+, \overline{W}^-$ are reverse tame by 13.13, and have stable π_1 at ∞ by 7.11, so that they are reverse π_1-tame.

(ii) The composite

$$p : e(\overline{W}^+) \xrightarrow{p_{\overline{W}^+}} \overline{W}^+ \longrightarrow \overline{W}$$

is a homotopy equivalence by 13.13. Reversing the role of the two ends of $\overline{W}$ similarly gives that the composite $e(\overline{W}^-) \longrightarrow \overline{W}^- \longrightarrow \overline{W}$ is a homotopy equivalence.

(iii) The homology identities are given as in 13.15 (ii).

(iv) These identities follow from (i), (ii) and $\overline{W} = \overline{W}^+ \cup \overline{W}^-$. □

Proposition 15.10 *Let W be a connected finite CW complex with a map $c : W \longrightarrow S^1$ such that $c_* : \pi_1(W) \longrightarrow \pi_1(S^1) = \mathbb{Z}$ is onto, and such that the infinite cyclic cover $\overline{W} = c^*\mathbb{R}$ of W is connected. The following conditions on (W, c) are equivalent:*

(i) *(W, c) is a CW band,*
(ii) *$\overline{W}^+$ and $\overline{W}^-$ are reverse tame,*
(iii) *$\overline{W}^+$ and $\overline{W}^-$ are forward tame,*
(iv) *$\overline{W}^+$ is forward and reverse tame,*
(v) *$\overline{W}^+$ and $\overline{W}^-$ are finitely dominated.*

If these conditions are satisfied

$$[\overline{W}^+]^{lf} \;=\; -[\overline{W}^-] \;,\; [\overline{W}^-]^{lf} \;=\; -[\overline{W}^+] \in \widetilde{K}_0(\mathbb{Z}[\pi])$$

where $\pi = \pi_1(\overline{W}) = \ker(c_* : \pi_1(W) \longrightarrow \pi_1(S^1))$.

Proof By 13.11 (ii) (W, c) is simple homotopy equivalent to a CW π_1-band (W', c'). By 9.6 and 9.8 the conditions (i)–(v) hold for (W, c) if and only if they hold for (W', c'). So there is no loss of generality in assuming that (W, c) is itself a π_1-band, with

$$\pi_1(\overline{W}^+ \cap \overline{W}^-) \;=\; \pi_1(\overline{W}^+) \;=\; \pi_1(\overline{W}^-) \;=\; \pi \;.$$

(i) $\Longleftrightarrow$ (v) Let $\widetilde{W}, \widetilde{W}^+, \widetilde{W}^-$ be the universal covers of $W, \overline{W}^+, \overline{W}^-$, so that

$$\widetilde{W} \;=\; \widetilde{W}^+ \cup \widetilde{W}^-$$

with $\widetilde{W}^+ \cap \widetilde{W}^-$ the universal cover of the finite CW complex $\overline{W}^+ \cap \overline{W}^-$. The cellular $\mathbb{Z}[\pi]$-module chain complexes fit into a short exact sequence

$$0 \longrightarrow C(\widetilde{W}^+ \cap \widetilde{W}^-) \longrightarrow C(\widetilde{W}^+) \oplus C(\widetilde{W}^-) \longrightarrow C(\widetilde{W}) \longrightarrow 0$$

with $C(\widetilde{W}^+ \cap \widetilde{W}^-)$ f.g. free. Thus $C(\widetilde{W})$ is finitely dominated if and only if $C(\widetilde{W}^+)$ and $C(\widetilde{W}^-)$ are finitely dominated. The space $\overline{W}$ is finitely dominated if and only if the $\mathbb{Z}[\pi]$-module chain complex $C(\widetilde{W})$ is finitely dominated (6.8), and similarly for $\overline{W}^+, \overline{W}^-$.

(ii) $\Longleftrightarrow$ (iii) $\Longleftrightarrow$ (iv) $\Longleftrightarrow$ (v) by 13.15.

The finiteness obstruction identities are given by 15.9. □

Definition 15.11 The *geometric fibring obstructions* of an ANR band (W, c) with respect to a choice of generating covering translation $\zeta : \overline{W} \longrightarrow \overline{W}$ are the torsions

$$\Phi^+(W, c) \;=\; \tau(T(\zeta) \longrightarrow W) \;,$$
$$\Phi^-(W, c) \;=\; \tau(T(\zeta^{-1}) \longrightarrow W) \in Wh(\pi_1(W))$$

of the homotopy equivalences

$$T(\zeta) \longrightarrow W \; ; \; (x,t) \longrightarrow p(x) \; ,$$
$$T(\zeta^{-1}) \longrightarrow W \; ; \; (x,t) \longrightarrow p(x)$$

with $p : \overline{W} \longrightarrow W$ the covering projection, using the canonical simple homotopy types on W (6.5) and $T(\zeta)$ and $T(\zeta^{-1})$ (14.3) to define the torsions.

□

Remark 15.12 (i) A CW band (W,c) is simple homotopy equivalent to the mapping torus $T(h)$ of a simple self homotopy equivalence $h : F \longrightarrow F$ of a finite CW complex F if and only if

$$\Phi^+(W,c) \;=\; \Phi^-(W,c) \;=\; 0 \in Wh(\pi_1(W))$$

by Ranicki [124, Chapter 20]. For any CW band (W,c)

$$\Phi^+(W \times S^1, c(pr_1)) \;=\; \Phi^-(W \times S^1, c(pr_1)) \;=\; 0 \in Wh(\pi_1(W) \times S^1) \; ,$$

with $W \times S^1$ simple homotopy equivalent to the mapping torus $T(h)$ of the simple self homotopy equivalence $h : F \longrightarrow F$ defined for any finite structure $(F, \phi : \overline{W} \times S^1 \longrightarrow F)$ on $\overline{W} \times S^1$ by

$$h \;=\; \phi(\zeta \times 1)\phi^{-1} \;:\; F \;\simeq\; \overline{W} \times S^1 \longrightarrow \overline{W} \times S^1 \;\simeq\; F \; .$$

(ii) The *torsion* of an n-dimensional geometric Poincaré complex W is

$$\tau(W) \;=\; \tau([W] \cap - : C(\widetilde{W})^{n-*} \longrightarrow C(\widetilde{W})) \in Wh(\pi_1(W)) \; ,$$

and W is *simple* if $\tau(W) = 0$. The fibring obstructions of an n-dimensional geometric Poincaré band (W,c) determine the torsion

$$\tau(W) \;=\; \Phi^+(W,c) + (-)^n \Phi^-(W,c)^* \in Wh(\pi_1(W)) \; .$$

Thus for a simple geometric Poincaré band (W,c) (e.g. a manifold band) the fibring obstructions are related by the duality

$$\Phi^+(W,c) \;=\; (-)^{n-1} \Phi^-(W,c)^* \in Wh(\pi_1(W)) \; ,$$

and $\Phi^+(W,c) \;=\; 0$ if and only if $\Phi^-(W,c) \;=\; 0$. In the manifold case $\Phi^+(W,c)$ is the fibring obstruction of Farrell [47] and Siebenmann [145], such that $\Phi^+(W,c) = 0$ if (and for $n \geq 6$ only if) $c : W \longrightarrow S^1$ is homotopic to the projection of a fibre bundle, i.e. if and only if W is homeomorphic to the mapping torus $T(h)$ of a self homeomorphism $h : F \longrightarrow F$ of a codimension 1 submanifold $F \subset W$. See Remark 24.17 below for the geometric interpretation of $\Phi^+(W,c) = \Phi^-(W,c) = 0$ for an ANR band (W,c). □

Proposition 15.13 *Let (W, c) be an ANR band. For any fundamental domain $(V; U, \zeta U)$ for $\overline{W}$ let*

$$\overline{W}^+ = \bigcup_{j=0}^{\infty} \zeta^j V \ , \ \overline{W}^- = \bigcup_{j=-\infty}^{-1} \zeta^j V \ .$$

(i) *There exists an integer $N^+ \geq 0$ such that $\bigcup_{j=0}^{N^+} \zeta^j V$ dominates $\overline{W}^+$* rel U.

(ii) *There exists an integer $N^- \geq 0$ such that $\bigcup_{j=-N^-}^{0} \zeta^j V$ dominates $\zeta\overline{W}^-$* rel ζU.

Proof (i) By 13.13 there exist a homotopy

$$h : \overline{W} \times I \longrightarrow \overline{W} \ ; \ (x, t) \longrightarrow h_t(x)$$

and an integer $N^+ \geq 0$ such that:

(a) $h_1(x) = x$ for $x \in \overline{W}$,

(b) $h_t(y) = y$ for $y \in \zeta^{-N^+}\overline{W}^- = \bigcup_{j=-\infty}^{-N^+-1} \zeta^j V, t \in I$,

(c) $h_0(\overline{W}^+) \subseteq \bigcup_{j=-N^+}^{-1} \zeta^j V$.

The maps

$$f : \overline{W}^+ \longrightarrow \bigcup_{j=0}^{N^+} \zeta^j V \ ; \ x \longrightarrow \zeta^{N^+} h_0 \zeta^{-N^+}(x) \ ,$$

$$g = \text{inclusion} : \bigcup_{j=0}^{N^+} \zeta^j V \longrightarrow \overline{W}^+$$

are such that there is defined a rel U homotopy

$$\zeta^{N^+} h_t \zeta^{-N^+} : gf \simeq 1 : \overline{W}^+ \longrightarrow \overline{W}^+ \ ,$$

so that $\bigcup_{j=0}^{N^+} \zeta^j V$ dominates $\overline{W}^+$ rel U.

(ii) As for (i), with the role of $\overline{W}^+, \overline{W}^-$ reversed. □

Definition 15.14 Let (W, c) be a band.

(i) The band (W, c) is *positively relaxed* if there exists a fundamental domain $(V; U, \zeta U)$ for $\overline{W}$ such that V dominates $\overline{W}^+$ rel U.

(ii) The band (W, c) is *negatively relaxed* if there exists a fundamental domain $(V; U, \zeta U)$ for $\overline{W}$ such that $\zeta^{-1}V$ dominates $\overline{W}^-$ rel U.

(iii) The band (W, c) is *relaxed* if there exists a fundamental domain $(V; U, \zeta U)$ for $\overline{W}$ which is both positively and negatively relaxed, i.e. if $N^+ = N^- = 0$ in 15.13. □

Example 15.15 (i) The mapping torus $T(h)$ of a self homotopy equivalence $h : F \longrightarrow F$ of a finite CW complex F is a relaxed CW band; in this case the infinite cyclic cover $\overline{T}(h)$ has a fundamental domain

$$(V; U, \zeta U) \simeq F \times (I; \{0\}, \{1\})$$

such that the inclusions $V \longrightarrow \overline{T}(h)^+$, $\zeta^{-1}V \longrightarrow \overline{T}(h)^-$ are homotopy equivalences rel U.

(ii) In 16.15 it is proved that if W is a finite CW complex with a map $c : W \longrightarrow S^1$ such that $c_* : \pi_1(W) \longrightarrow \pi_1(S^1)$ is onto, and which is homotopic to an approximate fibration, then (W, c) is a relaxed CW band. □

How do manifold bands arise? The obvious sources are:

(a) fibre bundles over S^1, e.g. complements of fibred knots,

(b) manifold tame ends, since these have open neighbourhoods which are infinite cyclic covers of manifold bands.

Here is another source:

(c) surgery theory.

Example 15.16 Surgery theory can be used to construct manifold bands (W, c) in dimensions $n \geq 5$ with prescribed fundamental group, as follows. For the sake of simplicity we shall only consider the untwisted case, with $c_* = \text{projection} : \pi_1(W) = \pi \times \mathbb{Z} \longrightarrow \mathbb{Z}$ for some finitely presented group π. Let $A = \mathbb{Z}[\pi]$, with Laurent polynomial extension $A[z, z^{-1}] = \mathbb{Z}[\pi \times \mathbb{Z}]$. As in Ranicki [128, 130] define a square matrix $\omega = (\omega_{ij})_{1 \leq i,j \leq k}$ with entries $\omega_{ij} \in A[z, z^{-1}]$ to be *Fredholm* if the $A[z, z^{-1}]$-module morphism

$$\begin{aligned} \omega \ : A[z, z^{-1}]^k &\longrightarrow A[z, z^{-1}]^k \ ; \\ (a_1, a_2, \dots, a_k) &\longrightarrow (\sum_{i=1}^k a_i \omega_{i1}, \sum_{i=1}^k a_i \omega_{i2}, \dots, \sum_{i=1}^k a_i \omega_{ik}) \end{aligned}$$

is injective and the cokernel is a f.g. projective A-module, or equivalently if ω becomes invertible over the Novikov rings $A((z))$ and $A((z^{-1}))$ (cf. 23.1 and 23.2 below). Let Ω be the set of Fredholm matrices in $A[z, z^{-1}]$, and let $\Lambda = \Omega^{-1}A[z, z^{-1}]$ be the (noncommutative) localization of $A[z, z^{-1}]$ inverting Ω. This type of localization is a generalization of the single-element inversion of 2.28 (i). The injection $A[z, z^{-1}] \longrightarrow \Lambda$ is a ring morphism with the universal property that a finite f.g. free $A[z, z^{-1}]$-module chain complex C is A-finitely dominated if and only if $H_*(\Lambda \otimes_{A[z,z^{-1}]} C) = 0$, by Ranicki

[126]. For a connected finite CW complex X with $\pi_1(X) = \pi \times \mathbb{Z}$ we have $\mathbb{Z}[\pi_1(X)] = A[z,z^{-1}]$ and $H_*(X;\Lambda) = H_*(\Lambda \otimes_{A[z,z^{-1}]} C(\widetilde{X}))$, with $\widetilde{X}$ the universal cover of X. The infinite cyclic cover $\overline{X} = \widetilde{X}/\pi$ of X is finitely dominated if and only if $H_*(X;\Lambda) = 0$. The realization theorems of Chapters 5 and 6 of Wall [165] for the L-groups have Λ-homology surgery versions for the Γ-groups of Cappell and Shaneson [20]. It is thus possible to realize every element $x \in L_{n+1}(\Lambda) = \Gamma_{n+1}(A[z,z^{-1}] \longrightarrow \Lambda)$ as the Λ-coefficient surgery obstruction $x = \sigma_*(f,b)$ of an $(n+1)$-dimensional normal map

$$(f,b) \ : \ (L; M \times S^1, N) \longrightarrow M \times S^1 \times (I; \{0\}, \{1\})$$

with M a closed $(n-1)$-dimensional manifold such that $\pi_1(M) = \pi$,

$$(f,b)| \ = \ 1 \ : \ M \times S^1 \longrightarrow M \times S^1$$

and with the restriction

$$(e,a) \ = \ (f,b)| \ : \ N \longrightarrow M \times S^1$$

a normal map inducing isomorphisms

$$e_* \ : \ \pi_1(N) \xrightarrow{\simeq} \pi_1(M \times S^1) \ = \ \pi \times \mathbb{Z} \ ,$$
$$e_* \ : \ H_*(N;\Lambda) \xrightarrow{\simeq} H_*(M \times S^1;\Lambda) \ = \ 0 \ .$$

Then (N,c) is an n-dimensional manifold band, with

$$c \ : \ N \xrightarrow{e} M \times S^1 \xrightarrow{\text{proj.}} S^1$$

inducing $c_* =$ projection : $\pi_1(N) = \pi \times \mathbb{Z} \longrightarrow \pi_1(S^1) = \mathbb{Z}$. (The construction of manifold bands in 10.3 (iii) is a special case). See [128] for a more detailed account, including the identification $L_{2*}(\Lambda) = W(A)$ with $W(A)$ the Witt group of nonsingular asymmetric forms over A of Quinn [113], and the computation $L_{2*+1}(\Lambda) = 0$ (implicit in [113]). □

We shall be more concerned with the geometric construction of bands. In Chapters 18–20 we shall relate the geometric and homotopy theoretic properties of ribbons and bands, using elementary versions of the geometric twist glueing of Siebenmann [145]:

Definition 15.17 Let X be a locally compact Hausdorff space with a proper map $d : X \longrightarrow \mathbb{R}$ and an end-preserving homeomorphism $h : X \longrightarrow X$. Suppose there exist homeomorphisms $f_\pm : U_\pm \longrightarrow X$ with $U_\pm \subseteq X$ open subspaces, $U_- \cap U_+ = \emptyset$, for which there exist $m > n > 0$ such that

$$d^{-1}(-\infty,-n) \subseteq U_- \ \ , \ \ d^{-1}(n,\infty) \subseteq U_+ \ ,$$

and such that the restrictions

$$f_-| \ : \ d^{-1}(-\infty,-m] \longrightarrow X \ \ , \ \ f_+| : d^{-1}[m,\infty) \longrightarrow X$$

are the inclusions. The *Siebenmann twist glueing* of X relative to h is the identification space

$$W_h(f_-,f_+) \ = \ X/\sim$$

with $\sim$ the equivalence relation generated by $x \sim f_+^{-1}hf_-(x)$ for $x \in U_-$. □

Note that $W_h(f_-,f_+)$ is homotopy equivalent to the mapping torus $T(h)$.

Siebenmann [145, p. 19] proved the following uniqueness result by an elementary method.

Proposition 15.18 *Let X be a locally compact Hausdorff space with a proper map $d : X \longrightarrow \mathbb{R}$. If the two ends of X have arbitrarily small neighbourhoods $U_\pm$ with homeomorphisms $f_\pm : U_\pm \longrightarrow X$ as above, then any two twist glueings $W_h(f_-,f_+), W_h(f'_-,f'_+)$ with respect to the same end-preserving homeomorphism $h : X \longrightarrow X$ are homeomorphic.* □

We shall be particularly concerned with the Siebenmann twist glueing construction in two special cases:

(i) Given a manifold ribbon (X,d) with $d : X \longrightarrow \mathbb{R}$ a 'manifold approximate fibration' (Chapter 16) we construct in Chapter 17 the 'geometric wrapping up' $(M,c) = (\widehat{X},\widehat{c})$, a manifold band with infinite cyclic cover $(\overline{M},\overline{c}) = (X,d)$. The construction in Chapter 17 uses elementary properties of approximate fibrations. In 18.6 M is identified with the Siebenmann twist glueing of X relative to $1 : X \longrightarrow X$.

(ii) The 'relaxation' of a manifold band (M,c) is constructed in Chapter 18 to be an h-cobordant manifold band (M',c') with $c' : M' \longrightarrow S^1$ a manifold approximate fibration. The construction in Chapter 18 uses elementary properties of 'bounded fibrations' (Chapter 16). In 18.7 M' is identified with the Siebenmann twist glueing of the infinite cyclic cover $\overline{M}$ of M relative to a generating covering translation $\zeta : \overline{M} \longrightarrow \overline{M}$.

Definition 15.19 Let (X,d), (X',d') denote spaces X, X' with maps $d : X \longrightarrow \mathbb{R}$, $d' : X' \longrightarrow \mathbb{R}$ (but not necessarily ribbons).

(i) A *homotopy equivalence* $f : (X,d) \longrightarrow (X',d')$ is a homotopy equivalence $f : X \longrightarrow X'$.

(ii) Suppose that the maps d, d' are proper. A *proper homotopy equivalence* $f : (X,d) \longrightarrow (X',d')$ is a proper homotopy equivalence $f : X \longrightarrow X'$ with a proper homotopy $d'f \simeq d : X \longrightarrow \mathbb{R}$. □

In Chapter 19 we shall develop a CW analogue of the Siebenmann twist glueing: given a connected CW ribbon (X,d) and a homotopy equivalence $h : (X,d) \longrightarrow (X,d)$ we use h to tie the two ends of X together to obtain a CW complex $X(h)$ with a map $d(h) : X(h) \longrightarrow S^1$ and with homotopy equivalences $(X,d) \simeq (\overline{X}(h), \overline{d}(h))$, $X(h) \simeq T(h)$. In the special case when $h : (X,d) \longrightarrow (X,d)$ is a proper homotopy equivalence with h a covering translation or the identity we obtain a relaxed CW band $(X[h], d[h])$ with a proper homotopy equivalence $(X,d) \simeq (\overline{X}[h], \overline{d}[h])$ which is simple in the sense of the infinite simple homotopy theory of Siebenmann [144]. In Chapter 20 the 'wrapping up' of a CW ribbon (X,d)

$$(\widehat{X}, \widehat{d}) = (X[1], d[1])$$

will be used to prove that $d : X \longrightarrow \mathbb{R}$ is proper homotopic to a proper bounded fibration. The 'relaxation' of a CW π_1-band (W,c) is defined in Chapter 20 to be the relaxed CW band

$$(W', c') = (\overline{W}[\zeta], \overline{c}[\zeta])$$

in the homotopy type of (W,c). In Chapters 24, 25 below we shall apply the homotopy theoretic twist glueing to study the fibring obstructions of relaxed CW bands (which are distinguished by the property that the $\widetilde{\text{Nil}}$-components vanish).

16

Approximate fibrations

We characterize forward and reverse tameness for open manifolds in terms of approximate lifting properties; the main result (16.13) is that an open manifold W of dimension $n \geq 5$ is forward tame and reverse tame if and only if there exists an open cocompact $X \subseteq W$ with a manifold approximate fibration $d : X \longrightarrow \mathbb{R}$. In Chapter 17 it will then be shown that an open manifold X of dimension $n \geq 5$ is the total space of a manifold approximate fibration $d : X \longrightarrow \mathbb{R}$ if and only if it is the infinite cyclic cover $X = \overline{M}$ of an n-dimensional manifold band (M, c).

Definition 16.1 Let B be a metric space.

(i) Let $\epsilon > 0$, and let X, Y be spaces equipped with maps $p : X \longrightarrow B$, $q : Y \longrightarrow B$. A map $f : X \longrightarrow Y$ is an *ϵ-homotopy equivalence over B* if there exists a map $g : Y \longrightarrow X$ together with homotopies

$$h : gf \simeq 1 : X \longrightarrow X \ , \ k : fg \simeq 1 : Y \longrightarrow Y$$

such that for all $x \in X, y \in Y$

$$d(p(x), qf(x)) < \epsilon \ , \ d(pg(y), q(y)) < \epsilon \ ,$$
$$\text{diameter}\, ph(\{x\} \times I) < \epsilon \ , \ \text{diameter}\, qk(\{y\} \times I) < \epsilon \ .$$

(ii) A *bounded homotopy equivalence* of spaces with maps to B is a homotopy equivalence which is an ϵ-homotopy equivalence over B for some $\epsilon > 0$.

□

Definition 16.2 Let B be a metric space.

(i) Let $\epsilon > 0$. An *ϵ-fibration* is a map $p : X \longrightarrow B$ such that for any space W and maps $f : W \longrightarrow X$, $F : W \times I \longrightarrow B$ satisfying

$$F(x, 0) = pf(x) \ \ (x \in W)$$

there exists a map $\widetilde{F} : W \times I \longrightarrow X$ such that

$$\widetilde{F}(x, 0) = f(x) \ , \ d(p\widetilde{F}(x, t), F(x, t)) < \epsilon \ \ (x \in W, t \in I) \ .$$

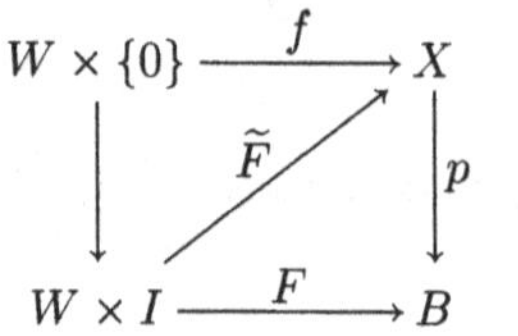

$$\begin{array}{ccc} W \times \{0\} & \xrightarrow{f} & X \\ \big\downarrow & \overset{\widetilde{F}}{\nearrow} & \big\downarrow p \\ W \times I & \xrightarrow{F} & B \end{array}$$

(ii) If $U \subseteq B$ then $p : X \longrightarrow B$ is an *ϵ-fibration over U* if the image of the homotopy F above is in U.

(iii) A *bounded fibration* is a map $p : X \longrightarrow B$ which is an ϵ-fibration for some $\epsilon > 0$. □

Proposition 16.3 *A map $p : X \longrightarrow \mathbb{R}$ from a metric space X is a bounded fibration if and only if it is bounded homotopy equivalent to the projection $Y \times \mathbb{R} \longrightarrow \mathbb{R}$ for some space Y.*

Proof Let p_i denote the projection onto the ith coordinate space ($i = 1, 2$). Assume first that X is bounded homotopy equivalent to $Y \times \mathbb{R}$ over $\mathbb{R}$ so that there exist maps $f : X \longrightarrow Y \times \mathbb{R}$ and $g : Y \times \mathbb{R} \longrightarrow X$ as in 16.1 (i) for some $\epsilon > 0$. Suppose we are given a lifting problem $F : W \times I \longrightarrow \mathbb{R}$ and $h : W \longrightarrow X$ such that $F(x, 0) = ph(x)$ for each $x \in W$. Convert this to a lifting problem in $Y \times \mathbb{R}$ by defining

$$G : W \times I \longrightarrow \mathbb{R} \; ; \; (x, t) \longrightarrow F(x, t) - F(x, 0) + p_2 f h(x)$$

with solution

$$\widetilde{G} : W \times I \longrightarrow Y \times \mathbb{R} \; ; \; (x, t) \longrightarrow (p_1 f h(x), G(x, t)) \; .$$

Note that G is ϵ-close to F and $pg\widetilde{G}$ is 2ϵ-close to F.

Let $H : 1 \simeq gf : X \longrightarrow X$ be a homotopy such that for all $x \in X$

$$\text{diameter}\, pH(\{x\} \times I) < \epsilon \; .$$

Define

$$F' : W \times [-1, 1] \longrightarrow X \; ; \; (x, t) \longrightarrow \begin{cases} g\widetilde{G}(x, t) & \text{if } 0 \le t \le 1 \; , \\ H(h(x), 1 + t) & \text{if } -1 \le t \le 0 \; . \end{cases}$$

Assume for the moment that W is paracompact. Then there exists a map $\alpha : W \longrightarrow (0, 1]$ such that for all $x \in W$

$$\text{diameter } pF'(\{x\} \times [-1, \alpha(x)]) < \epsilon$$

(see Dugundji [38, page 179]). Define

$$J : W \times I \longrightarrow W \times [-1, 1] \; ; \; (x, t) \longrightarrow \begin{cases} (x, t) & \text{if } \alpha(x) \le t \le 1 \; , \\ (x, t + \dfrac{t}{\alpha(x)} - 1) & \text{if } 0 \le t \le \alpha(x) \; . \end{cases}$$

Then $\widetilde{F} = F'J$ has the property that $\widetilde{F}(x,0) = h(x)$ and $p\widetilde{F}$ is 3ϵ-close to F. To avoid the paracompactness assumption, let

$$E(p) = \{(x,\lambda) \in X \times \mathbb{R}^I \mid p(x) = \lambda(0)\} .$$

Then $E(p)$ is a metric space (hence, paracompact) and, in the standard way, a universal lifting problem can be given in terms of $E(p)$, the bounded solution of which implies that p is a bounded fibration.

Conversely, assume that p is a bounded fibration. Define

$$f : X \longrightarrow X \times \mathbb{R} ; x \longrightarrow (x, p(x)) .$$

Note that $p = p_2 f$. We shall show that f is a bounded homotopy equivalence. In order to get an inverse, consider the following lifting problem. Let

$$h = \text{projection} : X \times \mathbb{R} \longrightarrow X ,$$
$$F : X \times \mathbb{R} \times I \longrightarrow \mathbb{R} ; (x,y,t) \longrightarrow (1-t)p(x) + ty .$$

Then there exists a map $\widetilde{F} : X \times \mathbb{R} \times I \longrightarrow X$ such that

$$\widetilde{F}(x,y,0) = x \quad ((x,y) \in X \times \mathbb{R})$$

and $p\widetilde{F}$ is ϵ-close to F for some $\epsilon > 0$. We claim that $\widetilde{F}_1 : X \times \mathbb{R} \longrightarrow X$ is a bounded homotopy inverse for f. First note that $\widetilde{F}_t f : 1 \simeq \widetilde{F}_1 f : X \longrightarrow X$ is a 2ϵ-homotopy over $\mathbb{R}$. The map

$$G : X \times \mathbb{R} \times I \longrightarrow X \times \mathbb{R} ; (x,y,t) \longrightarrow (\widetilde{F}(x,y,t), (1-t)y + tp_2 f \widetilde{F}(x,y,1))$$

is a 2ϵ-homotopy over $\mathbb{R}$

$$G : 1 \simeq f\widetilde{F}_1 : X \times \mathbb{R} \longrightarrow X \times \mathbb{R} . \qquad \square$$

Corollary 16.4 *If $p : X \longrightarrow \mathbb{R}$ is a proper bounded fibration such that $p^{-1}[0,\infty)$ and $p^{-1}(-\infty,0]$ are ANR's, then X is forward tame and reverse tame.*

Proof Apply 16.3 and 9.12. □

The following result says that the existence of a bounded fibration to $\mathbb{R}$ is a proper homotopy invariant.

Proposition 16.5 *If X is a metric space, $f : X \longrightarrow Y$ is a proper homotopy equivalence and $p : Y \longrightarrow \mathbb{R}$ is a bounded fibration then $pf : X \longrightarrow \mathbb{R}$ is properly homotopic to a bounded fibration $d : X \longrightarrow \mathbb{R}$.*

Proof Let $g : Y \longrightarrow X$ be a proper homotopy inverse for f and let $G : \mathrm{id}_X \simeq gf$ be a proper homotopy. Let $n_0 = 0$ and use the properness of p, f, g, and G to inductively choose $n_i \geq i$ $(i = 0, 1, 2, \ldots)$ so that:

(i) $pfg(p^{-1}[-n_{i-1}, n_{i-1}]) \subseteq [-n_i, n_i]$,
(ii) $pfg(p^{-1}[\pm n_i, \pm\infty)) \subseteq [\pm n_{i-1}, \pm\infty)$,
(iii) $pfG_t((pf)^{-1}[-n_{i-1}, n_{i-1}]) \subseteq [-n_i, n_i]$ for each t,
(iv) $pfG_t((pf)^{-1}[\pm n_i, \pm\infty)) \subseteq [\pm n_{i-1}, \pm\infty)$ for each t.

Let $\gamma : \mathbb{R} \longrightarrow \mathbb{R}$ be the PL homeomorphism such that

$$\gamma(\pm n_i) = \pm i \quad (i = 0, 1, 2, \ldots) .$$

Then pf is properly homotopic to $d = \gamma pf : X \longrightarrow \mathbb{R}$. One may check that γp is 3-close to γpfg and that $\gamma pfG : X \times I \longrightarrow \mathbb{R}$ is a 3-homotopy. In order to show that d is a bounded fibration, consider a lifting problem $h : Z \longrightarrow X$ and $H : Z \times I \longrightarrow \mathbb{R}$ such that

$$dh(z) = H(z, 0) \quad (z \in Z) .$$

It follows from standard arguments that we may assume that Z is paracompact (see the proof of 16.3). This lifting problem induces a lifting problem $fh : Z \longrightarrow Y$ and $H : Z \times I \longrightarrow \mathbb{R}$ for $\gamma p : Y \longrightarrow \mathbb{R}$. Suppose that $p : Y \longrightarrow \mathbb{R}$ is a b-fibration. Since γ is distance nonincreasing, γp is also a b-fibration. There is a solution $\widehat{H} : Z \times I \longrightarrow Y$ such that

$$fh(z) = \widehat{H}(z, 0) \quad (z \in Z)$$

and $\gamma p\widehat{H}$ is b-close to H. Use the paracompactness of Z to define a map $\phi : Z \longrightarrow (0, 1]$ such that $H(z, 0)$ is 1-close to $H(z, t)$ if $t \leq \phi(z)$. Define

$$\widetilde{H} : Z \times I \longrightarrow X ; (z, t) \longrightarrow \begin{cases} G(h(z), t/\phi(z)) & \text{if } 0 \leq t \leq 1/2 , \\ g\widehat{H}(z, t/\phi(z) - 1) & \text{if } 1/2 \leq t \leq 1 . \end{cases}$$

Then

$$h(z) = \widetilde{H}(z, 0) \quad (z \in Z)$$

and $d\widetilde{H}$ is $\max\{4, 3 + b\}$-close to H. Hence d is a bounded fibration. □

Note that the proof above just requires X to be properly dominated by a bounded fibration, rather than proper homotopy equivalent to one.

We shall be mainly concerned with bounded fibrations $p : X \longrightarrow B$ with p a proper map.

Definition 16.6 (i) An *approximate fibration* $p : X \longrightarrow B$ is a map which is an ϵ-fibration for every $\epsilon > 0$.

(ii) A *manifold approximate fibration* $p : W \longrightarrow B$ is an approximate fibration such that W and B are manifolds (either finite dimensional without boundary or Hilbert cube manifolds), and such that p is a proper map. □

Proposition 16.7 *A map $p : X \longrightarrow \mathbb{R}$ from a metric space X is an approximate fibration if and only if for every $\epsilon > 0$ it is ϵ-equivalent over $\mathbb{R}$ to the projection $Y \times \mathbb{R} \longrightarrow \mathbb{R}$ for some space Y.*

Proof The proof of 16.3 shows that a space which is ϵ-equivalent over $\mathbb{R}$ to a product with $\mathbb{R}$ is a 3ϵ-fibration. And conversely, an ϵ-fibration is 2ϵ-equivalent over $\mathbb{R}$ to a product with $\mathbb{R}$. □

More general versions of 16.3 and 16.7 are given in Hughes, Taylor and Williams [77, p. 47].

Proposition 16.8 *Let X be a metric space with a map $p : X \longrightarrow [0, \infty)$ which is ϵ-homotopy equivalent at ∞ to the projection $Y \times [0, \infty) \longrightarrow [0, \infty)$ for some space Y and some $\epsilon > 0$. Then there exists $u > 0$ such that p is an 3ϵ-fibration over (u, ∞).*

Proof Let $(f, g, X', Y') : X \longrightarrow Y \times [0, \infty)$ be an ϵ-equivalence at ∞ with $X' = p^{-1}([s, \infty))$ and $Y' = Y \times [t, \infty)$. For u much larger than s, t, the proof of 16.3 shows that p is a 3ϵ-fibration over (u, ∞). □

We shall use the sucking principle of Chapman [25, 26] to gain the control necessary to pass from bounded fibrations to approximate fibrations. This says that, for manifolds, there is essentially no difference between proper approximate fibrations and proper ϵ-fibrations for sufficiently small $\epsilon > 0$. At its simplest, the sucking principle takes on the following form.

Theorem 16.9 (Chapman) *For every $n \geq 5$ and $\epsilon > 0$ there exists $\delta > 0$ such that if W is an open n-dimensional manifold or a Hilbert cube manifold and $p : W \longrightarrow \mathbb{R}$ is a proper δ-fibration, then p is ϵ-homotopic to a manifold approximate fibration $p' : W \longrightarrow \mathbb{R}$.* □

The proof of 16.9 uses controlled engulfing.

Corollary 16.10 *If W is an open manifold of dimension ≥ 5 or a Hilbert cube manifold and $p : W \longrightarrow \mathbb{R}$ is a proper bounded fibration, then p is boundedly homotopic to a manifold approximate fibration.*

Proof Suppose p is an ϵ-fibration for some $\epsilon > 0$. Choose $\delta > 0$ by Theorem 16.9 so that any proper δ-fibration from an n-manifold to $\mathbb{R}$ is 1-homotopic to a manifold approximate fibration. Choose $L > 0$ so large that $\epsilon/L < \delta$ and define

$$\gamma \; : \; \mathbb{R} \longrightarrow \mathbb{R} \; ; \; x \longrightarrow x/L \; .$$

Then $\gamma p : M \longrightarrow \mathbb{R}$ is an (ϵ/L)-fibration and is 1-homotopic to a manifold approximate fibration $p' : M \longrightarrow \mathbb{R}$. It follows that p is L-homotopic to $\gamma^{-1}p'$ which is a manifold approximate fibration. □

Proposition 16.11 *For every $n \geq 5$ there exists $\epsilon > 0$ such that if $(W, \partial W)$ is an open n-dimensional manifold with compact boundary or a Hilbert cube manifold, and $p : W \longrightarrow [0, \infty)$ is a proper ϵ-fibration over (a, ∞) for some $a > 0$, then p is properly homotopic to a map p' such that*

$$p'| \; : \; (p')^{-1}(a+1, \infty) \longrightarrow (a+1, \infty)$$

is a manifold approximation fibration.

Proof The proof follows from Chapman's proof of 16.9 in [25, 26]. □

The next example shows that this version of Chapman's sucking principle fails for ANR's.

Example 16.12 We construct a non-manifold 2-dimensional CW band (W, c) such that $\overline{c} : \overline{W} \longrightarrow \mathbb{R}$ is not properly homotopic to an approximate fibration even though $\overline{c}$ is a proper bounded fibration.

Let D be the topologist's dunce cap, the space obtained from the standard 2-simplex σ by identifying its three edges, two with the same orientation and one with the opposite orientation. So D is a CW complex with one 0-cell, one 1-cell, and one 2-cell. Let x be the 0-cell and let y be a point in the interior of the 2-cell.

Let W be the space obtained from D by identifying x and y. Clearly, W can be given the structure of a finite 2-dimensional CW complex. (If D is obtained from the standard 2-simplex σ by identifying edges, and y is the barycentre of σ, the subdivision of σ obtained by starring from y induces a CW structure on D which in turn induces a CW structure on W. This CW structure on W has one 0-cell, four 1-cells, and three 2-cells.)

Fix a classifying map $c : W \longrightarrow S^1$ such that $c^{-1}(1) = \{x = y\}$. A $\mathbb{Z}$-equivariant lift $\overline{c} : \overline{W} \longrightarrow \mathbb{R}$ is a bounded fibration. This is because there is a strong deformation retraction of D to an arc in D joining x and y. This strong deformation retraction induces a bounded strong deformation retraction of $\overline{W}$ to a copy L of $\mathbb{R}$ such that $\overline{c}|_L : L \longrightarrow \mathbb{R}$ is a homeomorphism. It also follows from 17.14 below that $\overline{c}$ is a bounded fibration.

Write

$$\overline{W} = \bigcup_{i=-\infty}^{\infty} D_i$$

where D_i is a fundamental domain homeomorphic to D with vertices $z_i \in D_i$ such that $\overline{c}(z_i) = i$ and $D_i \cap D_{i+1} = \{z_{i+1}\}$.

Assume by way of contradiction that $\overline{c}$ is properly homotopic to an approximate fibration $d : \overline{W} \longrightarrow \mathbb{R}$. Since $\overline{W}$ and $\mathbb{R}$ are contractible, so is the homotopy fibre of d. This in turn implies that d is a cell-like map (Ferry [53, p. 337]). It can also be verified directly that d is a cell-like map. Since d is proper, there exists an integer i such that $d(z_i) < d(z_{i+1})$. Choose t such that $d(z_i) < t < d(z_{i+1})$.

Claim 1 $d^{-1}(t) \subseteq D_i$.
Proof Suppose on the contrary that there exists $u \in d^{-1}(t)$ such that $u \notin D_i$. Assume without loss of generality that $u \in \bigcup_{j<i} D_j$. By the connectivity of D_i, there exists $v \in D_i$ such that $d(v) = t$. Since $z_i \notin d^{-1}(t)$, $d^{-1}(t)$ has at least two components: one containing u, and another containing v. This contradicts the cell-likeness of $d^{-1}(t)$. □

Identify D_i with D and assume $i = 0$. Thus, we have a map $d : D \longrightarrow I$ such that $d(x) = 0$, $d(y) = 1$, and $d^{-1}(t)$ is cell-like for each $t \in (0,1)$. It follows that $d^{-1}(\frac{1}{2})$ is a cell-like subset of D which separates x from y. The following claim provides a contradiction.

Claim 2 *If $Z \subseteq D$ is a cell-like set and $x \notin Z$, then Z does not separate D.*
Proof Let $\pi : \sigma \longrightarrow D$ denote the quotient map. The proof is based on the following five items:

(i) Every proper 2-dimensional subpolyhedron P of D contains a free 1-dimensional face. For let Q be the subpolyhedron of σ which is the union of all 2-simplexes in $\pi^{-1}(P)$. It suffices to observe that that Q has a free 1-dimensional face not in $\partial\sigma$.

(ii) Every subpolyhedron P of D is aspherical. If $P = D$ this is true. On the other hand if P is proper, then (i) implies that P collapses to a 1-dimensional subpolyhedron which of course is aspherical. It also follows from Papakyriakopoulos [107, p. 19] that any subpolyhedron of D is aspherical.

(iii) If $P \subseteq Q \subseteq D \backslash \{x\}$ are subpolyhedra, P contracts to a point in Q and Q contracts to a point in $D\backslash\{x\}$, then there exists a contractible subpolyhedron P' of D such that $P \subseteq P' \subseteq Q$. By (ii) it suffices to find a simply-connected subpolyhedron P' with $P \subseteq P' \subseteq Q$. This would follow from standard plane topology if $Q \subseteq D\backslash\pi(\partial\sigma)$ (by adding appropriate complementary domains of P). In the more general case, pass to the universal cover U of $D\backslash V$ where V is the interior of a small regular neighbourhood of x. Lift P and Q to subpolyhedra $\widetilde{P}$, $\widetilde{Q}$ of U. The fundamental domain F for U is homeomorphic to a 2-cell so plane topology can be used in U to build P'. The idea is to work inductively starting at the outermost translates of F which meet $\widetilde{P}$ and add appropriate complementary domains in that translate to $\widetilde{P}$.

(iv) If $P \subseteq Q \subseteq D$ are subpolyhedra such that Q collapses to P across 2-simplexes and Q separates D, then P separates D. This can be verified by pulling P and Q back to σ and examining the several possible cases.

(v) No contractible 1-dimensional subpolyhedron P of D separates D. Since P collapses to a point and D is not separated by a point, a 1-dimensional analogue of (iv) is needed. As in (iv) this is verified by pulling back to σ and examining the several possible cases.

Given these five items we can finish the proof of Claim 5. It follows from (iii) that we can write

$$Z = \bigcap_{i=1}^{\infty} Z_i$$

where $Z_{i+1} \subseteq Z_i$ and Z_i is a contractible proper subpolyhedron of D. If Z separates D, then we may assume that each Z_i separates D. It follows from (i) and (iv) that Z_1 collapses to a 1-dimensional subpolyhedron P which does not separate D. Since Z_1 is contractible, so is P, which is a contradiction to (v). □

Theorem 16.13 *Let W be an open manifold of dimension ≥ 5 with compact boundary or a Hilbert cube manifold. The following conditions on W are equivalent:*

(i) *W is both forward and reverse tame,*
(ii) *W is forward tame and the end space $e(W)$ is finitely dominated,*
(iii) *there exist an open cocompact $X \subseteq W$ and a manifold approximate fibration $X \longrightarrow \mathbb{R}$.*

Proof (i) $\Longleftrightarrow$ (ii) follows from 9.15.
(i) $\Longrightarrow$ (iii) follows from 9.14, 16.8 and 16.11.
(iii) $\Longrightarrow$ (i) follows from 16.7 and 9.12. □

Theorem 16.13 is the existence part of the Teardrop Structure Theorem of Hughes, Taylor, Weinberger and Williams [76] in the simplest case (two strata, the lower stratum being a point). The next example shows that one cannot hope for a true analogue of 16.13 for ANR's, even if one only wants a proper bounded fibration.

Example 16.14 For the CW complex W of 16.12 the result of adding a point at $-\infty$ (thereby compactifying W^-) is an ANR $X = W \cup \{-\infty\}$ which satisfies all the hypotheses of 16.13, yet no open cocompact subset of X admits a proper approximate fibration to $\mathbb{R}$. □

Proposition 16.15 *If W is a connected finite CW complex with a map $c : W \longrightarrow S^1$ which is homotopic to an approximate fibration, and $c_* : \pi_1(W) \longrightarrow \pi_1(S^1)$ is onto, then (W, c) is a relaxed CW band.*

Proof We have to show that there exists a fundamental domain $(V; U, \zeta U)$ for $\overline{W} = c^*\mathbb{R}$ such that V dominates $\overline{W}^+$ rel U and $\zeta^{-1}V$ dominates $\overline{W}^-$ rel U. Assume first that c is an approximate fibration and consider the induced proper approximate fibration $\overline{c} : \overline{W} \longrightarrow \mathbb{R}$. Choose a fundamental domain $(V; U, \zeta U)$ such that $U \subseteq \overline{c}^{-1}(0, 1/2)$. Define a homotopy $F : [1/2, \infty) \times I \longrightarrow [1/2, \infty)$ such that $F(s, 0) = s, F(1/2, t) = 1/2$, and $F(s, 1) = 1/2$ for all $s, t \in [1/2, \infty) \times I$. Use the regular approximate homotopy lifting property of Coram and Duvall [36] to construct a homotopy $\widetilde{F} : \overline{c}^{-1}[1/2, \infty) \times I \longrightarrow \overline{c}^{-1}[1/2, \infty)$ such that $\widetilde{F}(x, 0) = x, \widetilde{F}(y, t) = y$ and $\widetilde{F}(x, 1) \in \overline{c}^{-1}[1/2, 1]$ for all $(x, t) \in \overline{c}^{-1}[1/2, \infty) \times I$ and $y \in \overline{c}^{-1}(1/2)$. Then $\widetilde{F}$ extends to a homotopy, also denoted $\widetilde{F}$, defined on all of $\overline{W}^+$, by setting $\widetilde{F}(x, t) = (x, t)$ for all $x \in \overline{W}^+ \cap \overline{c}^{-1}[0, 1/2]$ and $t \in I$. Thus, $\widetilde{F} : \overline{W}^+ \times I \longrightarrow \overline{W}^+$ satisfies $\widetilde{F}_0 = 1_{\overline{W}^+}$, $\widetilde{F}_t | U = \text{inclusion} : U \longrightarrow \overline{W}^+$ for each $t \in I$, and $\widetilde{F}_1(\overline{W}^+) \subseteq V$, showing that V dominates $\overline{W}^+$ rel U. A similar construction shows that $\zeta^{-1}V$ dominates $\overline{W}^-$ rel U.

If c is merely homotopic to an approximate fibration $c' : W \longrightarrow S^1$, then the associated infinite cyclic cover $\overline{W}' = c'^*\mathbb{R}$ of W has a fundamental domain $(V'; U', \zeta' U')$ with the required domination properties. Moreover, there is an isomorphism $h : \overline{W}' \longrightarrow \overline{W}$ of covering spaces so that $\zeta = h\zeta' h^{-1}$. It follows that $(hV; hU, \zeta hU)$ is a fundamental domain for $\overline{W}$ with the required domination properties. □

Proposition 16.16 *If $d : W \longrightarrow \mathbb{R}$ is a proper approximate fibration then the composite*

$$p : e(W^+) \longrightarrow W^+ \longrightarrow W$$

is a homotopy equivalence, with $W^+ = d^{-1}[0, \infty)$.

Proof W is forward tame by 16.4, and there exists a closed cocompact $V \subseteq W^+$ with a proper map $q : V \times [0, \infty) \longrightarrow W^+$ extending the inclusion. Chose $n > 0$ so large that $d^{-1}[n + 1, \infty) \subseteq U$. Define a homotopy

$$h : \mathbb{R} \times I \longrightarrow \mathbb{R} \; ; \; (x, t) \longrightarrow (1 - t)x + t \max\{N, x\} \, .$$

This homotopy can be approximately lifted to obtain a homotopy $\widetilde{h} : W \times I \longrightarrow \mathbb{R}$ with $\widetilde{h}_0 = \mathrm{id}_W$ and $d\widetilde{h}_t$ 1-close to $h_t d$ for each $t \in I$. In particular, $\widetilde{h}_1(W) \subseteq U$. Define

$$f : W \longrightarrow e(W^+) \; ; \; x \longrightarrow (t \longrightarrow q(\widetilde{h}_1(x), t)) \, .$$

Then f is a homotopy inverse for p. □

Corollary 16.17 *A manifold approximate fibration $d : W \longrightarrow \mathbb{R}$ is a ribbon.*
Proof The transversality conditions (i) in the definition of a ribbon (15.1) are given by manifold transversality. The homotopy conditions (ii) are given by 16.16. The homology conditions (iii) follow from the forward tameness of W (16.4) and 13.15 (ii). □

In Chapter 17 we shall prove that a manifold approximate fibration $d : W \longrightarrow \mathbb{R}$ of dimension ≥ 6 is proper homotopic to a lift $\overline{c} : \overline{M} \longrightarrow \mathbb{R}$ of the classifying map $c : M \longrightarrow S^1$ of a manifold band (M, c) such that $W = \overline{M}$.

17

Geometric wrapping up

Wrapping up is a geometric compactification procedure which for $n \geq 5$ associates to an n-dimensional manifold approximate fibration $(X, d : X \longrightarrow \mathbb{R})$ a relaxed n-dimensional manifold band $(M, c) = (\widehat{X}, \widehat{d} : \widehat{X} \longrightarrow S^1)$ with infinite cyclic cover $\overline{M} = X$, such that $\overline{c} : X \longrightarrow \mathbb{R}$ is an approximate fibration properly homotopic to d, with $X \times S^1$ homeomorphic to $M \times \mathbb{R}$. By 16.13 an open n-dimensional manifold W which is both forward and reverse tame has an open cocompact $X \subseteq W$ with a manifold approximate fibration $d : X \longrightarrow \mathbb{R}$, and the wrapping up provides a canonical collaring of the open $(n+1)$-dimensional manifold $W \times S^1$ with boundary $M = \widehat{X}$ (17.10).

We shall use wrapping up to prove that an ANR space X admits a proper bounded fibration $d : X \longrightarrow \mathbb{R}$ if and only if it is infinite simple homotopy equivalent to the infinite cyclic cover $\overline{M}$ of a CW band (M, c) (17.16). We also prove that an open manifold X admits a manifold approximate fibration $d : X \longrightarrow \mathbb{R}$ if and only if X is the infinite cyclic cover $\overline{M}$ of a manifold band (M, c) (17.18).

We begin in 17.1 with the wrapping up construction of a manifold band from a manifold approximate fibration over $\mathbb{R}$. Then in 17.11 we give some elementary consequences of the sucking principle (16.9). After observing in 17.12 that total spaces of proper bounded fibrations over $\mathbb{R}$ are finitely dominated, we present the main characterizations (17.16 and 17.18).

The next result of this chapter concerns bands. We know that finitely dominated infinite cyclic covers of ANR bands admit proper bounded fibrations to $\mathbb{R}$ (17.14), but might not admit any proper approximate fibration to $\mathbb{R}$ (16.12, 16.14). However, in 17.20 we show that if the ANR band is allowed to vary up to simple homotopy type, then it will have a finitely dominated infinite cyclic cover which admits a proper approximate fibration to $\mathbb{R}$.

The final part of this chapter is concerned with the homotopy theoretic analogues of the main characterizations; namely, when is a CW complex homotopy equivalent to a CW complex which admits a proper bounded fibration to $\mathbb{R}$, and when is an ANR (resp. manifold) homotopy equivalent to an ANR (resp. manifold) which admits an approximate fibration to $\mathbb{R}$?

Theorem 17.1 (Wrapping up) *Let $d : X \longrightarrow \mathbb{R}$ be an n-dimensional manifold approximate fibration, with $n \geq 5$ or $n = \infty$ (= Hilbert cube manifold). Then there exists an n-dimensional relaxed manifold band*

$$(M, c) \;=\; (\widehat{X}, \widehat{d} : \widehat{X} \longrightarrow S^1)$$

with infinite cyclic cover $\overline{M} = X$ such that:

(i) *the classifying map $c : M \longrightarrow S^1$ is a manifold approximate fibration, such that the lift $\overline{c} : \overline{M} \longrightarrow \mathbb{R}$ is a manifold approximate fibration properly homotopic to d,*
(ii) *the generating covering translation $\widehat{\zeta} : X \longrightarrow X$ for the infinite cyclic cover is isotopic to id_X,*
(iii) *the mapping torus $T(\widehat{\zeta})$ is homeomorphic to $\widehat{X} \times \mathbb{R}$,*
(iv) *$X \times S^1$ is homeomorphic to $\widehat{X} \times \mathbb{R}$.* □

Remark 17.2 (i) In general, $\widehat{X}$ is *not* in the canonical simple homotopy type of $X \times S^1$: the fibring obstructions of the wrapping up $(\widehat{X}, \widehat{d})$ are computed in 26.7 below to be

$$\Phi^+(\widehat{X}, \widehat{d}) \;=\; -[X^-] \;,\;\; \Phi^-(\widehat{X}, \widehat{d}) \;=\; [X^+]$$
$$\in \widetilde{K}_0(\mathbb{Z}[\pi]) \subseteq Wh(\pi \times \mathbb{Z}) \;\; (\pi = \pi_1(X)) \;,$$

so that the homeomorphism in 17.1 (iv) has torsion

$$\begin{aligned}\tau(X \times S^1 \cong \widehat{X} \times \mathbb{R}) \;&=\; \Phi^-(\widehat{X}, \widehat{d}) - \Phi^-(X \times S^1, p_{S^1}) \\ &=\; [X^+] \in \widetilde{K}_0(\mathbb{Z}[\pi]) \subseteq Wh(\pi \times \mathbb{Z}) \;.\end{aligned}$$

This does not contradict the topological invariance of Whitehead torsion, by which homeomorphisms of *compact* ANR's are simple. Nor does it contradict the fact that the *infinite* torsion of a homeomorphism of *non-compact* ANR's is zero.

(ii) Suppose that $n < \infty$, so that

$$[X^-] \;=\; (-)^{n-1}[X^+]^* \in \widetilde{K}_0(\mathbb{Z}[\pi]) \;,$$

and $[X^-] = 0$ if and only if $[X^+] = 0$. Thus $\Phi^+(\widehat{X}, \widehat{d}) = \Phi^-(\widehat{X}, \widehat{d}) = 0$ if and only if $[X^+] = 0$, and for $n \geq 6$ $\widehat{d} : \widehat{X} \longrightarrow S^1$ is homotopic to a fibre bundle projection if and only if X^+ can be collared. More precisely, $\widehat{d}$ is homotopic to a fibre bundle projection with fibre a closed $(n-1)$-dimensional manifold F if and only if X^+ can be collared with boundary F, in which case

$$(X,d) \cong (F \times \mathbb{R}, p_{\mathbb{R}}) \ , \ (\widehat{X}, \widehat{d}) \cong (F \times S^1, p_{S^1}) \ . \qquad \square$$

Example 17.3 (i) A manifold cobordism $(V;U,U')$ is *invertible* if there exists a manifold cobordism $(V';U',U)$ with rel ∂ homeomorphisms

$$\begin{aligned}(V;U,U') \cup (V';U',U) &\cong U \times (I;\{0\},\{1\}) \ ,\\ (V';U',U) \cup (V;U,U') &\cong U' \times (I;\{0\},\{1\}) \ .\end{aligned}$$

Stallings [155] used the collarings of U and U' and the infinite repetition trick to construct rel ∂ homeomorphisms

$$\begin{aligned}U \times ([0,\infty),\{0\}) &\cong (V;U,U') \cup (V';U',U) \cup (V;U,U') \cup \dots\\ &\cong (V \backslash U',U) \ ,\\ U' \times ([0,\infty),\{0\}) &\cong (V';U',U) \cup (V;U,U') \cup (V';U',U) \cup \dots\\ &\cong (V' \backslash U,U')\end{aligned}$$

as well as homeomorphisms

$$\begin{aligned}U \times \mathbb{R} &\cong \dots \cup (V;U,U') \cup (V';U',U) \cup (V;U,U') \cup \dots\\ &\cong U' \times \mathbb{R} \ .\end{aligned}$$

The s-cobordism theorem (6.6) shows that for $n \geq 6$ every n-dimensional manifold h-cobordism $(V;U,U')$ is invertible, with inverse $(V';U',U)$ any h-cobordism with torsion

$$\tau(V';U',U) = -\tau(V;U,U') \in Wh(\pi_1(V)) \ .$$

Connell [32] proved that for $n \geq 5$ an invertible n-dimensional manifold cobordism $(V;U,U')$ is necessarily an h-cobordism. The product of an n-dimensional h-cobordism $(V;U,U')$ with S^1 is an $(n+1)$-dimensional s-cobordism $(V \times S^1; U \times S^1, U' \times S^1)$, so that for $n \geq 5$ there exists a homeomorphism rel $U \times S^1$

$$(V \times S^1; U \times S^1, U' \times S^1) \cong U \times S^1 \times (I;\{0\},\{1\}) \ ,$$

as first observed by de Rham (Kervaire [83, p. 41]). Siebenmann [142, Thm. III] gave an explicit construction of a particular such homeomorphism, using only the existence of an inverse $(V';U,U')$ for $(V;U,U')$ and the infinite repetition trick. This was a direct precursor of wrapping up: the homeomorphism of open manifolds

$$X = U \times \mathbb{R} \cong X' = U' \times \mathbb{R}$$

was wrapped up to a homeomorphism of bands

$$\widehat{X} = U \times S^1 \cong \widehat{X}' = U' \times S^1 \ .$$

In fact, for $n \geq 5$ the following conditions on closed n-dimensional manifolds U, U' are equivalent:

(a) U, U' are h-cobordant,
(b) $U \times \mathbb{R}$, $U' \times \mathbb{R}$ are homeomorphic,
(c) $U \times S^1$, $U' \times S^1$ are homeomorphic.

(ii) Farrell and Hsiang [49] and Siebenmann [145] showed that for $n \geq 6$ the following conditions on an n-dimensional manifold band (W, c) are equivalent:

(a) the infinite cyclic cover $\overline{W} = c^*\mathbb{R}$ of W admits a fundamental domain $(V; U, \zeta U)$ such that there exists a homeomorphism rel U

$$U \times ([0,\infty), \{0\}) \;\cong\; (\overline{W}^+, U) \ ,$$

(b) $\overline{W}$ admits a fundamental domain $(V; U, \zeta U)$ such that the inclusion $U = U \times \{0\} \longrightarrow \overline{W}$ extends to a homeomorphism

$$U \times \mathbb{R} \;\cong\; \overline{W} \ ,$$

(c) $[\overline{W}^+] = 0 \in \widetilde{K}_0(\mathbb{Z}[\pi_1(\overline{W})])$.

(In order for (W, c) to admit a fundamental domain $(V; U, \zeta U)$ which is an h-cobordism it is necessary and sufficient that in addition the $\widetilde{\text{Nil}}$-components of the fibring obstructions vanish – see 24.16 below.) If (W, c) satisfies these conditions the manifold ribbon defined by the infinite cyclic cover

$$(X, d) \;=\; (\overline{W}, \overline{c})$$

is such that there exists a homeomorphism $(X, d) \cong (U \times \mathbb{R}, p_{\mathbb{R}})$. In particular, $d : X \longrightarrow \mathbb{R}$ is an approximate fibration, such that the wrapping up of (X, d) given by 17.1 is the n-dimensional manifold band $(\widehat{X}, \widehat{d}) = (U \times S^1, p_{S^1})$. The self homotopy equivalence

$$h \;:\; U \simeq \overline{W} \xrightarrow{\;\zeta\;} \overline{W} \simeq U$$

is such that the product

$$h \times 1 \;:\; \widehat{X} \;=\; U \times S^1 \longrightarrow \widehat{X} \;=\; U \times S^1$$

is homotopic to a self homeomorphism $\widehat{\zeta} : \widehat{X} \longrightarrow \widehat{X}$ with a homeomorphism

$$W \times S^1 \;\cong\; T(\widehat{\zeta} : \widehat{X} \longrightarrow \widehat{X}) \ .$$

The $(n+1)$-dimensional manifold band

$$(W, c) \times S^1 \;=\; (W \times S^1, W \times S^1 \xrightarrow{\text{proj.}} W \xrightarrow{\;c\;} S^1)$$

fibres, with fibre $\widehat{X}$ and monodromy $\widehat{\zeta}$. (This fact is well-known to the experts, cf. Ferry and Pedersen [57, p. 492].) □

The plan of the proof of 17.1 is as follows. The manifold $\widehat{X}$ is constructed using a variation of Chapman's wrapping up construction (which in turn is a variation of Siebenmann twist glueing (15.17); see Hughes and Prassidis [75] for the precise relation between the two constructions). The input needed is the approximate isotopy covering property for manifold approximate fibrations due to Hughes, allowing the standard shift map $\mathbb{R}\longrightarrow\mathbb{R}; s\longrightarrow s+1$ to be lifted to a covering translation $\widehat{\zeta} : X\longrightarrow X$ isotopic to the identity. Note that 17.1 (iv) is a direct consequence of (ii) and (iii).

In dealing with isotopies

$$G \,:\, X\times I \longrightarrow X \,;\, (x,s) \longrightarrow G_s(x)$$

we always assume that $G_0 = \mathrm{id.} : X\longrightarrow X$.

Theorem 17.4 (Approximate Isotopy Covering) *Let $p : M\longrightarrow B$ be a manifold approximate fibration where M is a manifold without boundary of dimension $n\geq 5$ or a Hilbert cube manifold ($n=\infty$). Let α be an open cover of B, and let $g : B\times I\longrightarrow B$ be an isotopy. Then there exists an isotopy $G : M\times I\longrightarrow M$ such that pG_t is α-close to $g_t p$ for each $t\in I$.* □

Comments on Proof. Note that 'Approximate Isotopy Covering' theorems are not well documented in the literature. However, the 'Controlled Isotopy Covering Theorem' in Hughes, Taylor and Williams [79] is derived from 'Controlled Straightening' in Hughes, Taylor and Williams [77], which in turn is derived from 'Approximate Straightening'. Now 'Approximate Straightening' is just the 'Approximation Theorem' of Hughes [72]. The point of all of this is that the Approximate Isotopy Covering Theorem follows easily from the Approximation Theorem of [72]. □

Proposition 17.5 *Let $a_1, a_2, a_3 > 0$ be real numbers such that $a_3 > a_1 + a_2$. Let $d : X\longrightarrow\mathbb{R}$ be a proper map, and let $\widehat{\zeta} : X\longrightarrow X$ be a homeomorphism with $d\widehat{\zeta}$ a_2-close to d. Also, let*

$$g \,:\, \mathbb{R} \longrightarrow \mathbb{R} \,;\, x \longrightarrow x + a_3$$

and let $G : X\longrightarrow X$ be a homeomorphism such that dG is a_1-close to gd. Then

$$\widehat{\zeta}G \,:\, X \longrightarrow X$$

is a covering translation of an infinite cyclic cover of $X/\widehat{\zeta}G$, with

$$U \;=\; d^{-1}(0) \;\;,\;\; V \;=\; \widehat{\zeta}Gd^{-1}(-\infty,0]\cap d^{-1}[0,\infty)$$

such that $(V; U, \widehat{\zeta}GU)$ is a fundamental domain.

Proof We shall show by induction on $n \in \mathbb{Z}_+$ that for each $x \in X$

$$d(\widehat{\zeta}G)^n x > dx + n(a_3 - a_1 - a_2) \ .$$

A similar argument would show

$$d(\widehat{\zeta}G)^{-n} x < dx - n(a_3 - a_1 - a_2)$$

and from these two inequalities it follows immediately that the orbit $\{(\widehat{\zeta}G)^n x \,|\, n \in \mathbb{Z}\}$ is closed and discrete in X (the fact that d is proper must be invoked here), so that the action

$$\mathbb{Z} \times X \longrightarrow X \ ; \ (n, x) \longrightarrow (\widehat{\zeta}G)^n x$$

is properly discontinuous.

Note that the estimates imply that $d\widehat{\zeta}G$ is $(a_1 + a_2)$-close to gd. It follows that

$$d\widehat{\zeta}Gx > gdx - a_1 - a_2 \ = \ dx + a_3 - a_2 - a_1 \ ,$$

establishing the inequality for $n = 1$. Assuming $n > 1$ and that

$$d(\widehat{\zeta}G)^{n-1} x > dx + (n-1)(a_3 - a_2 - a_1) \ ,$$

note that

$$\begin{aligned} d(\widehat{\zeta}G)^n x \ &= \ d\widehat{\zeta}G(\widehat{\zeta}G)^{n-1} x \\ &> \ dg(\widehat{\zeta}G)^{n-1} x - a_1 - a_2 \ = \ d(\widehat{\zeta}G)^{n-1} x + a_3 - a_2 - a_1 \\ &> \ dx + (n-1)(a_3 - a_2 - a_1) + a_3 - a_2 - a_1 \\ &= \ dx + n(a_3 - a_2 - a_1) \ . \end{aligned}$$

In order to establish the second statement, we need to verify that

$$X \ = \ \bigcup_{n \in \mathbb{Z}} (\widehat{\zeta}G)^n V \ .$$

Let $x \in X$ and $m = \min\{n \in \mathbb{Z} \,|\, 0 < d(\widehat{\zeta}G)^n x\}$. This minimum exists because the inequalities above show first of all that

$$\{n \in \mathbb{Z} \,|\, 0 < d(\widehat{\zeta}G)^n x\} \neq \emptyset \ ,$$

and second that

$$\{n \in \mathbb{Z} \,|\, 0 < d(\widehat{\zeta}G)^n x < k\}$$

is finite for each $k > 0$. It follows that $d(\widehat{\zeta}G)^{m-1} x \leq 0$ so that $(\widehat{\zeta}G)^m x \in V$.

□

Proposition 17.6 *Let $a_1, a_2, a_3 > 0$ be such that $a_3 > 17(a_1 + a_2)$, let $d : X \longrightarrow \mathbb{R}$ be a manifold approximate fibration with $\dim(X) \geq 5$, and let $\widehat{\zeta} : X \longrightarrow X$ be a homeomorphism such that $d\widehat{\zeta}$ is a_2-close to d. Also, let*

$$g_s \; : \; \mathbb{R} \longrightarrow \mathbb{R} \; ; \; t \longrightarrow t + a_3 s \quad (0 \leq s \leq 1) \; ,$$

and let $G_s : X \longrightarrow X$ be an isotopy such that dG_s is a_1-close to $g_1 d$. Then d is properly homotopic to a manifold approximate fibration $\widetilde{d} : X \longrightarrow \mathbb{R}$ such that there is defined a commutative diagram

$$\begin{array}{ccc} X & \xrightarrow{\widehat{\zeta} G_1} & X \\ \Big\downarrow{\scriptstyle \widetilde{d}} & & \Big\downarrow{\scriptstyle \widetilde{d}} \\ \mathbb{R} & \xrightarrow{g_1} & \mathbb{R} \end{array}$$

Proof Define a homotopy

$$h \; : \; d \; \simeq \; g_1 d G_1^{-1} \widehat{\zeta}^{-1} \; : \; X \longrightarrow \mathbb{R}$$

by

$$h_t(x) \;=\; \begin{cases} (1-2t)d(x) + 2td\widehat{\zeta}^{-1}(x) & \text{if } 0 \leq t \leq \frac{1}{2} \; , \\ g_{(2t-1)} d G_{(2t-1)}^{-1} \widehat{\zeta}^{-1}(x) & \text{if } \frac{1}{2} \leq t \leq 1 \; . \end{cases}$$

Since d is a_2-close to $d\widehat{\zeta}^{-1}$ and $g_s d$ is a_1-close to dG_s for each s, it follows that h is a $(2a_1 + a_2)$-homotopy.

Let $u : X \longrightarrow I$ be a map such that

$$\begin{aligned} u^{-1}(0) &= d^{-1}(-\infty, \tfrac{1}{4}(a_3 - a_2 - a_1)] \; , \\ u^{-1}(1) &= d^{-1}[\tfrac{3}{4}(a_3 - a_2 - a_1), \infty) \; . \end{aligned}$$

Note that $u(d^{-1}(0)) = 0$ and $u(\widehat{\zeta} G_1 d^{-1}(0)) = 1$. (The first equality is obvious; the second follows from the facts that dG_1 is a_1-close to $g_1 d$ and $d\widehat{\zeta}$ is a_2-close to d.) The map

$$d' \; : \; X \longrightarrow \mathbb{R} \; ; \; x \longrightarrow h_{u(x)}(x)$$

is such that

$$d'(x) \;=\; \begin{cases} d(x) & \text{if } x \in (-\infty, \frac{1}{8}(a_3 - a_2 - a_1)] \; , \\ g_1 d G_1^{-1} \widehat{\zeta}^{-1}(x) & \text{if } x \in [\frac{7}{8}(a_3 - a_2 - a_1), \infty) \; . \end{cases}$$

Since h is a bounded homotopy, d' is a bounded fibration. Moreover, d' is a manifold approximate fibration over

$$(-\infty, \tfrac{1}{8}(a_3 - a_2 - a_1)] \cup [\tfrac{7}{8}(a_3 - a_2 - a_1), \infty) \; .$$

It follows from a relative version of the sucking principle 16.9 (see [26]) that d' is boundedly homotopic rel

$$(-\infty, \tfrac{1}{16}(a_3 - a_2 - a_1)] \cup [\tfrac{15}{16}(a_3 - a_2 - a_1), \infty)$$

to a manifold approximate fibration $d'' : X \longrightarrow \mathbb{R}$. Let

$$V = \widehat{\zeta} G_1 d^{-1}(-\infty, 0] \cap d^{-1}[0, \infty) .$$

It follows from Proposition 17.5 that

$$X = \bigcup_{n \in \mathbb{Z}} (\widehat{\zeta} G_1)^n V .$$

Define $\widetilde{d} : X \longrightarrow \mathbb{R}$ by

$$\widetilde{d}|(\widehat{\zeta} G_1)^n V = g_1^n \circ d'' \circ (\widehat{\zeta} G_1)^{-n} .$$

Then $\widetilde{d}$ is a manifold approximate fibration boundedly close to d such that $\widetilde{d}\widehat{\zeta} G_1 = g_1 \widetilde{d}$. □

Proof of 17.1 Let $d : X \longrightarrow \mathbb{R}$ be a manifold approximate fibration as in the statement of 17.1. The isotopy

$$g : \mathbb{R} \times I \longrightarrow \mathbb{R} \; ; \; (s, t) \longrightarrow s + t$$

is such that $g_1 : \mathbb{R} \longrightarrow \mathbb{R}; s \longrightarrow s + 1$ is a generating covering translation of the universal covering $\mathbb{R} \longrightarrow S^1$. Choose $a_1 > 0$ such that $17a_1 < 1$. Apply 17.4 (Approximate Isotopy Covering) to get an isotopy $G : X \times I \longrightarrow X$ such that dG_s is a_1-close to $g_s d$. Apply 17.5 (with $\widehat{\zeta} = \mathrm{id}_X$, $a_3 = 1$, and a_2 as small as needed) to conclude that G_1 acts properly discontinuously on X with fundamental domain $(V; U, G_1 U)$ where

$$U = d^{-1}(0) \; , \; V = G_1 d^{-1}(-\infty, 0] \cap d^{-1}[0, \infty) .$$

Definition 17.7 The *wrapping up* of a manifold approximate fibration $d : X \longrightarrow \mathbb{R}$ is the manifold band

$$(\widehat{X}, \widehat{d}) = (X/G_1, X/G_1 \longrightarrow S^1)$$

with $\widehat{d} : X/G_1 \longrightarrow S^1$ classifying $\widehat{d}^* \mathbb{R} = X$. □

Lemma 17.8 *The manifold approximate fibration $d : X \longrightarrow \mathbb{R}$ is homotopic to a $\mathbb{Z}$-equivariant proper approximate fibration $\widetilde{d} : X \longrightarrow \mathbb{R}$.*

Proof Apply 17.6 (with $\widehat{\zeta} = \mathrm{id}_X$) to get a manifold approximate fibration $\widetilde{d} : X \longrightarrow \mathbb{R}$ such that $\widetilde{d} G_1 = g_1 \widetilde{d}$. Then $\widetilde{d}$ induces a manifold approximate fibration $c : \widehat{X} \longrightarrow S^1$ classifying $X \longrightarrow \widehat{X}$. (That c is a manifold approximate fibration follows from the fact that it is one locally (Coram [35]).) □

We now return to the proof of 17.1. It follows from 14.8 (i) that $\widehat{X} \times \mathbb{R} \cong T(G_1)$. On the other hand, since G_1 is isotopic to id_X there is a homeomorphism $T(G_1) \cong X \times S^1$. This completes the proof of 17.1. □

Remark 17.9 The wrapping up construction above is very similar to Chapman's original construction [25, 26]. However, our construction is technically easier and more conceptual because we make use of the Approximate Isotopy Covering Theorem 17.4. Chapman only had the technology to approximately cover compactly supported isotopies and the shift map $g_1 : \mathbb{R} \longrightarrow \mathbb{R}$ is far from compactly supported. Chapman had to truncate the shift map to get a compactly supported isotopy which he then approximately covered by a compactly supported isotopy on X. The upshot is that he constructed the fundamental domain V above, but the infinite cyclic cover $\overline{\widehat{X}}$ of $\widehat{X}$ was given as a proper open subset of X rather than equal to X. Using Chapman's approach it is far from obvious that the generating covering translation $\widehat{\zeta}$ is isotopic to the identity, but it is immediate in our approach. □

Theorem 17.10 *Let $(W, \partial W)$ be an open n-dimensional manifold with compact boundary and one end, with $n \geq 5$ or $n = \infty$ (= Hilbert cube manifold). If W is forward tame and reverse tame then the end space $e(W)$ is homotopy equivalent to an open cocompact submanifold $X \subset W$ with a proper map $d : X \longrightarrow \mathbb{R}$ such that*

$$(X, d) \;=\; (\overline{M}, \overline{c})$$

is the finitely dominated infinite cyclic cover of a relaxed n-dimensional manifold band $(M, c) = (\widehat{X}, \widehat{d})$ with $X \times S^1$ homeomorphic to $M \times \mathbb{R}$, and with a rel ∂ *homeomorphism*

$$(W, \partial W) \times S^1 \;\cong\; (N \backslash M, \partial W \times S^1)$$

for a compact $(n+1)$-dimensional manifold cobordism $(N; \partial W \times S^1, M)$.
Proof Combine 16.13 and 17.1. □

Let Q denote the Hilbert cube. Edwards proved that $X \times Q$ is a Hilbert cube manifold for any ANR X. (Recall our global assumption at the beginning of Chapter 1 that only locally compact, separable ANR's are to be considered.)

Proposition 17.11 (i) *For any ANR X, the following are equivalent:*

(a) *there exists a proper bounded fibration $d : X \longrightarrow \mathbb{R}$,*
(b) *for every $\epsilon > 0$ there exists a proper ϵ-fibration $d : X \longrightarrow \mathbb{R}$,*
(c) *there exists a proper approximate fibration $d : X \times Q \longrightarrow \mathbb{R}$.*

(ii) *For an open manifold X of dimension ≥ 5 or a Hilbert cube manifold, the conditions in (i) are equivalent to:*

(d) *there exists a manifold approximate fibration $d : X \longrightarrow \mathbb{R}$.*

Proof (a) $\Longrightarrow$ (b) By the usual method of shrinking $\mathbb{R}$.

(b) $\Longrightarrow$ (c) Let $d : X \longrightarrow \mathbb{R}$ be a proper ϵ-fibration for some $\epsilon > 0$. Then $p_1 \circ d : X \times Q \longrightarrow \mathbb{R}$ is also a proper ϵ-fibration. Since $X \times Q$ is a Hilbert cube manifold, if ϵ is sufficiently small, then Chapman's sucking principle implies that p is close to a proper approximate fibration.

(c) $\Longrightarrow$ (a) Since X and $X \times Q$ are proper homotopy equivalent, Proposition 16.5 can be applied.

(ii) (d) $\Longleftrightarrow$ (a) by 16.10. □

Proposition 17.12 *If $d : X \longrightarrow \mathbb{R}$ is a proper bounded fibration, then X is dominated by a compact space. Hence, if X has the homotopy type of a CW complex, then X is finitely dominated.*

Proof An ϵ-lift of a contraction of $\mathbb{R}$ to 0 gives a homotopy $K : X \times I \longrightarrow X$ such that $K_0 = \mathrm{id}_X$ and $K_1(X) \subseteq d^{-1}[-\epsilon, \epsilon]$. □

We now turn to the problem of deciding when there is a bounded or approximate fibration from a space to $\mathbb{R}$.

Lemma 17.13 (Sliding domination) *Let (W, c) be an ANR band, and let $\overline{c} : \overline{W} \longrightarrow \mathbb{R}$ be a $\mathbb{Z}$-equivariant lift of $c : W \longrightarrow S^1$. There exist a homotopy*

$$K_s \ : \ \overline{W} \times \mathbb{R} \longrightarrow \overline{W} \times \mathbb{R} \ \ (0 \leq s \leq 1)$$

and a constant $N > 1$ such that:

(i) $K_0 = \mathrm{id}_{\overline{W}}$,

(ii) *K_s is fibre-preserving over $\mathbb{R}$ (i.e. $p_2 K_s = p_2$ where $p_2 : \overline{W} \times \mathbb{R} \longrightarrow \mathbb{R}$ is the projection) for each s,*

(iii) *$K_s(x, t) = (x, t)$ if $\overline{c}(x) = t$ for each s (so that $K_s|\Gamma(\overline{c})$ is the inclusion, where $\Gamma(\overline{c})$ denotes the graph of $\overline{c}$),*

(iv) $K_1(\overline{W} \times \mathbb{R}) \subseteq \{(x, t) \,|\, t - N \leq \overline{c}(x) \leq t + N\}$.

Proof Let $\zeta : \overline{W} \longrightarrow \overline{W}$ denote the $(+1)$-generating covering translation. It follows from 15.10 that $\overline{W}^-$ is reverse tame. Hence there exist a homotopy

$$h \ : \ \overline{W} \times I \longrightarrow \overline{W} \ ; \ (x, t) \longrightarrow h_t(x)$$

and a constant $m_1 > 1$ such that:

(i) $h_0 = \mathrm{id}_{\overline{W}}$,
(ii) $h_t|\overline{c}^{-1}[-1,\infty)$ is the inclusion for each t,
(iii) $h_t\overline{c}^{-1}(-\infty,-1) \subseteq \overline{c}^{-1}(-\infty,-1)$ for each t,
(iv) $h_1\overline{W} \subseteq \overline{c}^{-1}[-m_1,\infty)$.

Compactness implies that there is a constant $m_2 > m_1$ such that for each t

$$h_t\overline{c}^{-1}[-m_1-1,\infty) \subseteq \overline{c}^{-1}[-m_2,\infty) .$$

Define a homotopy

$$H : \overline{W} \times \mathbb{R} \times I \longrightarrow \overline{W} \times \mathbb{R} \; ; \; (x,s,t) \longrightarrow H_s(x,t)$$

by

$$H_s(x,t) = (\widehat{\zeta}^{n+1}h_{s(t-n)}\widehat{\zeta}^{-1}h_1\widehat{\zeta}^{-1}h_{s(n+1-t)}\widehat{\zeta}^{1-n}(x),t)$$

for $n \leq t \leq n+1$. One can verify that:

(i) $H_0 = \mathrm{id}$,
(ii) H_s is fibre-preserving over $\mathbb{R}$ for each s,
(iii) $H_s(x,t) = (x,t)$ if $t \leq \overline{c}(x)$ for each s,
(iv) $H_s\{(x,t)\,|\,\overline{c}(x) \leq t\} \subseteq \{(x,t)\,|\,\overline{c}(x) \leq t\}$ for each s,
(v) $H_1(\overline{W} \times \mathbb{R}) \subseteq \{(x,t)\,|\,t - m_2 \leq \overline{c}(x)\}$.

Since $\overline{W}^+$ is also reverse tame (15.10), we can use the argument above to define a homotopy

$$G_s : \overline{W} \times \mathbb{R} \longrightarrow \overline{W} \times \mathbb{R} \quad (0 \leq s \leq 1)$$

and a constant $m_2' > 1$ such that:

(i) $G_0 = \mathrm{id}$,
(ii) G_s is fibre-preserving over $\mathbb{R}$ for each s,
(iii) $G_s(x,t) = (x,t)$ if $\overline{c}(x) \leq t$ for each s,
(iv) $G_s\{(x,t)\,|\,t \leq \overline{c}(x)\} \subseteq \{(x,t)\,|\,t \leq \overline{c}(x)\}$ for each s,
(v) $G_1(\overline{W} \times \mathbb{R}) \subseteq \{(x,t)\,|\,\overline{c}(x) \leq t + m_2'\}$.

Then let

$$K_s = G_s \circ H_s : \overline{W} \times \mathbb{R} \longrightarrow \overline{W} \times \mathbb{R} \quad (0 \leq s \leq 1)$$

and let $N = \max\{m_2, m_2'\}$. □

Proposition 17.14 *Let W be a compact ANR with a map $c : W \longrightarrow S^1$. The infinite cyclic cover $\overline{W} = c^*\mathbb{R}$ of W is finitely dominated (i.e. (W,c) is a band) if and only if $\overline{c} : \overline{W} \longrightarrow \mathbb{R}$ is a proper bounded fibration.*

Proof The 'if' statement follows from 17.12. Let $\widehat{\zeta} : \overline{W} \longrightarrow \overline{W}$ denote the $(+1)$-generating covering translation. Let K_s be given by 17.13. It remains to show that $\overline{c}$ has the $(N+1)$-homotopy lifting property. This estimate arises because $\overline{c}p_1K_1 : \overline{W} \times \mathbb{R} \longrightarrow \mathbb{R}$ is $(N+1)$-close to $p_2 : \overline{W} \times \mathbb{R} \longrightarrow \mathbb{R}$ where $p_1 : \overline{W} \times \mathbb{R} \longrightarrow \overline{W}$ and $p_2 : \overline{W} \times \mathbb{R} \longrightarrow \mathbb{R}$ are the projections (in fact, they are a distance at most N apart). Define

$$g : \overline{W} \longrightarrow \overline{W} \times \mathbb{R} \; ; \; x \longrightarrow (x, \overline{c}(x)) \; .$$

Thus g is the natural embedding of $\overline{W}$ onto the graph $\Gamma(\overline{c})$ of $\overline{c}$. A lifting problem for $\overline{c}$, say a homotopy $F : Z \times I \longrightarrow \mathbb{R}$ with an initial lift $f : Z \longrightarrow \overline{W}$ so that $F_0 = \overline{c}f$, induces a lifting problem for p_2 with homotopy F but with initial lift given by $gf : Z \longrightarrow \overline{W} \times \mathbb{R}$. Of course, p_2 is a fibration, so let $\widehat{F} : Z \times I \longrightarrow \overline{W} \times \mathbb{R}$ be a solution of this second problem so that $\widehat{F}_0 = gf$. It follows that $\widetilde{F} = p_1K_1\widehat{F}$ is an $(N+1)$-solution of the first problem. □

Corollary 17.15 *Let X be an open manifold of dimension $n \geq 5$ or a Hilbert cube manifold. If X is an infinite cyclic cover of a compact space, then there exists a manifold approximate fibration $d : X \longrightarrow \mathbb{R}$.*
Proof Apply 16.10 and 17.14. □

Proposition 17.16 (i) *For an ANR X the following are equivalent:*

(a) *there exists a proper bounded fibration $d : X \longrightarrow \mathbb{R}$,*
(b) *for every $\epsilon > 0$ there exists a proper ϵ-fibration $d : X \longrightarrow \mathbb{R}$,*
(c) *there exists a manifold approximate fibration $d : X \times Q \longrightarrow \mathbb{R}$,*
(d) *X is finitely dominated and $X \times Q$ is an infinite cyclic cover of a compact space,*
(e) *X is infinite simple homotopy equivalent to the finitely dominated infinite cyclic cover $\overline{W}$ of a CW band (W, c),*
(f) *X is proper homotopy equivalent to the finitely dominated infinite cyclic cover $\overline{W}$ of a CW band (W, c).*

(ii) *For an open manifold X of dimension $n \geq 5$ or a Hilbert cube manifold, the conditions of (i) are equivalent to:*

(g) *there exists a manifold approximated fibration $d : X \longrightarrow \mathbb{R}$,*
(h) *X is finitely dominated and is an infinite cyclic cover of a compact space.*

Proof (i) (a) $\Longleftrightarrow$ (b) $\Longleftrightarrow$ (c) by 17.11 (i).
(c) $\Longrightarrow$ (d) because $X \times Q$ is a Hilbert cube manifold, so 17.1 implies that $X \times Q$ is the finitely dominated infinite cyclic cover of a Hilbert cube

manifold band.

(d) $\Longrightarrow$ (e) Let K be a compact space with infinite cyclic cover $\overline{K} = X \times Q$. Since K is a compact Hilbert cube manifold, X is homeomorphic to $Y \times Q$ for some finite CW complex Y (by Chapman's Triangulation Theorem). Then Y is a CW band with finitely dominated infinite cyclic cover $\overline{Y}$ such that $X \times Q$ and $\overline{Y} \times Q$ are homeomorphic. By the work of Chapman, this is what it means for X and $\overline{Y}$ to be infinite simple homotopy equivalent.

(e) $\Longrightarrow$ (f) is obvious.

(f) $\Longrightarrow$ (a) by 16.5 and 17.14.

(ii) (a) $\Longleftrightarrow$ (g) by 17.11 (ii).

(g) $\Longrightarrow$ (h) by 17.1.

(h) $\Longrightarrow$ (a) by 17.14. □

Remark 17.17 (i) If the conditions of 17.16 (i) are satisfied, then the compact space of which X is an infinite cyclic cover is a compact ANR and, hence, of the homotopy type of a finite CW complex (by the theorem of West [168]).

(ii) It follows from 15.9 that the conditions of 17.16 (i) imply that X is proper homotopy equivalent to an ANR ribbon. For a CW ribbon (X, d) the converse is established in 20.3 (ii): X is proper homotopy equivalent to the finitely dominated infinite cyclic cover of a CW band. □

Corollary 17.18 *An open manifold X of dimension ≥ 5 is the total space of a manifold approximate fibration $X \longrightarrow \mathbb{R}$ if and only if it is the infinite cyclic cover $X = \overline{M}$ of a compact manifold band (M, c).* □

Theorem 17.19 (i) *For a strongly locally finite CW complex X with a finite number of ends, the following are equivalent:*

(a) *X is forward and reverse tame,*

(b) *there exists a CW band (W, c) such that X and $\overline{W}^+$ are proper homotopy equivalent at ∞.*

(ii) *For a manifold X with a finite number of ends of dimension ≥ 5 with compact boundary or a Hilbert cube manifold, the conditions above are equivalent to:*

(c) *there exists a manifold band (W, c) such that $\overline{W}^+$ is homeomorphic to a closed cocompact subspace of X,*

(d) *there exists a manifold band (W, c) such that $\overline{W}$ is homeomorphic to an open cocompact subspace of X.*

Proof (i) (a) $\Longrightarrow$ (b) By 16.13 and 17.1 $X \times Q$ has an open cocompact subspace $\overline{U}$ which is the finitely dominated infinite cyclic cover of a Hilbert cube manifold band (U, c_1). Since U is homeomorphic to $W \times Q$ for some finite CW complex W, $(W, c = c_1 \circ \text{inclusion})$ is a CW band such that X and $\overline{W}^+$ are proper homotopy equivalent at ∞.
(b) $\Longrightarrow$ (a) by 15.9 (i), 9.6 and 9.8.
(ii) (a) $\Longrightarrow$ (d) by 16.13 and 17.1.
(d) $\Longrightarrow$ (c) and (c) $\Longrightarrow$ (b) are obvious. □

Proposition 17.20 *Every ANR band (X, c) is simple homotopy equivalent to one such that $\overline{c} : \overline{X} \longrightarrow \mathbb{R}$ is proper homotopic to a proper approximate fibration.* □

The proof of 17.20 will be based on the following two lemmas.

Lemma 17.21 *Suppose M is a finitely dominated manifold such that ∂M is also finitely dominated. Then there exist a compact subset $C \subseteq M$ and a homotopy $h : \mathrm{id}_M \simeq h_1 : M \times I \longrightarrow M$ such that:*

(i) $h_1(M) \subseteq C$,
(ii) *if $x \in \partial M$ (resp.* $\mathrm{int}(M)$*) then $h(x \times I) \subseteq \partial M$ (resp.* $\mathrm{int}(M)$*).*

Proof This is a standard construction using a collar of ∂M in M. □

Lemma 17.22 *Let $(N, \partial N)$ be a compact n-dimensional manifold with boundary such that $\pi_1(\partial N) \longrightarrow \pi_1(N)$ is a split injection. If $(b, \partial b) : (N, \partial N) \longrightarrow S^1$ is a map such that (N, b) is a band then the boundary $(\partial N, \partial b)$ is also a band.*

Proof Let $\widetilde{N}$ be the universal cover of N, and let $\widetilde{\partial N}$ be the corresponding cover of ∂N. We need to show that the infinite cyclic cover $\overline{\partial N} = (\partial b)^* \mathbb{R}$ of ∂N is finitely dominated, which by 6.9 (i) is equivalent to the $\mathbb{Z}[\pi_1(\overline{N})]$-finite domination of the cellular $\mathbb{Z}[\pi_1(N)]$-module chain complex $C(\widetilde{\partial N})$. The infinite cyclic cover $\overline{N} = b^* \mathbb{R}$ of N is finitely dominated, so that $C(\widetilde{N})$ is $\mathbb{Z}[\pi_1(\overline{N})]$-finitely dominated, and so is the n-dual $\mathbb{Z}[\pi_1(N)]$-module chain complex $C(\widetilde{N})^{n-*}$. By the exactness of

$$0 \longrightarrow C(\widetilde{\partial N}) \longrightarrow C(\widetilde{N}) \longrightarrow C(\widetilde{N}, \widetilde{\partial N}) \longrightarrow 0$$

and the Poincaré–Lefschetz $\mathbb{Z}[\pi_1(N)]$-module chain equivalence

$$C(\widetilde{N}, \widetilde{\partial N}) \simeq C(\widetilde{N})^{n-*}$$

there is defined a $\mathbb{Z}[\pi_1(N)]$-module chain equivalence

$$C(\widetilde{\partial N}) \simeq \mathcal{C}(C(\widetilde{N}) \longrightarrow C(\widetilde{N})^{n-*})_{*+1} ,$$

so that $C(\widetilde{\partial N})$ is indeed $\mathbb{Z}[\pi_1(\overline{N})]$-finitely dominated. □

Proof of 17.20 By West's result on the homotopy finiteness of compact ANR's there is no loss of generality in assuming that (X, c) is a CW band. Let N be a regular neighbourhood of X in some Euclidean space of sufficiently high dimension that the inclusion $\partial N \longrightarrow N$ induces an isomorphism $\pi_1(\partial N) \cong \pi_1(N)$ and $\dim(N) > 5$.

Thus, X is simple homotopy equivalent to N and there is a map

$$b : N \simeq X \xrightarrow{c} S^1$$

inducing a finitely dominated infinite cyclic cover $\overline{N}$. It follows from Proposition 17.14 that the induced map $\overline{b} : \overline{N} \longrightarrow \mathbb{R}$ is a proper bounded fibration. Lemma 17.22 implies that $\overline{\partial N}$ is also finitely dominated, so $\overline{\partial b} : \overline{\partial N} \longrightarrow \mathbb{R}$ is also a proper bounded fibration.

The rest of the proof consists of applying a stratified sucking principle from Hughes [74] to show that $\overline{b}$ is boundedly homotopic to an approximate fibration. (Note that 16.10 cannot be used because $\overline{N}$ has a boundary.) The idea is that $\overline{N}$ is a stratified space with strata $\overline{\partial N}$ and $\mathrm{int}(\overline{N})$. The proof of Proposition 17.14 actually shows that $(\overline{b}, \overline{\partial b}) : (\overline{N}, \overline{\partial N}) \longrightarrow \mathbb{R}$ is a proper *stratified* bounded fibration. This is because Lemma 17.22 shows that $(\overline{N}, \overline{\partial N})$ is finitely dominated in a stratified sense. Now use the stratified sucking theorem [74]. □

Remark 17.23 (i) The wrapping up $(\widehat{X}, \widehat{d})$ of a manifold approximate fibration (X, d) can also be constructed by the end obstruction theory of Siebenmann [140] (quoted in 10.2) and the projective surgery theory of Pedersen and Ranicki [109], as follows.

The *total projective surgery obstruction groups* $\mathbb{S}^p_*(K)$ of [109] are defined for any space K to fit into the algebraic surgery exact sequence

$$\begin{aligned} \dots \longrightarrow H_m(K; \mathbb{L}_\cdot) \xrightarrow{A^p} L^p_m(\mathbb{Z}[\pi_1(K)]) \longrightarrow \mathbb{S}^p_m(K) \\ \longrightarrow H_{m-1}(K; \mathbb{L}_\cdot) \longrightarrow \dots , \end{aligned}$$

with $\mathbb{L}_\cdot$ the 1-connective simply-connected surgery spectrum such that $\pi_*(\mathbb{L}_\cdot) = L_*(\mathbb{Z})$, and A^p the assembly map in projective L-theory.

The *total projective surgery obstruction* $s^p(K) \in \mathbb{S}^p_m(K)$ of a finitely dominated m-dimensional Poincaré space K is such that $s^p(K) = 0$ if (and for $m \geq 5$ only if) $K \times S^1$ is homotopy equivalent to a compact $(m+1)$-dimensional manifold L. (See Ranicki [125] for a detailed exposition of the total surgery obstruction.) If $s^p(K) = 0$ then the composite $c : L \simeq K \times S^1 \longrightarrow S^1$ defines an $(m+1)$-dimensional manifold band (L, c), such that K is homotopy equivalent to the infinite cyclic cover $\overline{L} = c^*\mathbb{R}$ of L. It was shown in [109] that L can be chosen such that $\overline{L} \times S^1$ is homeomorphic to $L \times \mathbb{R}$.

Given an n-dimensional manifold approximate fibration $(X, d : X \longrightarrow \mathbb{R})$ there is defined an open n-dimensional manifold with compact boundary

$$(W, \partial W) \;=\; (X^+, X^+ \cap X^-)$$

with one end which is both forward and reverse tame, such that $e(W)$ is a finitely dominated $(n-1)$-dominated Poincaré space homotopy equivalent to X (16.17). The projective class at ∞ of $W \times S^1$ is

$$[W \times S^1]_\infty \;=\; [W \times S^1] \;=\; 0 \in \widetilde{K}_0(\mathbb{Z}[\pi][z, z^{-1}]) \;\; (\pi = \pi_1(X)) \,,$$

so that for $n \geq 5$ there exists a compact $(n+1)$-dimensional manifold cobordism $(N; \partial W \times S^1, M)$ with a rel ∂ homeomorphism

$$(N \backslash M, \partial W \times S^1) \;\cong\; (W, \partial W) \times S^1$$

and a homotopy equivalence $M \simeq e(W) \times S^1$, such that $(N; \partial W \times S^1, M)$ is unique up to adjoining h-cobordisms to M (10.2). By [109] it is possible to choose M to be such that $\overline{M} \times S^1$ is homeomorphic to $M \times \mathbb{R}$ and also

$$[\overline{M}^+] \;=\; [W] \in \widetilde{K}_0(\mathbb{Z}[\pi]) \,.$$

Then $(M, c : M \longrightarrow S^1) = (\widehat{X}, \widehat{d})$ is the wrapping up of (X, d) with $X = \overline{M}$, and X is a finitely dominated $(n-1)$-dimensional geometric Poincaré complex such that $s^p(X) = 0 \in \mathbb{S}^p_{n-1}(X)$. In fact, this type of wrapping up is the method used by Freedman and Quinn [60, p. 225] to classify tame ends of 4-dimensional manifolds with good fundamental group.

(ii) Let $(W, \partial W)$ be an open n-dimensional manifold with compact boundary, such that $n \geq 5$ and W is both forward and reverse tame. By 17.10 there exists an open cocompact $X \subseteq W$ with a manifold approximate fibration $X \longrightarrow \mathbb{R}$, such that the end space $e(W)$ is a finitely dominated $(n-1)$-dimensional Poincaré space homotopy equivalent to X. The $\mathbb{S}$-groups are homotopy invariant, so that

$$s^p(e(W)) \;=\; s^p(X) \;=\; 0 \in \mathbb{S}^p_{n-1}(e(W)) \,,$$

with $s^p(X) = 0$ as in (i). The product $e(W) \times S^1$ is homotopy equivalent to a compact n-dimensional manifold, namely the wrapping up $\widehat{X}$ of X, with $X \times S^1 \cong \widehat{X} \times \mathbb{R}$. □

We now determine when a space is homotopy equivalent to a space which admits a manifold approximate fibration or proper bounded fibration to $\mathbb{R}$.

Proposition 17.24 *The following conditions on a CW complex X are equivalent for $n \geq 5$:*

(i) *X is homotopy equivalent to the infinite cyclic cover $\overline{M} = c^*\mathbb{R}$ of an n-dimensional manifold approximate fibration $c : M \longrightarrow S^1$,*

(ii) *X is homotopy equivalent to an open n-dimensional manifold W with a manifold approximate fibration $d : W \longrightarrow \mathbb{R}$,*

(iii) *$X \times S^1$ is homotopy equivalent to a closed n-dimensional manifold,*

(iv) *X is a finitely dominated $(n-1)$-dimensional geometric Poincaré complex with Pedersen–Ranicki [109] total projective surgery obstruction $s^p(X) = 0 \in \mathbb{S}^p_{n-1}(X)$.*

Proof (i) $\Longrightarrow$ (ii) Let $d = \overline{c} : W = \overline{M} \longrightarrow \mathbb{R}$.

(ii) $\Longrightarrow$ (i), (ii) $\Longrightarrow$ (iii) The wrapping up $(\widehat{W}, \widehat{d}) = (M, c)$ of (W, d) (17.1) is such that $c : M \longrightarrow S^1$ is an approximate fibration, with homotopy equivalences $X \simeq W = \overline{M}$, $X \times S^1 \simeq W \times S^1 \simeq M$.

(iii) $\Longrightarrow$ (ii) Let $X \times S^1 \simeq N$ for a closed n-dimensional manifold N. The infinite cyclic cover of N classified by $N \simeq X \times S^1 \longrightarrow S^1$ is an open n-dimensional manifold $W = \overline{N}$ proper homotopy equivalent to $X \times \mathbb{R}$. There exists a proper map $d : W \longrightarrow \mathbb{R}$ which is boundedly homotopy equivalent to the projection $X \times \mathbb{R} \longrightarrow \mathbb{R}$. Also, W is homotopy equivalent to X. Now 16.3 implies that d is a bounded fibration and 16.10 implies that W admits a manifold approximate fibration to $\mathbb{R}$.

(iii) $\Longleftrightarrow$ (iv) by [109]. □

Remark 17.25 The equivalence (i) $\Longleftrightarrow$ (iii) in 17.24 was first obtained by Chapman [26]. □

Proposition 17.26 *Let X be a CW complex. The following conditions are equivalent:*

(i) *X is finitely dominated,*

(ii) *$X \times S^1$ is homotopy equivalent to a finite CW complex,*

(iii) *X is homotopy equivalent to a CW complex which admits a proper bounded fibration to $\mathbb{R}$,*

(iv) *X is homotopy equivalent to a CW complex which admits a proper approximate fibration to $\mathbb{R}$.*

Proof (i) $\Longleftrightarrow$ (ii) by 6.7 (ii).

(i) $\Longrightarrow$ (iii) Let X be a finitely dominated CW complex. Then $X \times S^1$ is homotopy equivalent to a CW band with X homotopy equivalent to its infinite cyclic cover. Apply 17.14 to this infinite cyclic cover.

(iii) $\Longrightarrow$ (iv) by 17.16 and 17.20.

(iv) $\Longrightarrow$ (i) by 17.12. □

18

Geometric relaxation

By definition (15.14), a band $(M, c : M \longrightarrow S^1)$ is relaxed if there exists a fundamental domain $(V; U, \zeta U)$ for the infinite cyclic cover $\overline{M} = c^*\mathbb{R}$ of M such that V dominates $\overline{W}^+$ rel U, and $\zeta^{-1}V$ dominates $\overline{W}^-$ rel U. By 16.15 a CW band (M, c) with c an approximate fibration is relaxed. We shall now associate to a manifold band (M, c) with $\dim(M) \geq 5$ an h-cobordant relaxed manifold band (M', c') with $c' : M' \longrightarrow S^1$ a manifold approximate fibration, using approximate lifting properties. We shall relate this to the original construction of Siebenmann [145], which obtained the relaxation by ζ-twist glueing the two ends of the infinite cyclic cover $\overline{M}$.

Proposition 18.1 *For any manifold band* (M, c) *with* $\dim(M) \geq 5$ *a generating covering translation* $\zeta : \overline{M} \longrightarrow \overline{M}$ *is isotopic to a generating covering translation* $\zeta' : \overline{M} \longrightarrow \overline{M}$ *of an infinite cyclic covering* $\overline{M} \longrightarrow \overline{M}/\zeta' = M'$ *such that* (M', c') *is a relaxed manifold band, with* $c' : M' \longrightarrow S^1$ *a manifold approximate fibration.*

Proof By Proposition 17.14 $\overline{c} : \overline{M} \longrightarrow \mathbb{R}$ is a bounded fibration. Thus there is a bounded homotopy $h : \overline{c} \simeq d$ where $d : \overline{M} \longrightarrow \mathbb{R}$ is a manifold approximate fibration (16.10). Let $a_2 > 0$ be the bound on the homotopy h. In particular, $d\zeta$ is $(2a_2 + 1)$-close to d where $\zeta : \overline{M} \longrightarrow \overline{M}$ is the generating covering translation. Choose $a_3 > 17(2a_2 + 1)$. Let $g_s : \mathbb{R} \longrightarrow \mathbb{R}$ be the isotopy $g_s : t \longrightarrow t + a_3 s$. Choose $a_1 > 0$ such that $a_3 > 17(a_1 + 2a_2 + 1)$ and let $G_s : \overline{M} \longrightarrow \overline{M}$ be an isotopy with dG_s a_1-close to $g_s d$ (17.4, Approximate Isotopy Covering). Let $\zeta' = \zeta G_1 : \overline{M} \longrightarrow \overline{M}$. It follows from 17.5 that ζ' acts properly discontinuously on $\overline{M}$ so that $\overline{M} \longrightarrow \overline{M}/\zeta'$ is an infinite cyclic covering. It follows from Proposition 17.6 that d is properly homotopic to a manifold approximate fibration $\tilde{d} : \overline{M} \longrightarrow \mathbb{R}$ such that $\tilde{d}\zeta' = g_1 \tilde{d}$. If S^1 is identified with $\mathbb{R}/\mathbb{Z}$, then $\tilde{d}$ induces a manifold approximate fibration

$$c' : M' = \overline{M}/\zeta' \longrightarrow S^1 \; ; \; [x] \longrightarrow [\tilde{d}(x)/a_3] \; .$$

That c' is a manifold approximate fibration follows from the fact that it is one locally (Coram [35]). □

Definition 18.2 The *relaxation* of a manifold band (M, c) with $\dim(M) \geq 5$ is the relaxed manifold band of 18.1

$$(M', c') = (\overline{M}/\zeta', \overline{M}/\zeta' \longrightarrow S^1) . \qquad \square$$

Relaxation Theorem 18.3 *The relaxation (M', c') of a manifold band (M, c) with $\dim(M) \geq 5$ has the following properties:*

(i) $\overline{M}' = \overline{M}$,
(ii) *the generating covering translation $\zeta' : \overline{M}' \longrightarrow \overline{M}'$ is isotopic to the generating covering translation $\zeta : \overline{M} \longrightarrow \overline{M}$,*
(iii) *the classifying map $c' : M' \longrightarrow S^1$ is a manifold approximate fibration,*
(iv) *there exist homeomorphisms*

$$M \times \mathbb{R} \cong T(\zeta) \cong T(\zeta') \cong M' \times \mathbb{R} ,$$

(v) *the homeomorphism of (iv) determines the* relaxation h-cobordism $(W; M, M')$ *with well-defined torsion in $Wh(\pi_1(M))$ which depends only on the homotopy class of the classifying map $c : M \longrightarrow S^1$. The homotopy equivalence $f : M \longrightarrow M'$ induced by the h-cobordism is such that $c \simeq c'f$.*

The following conditions are equivalent:

(a) *$c : M \longrightarrow S^1$ is homotopic to a manifold approximate fibration,*
(b) *$(W; M, M')$ is a trivial h-cobordism,*
(c) *$f : M \longrightarrow M'$ is homotopic to a homeomorphism.*

Proof (M', c') is a relaxed manifold band by 16.15.

(i),(ii),(iii) obvious.

(iv) It follows from 14.8 (ii) that $T(\zeta) \cong M \times \mathbb{R}$ and $T(\zeta') \cong M' \times \mathbb{R}$. Since ζ' is isotopic to ζ we have $T(\zeta') \cong T(\zeta)$.

(v) We begin with an explicit description of the homeomorphism

$$M \times \mathbb{R} \cong M' \times \mathbb{R} .$$

Consider

$$T(\zeta) = (\overline{M} \times \mathbb{R})/\sim \ , \ T(\zeta') = (\overline{M} \times \mathbb{R})/\approx$$

where

$$(x, s + n) \sim (\zeta^n x, s) \ , \ (x, s + n) \approx ((\zeta')^n x, s) \ , \ n \in \mathbb{Z} .$$

The natural projections $\pi_1 : T(\zeta) \longrightarrow S^1 = \mathbb{R}/\mathbb{Z}$ and $\pi_2 : T(\zeta') \longrightarrow S^1 = \mathbb{R}/\mathbb{Z}$

are $[x,s]\longrightarrow[s]$. Define a homeomorphism

$$h_1 \ :\ M\times\mathbb{R} \ = \ \overline{M}/\zeta\times\mathbb{R} \xrightarrow{\simeq} T(\zeta) \ = \ (\overline{M}\times\mathbb{R})/\!\sim \ ; \ ([x],s) \longrightarrow [x, s-\overline{c}(x)] \ .$$

The composition $M\xrightarrow{\times 0}M\times\mathbb{R}\xrightarrow{h_1}T(\zeta)$ is $[x]\longrightarrow[x,-\overline{c}(x)]$ for $x\in\overline{M}$, and the diagram

$$\begin{array}{ccccc} M & \xrightarrow{\times 0} & M\times\mathbb{R} & \xrightarrow{h_1} & T(\zeta) \\ {\scriptstyle c}\big\downarrow & & & & \big\downarrow \\ S^1 & & \xrightarrow{\quad -1 \quad} & & S^1 \end{array}$$

commutes where $-1 : S^1\longrightarrow S^1; [s]\longrightarrow[-s]$. The isotopy $\zeta\cong\zeta'$ induces a homeomorphism

$$h_2 \ : \ T(\zeta) \xrightarrow{\simeq} T(\zeta') \ ; \ [x,s] \longrightarrow [\zeta G_{n+1-s}\zeta^{n-1}x, s-n]$$
$$(n\leq s\leq n+1 \ , \ n\in\mathbb{Z})$$

which commutes with the natural projections $T(\zeta)\longrightarrow S^1$ and $T(\zeta')\longrightarrow S^1$. Define a homeomorphism

$$h_3 \ : \ T(\zeta') \ = (\overline{M}\times\mathbb{R})/\approx \ \xrightarrow{\simeq} \ M'\times\mathbb{R} \ = \ \overline{M}/\zeta'\times\mathbb{R} \ ;$$
$$[x,s] \longrightarrow ([x], s+\overline{c}'(x)) \ .$$

As above, the diagram

$$\begin{array}{ccccc} M' & \xrightarrow{\times 0} & M'\times\mathbb{R} & \xrightarrow{h_3^{-1}} & T(\zeta') \\ {\scriptstyle c'}\big\downarrow & & & & \big\downarrow \\ S^1 & & \xrightarrow{\quad -1 \quad} & & S^1 \end{array}$$

commutes. The homeomorphism

$$h \ = \ h_3h_2h_1 \ : \ M\times\mathbb{R} \xrightarrow{\simeq} M'\times\mathbb{R}$$

determines an h-cobordism $(W;M,M')$ by choosing $L>0$ so large that

$$h(M\times(-\infty,0])\subseteq M'\times(-\infty,L)$$

and letting

$$W \ = \ M'\times(-\infty,L]\cap h(M\times[0,\infty)) \ ,$$

that is, $W\subseteq M'\times\mathbb{R}$ is the region between $h(M\times\{0\})$ and $M'\times\{L\}$ in $M'\times\mathbb{R}$. In the h-cobordism $(W;M,M')$ we identify M with $h(M\times\{0\})$ and

M' with $M' \times \{L\}$. With these identifications there is a map $\tilde{c} : W \longrightarrow S^1$ such that $\tilde{c}|M = c$ and $\tilde{c}|M' = c'$. This map is constructed by using the homotopy extension property to adjust the map $\pi_2 h_3^{-1}| : W \longrightarrow S^1$. For

$$\pi_2 h_3^{-1}|h(M \times \{0\}) \;=\; -ch^{-1}|h(M \times \{0\})$$

and $\pi_2 h_3^{-1}|M' \times \{L\}$ is homotopic to the composition

$$M' \times \{L\} \longrightarrow M' \times \{0\} \xrightarrow{\pi_2 h_3^{-1}} S^1 \;;\; (x, L) \longrightarrow -c'(x)\;.$$

The homotopy equivalence $f : M \longrightarrow M'$ is such that

$$c'f \;\simeq\; \tilde{c} \circ i \;=\; c \;:\; M \longrightarrow S^1$$

with $i =$ inclusion $: M \longrightarrow W$. We now show that the torsion of $(W; M, M')$ is well-defined, by which we mean that it is independent of the choices made in constructing M'. So suppose $G'_s : \overline{M} \longrightarrow \overline{M}$ is another isotopy such that $\overline{c}G'_s$ is $(a_1 + 2a_2)$-close to $g_s\overline{c}$. By a one-parameter and relative version of Approximate Isotopy Covering 17.4 (see Hughes, Taylor and Williams [77]) there is a two-parameter isotopy $\Gamma_{s,t} : \overline{M} \longrightarrow \overline{M}$ such that $\Gamma_{0,t} = \mathrm{id}_{\overline{M}}$, $\overline{c}\Gamma_{s,t}$ is $(a_1 + 2a_2)$-close to $g_s\overline{c}$, $\Gamma_{s,0} = G_s$ and $\Gamma_{s,1} = G'_s$. The homeomorphism

$$\beta \;=\; (\zeta \times \mathrm{id}_I)\Gamma_{1,-} \;:\; \overline{M} \times 1 \times I \xrightarrow{\simeq} \overline{M} \times 1 \times I$$

defines a properly discontinuous action of $\mathbb{Z}$ on $\overline{M} \times I$ (Proposition 17.5). We claim that the natural projection $(\overline{M} \times I)/\beta \longrightarrow I$ is a locally trivial bundle projection. First note that since $\overline{M} \times I \longrightarrow (\overline{M} \times I)/\beta$ is a covering and the composition $\overline{M} \times I \longrightarrow (\overline{M} \times I)/\beta \longrightarrow I$ is locally trivial, it follows easily that $(\overline{M} \times I)/\beta \longrightarrow I$ is a Serre fibration. Because $(\overline{M} \times I)/\beta$ and I are finite dimensional ANR's, $(\overline{M} \times I)/\beta \longrightarrow I$ is also a Hurewicz fibration (Ungar [162]). The fibres are manifolds of dimension greater than 4, so that it is locally trivial (Chapman and Ferry [29]). Since $\beta_0 = \zeta G_1$, there is a trivializing homeomorphism $\alpha : M' \times I \longrightarrow (\overline{M} \times I)/\beta$ such that $\alpha_0 = \mathrm{id}_{M'}$. Let

$$H \;:\; (\overline{M} \times I)/\beta \times \mathbb{R} \xrightarrow{\simeq} M \times I \times \mathbb{R}$$

be the homeomorphism given by the composite

$$(\overline{M} \times I)/\beta \times \mathbb{R} \;\cong\; T(\beta) \;\cong\; T(\zeta \times \mathrm{id}_I) \;\cong\; M \times I \times \mathbb{R}.$$

Since H_0 is isotopic to $H_1 \circ (\alpha_1 \times \mathrm{id}_{\mathbb{R}})$, it follows that the h-cobordisms determined by H_0 and $H_1 \circ (\alpha_1 \times \mathrm{id}_{\mathbb{R}})$ are homeomorphic (using the Isotopy Extension Theorem of Edwards and Kirby [41]), and have the same torsion. Hence, the h-cobordisms determined by H_0 and H_1 have the same torsion. These are the h-cobordisms given by the two sets of data so we have established well-definedness. This also shows that the torsion depends only on the homotopy class of c, for if $c \simeq c_1$, then both c and c_1 induce data for constructing the relaxation of (M, c) yielding the same torsion.

(a) $\Longrightarrow$ (b) We now assume that c is homotopic to a manifold approximate fibration and show that the h-cobordism is trivial. By the preceding paragraph we may assume that c itself is a manifold approximate fibration. Therefore, in the construction of M' we may take $d = \overline{c}$, $a_3 > 17$ and a_1, a_2 as small as we like. Then we shall have $\overline{c}G_s$ a_1-close to $g_s\overline{c}$. Proposition 17.5 can be used to show that ζG_s acts properly discontinuously on $\overline{M}$ for each s. Then the homeomorphism

$$\gamma = (\zeta \times \mathrm{id}_I)G : \overline{M} \times I \longrightarrow \overline{M} \times I$$

induces a properly discontinuous action on $\overline{M} \times I$ with $(\overline{M} \times I)/\gamma \cong M \times I$ by the argument above. Since $\gamma_0 = \zeta$ and $\gamma_1 = \zeta G_1$, it follows as above that γ may be used to show that the h-cobordism between $M = \overline{M}/\zeta$ and $M' = \overline{M}/\zeta G_1$ is trivial.

(b) $\Longrightarrow$ (c) If $h : M \times I \longrightarrow W$ is a homeomorphism with $h_0 = \mathrm{id}_M$ then $M = M \times \{1\} \xrightarrow{h|} M' \subseteq W$ is a homeomorphism homotopic to f.

(c) $\Longrightarrow$ (a) Since $c \simeq c'f$, if f is homotopic to a homeomorphism $h : M \longrightarrow M'$ then c is homotopic to the manifold approximate fibration $c'f$.

□

Remark 18.4 (i) For any $(n+1)$-dimensional h-cobordism $(W; M, M')$ the torsion of the homotopy equivalence $f : M \longrightarrow W \longrightarrow M'$ is

$$\begin{aligned}\tau(f) &= \tau(M \longrightarrow W) - \tau(M' \longrightarrow W) \\ &= \tau(M \longrightarrow W) + (-)^n \tau(M \longrightarrow W)^* \in Wh(\pi_1(M)) .\end{aligned}$$

It will be shown in 26.13 that for the relaxation h-cobordism $(W; M, M')$ of 18.3 (v) $\tau(M \longrightarrow W)$ and $\tau(M' \longrightarrow W)$ are in complementary direct summands of the Whitehead group of $\pi_1(M) = \pi_1(\overline{M}) \times_{\zeta_*} \mathbb{Z}$, namely the two copies of the reduced nilpotent class group $\widetilde{\mathrm{Nil}}_0$, and that there is a Poincaré duality $\tau(M' \longrightarrow W) = (-)^{n-1}\tau(M \longrightarrow W)^*$. Thus the conditions (a), (b), (c) in 18.3 are also equivalent to:

(d) $\tau(f) = 0$.

In particular, f is simple if and only if f is homotopic to a homeomorphism.

(ii) It will follow from 26.10 (ii) that an n-dimensional manifold band (M, c) with $n \geq 5$ is relaxed if and only if the homotopy equivalence $f : M \longrightarrow M'$ in 18.3 (v) is simple. Combining this with (i) and 18.3 (iii) gives:

a manifold band (M, c) *with* $\dim M \geq 5$ *is relaxed if and only if the map* $c : M \longrightarrow S^1$ *is homotopic to a manifold approximate fibration.* □

Lemma 18.5 *For a manifold approximate fibration $d : X \longrightarrow \mathbb{R}$ with $\dim(X) \geq 5$, there exists an isotopy of open embeddings*

$$H \; : \; d^{-1}(-\infty, 1) \times I \longrightarrow X$$

with $H_0 : d^{-1}(-\infty,1) \longrightarrow X$ the inclusion, $H_1 : d^{-1}(-\infty,1) \longrightarrow X$ a homeomorphism, and for every $t \in I$ $H_t| : d^{-1}(-\infty, 0] \longrightarrow X$ the inclusion.

Proof Define

$$X^* \;=\; \{(x,t) \in X \times I \,|\, d(x) < \frac{1}{1-t}\} \;,$$

$$d^* \,:\, X^* \longrightarrow \mathbb{R} \times I\,;\,(x,t) \longrightarrow \begin{cases} (d(x),t) & \text{if } d(x) \leq 0\;, \\ \left(\dfrac{d(x)}{1-(1-t)d(x)}, t\right) & \text{if } 0 \leq d(x) < \dfrac{1}{1-t} \end{cases}$$

so that $d^*| : (d^*)^{-1}(\mathbb{R} \times \{t\}) \longrightarrow \mathbb{R} \times \{t\}$ is a manifold approximate fibration for each $t \in I$ and

$$d^*| \;=\; d| \times \mathrm{id}_I \;:\; (d^*)^{-1}((-\infty,0] \times I) \;=\; d^{-1}(-\infty,0] \times I \longrightarrow (-\infty,0] \times I\;.$$

It follows from the argument of Hughes [73, p. 75] that the composition

$$\pi \;:\; X^* \xrightarrow{d^*} \mathbb{R} \times I \xrightarrow{\text{proj.}} I$$

is a fibre bundle with trivial subbundle $\pi| \,:\, d^{-1}(-\infty,0] \times I \longrightarrow I$. Since $\pi^{-1}(0) = d^{-1}(-\infty,1)$ a trivializing homeomorphism

$$H \;:\; d^{-1}(-\infty,1) \times I \longrightarrow X^* \subseteq X \times I$$

with

$$\begin{aligned} H_0 \;&=\; \text{identity} \;:\; \pi^{-1}(0) \longrightarrow d^{-1}(-\infty,1)\;, \\ H| \;&=\; \text{inclusion} \;:\; d^{-1}(-\infty,0] \times I \longrightarrow X^* \end{aligned}$$

gives the desired isotopy of open embeddings. □

Proposition 18.6 *Let $d : X \longrightarrow \mathbb{R}$ be a manifold approximate fibration with $\dim(X) \geq 5$. The wrapping up $\widehat{X}$ of 17.1 is homeomorphic to any Siebenmann twist glueing $W_1(f_-,f_+)$ of X relative to the identity $1 : X \longrightarrow X$ (15.17).*

Proof The isotopy

$$g_s \;:\; \mathbb{R} \longrightarrow \mathbb{R}\;;\; t \longrightarrow t+s \;\; (0 \leq s \leq 1)$$

can be covered up to a_1 by an isotopy $G_s : X \longrightarrow X$ where $a_1 > 0$ is such that $17a_1 < 1$. Then G_1 is a covering translation and $\widehat{X} = X/G_1$ is the wrapping up X (17.7). Define a fundamental domain $(V; U, G_1U)$ by

$$U \;=\; d^{-1}(0) \;\;,\;\; V \;=\; G_1 d^{-1}(-\infty,0] \cap d^{-1}[0,\infty)\;.$$

Let

$$U_- = d^{-1}(-\infty, \tfrac{1}{4}) \ , \ U_+ = d^{-1}(\tfrac{3}{4}, \infty) \subset X .$$

By Edwards and Kirby [41] there exists an isotopy $H : X \times I \longrightarrow X \times I$ supported on $d^{-1}[-2,3]$ such that

$$H_s|d^{-1}([-1,2]) = G_s| \ (0 \le s \le 1) .$$

By 18.5 $H_s|d^{-1}(-\infty, 0]$ extends to an isotopy $k_s^- : U_- \longrightarrow X$ $(0 \le s \le 1)$ of open embeddings such that $k_1^- U_- = X$ and

$$H_s| = k_s| = \text{inclusion} : d^{-1}(-\infty, -2] \longrightarrow X \ \ (0 \le s \le 1) .$$

Since we can choose $a_1 > 0$ as small as desired we can assume that $G_1 d^{-1}(0) \subseteq d^{-1}(\frac{7}{8}, \infty)$. By 18.5 again, there is an isotopy $k_s^+ : U_+ \longrightarrow X$ of open embeddings such that $k_1^+ U_+ = X$ and $k_s^+| = \text{inclusion} : d^{-1}([\frac{7}{8}, \infty)) \longrightarrow X$ for $0 \le s \le 1$. Let $f_- = k_1^-$, $f_+ = k_1^+$ and form the Siebenmann twist glueing $W_1(f_-, f_+)$. We now demonstrate that the inclusion : $V \longrightarrow X$ induces a homeomorphism

$$\widehat{X} = X/G_1 \xrightarrow{\simeq} W_1(f_-, f_+) = X/\sim .$$

The map is well-defined for if $x \in U$ then $G_1(x) = f_+^{-1} f_-(x)$. The map is onto because every $x \in X$ is $\sim$-related to a point in V. This also shows that $X/\sim = V/\sim$, from which it follows that the map is one-to-one. Uniqueness follows from 15.18. □

Proposition 18.7 *Let* (W, c) *be a manifold band,* $\dim(W) \ge 5$*, with generating covering translation* $\zeta : \overline{W} \longrightarrow \overline{W}$*. Any Siebenmann twist glueing* $W_\zeta(f_-, f_+)$ *of* $\overline{W}$ *relative to* ζ *is homeomorphic to the relaxation* W' *of 18.2.*

Proof Let $\overline{c} : \overline{W} \longrightarrow \mathbb{R}$ be the bounded fibration induced by $c : W \longrightarrow S^1$. Let $a_1, a_2, a_3 > 0$ be as in 18.1, so that there is an a_2-homotopy from $\overline{c}$ to a manifold approximate fibration $d : \overline{W} \longrightarrow \mathbb{R}$, and recall that a_1 is allowed to be chosen as small as desired. The isotopy

$$g_s : \mathbb{R} \longrightarrow \mathbb{R} \ ; \ t \longrightarrow t + a_3 s \ \ (0 \le s \le 1)$$

can be covered up to a_1 by an isotopy $G_s : \overline{W} \longrightarrow \overline{W}$. Then $\zeta' = \zeta G_1$ is a covering translation and $W' = \overline{W}/\zeta'$ is the relaxation of W. The infinite cyclic cover $\overline{W}'$ of W' has a fundamental domain $(V; U, \zeta' U)$ with

$$U = d^{-1}(0) \ , \ V = \zeta' d^{-1}(-\infty, 0] \cap d^{-1}[0, \infty) .$$

Let

$$U_- = d^{-1}(-\infty, 1) \ , \ U_+ = d^{-1}(2, \infty) \subset \overline{W} .$$

By Edwards and Kirby [41] there exists an isotopy $H : \overline{W} \times I \longrightarrow \overline{W} \times I$ supported on $d^{-1}[-2, N+1]$ such that $H_s|d^{-1}[-1, N] = G_s|$ for $0 \le s \le 1$

where N is so large that $V \subseteq d^{-1}(-\infty, N)$. By 18.5 $H_s|d^{-1}(-\infty, 0]$ extends to an isotopy $k_s^- : U_- \longrightarrow \overline{W}$ of open embeddings such that $k_1^- U_- = \overline{W}$ and

$$H_s| = k_s| = \text{inclusion} : d^{-1}(-\infty, -2] \longrightarrow \overline{W} \quad (0 \leq s \leq 1) .$$

Since $a_3 > 17$ and we can choose $a_1 > 0$ as small as desired we can assume that $G_1 d^{-1}(0) \subseteq d^{-1}(3, \infty)$. By 18.5 again, there is an isotopy $k_s^+ : U_+ \longrightarrow \overline{W}$ of open embeddings such that $k_1^+ U_+ = \overline{W}$ and $k_s^+| =$ inclusion : $d^{-1}[4, \infty) \longrightarrow \overline{W}$ for $0 \leq s \leq 1$. Let $f_- = k_1^-$, $f_+ = k_1^+$ and form the Siebenmann twist glueing $W_\zeta(f_-, f_+)$. We now demonstrate that inclusion : $V \longrightarrow \overline{W}$ induces a homeomorphism

$$W' = \overline{W}/\zeta' \xrightarrow{\simeq} W_\zeta(f_-, f_+) = \overline{W}/\sim .$$

The map is well-defined for if $x \in U$ then $\zeta G_1(x) = f_+^{-1} \zeta f_-(x)$. To show that the map is onto it suffices to show that every $x \in U_-$ is $\sim$-related to a point in V. This is clear for $d(x) > -1$, so we consider the case $d(x) \leq -1$. For such an x we have $x \sim f_+^{-1} \zeta f_-(x) = f_+^{-1} \zeta(x)$, so it suffices to show that $f_+^{-1} \zeta(x) \in V$. For this we must make sure that a_1 is chosen small enough that $a_3 - a_1 > 2a_2$, in which case

$$d^{-1}(-\infty, 2a_2] \subseteq d^{-1}(-\infty, a_3 - a_1] \subseteq G_1 d^{-1}(-\infty, 0] .$$

Since $d(x) \leq -1$ it follows that $\overline{c}(x) \leq a_2 - 1$ so $\overline{c}\zeta(x) \leq a_2$ and $d\zeta(x) \leq 2a_2$. Thus $\zeta(x) \in G_1 d^{-1}(-\infty, 0]$, from which it follows that $f_+^{-1} \zeta(x) \in V$. This also shows that $\overline{W}/\sim = V/\sim$, from which it follows that the map is one-to-one. Uniqueness follows from 15.18. □

19

Homotopy theoretic twist glueing

Given a *CW* ribbon $(X, d : X \longrightarrow \mathbb{R})$ and a self homotopy equivalence $h : (X, d) \longrightarrow (X, d)$ we use h to identify the two copies of X in $e(X) \simeq X \amalg X$, constructing an infinite *CW* complex $X(h)$ equipped with a map $d(h) : X(h) \longrightarrow S^1$. The 'homotopy theoretic twist glueing' $X(h)$ is homotopy equivalent to the mapping torus $T(h)$. The induced infinite cyclic cover $\overline{X}(h) = d(h)^*\mathbb{R}$ of $X(h)$ is related to (X, d) by a homotopy equivalence

$$F(h) \; : \; (X, d) \xrightarrow{\simeq} (\overline{X}(h), \overline{d}(h))$$

such that the generating covering translation $\zeta_{\overline{X}(h)} : \overline{X}(h) \longrightarrow \overline{X}(h)$ fits into a homotopy commutative square

$$\begin{array}{ccc} X & \xrightarrow{F(h)} & \overline{X}(h) \\ {\scriptstyle h}\downarrow & & \downarrow{\scriptstyle \zeta_{\overline{X}(h)}} \\ X & \xrightarrow{F(h)} & \overline{X}(h) \end{array}$$

The construction of $(X(h), d(h))$ from $(X, d), h$ is a homotopy theoretic version of Siebenmann twist glueing (15.17). If $h : (X, d) \longrightarrow (X, d)$ is a proper homotopy equivalence which is either an end-preserving covering translation or the identity we refine the construction of $(X(h), d(h))$, $F(h)$ to obtain a relaxed *CW* π_1-band $(X[h], d[h])$ in the homotopy type of $(X(h), d(h))$ with an infinite simple homotopy equivalence

$$F[h] \; : \; (X, d) \xrightarrow{\simeq} (\overline{X}[h], \overline{d}[h]) \; .$$

In Chapter 20 we shall show that the wrapping up $(\widehat{X}, \widehat{d})$ of a manifold ribbon (X, d) constructed in Chapter 17 has the simple homotopy type of the 1-twist glueing $(X[1], d[1])$, and that the relaxation (M', c') of a manifold band (M, c) constructed in Chapter 18 has the simple homotopy type of the ζ-twist glueing $(X[\zeta], d[\zeta])$ of the manifold ribbon $(X, d) = (\overline{M}, \overline{c})$.

The homotopy theoretic twist glueings fit into various homotopy pushouts. The chain homotopy analogue of the homotopy pushout property of the double mapping cylinder (12.1) is given by:

Definition 19.1 Let A be a ring.

(i) The *algebraic mapping pushout* of A-module chain maps $f : P \longrightarrow Q$, $g : P \longrightarrow R$ is the algebraic mapping cone

$$\mathcal{M}(f,g) = \mathcal{C}(\begin{pmatrix} f \\ g \end{pmatrix} : P \longrightarrow Q \oplus R) .$$

(ii) The *algebraic mapping pullback* of A-module chain maps $h : Q \longrightarrow S$, $k : R \longrightarrow S$ is the algebraic mapping cone

$$\mathcal{P}(h,k) = \mathcal{C}((h \quad k) : Q \oplus R \longrightarrow S)_{*+1} .$$

(iii) A square of A-module chain complexes and chain maps

$$\begin{array}{ccc} P & \xrightarrow{f} & Q \\ {\scriptstyle g}\downarrow & & \downarrow{\scriptstyle h} \\ R & \xrightarrow{k} & S \end{array}$$

is *chain homotopy cartesian* if there is given a chain homotopy

$$e : hf \simeq kg : P \longrightarrow S$$

such that the A-module chain map

$$\mathcal{M}(f,g) = \mathcal{C}(\begin{pmatrix} f \\ g \end{pmatrix} : P \longrightarrow Q \oplus R) \longrightarrow S$$

defined by

$$\mathcal{M}(f,g)_n = P_{n-1} \oplus Q_n \oplus R_n \longrightarrow S_n \ ; \ (x,y,z) \longrightarrow e(x) + h(y) - k(z)$$

is a chain equivalence, or equivalently such that the A-module chain map

$$P \longrightarrow \mathcal{P}(h,k) = \mathcal{C}((h \quad -k) : Q \oplus R \longrightarrow S)_{*+1}$$

defined by

$$P_n \longrightarrow \mathcal{P}(h,k)_n = Q_n \oplus R_n \oplus S_{n+1} \ ; \ x \longrightarrow (f(x), g(x), e(x))$$

is a chain equivalence. □

Example 19.2 A homotopy pushout of spaces induces a chain homotopy cartesian square on the chain level. □

A chain homotopy cartesian square has the universal property of a chain homotopy pushout (analogous to 12.2), and also the universal property of a chain homotopy pullback:

Proposition 19.3 *A chain homotopy cartesian square as in 19.1 has the following universal properties:*

(i) *Given an A-module chain complex T, chain maps $u : Q \longrightarrow T$, $v : R \longrightarrow T$ and a chain homotopy $uf \simeq vg : P \longrightarrow T$ there exists a chain map*

$$(u, v) \ : \ S \longrightarrow T$$

such that there is defined a chain homotopy commutative diagram

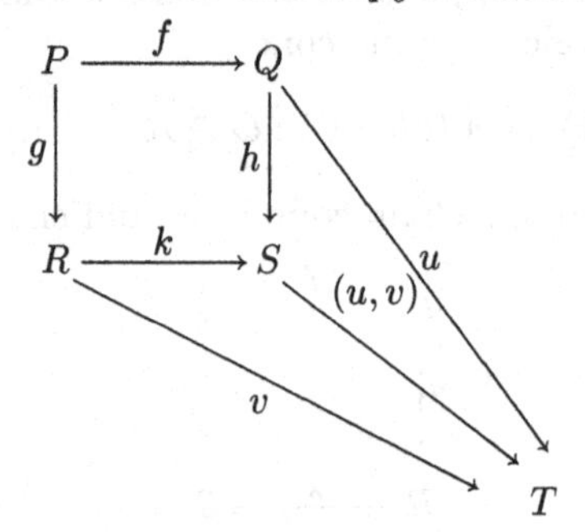

(ii) *Given an A-module chain complex T, chain maps $u : T \longrightarrow Q$, $v : T \longrightarrow R$ and a chain homotopy $hu \simeq kv : T \longrightarrow S$ there exists a chain map*

$$(u, v) \ : \ T \longrightarrow P$$

such that there is defined a chain homotopy commutative diagram

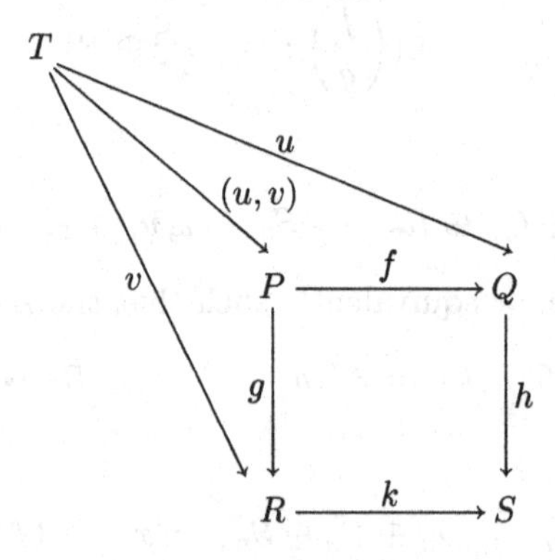

□

Definition 19.4 (i) A *homotopy cobordism* (V, U_1, U_2, f_1, f_2) is a diagram of spaces and maps

$$U_1 \xrightarrow{f_1} V \xleftarrow{f_2} U_2$$

which may be visualized as

$$U_1 \xrightarrow{f_1} \quad V \quad \xleftarrow{f_2} U_2$$

If the spaces are *CW* complexes and the maps are cellular this is a *CW cobordism.*

(ii) The *union* of homotopy cobordisms

$$c = (V, U_1, U_2, f_1, f_2) \ , \ c' = (V', U_1', U_2', f_1', f_2')$$

with $U_2 = U_1'$ is the homotopy cobordism

$$c \cup c' = (V'', U_1'', U_2'', f_1'', f_2'')$$

with

$$f_1'' : U_1'' = U_1 \xrightarrow{f_1} V \longrightarrow V'' = \mathcal{M}(f_2, f_1') \ ,$$

$$f_2'' : U_2'' = U_2' \xrightarrow{f_2'} V' \longrightarrow V'' = \mathcal{M}(f_2, f_1') \ ,$$

so that there is defined a homotopy commutative diagram

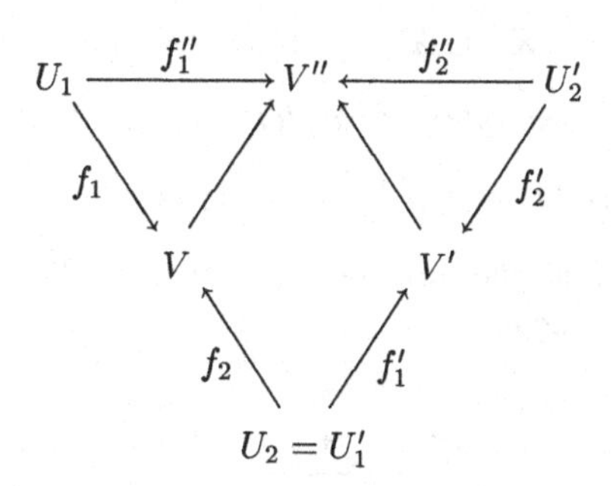

with the square a homotopy pushout.

$$U_1'' = U_1 \xrightarrow{f_1} \quad V \quad \xleftarrow{f_2} U_2 = U_1' \xrightarrow{f_1'} \quad V' \quad \xleftarrow{f_2'} U_2'' = U_2'$$

$$V''$$

□

Example 19.5 (i) A manifold cobordism $(V; U_1, U_2)$ determines a homotopy cobordism (V, U_1, U_2, f_1, f_2) with $f_1 : U_1 \longrightarrow V$, $f_2 : U_2 \longrightarrow V$ the inclusions. The union of manifold cobordisms $(V; U_1, U_2)$, $(V'; U_1', U_2')$ with

$U_2 = U_1'$

$$(V;U_1,U_2)\cup(V';U_1',U_2') = (V\cup_{U_2=U_1'}V';U_1,U_2')$$

is a manifold cobordism which is rel ∂ homotopy equivalent to the union homotopy cobordism $(V,U_1,U_2,f_1,f_2)\cup(V',U_1',U_2',f_1',f_2')$.

(ii) A fundamental domain $(V;U,\zeta U)$ for an infinite cyclic cover $\overline{W}$ of a space W with generating covering translation $\zeta:\overline{W}\longrightarrow\overline{W}$ determines a homotopy cobordism (V,U,U,f^+,f^-) with

$$f^+ : U\longrightarrow V\ ;\ x\longrightarrow x\ ,$$
$$f^- : U\longrightarrow V\ ;\ y\longrightarrow \zeta y\ . \qquad \square$$

Proposition 19.6 *Let (X,d) be a CW complex ribbon, and write the inclusions as*

$$j^+ : U = d^{-1}(0)\longrightarrow X^+ = d^{-1}[0,\infty)\ ,$$
$$j^- : U\longrightarrow X^- = d^{-1}(-\infty,0]\ ,$$
$$q^+ : X^+\longrightarrow X\ ,\ q^- : X^-\longrightarrow X\ ,$$

so that

$$X = X^+\cup X^-\ ,\ U = X^+\cap X^-\ ,$$

with $U\subset X$ a finite subcomplex. Also, let

$$\pi = \pi_1(U) = \pi_1(X^+) = \pi_1(X^-) = \pi_1(X)$$

and let $\widetilde{U},\widetilde{X}^+,\widetilde{X}^-,\widetilde{X}$ be the universal covers of U,X^+,X^-,X.

(i) *The commutative square*

$$\begin{array}{ccc} U & \xrightarrow{j^+} & X^+ \\ {\scriptstyle j^-}\big\downarrow & & \big\downarrow{\scriptstyle q^+} \\ X^- & \xrightarrow{q^-} & X \end{array}$$

is a homotopy pushout, with the natural map defining a homotopy equivalence

$$\mathcal{M}(j^+,j^-)\xrightarrow{\simeq} X\ .$$

(ii) *X^+ and X^- dominate X.*

(iii) *The finite f.g. free $\mathbb{Z}[\pi]$-module chain complex $C(\widetilde{U})$ dominates $C(\widetilde{X}^+)$ and $C(\widetilde{X}^-)$.*

(iv) *X^+, X^- and X are finitely dominated.*

Proof (i) This is a standard property of CW complexes: pushout squares are homotopy pushouts.

(ii) Use the homotopy equivalences $e(X^{\pm}) \simeq X$ to define dominations of X by X^+ and X^-

$$(X^+, p^+, q^+, q^+p^+ \simeq 1_X)\ ,$$
$$(X^-, p^-, q^-, q^-p^- \simeq 1_X)$$

with

$$q^{\pm} \ :\ X^{\pm} \longrightarrow X$$

the inclusions and

$$p^{\pm} \ :\ X \simeq e(X^{\pm}) \longrightarrow X^{\pm}\ .$$

(iii) The $\mathbb{Z}[\pi]$-module chain homotopy cartesian square

$$\begin{array}{ccc}
C(\widetilde{U}) & \xrightarrow{\ j^+\ } & C(\widetilde{X}^+) \\
{\scriptstyle j^-}\big\downarrow & & \big\downarrow{\scriptstyle q^+} \\
C(\widetilde{X}^-) & \xrightarrow{\ q^-\ } & C(\widetilde{X})
\end{array}$$

is a chain homotopy pullback by 19.3. The $\mathbb{Z}[\pi]$-module chain maps

$$i^+ \ :\ C(\widetilde{X}^+) \xrightarrow{(1,p^-q^+)} \mathcal{P}(q^+,q^-) \simeq C(\widetilde{U})\ ,$$
$$i^- \ :\ C(\widetilde{X}^-) \xrightarrow{(p^+q^-,1)} \mathcal{P}(q^+,q^-) \simeq C(\widetilde{U})$$

fit into chain homotopy commutative diagrams

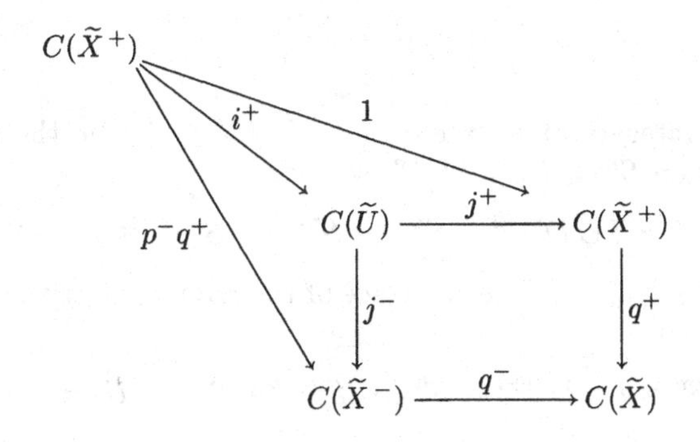

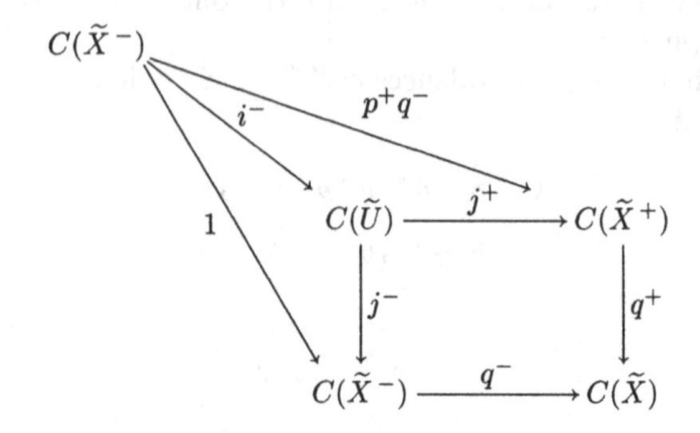

so that there are defined chain homotopy dominations of $C(\widetilde{X}^+)$ and $C(\widetilde{X}^-)$

$$(C(\widetilde{U}),i^+,j^+,j^+i^+\simeq 1_{C(\widetilde{X}^+)})\ ,\ (C(\widetilde{U}),i^-,j^-,j^-i^-\simeq 1_{C(\widetilde{X}^-)})\ .$$

(iv) It follows from (iii) and 6.8 (i) that X^+ and X^- are finitely dominated. Combining this with (ii) gives that X is finitely dominated also. □

Remark 19.7 In general, the finite subcomplex $U = X^+\cap X^-$ of a CW ribbon (X,d) does not dominate X^+ and X^-. In particular, this fails to be the case for the example in (i) below. In Chapter 20 below it is proved that every CW ribbon (X,d) is infinite simple homotopy equivalent to the infinite cyclic cover $(\overline{W},\overline{c})$ of a relaxed CW π_1-band (W,c). In (ii) below it is shown that every relaxed CW π_1-band (W,c) is simple homotopy equivalent to a relaxed CW π_1-band (W_1,c_1) with a π_1-fundamental domain $(V_1,U_1,U_1,f_1^+,f_1^-)$ for $\overline{W}_1$ such that $U_1=\overline{W}_1^+\cap\overline{W}_1^-$ dominates $\overline{W}_1^+$ and $\overline{W}_1^-$ rel U_1. Thus every CW ribbon (X,d) is infinite simple homotopy equivalent to one (also denoted (X,d)) such that U dominates X^+ and X^-, with dominations

$$(U,i^+:X^+\longrightarrow U,j^+,j^+i^+\simeq 1_{X^+})\ ,\ (U,i^-:X^-\longrightarrow U,j^-,j^-i^-\simeq 1_{X^-})$$

inducing the chain homotopy dominations

$$(C(\widetilde{U}),i^+,j^+,j^+i^+\simeq 1_{C(\widetilde{X}^+)})\ ,\ (C(\widetilde{U}),i^-,j^-,j^-i^-\simeq 1_{C(\widetilde{X}^-)})$$

of 19.6.

(i) For any integers $m,n\geq 1$ let (V_1,U_1,U_2,f_1,f_2) be the trace of the trivial surgery on $S^m\subset U_1=S^{m+n}$, so that

$$V_1\ =\ S^{m+n}\times I\cup D^{m+1}\times S^n\ \simeq\ S^{m+n}\vee S^n\ ,\ U_2\ =\ S^m\times S^n\ .$$

Also, let (V_2,U_2,U_3,f_2',f_3) be the trace of the trivial surgery on $S^m\subset U_2=S^m\times S^n$, so that

$$V_2\ =\ S^m\times S^n\times I\cup D^{m+1}\times S^n\ \simeq\ S^{m+n}\vee S^m\ ,\ U_3=\ U_1\ =\ S^{m+n}\ .$$

Then

$$(V_1, U_1, U_2, f_1, f_2) \cup (V_2, U_2, U_3, f_2', f_3) \simeq (S^{m+n} \times I, S^{m+n} \times \{0\}, S^{m+n} \times \{1\}, i_0, i_1) ,$$

$$(V_2, U_2, U_3, f_2', f_3) \cup (V_1, U_1, U_2, f_1, f_2) \simeq (S^{m+n} \vee S^m \vee S^n, S^m \times S^n, S^m \times S^n, q \vee p_1 \vee 0, q \vee 0 \vee p_2)$$

with i_0, i_1 the inclusions, $p_1 : S^m \times S^n \longrightarrow S^m$, $p_2 : S^m \times S^n \longrightarrow S^n$ the projections and $q : S^m \times S^n \longrightarrow S^{m+n}$ a degree 1 map. The infinite cyclic cover of the CW band

$$(W, c) = (S^{m+n} \times S^1, \text{projection} : S^{m+n} \times S^1 \longrightarrow S^1)$$

is thus a CW ribbon

$$(X, d) = (\overline{W}, \overline{c}) = (S^{m+n} \times \mathbb{R}, \text{projection} : S^{m+n} \times \mathbb{R} \longrightarrow \mathbb{R})$$

with a fundamental domain (V, U, U, f^+, f^-) such that up to homotopy

$$f^+ : U = S^m \times S^n \xrightarrow{q \vee p_1 \vee 0} S^{m+n} \vee S^m \vee S^n \simeq V ,$$

$$f^- : U = S^m \times S^n \xrightarrow{q \vee 0 \vee p_2} S^{m+n} \vee S^m \vee S^n \simeq V ,$$

$$g^+ : V \simeq S^{m+n} \vee S^m \vee S^n \longrightarrow S^{m+n} \vee S^m \simeq X^+ ,$$

$$g^- : V \simeq S^{m+n} \vee S^m \vee S^n \longrightarrow S^{m+n} \vee S^n \simeq X^- .$$

(ii) For any CW π_1-band (W, c) and any π_1-fundamental domain (V, U, U, f^+, f^-) for the infinite cyclic cover $\overline{W} = c^*\mathbb{R}$ of W let

$$V_1 = \mathcal{M}(f^+, f^-)$$

be the double mapping cylinder, so that there is defined a homotopy pushout

$$\begin{array}{ccc} U & \xrightarrow{f^+} & V \\ {\scriptstyle f^-}\downarrow & & \downarrow{\scriptstyle f_1^-} \\ V & \xrightarrow{f_1^+} & V_1 \end{array}$$

The CW cobordism $(V_1, V, V, f_1^+, f_1^-)$ is a π_1-fundamental domain for the CW π_1-band defined by the mapping coequalizer

$$(W_1, c_1) = \mathcal{W}(f_1^+, f_1^-) .$$

The application of 13.18 with

$$i^+ = f^+ , \; i^- = f^- : K = U \longrightarrow X = Y = V ,$$

$$j^+ = j^- = 1 : L = V \longrightarrow X = Y = V$$

shows that (W_1, c_1) is simple homotopy equivalent to (W, c). If (W, c) is a relaxed CW π_1-band and V dominates $\overline{W}^+$ rel U and V dominates $\overline{W}^-$ rel ζU let

$$k^+ : \overline{W}^+ \longrightarrow V \ , \ k^- : \zeta\overline{W}^- \longrightarrow V$$

be maps such that

$$\begin{aligned} g^+k^+ &\simeq 1 : \overline{W}^+ \longrightarrow \overline{W}^+ \ \mathrm{rel}\, U \ , \\ g^-k^- &\simeq 1 : \zeta\overline{W}^- \longrightarrow \zeta\overline{W}^- \ \mathrm{rel}\, \zeta U \ , \end{aligned}$$

with $g^+ : V \longrightarrow \overline{W}^+$, $g^- : V \longrightarrow \zeta\overline{W}^-$ the inclusions. The homotopy commutative square

$$\begin{array}{ccc} U & \xrightarrow{f^+} & V \\ {\scriptstyle f^-}\downarrow & & \downarrow{\scriptstyle k^-g^-} \\ V & \xrightarrow{k^+g^+} & V \end{array}$$

is a homotopy pushout, so that

$$\begin{aligned} f_1^+ &: U_1 = V \xrightarrow{k^+g^+} V \simeq V_1 \ , \ f_1^- : U_1 = V \xrightarrow{k^-g^-} V \simeq V_1 \ , \\ g_1^+ &: V \xrightarrow{g^+} \overline{W}^+ \simeq \overline{W}_1^+ \ , \ g_1^- : V \xrightarrow{g^-} \overline{W}^- \simeq \overline{W}_1^- \ . \end{aligned}$$

(Warning: the homotopy equivalence $V \simeq V_1$ is not simple in general – in the notation of 22.5 $\tau(V \simeq V_1) = \phi^+ \in Wh(\pi_1(\overline{W}))$.) The fundamental domain $(V_1, U_1, U_1, f_1^+, f_1^-)$ for the infinite cyclic cover $(\overline{W}_1, \overline{c}_1)$ of (W_1, c_1) is such that U_1 dominates $\overline{W}_1^+$ and $\overline{W}_1^-$, with

$$\begin{aligned} g_1^+f_1^+ &: U_1 = V \xrightarrow{g^+} \overline{W}^+ \simeq \overline{W}_1^+ \ , \\ g_1^-f_1^- &: U_1 = V \xrightarrow{g^-} \overline{W}^- \simeq \overline{W}_1^- \ . \end{aligned}$$

□

Definition 19.8 Given a CW ribbon (X, d) and a homotopy equivalence $h : X \longrightarrow X$ define the *h-twist glueing* $(X(h), d(h))$ of X to be the (infinite) CW complex

$$\begin{aligned} X(h) &= \mathcal{W}(f^+(h), f^-(h)) \\ &= U \times I \cup_{f^+(h) \amalg f^-(h)} V(h) \end{aligned}$$

with

$$\begin{aligned} &(V(h), U, U, f^+(h), f^-(h)) \\ &\quad = (X^+, U, X, j^+, p^+h) \cup (X^-, X, U, p^-, j^-) \end{aligned}$$

the union cobordism and

$$d(h) \; : \; X(h) \longrightarrow \mathcal{W}(1_{\{\text{pt.}\}}, 1_{\{\text{pt.}\}}) \; = \; S^1$$

the canonical map. The double mapping cylinder

$$V(h) \; = \; \mathcal{M}(p^+h, p^-)$$

fits into the homotopy pushout square

$$\begin{array}{ccc} X & \xrightarrow{p^-} & X^- \\ {\scriptstyle p^+h}\big\downarrow & & \big\downarrow{\scriptstyle k^-(h)} \\ X^+ & \xrightarrow{k^+(h)} & V(h) \end{array}$$

and the maps $f^+(h), f^-(h)$ are the composites

$$f^+(h) \; : \; U \xrightarrow{j^+} X^+ \xrightarrow{k^+(h)} V(h) \; ,$$

$$f^-(h) \; : \; U \xrightarrow{j^-} X^- \xrightarrow{k^-(h)} V(h) \; .$$

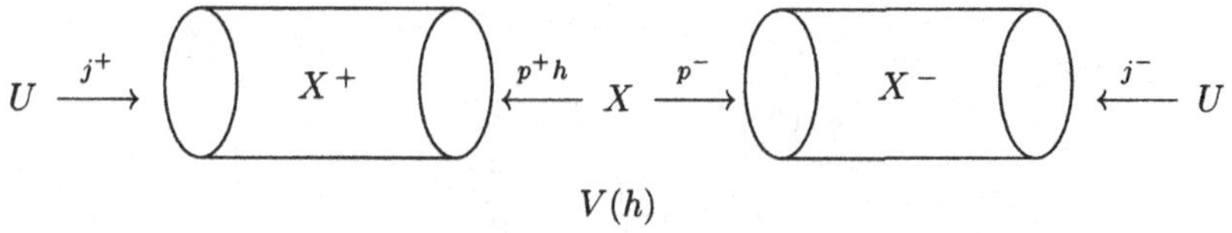

$$U \xrightarrow{j^+} \boxed{\;X^+\;} \xleftarrow{p^+h} X \xrightarrow{p^-} \boxed{\;X^-\;} \xleftarrow{j^-} U$$

$V(h)$

□

Proposition 19.9 (i) *The maps $f^{\pm}(h) : U \longrightarrow V(h)$ induce isomorphisms*

$$f^{\pm}(h)_* \; : \; \pi_1(U) \; = \; \pi_1(X) \xrightarrow{\simeq} \pi_1(V(h))$$

such that

$$(f^+(h)_*)^{-1} f^-(h)_* \; = \; h_* \; : \; \pi_1(U) \; = \; \pi_1(X) \xrightarrow{\simeq} \pi_1(X) \; .$$

The h-twist glueing $X(h)$ is homotopy equivalent to the mapping torus $T(h)$, with $d(h)$ homotopic to the canonical map $T(h) \longrightarrow S^1$:

$$d(h) \; : \; X(h) \; \simeq \; T(h) \longrightarrow S^1 \; .$$

The infinite cyclic cover of $X(h)$

$$\overline{X}(h) \; = \; \mathbb{Z} \times V(h)/\{(j, f^-(h)(x)) = (j+1, f^+(h)(x)) \,|\, j \in \mathbb{Z}, x \in U\}$$

is homotopy equivalent to X, with the generating covering translation

$$\zeta_{\overline{X}(h)} \; : \; \overline{X}(h) \longrightarrow \overline{X}(h) \; ; \; (j, v) \longrightarrow (j+1, v)$$

such that

$$\zeta_{\overline{X}(h)} \simeq h : \overline{X}(h) \simeq X \longrightarrow \overline{X}(h) \simeq X .$$

The restrictions

$$\begin{aligned} \zeta^+_{\overline{X}(h)} &= \zeta_{\overline{X}(h)}| : \overline{X}(h)^+ \longrightarrow \overline{X}(h)^+ , \\ \zeta^-_{\overline{X}(h)} &= \zeta^{-1}_{\overline{X}(h)}| : \overline{X}(h)^- \longrightarrow \overline{X}(h)^- \end{aligned}$$

are such that

$$\begin{aligned} \zeta^+_{\overline{X}(h)} &: \overline{X}(h)^+ \simeq X^+ \xrightarrow{p^+hq^+} X^+ \simeq \overline{X}(h)^+ , \\ \zeta^-_{\overline{X}(h)} &: \overline{X}(h)^- \simeq X^- \xrightarrow{p^-h^{-1}q^-} X^- \simeq \overline{X}(h)^- . \end{aligned}$$

(ii) *The maps*

$$\begin{aligned} g^+(h) &: V(h) \longrightarrow \overline{X}(h)^+ \simeq X^+ , \\ g^-(h) &: V(h) \longrightarrow \overline{X}(h)^- \simeq X^- \end{aligned}$$

fit into a homotopy pushout square

$$\begin{array}{ccc} V(h) & \xrightarrow{g^+(h)} & X^+ \\ {\scriptstyle g^-(h)}\big\downarrow & & \big\downarrow{\scriptstyle q^+} \\ X^- & \xrightarrow{hq^-} & X \end{array}$$

and homotopy commutative diagrams

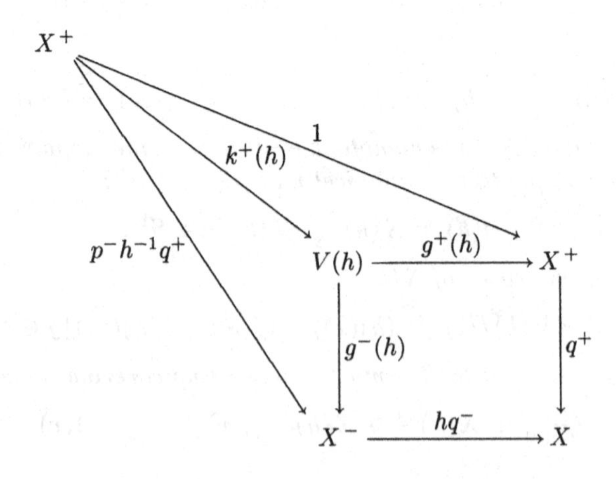

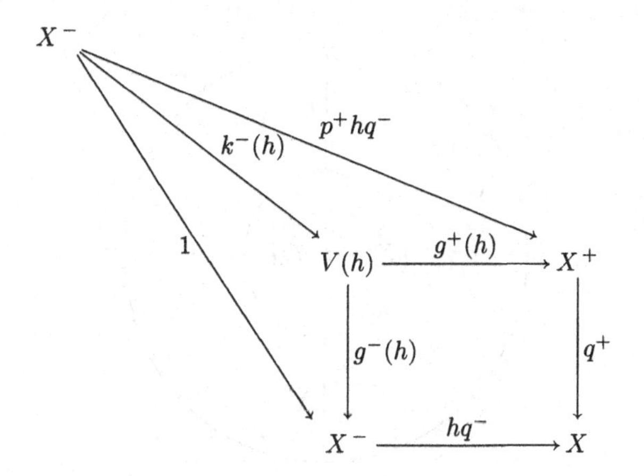

In particular, $V(h)$ dominates X^+ and X^- rel U.

(iii) *There are defined homotopy pushout squares*

$$\begin{array}{ccc} U & \xrightarrow{f^+(h)} & V(h) \\ {\scriptstyle j^-}\big\downarrow & & \big\downarrow{\scriptstyle g^-(h)} \\ X^- & \xrightarrow{p^-h^{-1}q^-} & X^- \end{array} \qquad \begin{array}{ccc} U & \xrightarrow{f^-(h)} & V(h) \\ {\scriptstyle j^+}\big\downarrow & & \big\downarrow{\scriptstyle g^+(h)} \\ X^+ & \xrightarrow{p^+hq^+} & X^+ \end{array}$$

(iv) *$V(h)$ is finitely dominated, with finiteness obstruction*

$$[V(h)] = (h_* - 1)[X^+] = (1 - h_*)[X^-] \in \widetilde{K}_0(\mathbb{Z}[\pi_1(X)]) .$$

Proof (i) The map

$$(q^+j^+, q^+g^+(h)) : \mathcal{W}(f^+(h), f^-(h)) = X(h) \longrightarrow \mathcal{W}(1_X, h) = T(h)$$

is a homotopy equivalence – this is a special case of 13.18, since the maps

$$j^+ : U \longrightarrow X^+ , \quad j^- : U \longrightarrow X^- ,$$

$$p^+h : X \longrightarrow X^+ , \quad p^- : X \longrightarrow X^-$$

are such that up to homotopy

$$f^+(h) : U \xrightarrow{j^+} X^+ \xrightarrow{k^+(h)} \mathcal{M}(p^+h, p^-) = V(h) ,$$

$$f^-(h) : U \xrightarrow{j^-} X^- \xrightarrow{k^-(h)} \mathcal{M}(p^+h, p^-) = V(h) ,$$

$$h : X \xrightarrow{p^+h} X^+ \xrightarrow{q^+} \mathcal{M}(j^+, j^-) = X ,$$

$$1 : X \xrightarrow{p^-} X^- \xrightarrow{q^-} \mathcal{M}(j^+, j^-) = X .$$

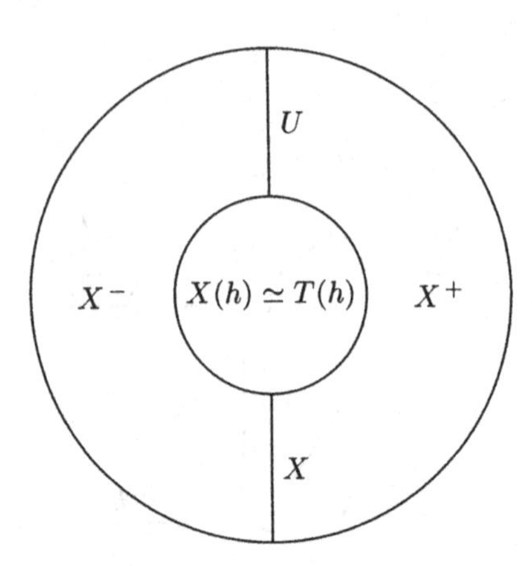

The infinite cyclic cover $\overline{X}(h)$ of $X(h)$ has a fundamental domain

$$(\{0\}\times X^- \cup_{\{0\}\times f^-(h)(U)=\{1\}\times f^+(h)(U)} \{1\}\times X^+ ;\ \{0\}\times X,\ \{1\}\times X)$$

which is (homotopy equivalent to) a fundamental domain $(\mathcal{M}(h : X \longrightarrow X), X\times\{0\}, X\times\{1\})$ of the canonical infinite cyclic cover $\overline{T}(h)$ of the mapping torus $T(h)$, so that

$$\zeta_{\overline{X}(h)} \simeq \zeta_{\overline{T}(h)} \simeq h \ :\ \overline{X}(h) \simeq \overline{T}(h) \simeq X \longrightarrow \overline{X}(h) \simeq \overline{T}(h) \simeq X\ .$$

(ii) The square is a homotopy pushout because it is homotopy equivalent to the pushout square

$$\begin{array}{ccc} V(h) & \longrightarrow & \overline{X}(h)^+ \\ \big\downarrow & & \big\downarrow \\ \overline{X}(h)^- & \longrightarrow & \overline{X}(h) \end{array}$$

(iii) The squares are homotopy pushouts because they are homotopy equivalent to the pushout squares

$$\begin{array}{ccc} U & \xrightarrow{f^+(h)} & V(h) \\ \big\downarrow & & \big\downarrow \\ \overline{X}(h)^- & \xrightarrow{\zeta^-_{\overline{X}(h)}} & \overline{X}(h)^- \end{array} \qquad \begin{array}{ccc} U & \xrightarrow{f^-(h)} & V(h) \\ \big\downarrow & & \big\downarrow \\ \overline{X}(h)^+ & \xrightarrow{\zeta^+_{\overline{X}(h)}} & \overline{X}(h)^+ \end{array}$$

(iv) Let $\widetilde{V}(h), \widetilde{X}, \widetilde{X}^+, \widetilde{X}^-, \widetilde{X}(h)$ be the universal covers of $V(h), X, X^+$, $X^-, X(h)$ respectively, and let

$$\pi_1(V(h)) = \pi_1(X) = \pi_1(X^+) = \pi_1(X^-) = \pi .$$

The $\mathbb{Z}[\pi]$-module chain complexes $C(\widetilde{X})$, $h_*C(\widetilde{X}^+)$, $C(\widetilde{X}^-)$ in the chain homotopy cartesian square

$$\begin{array}{ccc} C(\widetilde{X}) & \xrightarrow{p^-} & C(\widetilde{X}^-) \\ {\scriptstyle p^+h}\big\downarrow & & \big\downarrow{\scriptstyle k^-(h)} \\ h_*C(\widetilde{X}^+) & \xrightarrow{k^+(h)} & C(\widetilde{V}(h)) \end{array}$$

are finitely dominated, so that $C(\widetilde{V}(h))$ is also a finitely dominated $\mathbb{Z}[\pi]$-module chain complex, and $V(h)$ is a finitely dominated CW complex (by 6.8). It now follows from

$$\begin{aligned} [X] &= [X^+] + [X^-] \\ &= h_*[X] = h_*[X^+] + h_*[X^-] \in \widetilde{K}_0(\mathbb{Z}[\pi]) \end{aligned}$$

that the finiteness obstruction of $V(h)$ is given by

$$\begin{aligned} [V(h)] &= h_*[X^+] + [X^-] - [X] \\ &= (h_* - 1)[X^+] = (1 - h_*)[X^-] \in \widetilde{K}_0(\mathbb{Z}[\pi]) . \end{aligned}$$

□

Example 19.10 If $h_* = 1 : \pi_1(X) \longrightarrow \pi_1(X)$ then

$$[V(h)] = [X^+] - [X^+] = 0 \in \widetilde{K}_0(\mathbb{Z}[\pi_1(X)]) ,$$

and $V(h)$ is homotopy finite. □

Proposition 19.11 (i) *Let $h = 1 : X \longrightarrow X$. The cellular chain complex $C(\widetilde{V}(1))$ of the universal cover $\widetilde{V}(1)$ of $V(1)$ is equipped with a $\mathbb{Z}[\pi_1(X)]$-module chain equivalence*

$$C(\widetilde{U}) \simeq C(\widetilde{V}(1))$$

such that

$$f^+(1) : C(\widetilde{U}) \xrightarrow{i^+j^+} C(\widetilde{U}) \simeq C(\widetilde{V}(1)) ,$$

$$f^-(1) : C(\widetilde{U}) \xrightarrow{i^-j^-} C(\widetilde{U}) \simeq C(\widetilde{V}(1)) .$$

The chain equivalence determines a particular simple chain homotopy type on $C(\widetilde{V}(1))$, and hence particular simple homotopy types on the spaces $V(1)$,

$(X(1), d(1))$.

(ii) *If* $h : X \longrightarrow X$ *is a covering translation with*

$$h(X^+) \subset X^+ \quad , \quad h^{-1}(X^-) \subset X^-$$

let

$$\begin{aligned} U &= X^+ \cap X^- \quad , \quad V = X^+ \cap h(X^-) \subset X \ , \\ f^+ &: U \longrightarrow V \ ; \ x \longrightarrow x \ , \\ f^- &: U \longrightarrow V \ ; \ x \longrightarrow h(x) \ . \end{aligned}$$

Then (V, U, U, f^+, f^-) *is a* π_1*-fundamental domain for* X *regarded as the infinite cyclic cover of the CW* π_1*-band* X/h, *with* f^+, f^- *inducing isomorphisms*

$$f^+_* \ , \ f^-_* \ : \ \pi_1(U) \xrightarrow{\simeq} \pi_1(V) \ = \ \pi_1(X)$$

such that $f^-_* = h_* f^+_*$.

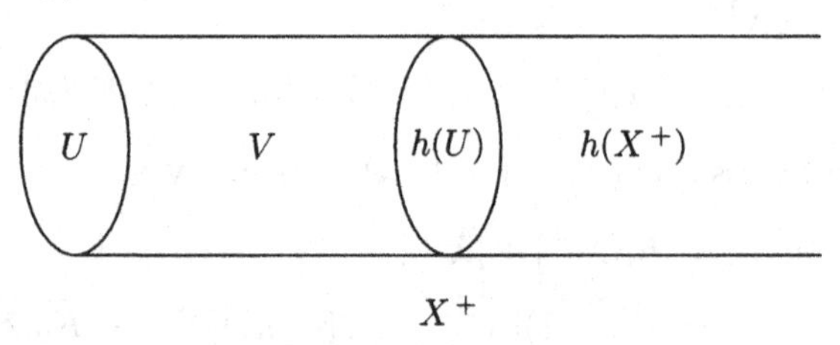

The cellular chain complex $C(\widetilde{V}(h))$ *of the universal cover* $\widetilde{V}(h)$ *of* $V(h)$ *is equipped with a* $\mathbb{Z}[\pi_1(X)]$*-module chain equivalence*

$$C(\widetilde{V}) \ \simeq \ C(\widetilde{V}(h))$$

which determines a particular simple chain homotopy type on $C(\widetilde{V}(h))$. *This determines a particular simple homotopy type on the space* $V(h)$, *and also on* $(X(h), d(h))$.

Proof (i) Both $C(\widetilde{U})$ and $C(\widetilde{V}(1))$ fit into chain homotopy cartesian squares

$$\begin{array}{ccc} C(\widetilde{X}) & \xrightarrow{p^-} & C(\widetilde{X}^-) \\ {\scriptstyle p^+}\downarrow & & \downarrow{\scriptstyle i^-} \\ C(\widetilde{X}^+) & \xrightarrow{i^+} & C(\widetilde{U}) \end{array} \qquad \begin{array}{ccc} C(\widetilde{X}) & \xrightarrow{p^-} & C(\widetilde{X}^-) \\ {\scriptstyle p^+}\downarrow & & \downarrow{\scriptstyle k^-(1)} \\ C(\widetilde{X}^+) & \xrightarrow{k^+(1)} & C(\widetilde{V}(1)) \end{array}$$

(ii) The maps

$$\begin{aligned} g^+ &: V \longrightarrow X^+ \ ; \ x \longrightarrow x \ , \\ g^- &: V \longrightarrow X^- \ ; \ x \longrightarrow h^{-1}(x) \end{aligned}$$

fit into a pushout square

$$\begin{array}{ccc} V & \xrightarrow{g^+} & X^+ \\ {\scriptstyle g^-}\big\downarrow & & \big\downarrow{\scriptstyle q^+} \\ X^- & \xrightarrow{hq^-} & X \end{array}$$

The restrictions

$$\begin{aligned} h^+ &= h| : X^+ \longrightarrow X^+ , \\ h^- &= h^{-1}| : X^- \longrightarrow X^- \end{aligned}$$

are proper homotopy equivalences at ∞ such that there are defined commutative diagrams

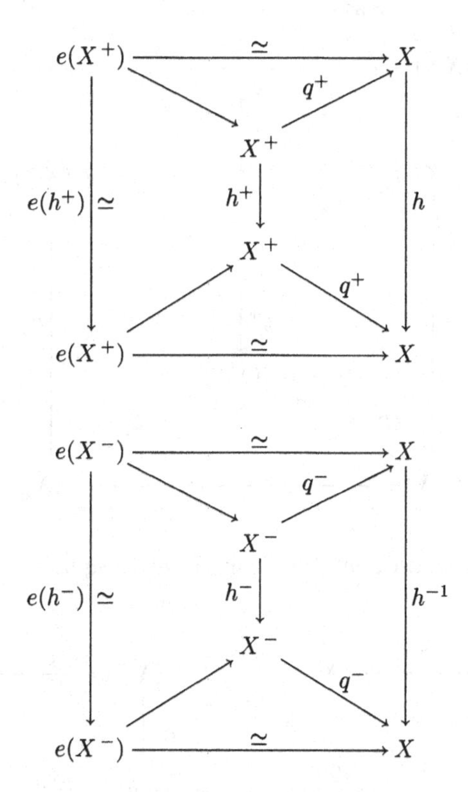

so that

$$\begin{aligned} p^+h &\simeq h^+p^+ : X \longrightarrow X^+ , \\ p^-h^{-1} &\simeq h^-p^- : X \longrightarrow X^- . \end{aligned}$$

The $\mathbb{Z}[\pi]$-module chain maps

$$\begin{aligned} k^+ &= f^+i^+ : C(\widetilde{X}^+) \longrightarrow C(\widetilde{V}) , \\ k^- &= f^-i^- : C(\widetilde{X}^-) \longrightarrow C(\widetilde{V}) \end{aligned}$$

fit into chain homotopy commutative diagrams

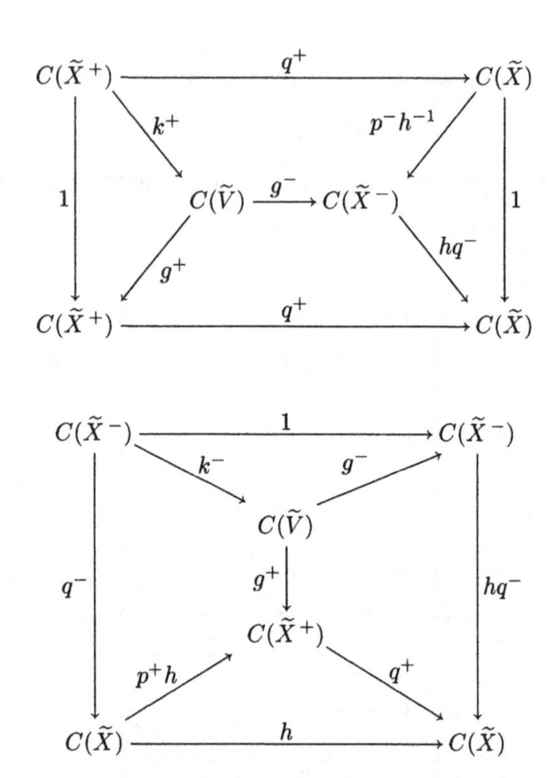

It follows that the chain homotopy commutative squares

$$\begin{array}{ccc} C(\widetilde{X}^+) & \xrightarrow{\quad q^+ \quad} & C(\widetilde{X}) \\ {\scriptstyle k^+}\big\downarrow & & \big\downarrow{\scriptstyle p^-h^{-1}} \\ C(\widetilde{V}) & \xrightarrow{\quad g^- \quad} & C(\widetilde{X}^-) \end{array} \qquad \begin{array}{ccc} C(\widetilde{X}^-) & \xrightarrow{\quad k^- \quad} & C(\widetilde{V}) \\ {\scriptstyle q^-}\big\downarrow & & \big\downarrow{\scriptstyle g^+} \\ C(\widetilde{X}) & \xrightarrow{\quad p^+h \quad} & C(\widetilde{X}^+) \end{array}$$

are chain homotopy cartesian. Moreover, it follows from the chain homotopy commutative diagram

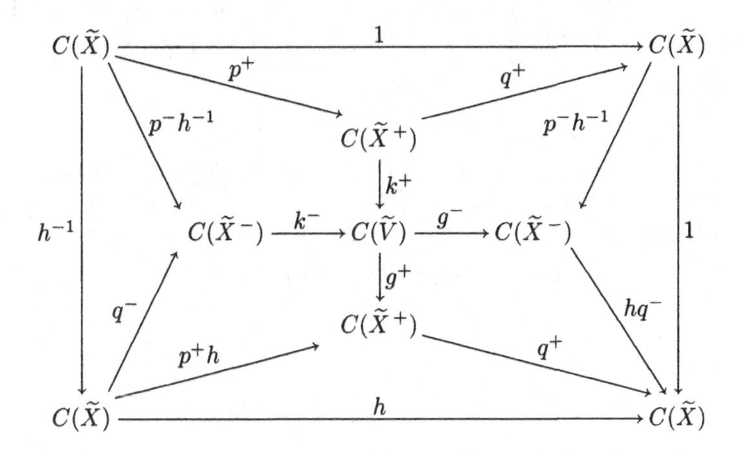

that the chain homotopy commutative square

$$\begin{array}{ccc}
C(\widetilde{X}) & \xrightarrow{\;p^+\;} & C(\widetilde{X}^+) \\
{\scriptstyle p^-h^{-1}}\big\downarrow & & \big\downarrow{\scriptstyle k^+} \\
C(\widetilde{X}^-) & \xrightarrow{\;k^-\;} & C(\widetilde{V})
\end{array}$$

is chain homotopy cartesian, so that there is defined a chain equivalence

$$C(\widetilde{V}) \simeq C(\widetilde{V}(h))$$

such that

$$\begin{aligned}
k^+(h) &: C(\widetilde{X}^+) \xrightarrow{k^+} C(\widetilde{V}) \simeq C(\widetilde{V}(h)) , \\
k^-(h) &: C(\widetilde{X}^-) \xrightarrow{k^-} C(\widetilde{V}) \simeq C(\widetilde{V}(h)) .
\end{aligned}$$

□

Proposition 19.12 *Let (X,d) be a CW ribbon, and let $h:(X,d)\longrightarrow(X,d)$ be a proper homotopy equivalence such that $h: X\longrightarrow X$ is a cellular end-preserving homeomorphism which is either a covering translation or the identity, with*

$$h(X^+)\subseteq X^+ \;,\; h^{-1}(X^-)\subseteq X^- .$$

Let

$$(V,U,U,f^+,f^-)$$
$$= \begin{cases} \textit{a fundamental domain for } X & \textit{if } h \textit{ is a covering translation} , \\ (U,U,U,1,1) & \textit{if } h=1 . \end{cases}$$

Choose a finite CW complex $V[h]$ in the simple homotopy type of $V(h)$ given by 19.11, and let $(V[h], U, U, f^+[h], f^-[h])$ be the finite CW cobordism defined by

$$f^{\pm}[h] \; : \; U \xrightarrow{f^{\pm}(h)} V(h) \; \simeq \; V[h] \; .$$

(i) *The finite CW cobordism $(V[h], U, U, f^+[h], f^-[h])$ is a π_1-fundamental domain for the infinite cyclic cover $\overline{X}[h] = d[h]^*\mathbb{R}$ of a relaxed CW π_1-band $(X[h], d[h])$ with*

$$\begin{aligned} X[h] \; &= \; \mathcal{W}(f^+[h], f^-[h]) \\ &= \; U \times I \cup_{f^+[h] \amalg f^-[h]} V[h] \end{aligned}$$

and

$$d[h] \; : \; X[h] \; \simeq \; X(h) \xrightarrow{d(h)} S^1 \; .$$

The chain homotopy idempotents given by 19.6

$$r^+ \; = \; i^+ j^+ \;\; , \;\; r^- \; = \; i^- j^- \; : \; C(\widetilde{U}) \longrightarrow C(\widetilde{U})$$

are such that

$$f^{\pm}[h] \; \simeq \; f_h^{\pm} r^{\pm} \; : \; C(\widetilde{U}) \longrightarrow C(\widetilde{V}[h]) \; ,$$

with

$$f_h^{\pm} \; : \; C(\widetilde{U}) \xrightarrow{f^{\pm}} C(\widetilde{V}) \longrightarrow C(\widetilde{V}[h]) \; .$$

(ii) *There is defined an end-preserving proper homotopy equivalence*

$$F[h] \; : \; (X, d) \longrightarrow (\overline{X}[h], \overline{d}[h])$$

such that the generating covering translation $\zeta_{\overline{X}[h]} : \overline{X}[h] \longrightarrow \overline{X}[h]$ fits into a homotopy commutative square

$$\begin{array}{ccc} X & \xrightarrow{F[h]} & \overline{X}[h] \\ {\scriptstyle h}\big\downarrow & & \big\downarrow{\scriptstyle \zeta_{\overline{X}[h]}} \\ X & \xrightarrow{F[h]} & \overline{X}[h] \end{array}$$

If $h : X \longrightarrow X$ is a covering translation then $d : X \longrightarrow \mathbb{R}$ is proper homotopic to a $\mathbb{Z}$-equivariant lift of a map $d/h : X/h \longrightarrow S^1$ classifying $(d/h)^*\mathbb{R} = X$, and $F[h]$ is homotopic to a $\mathbb{Z}$-equivariant lift of a homotopy equivalence $(X/h, d/h) \longrightarrow (X[h], d[h])$ of CW bands.
If $h = 1 : X \longrightarrow X$ the composite

$$G \; : \; X \times S^1 \; = \; T(1_X) \xrightarrow{T(F[1])} T(\zeta_{\overline{X}[1]}) \xrightarrow{\text{proj.}} \overline{X}[1]/\zeta_{\overline{X}[1]} \; = \; X[1]$$

is a homotopy equivalence such that

$$F[1] \; : \; X \xrightarrow{\text{incl.}} X \times \mathbb{R} \xrightarrow{\overline{G}} \overline{X}[1] \; .$$

Proof (i) The chain homotopy $f^{\pm}[h] \simeq f_h^{\pm}r^{\pm}$ is given by the chain homotopy commutative diagram

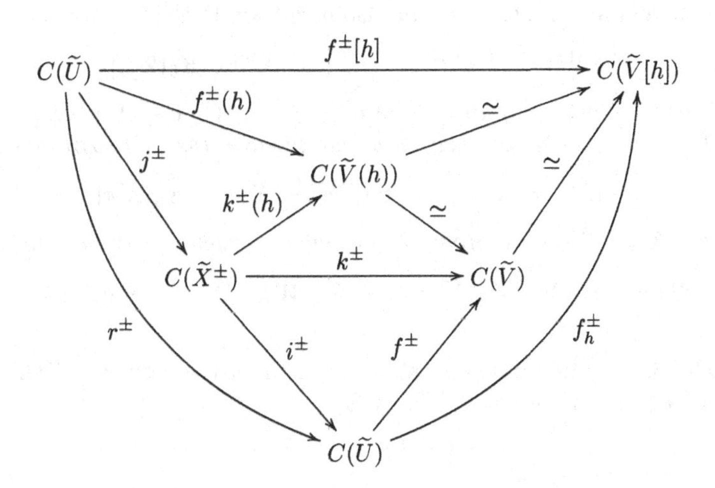

(ii) The homotopy equivalence $X[h] \simeq X(h)$ is induced by the rel ∂ homotopy equivalence

$$(V[h], U, U, f^+[h], f^-[h]) \simeq (V(h), U, U, f^+(h), f^-(h)) .$$

The composite

$$X \longrightarrow X \times \mathbb{R} \longrightarrow \overline{T}(h) \longrightarrow \overline{X}(h)$$

of

$$X \longrightarrow X \times \mathbb{R} \; ; \; x \longrightarrow (x, d(x)) \, ,$$
$$X \times \mathbb{R} \longrightarrow \overline{T}(h) \; ; \; (x, a+b) \longrightarrow (h^a(x), b) \;\; (a \in \mathbb{Z}, \, b \in [0,1))$$

and a $\mathbb{Z}$-equivariant lift $\overline{T}(h) \longrightarrow \overline{X}(h)$ of the homotopy equivalence $T(h) \longrightarrow X(h)$ of 19.9 (i) is a proper map

$$F[h] \; : \; X \longrightarrow \overline{X}(h) \, .$$

Now $F[h]$ induces isomorphisms of the fundamental groups, the fundamental groups at ∞, the homology groups of the universal covers

$$\widetilde{F}[h]_* \; : \; H_*(\widetilde{X}) \xrightarrow{\simeq} H_*(\widetilde{X}(h))$$

and also the locally π-finite homology groups of the universal covers

$$\widetilde{F}[h]_* \; : \; H_*^{lf,\pi}(\widetilde{X}) \xrightarrow{\simeq} H_*^{lf,\pi}(\widetilde{X}(h)) \, ,$$

so that $F[h]$ is a proper homotopy equivalence by 5.7. □

Proposition 19.13 *Let (X,d) be a CW ribbon with $\pi_1(X) = \pi$, and let $h : (X,d) \longrightarrow (X,d)$ be a proper homotopy equivalence.*

(i) *The finiteness obstruction of $V(h)$ is the image of the infinite torsion $\tau^{lf}(h) \in Wh^{lf}(X)$ (11.1) under the isomorphism $Wh^{lf}(X) \cong \widetilde{K}_0(\mathbb{Z}[\pi])$:*

$$[V(h)] = [\tau^{lf}(h)] = (h_* - 1)[X^+] \in \widetilde{K}_0(\mathbb{Z}[\pi]) .$$

Thus $V(h)$ is homotopy finite if and only if $\tau^{lf}(h) = 0 \in Wh^{lf}(X)$.

(ii) *If h is a covering translation or the identity (as in 19.12) then*

$$\tau^{lf}(h) = [V(h)] = 0 \in Wh^{lf}(X) = \widetilde{K}_0(\mathbb{Z}[\pi])$$

and $F[h] : X \longrightarrow \overline{X}[h]$ is a proper homotopy equivalence with infinite torsion

$$\tau^{lf}(F) = [X^+] - [\overline{X}^+[h]] = 0 \in Wh^{lf}(X) = \widetilde{K}_0(\mathbb{Z}[\pi]) . \qquad \square$$

Remark 19.14 The fundamental group of the h-twist glueing $X(h)$ is the α-twisted extension of $\pi = \pi_1(X)$ by $\mathbb{Z}$

$$\pi_1(X(h)) = \pi \times_\alpha \mathbb{Z}$$

with $\alpha = h_* : \pi \longrightarrow \pi$. The splitting theorem of Farrell and Hsiang [48] expresses the Whitehead group as

$$Wh(\pi \times_\alpha \mathbb{Z}) = Wh(\pi, \alpha) \oplus \widetilde{\mathrm{Nil}}_0(\mathbb{Z}[\pi], \alpha) \oplus \widetilde{\mathrm{Nil}}_0(\mathbb{Z}[\pi], \alpha^{-1})$$

with $Wh(\pi, \alpha)$ the class group of α-twisted automorphisms of f.g. projective $\mathbb{Z}[\pi]$-modules, the relative group of Siebenmann [145] in the exact sequence

$$Wh(\pi) \xrightarrow{1-\alpha} Wh(\pi) \longrightarrow Wh(\pi, \alpha) \longrightarrow \widetilde{K}_0(\mathbb{Z}[\pi]) \xrightarrow{1-\alpha} \widetilde{K}_0(\mathbb{Z}[\pi]) ,$$

and $\widetilde{\mathrm{Nil}}_0(\mathbb{Z}[\pi], \alpha)$ the reduced nilpotent class group of α-twisted endomorphisms of f.g. projective $\mathbb{Z}[\pi]$-modules. (See Chapter 21 for a slightly more detailed account.) The fibring obstructions of the relaxed *CW* band $X[h]$ of 19.12 will be shown in Chapter 26 to be such that

$$\Phi^{\pm}(X[h]) \in Wh(\pi, \alpha) \subseteq Wh(\pi \times_\alpha \mathbb{Z}) ,$$

with image

$$[\Phi^{\pm}(X[h])] = \pm[X^{\mp}] \in \ker(1 - \alpha : \widetilde{K}_0(\mathbb{Z}[\pi]) \longrightarrow \widetilde{K}_0(\mathbb{Z}[\pi])) .$$

(Actually, only the case $\alpha = 1$ is considered in Chapter 26, but the result holds for arbitrary α.) $\square$

Example 19.15 Let $(X,d) = (\overline{W}, \overline{c})$ be the manifold ribbon defined by the infinite cyclic cover of an n-dimensional manifold band (W,c), and let $h : X \longrightarrow X$ be an end-preserving homeomorphism which is either a covering translation or the identity, as in 19.12. For $n \geq 6$ the relaxed *CW* π_1-band $(X[h], d[h])$ is realized by a relaxed manifold band (also denoted by

$(X[h],d[h]))$, constructed by the original geometric twist glueing construction of Siebenmann [145], as follows. Let $(V;U,\zeta U)$ be a manifold fundamental domain for the infinite cyclic cover $\overline{W}$. Use engulfing (Siebenmann, Guillou and Hähl [149]) to realize the maps $p^+ : X \longrightarrow X^+$, $p^- : X \longrightarrow X^-$ by embeddings

$$[p^+] : X \subset X^+ \ , \ [p^-] : X \subset X^-$$

as disjoint open neighbourhoods of the ends. The finite CW fundamental domain $(V[h],U,U,f^+[h],f^-[h])$ for $\overline{X}[h]$ of 19.12 is realized by the compact n-dimensional manifold cobordism (also denoted by $(V[h],U,U,f^+[h],$ $f^-[h]))$ which fits into pushout squares

$$\begin{array}{ccc} X & \xrightarrow{[p^-]} & X^- \\ {\scriptstyle [p^+]h}\downarrow & & \downarrow \\ X^+ & \longrightarrow & V[h] \end{array} \qquad \begin{array}{ccc} V[h] & \longrightarrow & X^- \\ \downarrow & & \downarrow{\scriptstyle \zeta[h]q^-} \\ X^+ & \xrightarrow{q^+} & X \end{array}$$

with

$$f^+[h] : U \longrightarrow X^+ \longrightarrow V[h] \ , \ f^-[h] : U \longrightarrow X^- \longrightarrow V[h]$$

and $\zeta[h] : X \longrightarrow X$ a covering translation isotopic to h. The h-twist glueing

$$X[h] = \mathcal{W}(f^+[h],f^-[h]) = U \times I \cup_{f^+[h] \amalg f^-[h]} V[h]$$

is a relaxed manifold band such that

$$\begin{aligned} &\zeta[h] = \zeta_{\overline{X}[h]} \simeq h : \overline{X}[h] = X \longrightarrow \overline{X}[h] = X \ , \\ &\overline{X}[h]^+ = X^+ \ , \ \overline{X}[h]^- = X^- \ , \\ &\overline{X}[h]^+ \cap \overline{X}[h]^- = X^+ \cap X^- = U \ , \\ &V[h] = X^+ \cap \zeta[h](X^-) \simeq X^+ \cap h(X^-) \simeq V(h) \ , \\ &f^+(h) : U \xrightarrow{f^+[h]} V[h] \simeq V(h) \ , \\ &f^-(h) : U \xrightarrow{f^-[h]} V[h] \simeq V(h) \ . \end{aligned}$$

□

20

Homotopy theoretic wrapping up and relaxation

We now use the homotopy theoretic twist glueing of Chapter 19 to develop the homotopy theoretic analogues of the geometric wrapping up and relaxation techniques of Chapters 17, 18. We prove that every CW ribbon (X, d) is infinite simple proper homotopy equivalent to the infinite cyclic cover $(\overline{W}, \overline{c})$ of a relaxed CW π_1-band (W, c), so that (X, d) is both forward and reverse tame.

Definition 20.1 The *wrapping up* of a CW ribbon (X, d) is the relaxed 1-twist glueing CW π_1-band given by 19.12

$$(\widehat{X}, \widehat{d}) \;=\; (X[1], d[1]) \;. \qquad \square$$

Remark 20.2 If (X, d) is a CW ribbon such that the chain homotopy dominations $(C(\widetilde{U}), i^{\pm}, j^{\pm}, j^{\pm}i^{\pm} \simeq 1_{C(\widetilde{X}^{\pm})})$ of 19.6 are realized by dominations $(U, i^{\pm}, j^{\pm}, j^{\pm}i^{\pm} \simeq 1_{X^{\pm}})$ (as in 19.7 (ii)) with a homotopy

$$i^+p^+ \;\simeq\; i^-p^- \;:\; X \longrightarrow U \;=\; X^+ \cap X^-$$

then the homotopy commutative square

$$\begin{array}{ccc} X & \xrightarrow{\;p^+\;} & X^+ \\ {\scriptstyle p^-}\big\downarrow & & \big\downarrow{\scriptstyle i^+} \\ X^- & \xrightarrow{\;i^-\;} & U \end{array}$$

is a homotopy pushout such that the homotopy equivalence

$$V(1) \;=\; \mathcal{M}(p^+, p^-) \xrightarrow{\;\simeq\;} U$$

is simple, with

$$i^+j^+ \;:\; U \xrightarrow{f^+(1)} V(1) \longrightarrow U \;,$$

$$i^-j^- \;:\; U \xrightarrow{f^-(1)} V(1) \longrightarrow U \;.$$

The wrapping up $(\widehat{X}, \widehat{d})$ is a relaxed CW π_1-band, which can be taken to be the mapping coequalizer (13.7) of the homotopy idempotents $i^{\pm}j^{\pm} : U \longrightarrow U$ of the finite CW complex U

$$(X[1], d[1]) \;=\; \mathcal{W}(i^+j^+, i^-j^-) \,. \qquad \square$$

Proposition 20.3 *The wrapping up of a CW ribbon (X, d) is a relaxed CW π_1-band*

$$(W, c) \;=\; (\widehat{X}, \widehat{d})$$

such that:

(i) *the covering translation of the infinite cyclic cover $\overline{W} = \widehat{d}^*(\mathbb{R})$ of W is homotopic to the identity,*

$$\zeta_{\overline{W}} \;\simeq\; 1 \;:\; \overline{W} \longrightarrow \overline{W} \,,$$

(ii) *there is defined an infinite simple homotopy equivalence of CW ribbons*

$$F \;:\; (X, d) \xrightarrow{\;\simeq\;} (\overline{W}, \overline{c})$$

with

$$X^+ \cap X^- \;=\; \overline{W}^+ \cap \overline{W}^-$$

and the restrictions $F| \,:\, X^{\pm} \longrightarrow \overline{W}^{\pm}$ are proper homotopy equivalences rel $X^+ \cap X^-$,

(iii) *there is defined an infinite simple homotopy equivalence of CW ribbons*

$$G \;:\; (X \times S^1, d(pr_1)) \xrightarrow{\;\simeq\;} (W \times \mathbb{R}, pr_2) \,.$$

Proof (i) This is the special case $h = 1$ of the homotopy $\zeta_{\overline{X}[h]} \simeq h$ given by 19.9 (i).

(ii) Let $F = F[1]$, with $F[1]$ as defined in 19.12 (ii), such that $\tau^{lf}(F) = 0$ by 19.13 (ii).

(iii) Use the homotopy equivalence

$$T(F) \;:\; T(1_X) \;=\; X \times S^1 \xrightarrow{\;\simeq\;} T(\zeta_{\overline{W}})$$

and the homotopy equivalence defined by the projection

$$p \;:\; T(\zeta_{\overline{W}}) \xrightarrow{\;\simeq\;} \overline{W}/\zeta_{\overline{W}} \;=\; W \;;\; (x, t) \longrightarrow x$$

to define a proper map

$$G \;:\; X \times S^1 \longrightarrow W \times \mathbb{R} \;;\; (x, s) \longrightarrow (pT(F)(x, s), d(x))$$

which is a homotopy equivalence. Moreover, G induces isomorphisms of fundamental groups at ∞ and the locally π-finite homology groups of the universal covers (with $\pi = \pi_1(X) \times \mathbb{Z}$), so that G is a proper homotopy equivalence by 5.7. The isomorphism

$$Wh^{lf}(X \times S^1) \cong \widetilde{K}_0(\mathbb{Z}[\pi_1(X) \times \mathbb{Z}])$$

sends $\tau^{lf}(G) \in Wh^{lf}(X \times S^1)$ to

$$[X^+ \times S^1] - [W \times [0,\infty)] = 0 \in \widetilde{K}_0(\mathbb{Z}[\pi_1(X) \times \mathbb{Z}]) ,$$

so that G is infinite simple. □

Theorem 20.4 (i) *If (X,d) is a CW ribbon then $d : X \longrightarrow \mathbb{R}$ is proper homotopic to a proper bounded fibration, and X is forward and reverse tame.*

(ii) *If (X,d) is a manifold ribbon with X an open manifold of dimension ≥ 5 or a Hilbert cube manifold then $d : X \longrightarrow \mathbb{R}$ is proper homotopic to a manifold approximate fibration. The homotopy theoretic wrapping up $(\widehat{X},\widehat{d})$ of 20.1 is realized (within its simple homotopy type) by the geometric wrapping up of 17.1, with the infinite simple homotopy equivalence G of 20.3 realized by the homeomorphism $X \times S^1 \cong \widehat{X} \times \mathbb{R}$ of 17.1.*

Proof (i) Immediate from 17.16 (i) and 19.12 (ii).

(ii) It is immediate from (i) and 16.10 that d is proper homotopic to a manifold approximate fibration $d' : X \longrightarrow \mathbb{R}$. By 18.6 the geometric wrapping up of 17.1 is homeomorphic to any Siebenmann 1-twist glueing of (X,d). Thus it suffices to show that there is a Siebenmann 1-twist glueing $W_1(f_-,f_+)$ of X with a natural homotopy equivalence $\widehat{X} \longrightarrow W_1(f_-,f_+)$. As in 18.6 if $U_- = (d')^{-1}(-\infty,-1)$ and $U_+ = (d')^{-1}(1,\infty)$ then there are homeomorphisms $f_\pm : U_\pm \longrightarrow X$ which are isotopic to the identity through open embeddings. The maps

$$[p_\pm] : X \xrightarrow{(f_\pm)^{-1}} U_\pm \subset X^\pm$$

realize $p_\pm : X \simeq e(X^\pm) \longrightarrow X^\pm$ up to homotopy. It is then clear that the Siebenmann twist glueing $W_1(f_-,f_+)$ is homotopy equivalent to the 1-twist glueing $\mathcal{W}(f^+(1), f^-(1))$ with $[p_\pm]$ replacing $p_\pm$ in 19.8 (cf. 19.15). □

Remark 20.5 Let (X,d) be an n-dimensional manifold ribbon with $\pi_1(X) = \pi$.

(i) X is a finitely dominated $(n-1)$-dimensional geometric Poincaré complex with finiteness obstruction

$$[X] = [X^+] + [X^-] = [X^+] + (-)^{n-1}[X^+]^* \in \widetilde{K}_0(\mathbb{Z}[\pi]) .$$

The inverse image of $0 \in \mathbb{R}$ is a compact $(n-1)$-dimensional manifold

$$U = d^{-1}(0) \subset X$$

such that the inclusion defines a degree 1 normal map $(f,b) : U \longrightarrow X$. For

$n \geq 6$ the finiteness obstruction $[X^+] \in \widetilde{K}_0(\mathbb{Z}[\pi])$ is the codimension 1 splitting obstruction to making (f,b) normal bordant to a homotopy equivalence by codimension 1 surgeries on U inside X. For any n the projective surgery obstruction (Pedersen and Ranicki [109]) of (f,b) is

$$\sigma_*^p(f,b) \;=\; 0 \in L_{n-1}^p(\mathbb{Z}[\pi]) \;,$$

since (f,b) extends to a finitely dominated normal bordism

$$(g,c) \;:\; (X^+;U,e(X^+)) \longrightarrow X \times (I;\{0\},\{1\})$$

such that $(g,c)| : e(X^+) \longrightarrow X$ is a homotopy equivalence. If X is homotopy finite then (f,b) has finite surgery obstruction

$$\begin{aligned}\sigma_*^h(f,b) \;&=\; [X^+] \in \mathrm{im}(\widehat{H}^n(\mathbb{Z}_2;\widetilde{K}_0(\mathbb{Z}[\pi])) \longrightarrow L_{n-1}^h(\mathbb{Z}[\pi])) \\ &=\; \ker(L_{n-1}^h(\mathbb{Z}[\pi]) \longrightarrow L_{n-1}^p(\mathbb{Z}[\pi])) \;,\end{aligned}$$

in accordance with the Rothenberg-type exact sequence of Ranicki [118]

$$\begin{aligned}\dots \longrightarrow L_n^p(\mathbb{Z}[\pi]) &\longrightarrow \widehat{H}^n(\mathbb{Z}_2;\widetilde{K}_0(\mathbb{Z}[\pi])) \\ &\longrightarrow L_{n-1}^h(\mathbb{Z}[\pi]) \longrightarrow L_{n-1}^p(\mathbb{Z}[\pi]) \longrightarrow \dots \;.\end{aligned}$$

(ii) The construction of the normal map $(f,b) : U \longrightarrow X$ in (i) goes back to the special case considered by Browder [12], with X an open n-dimensional PL manifold homeomorphic to $M \times \mathbb{R}$ for a compact $(n-1)$-dimensional PL manifold M. For $n \geq 6$ the finiteness obstruction (= codimension 1 splitting obstruction) $[X^+] \in \widetilde{K}_0(\mathbb{Z}[\pi])$ is such that $[X^+] = 0$ if and only if X is PL homeomorphic to $N \times \mathbb{R}$ for a compact $(n-1)$-dimensional PL manifold N (Novikov [104], Golo [64], Bryant and Pacheco [17]). In the case considered in [12] $\pi_1(X) = \pi_1(M) = \{1\}$, so the obstruction takes its value in $\widetilde{K}_0(\mathbb{Z}) = 0$ – it was proved in [12] that such X is indeed PL homeomorphic to $N \times \mathbb{R}$.

(iii) The finitely dominated n-dimensional geometric Poincaré cobordism $(X^+;U,e(X^+))$ of (i) gives the identity of projective symmetric signatures

$$\sigma_p^*(X) \;=\; \sigma^*(U) \in L_p^{n-1}(\mathbb{Z}[\pi]) \;.$$

This is a generalization of the identity

$$\mathrm{signature}(X) \;=\; \mathrm{signature}(U) \in L^{4k}(\mathbb{Z}) \;=\; \mathbb{Z}$$

obtained by Novikov [103] in the case $n = 4k+1$, for any open $(4k+1)$-dimensional manifold X with a proper map $d : X \longrightarrow \mathbb{R}$ transverse regular at $0 \in \mathbb{R}$. The identity was used in [103] to prove the homotopy invariance of the codimension 1 component $\mathcal{L}_k(M^{4k+1}) \in H^{4k}(M;\mathbb{Q})$ of the $\mathcal{L}$-genus of a $(4k+1)$-dimensional manifold M^{4k+1}. Novikov [104] proved the topological invariance of the $\mathcal{L}$-genus $\mathcal{L}_*(M) \in H^{4*}(M;\mathbb{Q})$ and the rational Pontrjagin classes $p_*(M) \in H^{4*}(M;\mathbb{Q})$ for all manifolds M, using the tori T^i and the signature properties of open manifolds with a proper map to $\mathbb{R}^i$, generalizing the method of Browder [12]. This proof was interpreted in Ranicki [125, Appendix C] in terms of the lower L-theory of Ranicki [124]. Gromov [65] used a signature identity of the above type in the case $n \equiv 0 \,(\mathrm{mod} 2)$ with coefficients in a flat bundle, replacing the tori in [104] by surfaces of higher genus. See Ranicki [127, 4.2] for the relationship of the proofs to each other. □

Remark 20.6 (i) In 26.7 it will be proved that the wrapping up $(\widehat{X}, \widehat{d})$ of a CW ribbon (X, d) has fibring obstructions

$$\Phi^+(\widehat{X}, \widehat{d}) = -\overline{B}'([X^-]) \ , \ \Phi^-(\widehat{X}, \widehat{d}) = \overline{B}'([X^+]) \in Wh(\pi_1(X) \times \mathbb{Z})$$

with

$$\overline{B}' : \widetilde{K}_0(\mathbb{Z}[\pi_1(X)]) \longrightarrow Wh(\pi_1(X) \times \mathbb{Z}) \ ;$$
$$[P] \longrightarrow \tau(-z : P[z, z^{-1}] \longrightarrow P[z, z^{-1}])$$

the geometrically significant split injection of Ranicki [122]. Thus $\widehat{X}$ is not in general in the canonical (finite) simple homotopy type of $X \times S^1$ (cf. 17.2).

(ii) In 27.4 below it will be shown that a CW band (W, c) is simple homotopy equivalent to the wrapping up $(\widehat{X}, \widehat{d})$ of a CW ribbon (X, d) if and only if (W, c) is relaxed and there exists a homotopy $\zeta \simeq 1 : \overline{W} \longrightarrow \overline{W}$. □

Definition 20.7 The *relaxation* of a CW π_1-band (W, c) is the ζ-twist glueing of the infinite cyclic cover CW ribbon $(\overline{W}, \overline{c})$ given by 19.12, the relaxed CW π_1-band

$$(W', c') = (\overline{W}[\zeta], \overline{c}[\zeta]) \ .$$

The relaxation (W', c') is related to (W, c) by a homotopy equivalence

$$(W, c) \simeq (W', c')$$

(which is not simple in general) with

$$\zeta_{\overline{W}'} \simeq \zeta : \overline{W}' \simeq \overline{W} \longrightarrow \overline{W}' \simeq \overline{W} \ .$$ □

Proposition 20.8 *Given a CW π_1-band (W, c) let*

$$(X, d) = (\overline{W}, \overline{c})$$

be the CW ribbon defined by the infinite cyclic cover, and write the wrapping up as

$$(\widehat{X}, \widehat{d}) = (\widehat{W}, \widehat{c}) \ .$$

The mapping torus of the simple self homotopy equivalence of the wrapping up

$$\widehat{\zeta} : \widehat{W} \xrightarrow{\simeq} \overline{W} \times S^1 \xrightarrow{\zeta \times 1} \overline{W} \times S^1 \xrightarrow{\simeq} \widehat{W}$$

is such that there is defined a simple homotopy equivalence

$$T(\widehat{\zeta}) \xrightarrow{\simeq} W \times S^1 .$$

Proof The homotopy equivalence $X \times S^1 \simeq \widehat{X}$ is a restriction of the proper homotopy equivalence $G : X \times S^1 \simeq \widehat{X} \times \mathbb{R}$ of 20.3 (iii). The simple homotopy equivalence $T(\widehat{\zeta}) \simeq_s W \times S^1$ is the composite of the evident simple homotopy equivalence $T(\widehat{\zeta}) \simeq_s T(\zeta) \times S^1$ and the simple homotopy equivalence proj. $\times 1 : T(\zeta) \times S^1 \simeq_s W \times S^1$. □

Remark 20.9 It will be proved in Chapter 26 that the relaxation W' has fibring obstructions

$$\Phi^{\pm}(W') = \Phi^{\pm}(W)' \in Wh(\pi_1(W))$$

with $\Phi^{\pm}(W)'$ obtained from $\Phi^{\pm}(W)$ by setting the $\widetilde{\text{Nil}}$-components to 0, and

$$\tau(W \simeq W') = \Phi^{\pm}(W') - \Phi^{\pm}(W) \in Wh(\pi_1(W))$$

the sum of the $\widetilde{\text{Nil}}$-components. □

Example 20.10 The geometric relaxation (W', c') (18.2) of a manifold band (W, c) with $\dim(W) \geq 6$ is a relaxed manifold band in the simple homotopy type of the homotopy theoretic relaxation (W', c') of 20.7. The simple homotopy equivalence $\widehat{\zeta} : \widehat{W} \longrightarrow \widehat{W}$ of 20.8 is realized by a homeomorphism and the simple homotopy equivalence $W \times S^1 \simeq_s T(\widehat{\zeta})$ is realized by a homeomorphism $W \times S^1 \cong T(\widehat{\zeta})$ with a lift to a $\mathbb{Z}$-equivariant homeomorphism

$$\overline{W} \times S^1 \cong \overline{T}(\widehat{\zeta}) = \widehat{W} \times \mathbb{R} .$$

The composite

$$W \times S^1 \xrightarrow{\text{proj.}} W \xrightarrow{c} S^1$$

is homotopic to the projection of a fibre bundle over S^1 with fibre $\widehat{W}$ and monodromy $\widehat{\zeta} : \widehat{W} \longrightarrow \widehat{W}$. □

Wrapping up and relaxation are related by:

Proposition 20.11 *Let (W,c) be a CW π_1-band. A fundamental domain (V,U,U,f^+,f^-) for the infinite cyclic cover $\overline{W}=c^*\mathbb{R}$ of W and corresponding fundamental domains*

$$\begin{aligned}(\widehat{V},U,U,\widehat{f}^+,\widehat{f}^-) &= (V[1],U,U,f[\zeta]^+,f[\zeta]^-)\ ,\\ (V',U,U,f'^+,f'^-) &= (V[\zeta],U,U,f[\zeta]^+,f[\zeta]^-)\end{aligned}$$

for the wrapping up and relaxation

$$(\widehat{W},\widehat{c}) = (W[1],c[1])\ ,\ (W',c') = (W[\zeta],c[\zeta])$$

are related by simple rel ∂ *homotopy equivalences*

$$\begin{aligned}(V,U,U,f^+,f^-)\cup(\widehat{V},U,U,\widehat{f}^+,\widehat{f}^-) &\simeq (V',U,U,f'^+,f'^-)\ ,\\ (\widehat{V},U,U,\widehat{f}^+,\widehat{f}^-)\cup(V,U,U,f^+,f^-) &\simeq (V',U,U,f'^+,f'^-)\ .\end{aligned}$$

Proof The homotopy pushout property of the square

$$\begin{array}{ccc}\overline{W} & \xrightarrow{p^+} & \overline{W}^+\\ {\scriptstyle p^-}\big\downarrow & & \big\downarrow{\scriptstyle \widehat{k}^+}\\ \overline{W}^- & \xrightarrow{\widehat{k}^-} & \widehat{V}\end{array}$$

and the homotopy commutative diagram

$$\begin{array}{ccc}\overline{W} & \xrightarrow{p^+} & \overline{W}^+\\ {\scriptstyle p^-}\big\downarrow & & \big\downarrow{\scriptstyle k'^+\zeta^+}\\ \overline{W}^- & \xrightarrow{k'^-} & V'\end{array}$$

give a map

$$\ell : \widehat{V} \longrightarrow V'$$

such that

$$\begin{aligned}\ell\widehat{k}^+ &\simeq k'^+\zeta^+ : \overline{W}^+ \longrightarrow V'\ ,\\ \ell\widehat{k}^- &\simeq k'^- : \overline{W}^- \longrightarrow V'\ .\end{aligned}$$

The homotopy commutative diagram

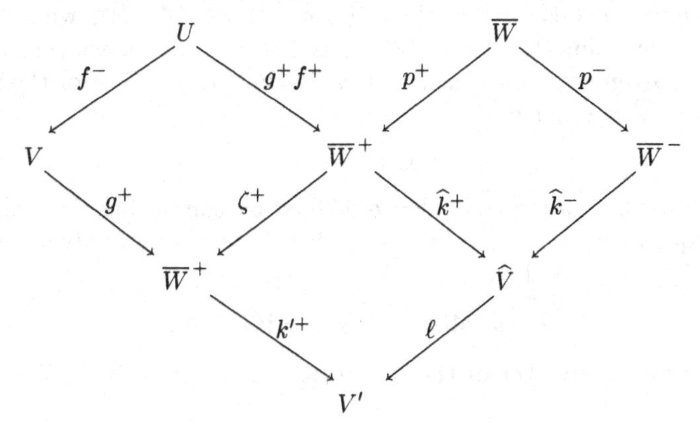

is such that each parallelogram is a homotopy pushout, and

$$f'^+ \; : \; U \xrightarrow{f^+} V \xrightarrow{g^+} \overline{W}^+ \xrightarrow{k'^+} V' \; ,$$

$$f'^- \; : \; U \xrightarrow{f^-} V \xrightarrow{g^-} \overline{W}^- \xrightarrow{\widehat{k}^-} \widehat{V} \xrightarrow{\ell} V' \; .$$

It follows that there is defined a homotopy pushout square

$$\begin{array}{ccc} U & \xrightarrow{\widehat{f}^+} & \widehat{V} \\ {\scriptstyle f^-}\big\downarrow & & \big\downarrow{\scriptstyle \ell} \\ V & \xrightarrow{k'^+} & V' \end{array}$$

with

$$\widehat{f}^+ \; = \; \widehat{k}^+ g^+ f^+ \; : \; \overline{W}^+ \longrightarrow \widehat{V} \; .$$

The corresponding map

$$V \cup \widehat{V} \; = \; \mathcal{M}(\widehat{f}^+, f^-) \longrightarrow V'$$

defines a simple rel ∂ homotopy equivalence

$$(V, U, U, f^+, f^-) \cup (\widehat{V}, U, U, \widehat{f}^+, \widehat{f}^-) \; \simeq \; (V', U, U, f'^+, f'^-) \; .$$

Similarly for the simple rel ∂ homotopy equivalence

$$(\widehat{V}, U, U, \widehat{f}^+, \widehat{f}^-) \cup (V, U, U, f^+, f^-) \; \simeq \; (V', U, U, f'^+, f'^-) \; . \qquad \square$$

A CW π_1-band (W, c) and its relaxation (W', c') are related by the following 'CW h-cobordism' $(Z; W, W')$. For manifold W it is possible to realize $(Z; W, W')$ by a manifold h-cobordism.

Definition 20.12 A CW π_1-band (W,c) is related to the relaxation (W',c') by the *relaxation CW h-cobordism* $((Z,d);(W,c),(W',c'))$, with (Z,d) a CW band containing (W,c) and (W',c') as deformation retracts, constructed as follows. Given a π_1-fundamental domain (V,U,U,f^+,f^-) for the infinite cyclic cover $\overline{W}=c^*\mathbb{R}$ let

$$h\ :\ V\cup\widehat{V}\longrightarrow\widehat{V}\cup V'$$

be the simple homotopy equivalence defined by composing the simple homotopy equivalences $V\cup\widehat{V}\simeq V'$, $V'\simeq\widehat{V}\cup V'$ given by applying 20.11 to (W,c) and (W',c'), and let

$$X\ =\ \mathcal{M}(h:V\cup\widehat{V}\longrightarrow\widehat{V}\cup V')\ .$$

The mapping coequalizer of the two inclusions $e^+,e^-\ :\ \widehat{V}\longrightarrow X$ is a CW band

$$Z\ =\ \mathcal{W}(e^+,e^-)$$

such that $(Z;W,W')$ is a CW triad, with the inclusions

$$W\ =\ \mathcal{W}(f^+,f^-)\longrightarrow Z\ \ ,\ \ W'\ =\ \mathcal{W}(f'^+,f'^-)\longrightarrow Z$$

homotopy equivalences. The infinite cyclic cover $\overline{Z}$ of Z has a π_1-fundamental domain $(X,\widehat{V},\widehat{V},e^+,e^-)$ which restricts to the π_1-fundamental domains (V,U,U,f^+,f^-), (V',U,U,f'^+,f'^-) of the infinite cyclic covers $\overline{W}=c^*\mathbb{R}$, $\overline{W}'=c'^*\mathbb{R}$ of W, W'.

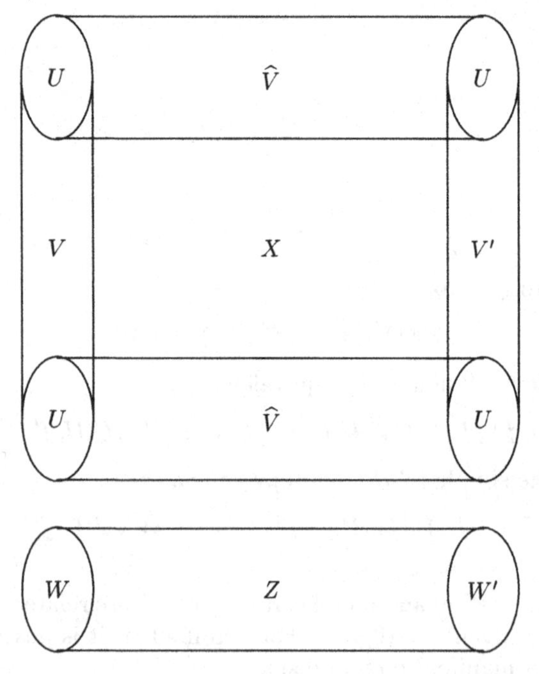

□

Remark 20.13 (i) The CW structure of the relaxation h-cobordism $(W;M,M')$ (18.3) of a manifold band (M,c) is the relaxation CW h-cobordism of 20.12, up to simple homotopy equivalence – see 26.14 below for a more detailed discussion.

(ii) In Chapter 26 we shall identify the torsions of the relaxation CW h-cobordism $\tau(W\longrightarrow Z)$, $\tau(W'\longrightarrow Z)\in Wh(\pi_1(W))$ with the nilpotent components of the fibring obstructions $\Phi^{\pm}(W)\in Wh(\pi_1(W))$, so that

$$\begin{aligned}\Phi^+(W)-\Phi^+(W') &= \Phi^-(W)-\Phi^-(W')\\ &= \tau(W\longrightarrow W')\\ &= \tau(t^+)-\tau(t^-)\in Wh(\pi_1(W))\end{aligned}$$

and $\Phi^{\pm}(W')$ is obtained from $\Phi^{\pm}(W)$ by setting the nilpotent components to zero. □

Example 20.14 Let (W,c) be an n-dimensional manifold band, with $n\geq 6$.

(i) The following conditions are equivalent:

(a) the reduced projective class

$$[\overline{W}^+] = (-)^{n-1}[\overline{W}^-]^*\in \widetilde{K}_0(\mathbb{Z}[\pi_1(\overline{W})])$$

is such that $[\overline{W}^+]=0$,

(b) there are defined homeomorphisms

$$\overline{W}\cong M\times\mathbb{R}\ ,\ \widehat{W}\cong M\times S^1$$

for a closed $(n-1)$-dimensional manifold M ,

(c) the homotopy equivalence given by 20.3

$$h = G^{-1}| : \widehat{W}\xrightarrow{\simeq}\overline{W}\times S^1$$

splits along $\overline{W}\times\{*\}\subset\overline{W}\times S^1$, i.e. the codimension 1 submanifold

$$M = h^{-1}(\overline{W}\times\{*\})\subset\widehat{W}$$

is such that the restriction $h|:M\longrightarrow\overline{W}$ is a homotopy equivalence.

(See Browder [12] for the first application of surgery to prove that $\overline{W}\cong M\times\mathbb{R}$ in the unobstructed case $\pi_1(\overline{W})=\{1\}$.)

(ii) The manifold ζ-twist glueing $(W',c')=(W[\zeta],c[\zeta])$ is the relaxation of (W,c) in the sense of Siebenmann [145]. The infinite cyclic cover $\overline{W}'$ of W' is homeomorphic to the infinite cyclic cover $\overline{W}$ of W, and $\overline{W}'$ has a fundamental domain $(V';U,\zeta'U)$ such that V' dominates $\overline{W}'^+$ rel U and V' dominates $\zeta'\overline{W}'^-$ rel $\zeta'U$. The relaxation W' is h-cobordant to W, and (equivalently) $W\times S^1$ is homeomorphic to $W'\times S^1$. Such an h-cobordism

$(Z; W, W')$ was obtained by Farrell [46] and Wall [165, 12.9] using a winding trick, and also by Cappell [19, II.1,VI] using the closely related nilpotent normal cobordism construction. □

Part Three: The algebraic theory

21

Polynomial extensions

The cellular chain complex of an infinite cyclic cover of a CW complex is a chain complex over a polynomial extension ring. In Chapter 21 we recall from Ranicki [122, 123, 124] the chain complex treatments of the mapping torus and of the splitting theorems of Bass, Heller and Swan [5] and Bass [4],

$$\begin{aligned} Wh(A[z]) &= Wh(A) \oplus \widetilde{\mathrm{Nil}}_0(A) , \\ Wh(A[z, z^{-1}]) &= Wh(A) \oplus \widetilde{K}_0(A) \oplus \widetilde{\mathrm{Nil}}_0(A) \oplus \widetilde{\mathrm{Nil}}_0(A) , \end{aligned}$$

including the algebraic analogue for finite based f.g. free $A[z, z^{-1}]$-module chain complexes of the geometric transversality construction of fundamental domains for infinite cyclic covers of finite CW complexes.

Definition 21.1 (i) The *polynomial extension* of a ring A is the ring

$$A[z] = \{\sum_{j=0}^{\infty} a_j z^j \,|\, a_j \in A, \{j \geq 0 \,|\, a_j \neq 0\} \text{ finite}\} .$$

Similarly for $A[z^{-1}]$, which is isomorphic to $A[z]$.

(ii) The *Laurent polynomial extension* of A is the ring

$$A[z, z^{-1}] = \{ \sum_{j=-\infty}^{\infty} a_j z^j \,|\, a_j \in A, \{j \in \mathbb{Z} \,|\, a_j \neq 0\} \text{ finite}\}$$

obtained from $A[z]$ by inverting z. □

Remark 21.2 (i) Given a ring A and an automorphism $\alpha : A \longrightarrow A$ let z be an indeterminate over A such that

$$az = z\alpha(a) \quad (a \in A) .$$

Given an A-module P let $\alpha_! P$ be the A-module with elements $\alpha_! x$ $(x \in P)$

and

$$\alpha_! x + \alpha_! y \ = \ \alpha_!(x+y) \ , \ a(\alpha_! x) \ = \ \alpha_!(\alpha^{-1}(a)x) \in \alpha_! P \ .$$

The α-twisted polynomial extensions $A_\alpha[z]$, $A_\alpha[z,z^{-1}]$ are defined by analogy with $A[z]$, $A[z,z^{-1}]$. There are natural direct sum decompositions

$$\begin{aligned} Wh(A_\alpha[z]) &= Wh(A) \oplus \widetilde{\mathrm{Nil}}_0(A,\alpha) \ , \\ Wh(A_\alpha[z,z^{-1}]) &= Wh(A,\alpha) \oplus \widetilde{\mathrm{Nil}}_0(A,\alpha) \oplus \widetilde{\mathrm{Nil}}_0(A,\alpha^{-1}) \ . \end{aligned}$$

Here, $Wh(A,\alpha)$ is the Grothendieck group of equivalence classes of pairs (P,f) with P a f.g. projective A-module and $f : P \longrightarrow \alpha_! P$ an isomorphism, and $\widetilde{\mathrm{Nil}}_0(A,\alpha)$ is the reduced nilpotent class group of equivalence classes of pairs (P,ν) with P a f.g. projective A-module and $\nu : P \longrightarrow \alpha_! P$ a nilpotent morphism (Farrell and Hsiang [48], Siebenmann [145]). The algebraic results of Chapters 21, 22 apply equally well in the α-twisted case, but for the sake of simplicity we shall only consider the special case

$$\alpha \ = \ 1 \ : \ A \longrightarrow A \ , \ A_\alpha[z] \ = \ A[z] \ , \ A_\alpha[z,z^{-1}] \ = \ A[z,z^{-1}]$$

with $\alpha_! P = P$.

(ii) If $\overline{W}$ is a connected infinite cyclic cover of a connected space W and the generating covering translation $\zeta : \overline{W} \longrightarrow \overline{W}$ induces the automorphism $\alpha = \zeta_* : \pi_1(\overline{W}) \longrightarrow \pi_1(\overline{W})$ then

$$\pi_1(W) \ = \ \pi_1(\overline{W}) \times_\alpha \mathbb{Z} \ , \ \mathbb{Z}[\pi_1(W)] \ = \ \mathbb{Z}[\pi_1(\overline{W})]_\alpha[z,z^{-1}] \ .$$

In dealing with infinite cyclic covers we again make the simplifying assumption $\alpha = 1$, so that

$$\pi_1(W) \ = \ \pi_1(\overline{W}) \times \mathbb{Z} \ , \ \mathbb{Z}[\pi_1(W)] \ = \ \mathbb{Z}[\pi_1(\overline{W})][z,z^{-1}] \ . \qquad \square$$

Convention 21.3 In dealing with $A[z]$- and $A[z,z^{-1}]$-modules M we shall always denote the action of z on M by ζ, that is

$$\zeta \ : \ M \longrightarrow M \ ; \ x \longrightarrow zx \ . \qquad \square$$

Thus if X is a connected CW complex with $\pi_1(X) = \pi \times \mathbb{Z}$ and $\zeta : \widetilde{X} \longrightarrow \widetilde{X}$ is the action of $z = 1 \in \mathbb{Z} \subset \pi \times \mathbb{Z}$ on the universal cover $\widetilde{X}$ then the induced $\mathbb{Z}[\pi]$-module chain map $\zeta : C(\widetilde{X}) \longrightarrow C(\widetilde{X})$ is the action of $z \in \mathbb{Z}[\pi \times \mathbb{Z}] = \mathbb{Z}[\pi][z,z^{-1}]$ on $C(\widetilde{X})$.

Definition 21.4 A CW band (W,c) is *untwisted* if (on each component) the induced surjection of fundamental groups

$$c_* \ : \ \pi_1(W) \longrightarrow \pi_1(S^1) \ = \ \mathbb{Z}$$

splits, so that

$$\pi_1(W) = \pi_1(\overline{W}) \times \mathbb{Z} \ , \ \mathbb{Z}[\pi_1(W)] = \mathbb{Z}[\pi_1(\overline{W})][z, z^{-1}] \ . \qquad \square$$

We shall be mainly concerned with untwisted CW bands from now on, in order to only have to consider untwisted polynomial extensions.

Definition 21.5 The *algebraic mapping coequalizer* of A-module chain maps $f^+, f^- : D \longrightarrow E$ is the $A[z, z^{-1}]$-module chain complex

$$\mathcal{W}(f^+, f^-) = \mathcal{C}(f^+ - z^{-1}f^- : D[z, z^{-1}] \longrightarrow E[z, z^{-1}]) \ . \qquad \square$$

Example 21.6 Let $f^+, f^- : U \longrightarrow V$ be maps of connected CW complexes such that

$$f^+_* = f^-_* : \pi = \pi_1(U) \longrightarrow \pi_1(V)$$

is an isomorphism, and let $\widetilde{f}^+, \widetilde{f}^- : C(\widetilde{U}) \longrightarrow C(\widetilde{V})$ be the induced $\mathbb{Z}[\pi]$-module chain maps of the cellular chain complexes of the universal covers $\widetilde{U}, \widetilde{V}$ of U, V. The mapping coequalizer (13.7)

$$\mathcal{W}(f^+, f^-) = U \times I \cup_{f^+ \cup f^-} V$$

has fundamental group

$$\pi_1(\mathcal{W}(f^+, f^-)) = \pi \times \mathbb{Z}$$

and the cellular $\mathbb{Z}[\pi \times \mathbb{Z}]$-module chain complex of the universal cover $\widetilde{\mathcal{W}}(f^+, f^-)$ of $\mathcal{W}(f^+, f^-)$ is given by the algebraic mapping coequalizer

$$\begin{aligned} C(\widetilde{\mathcal{W}}(f^+, f^-)) &= \mathcal{W}(\widetilde{f}^+, \widetilde{f}^-) \\ &= \mathcal{C}(\widetilde{f}^+ - z^{-1}\widetilde{f}^- : C(\widetilde{U})[z, z^{-1}] \longrightarrow C(\widetilde{V})[z, z^{-1}]) \end{aligned}$$

(assuming that f^+, f^- are the inclusions of disjoint subcomplexes). $\square$

By analogy with the geometric mapping torus:

Definition 21.7 Let $h : C \longrightarrow C$ be an A-module chain map.

(i) The *algebraic mapping torus* of h is the $A[z, z^{-1}]$-module chain complex

$$T(h) = \mathcal{W}(1, h) = \mathcal{C}(1 - z^{-1}h : C[z, z^{-1}] \longrightarrow C[z, z^{-1}]) \ .$$

(ii) The *modified algebraic mapping torus* of h is the $A[z, z^{-1}]$-module chain complex

$$T'(h) = \mathcal{C}(1 - zh : C[z, z^{-1}] \longrightarrow C[z, z^{-1}]) \ . \qquad \square$$

Example 21.8 (i) The canonical infinite cyclic cover $\overline{T}(h)$ of a mapping torus $T(h : X \longrightarrow X) = \mathcal{W}(1,h)$ is such that there is defined a pushout square

$$\begin{array}{ccc} X & \longrightarrow & \overline{T}^{+}(h) \\ \simeq \big\downarrow & & \big\downarrow \simeq \\ \overline{T}^{-}(h) & \longrightarrow & \overline{T}(h) \end{array}$$

as in 14.6 (vii). If X is a CW complex and $h : X \longrightarrow X$ is a cellular map the cellular $\mathbb{Z}[z,z^{-1}]$-module chain complexes are such that

$$\begin{aligned} C(\overline{T}(h)) &= \mathcal{C}(1 - z^{-1}h : C(X)[z,z^{-1}] \longrightarrow C(X)[z,z^{-1}]) \\ &\simeq C(\overline{T}^{+}(h)) = \mathcal{C}(1 - z^{-1}h : zC(X)[z] \longrightarrow C(X)[z]) , \\ C(\overline{T}^{-}(h)) &= \mathcal{C}(1 - z^{-1}h : z^{-1}C(X)[z^{-1}] \longrightarrow z^{-1}C(X)[z^{-1}]) \simeq C(X) . \end{aligned}$$

(ii) If X is a connected finite CW complex with universal cover $\widetilde{X}$ and $h : X \longrightarrow X$ is a cellular map such that $h_* = 1 : \pi_1(X) \longrightarrow \pi_1(X)$ then

$$\pi_1(T(h)) = \pi_1(X) \times \mathbb{Z} \ , \ \mathbb{Z}[\pi_1(T(h))] = \mathbb{Z}[\pi_1(X)][z,z^{-1}]$$

and the cellular $\mathbb{Z}[\pi_1(T(h))]$-module chain complex of the universal cover $\widetilde{T(h)}$ of $T(h)$ is the algebraic mapping torus of the induced $\mathbb{Z}[\pi_1(X)]$-module chain map $\widetilde{h} : C(\widetilde{X}) \longrightarrow C(\widetilde{X})$

$$\begin{aligned} C(\widetilde{T(h)}) &= T(\widetilde{h} : C(\widetilde{X}) \longrightarrow C(\widetilde{X})) \\ &= \mathcal{C}(1 - z^{-1}\widetilde{h} : C(\widetilde{X})[z,z^{-1}] \longrightarrow C(\widetilde{X})[z,z^{-1}]) . \end{aligned}$$

□

Proposition 21.9 (Ranicki [123, 124]) (i) *An A-module chain homotopy $e : h \simeq h' : C \longrightarrow C$ induces an $A[z,z^{-1}]$-module chain equivalence*

$$T(h) \longrightarrow T(h') \ ; \ (x,y) \longrightarrow (x + e(y), y) .$$

(ii) *For any A-module chain maps $f : C \longrightarrow D$, $g : D \longrightarrow C$ there is defined an $A[z,z^{-1}]$-module chain equivalence*

$$T(gf) \longrightarrow T(fg) \ ; \ (x,y) \longrightarrow (f(x), f(y)) .$$

(iii) *If C is a finitely dominated A-module chain complex and $h : C \longrightarrow C$ is any chain map the algebraic mapping torus $T(h : C \longrightarrow C)$ has a canonical simple chain homotopy type (as in 6.3). If*

$$(D, f : C \longrightarrow D, g : D \longrightarrow C, gf \simeq 1 : C \longrightarrow C)$$

is a finite domination of C then for any choice of basis for D the algebraic *mapping torus $T(fhg : D \longrightarrow D)$ is a finite based f.g. free $A[z,z^{-1}]$-module chain complex in the canonical simple chain homotopy type.* □

Definition 21.10 The *Whitehead group* of $A[z,z^{-1}]$ is

$$Wh(A[z,z^{-1}]) = \begin{cases} Wh(\pi \times \mathbb{Z}) \\ \mathrm{coker}(K_1(\mathbb{Z}[z,z^{-1}]) \longrightarrow K_1(A[z,z^{-1}])) \\ \qquad = K_1(A[z,z^{-1}])/\{\pm z\} \end{cases}$$

$$\begin{cases} \text{if } A = \mathbb{Z}[\pi] \text{ is a group ring ,} \\ \text{otherwise .} \end{cases}$$ □

The splitting theorem for $Wh(A[z,z^{-1}])$ involves the following K-group of nilpotent endomorphisms.

Definition 21.11 (i) The *nilpotent class group* $\mathrm{Nil}_0(A)$ is the abelian group generated by pairs (P,ν) with P a f.g. projective A-module and $\nu : P \longrightarrow P$ a nilpotent endomorphism, with one relation

$$(P,\nu) = (P',\nu') + (P'',\nu'') \in \mathrm{Nil}_0(A)$$

for each exact sequence

$$0 \longrightarrow P' \xrightarrow{f} P \xrightarrow{f'} P'' \longrightarrow 0$$

with

$$\nu f = f\nu' : P' \longrightarrow P \ , \ \nu'' f' = f'\nu : P \longrightarrow P'' \ .$$

(ii) The *reduced nilpotent class group* is

$$\widetilde{\mathrm{Nil}}_0(A) = \mathrm{coker}(K_0(A) \longrightarrow \mathrm{Nil}_0(A))$$

with

$$K_0(A) \longrightarrow \mathrm{Nil}_0(A) \ ; \ [P] \longrightarrow [P,0] \ .$$

The direct sum decomposition

$$\mathrm{Nil}_0(A) = K_0(A) \oplus \widetilde{\mathrm{Nil}}_0(A)$$

is such that the projective class $[P] \in K_0(A)$ a component of the nilpotent class $[P,\nu] \in \mathrm{Nil}_0(A)$. □

See Ranicki [124, Chapter 9] for the chain complex treatment of the nilpotent class group $\mathrm{Nil}_0(A)$ of a ring A, including the definition of the nilpotent class $[P,\nu] \in \mathrm{Nil}_0(A)$ of a finitely dominated A-module chain complex P with a chain homotopy nilpotent self chain map $\nu : P \longrightarrow P$.

Let $i^+ : A \longrightarrow A[z]$, $i : A \longrightarrow A[z,z^{-1}]$ be the inclusions.

Proposition 21.12 (Bass [4]) (i) *The torsion group of* $A[z]$ *is such that*

$$K_1(A[z]) = K_1(A) \oplus \widetilde{\mathrm{Nil}}_0(A) ,$$

with an isomorphism

$$K_1(A) \oplus \widetilde{\mathrm{Nil}}_0(A) \longrightarrow K_1(A[z]) ;$$
$$(\tau, [P, \nu]) \longrightarrow i_!^+\tau + \tau(1 - z\nu : P[z] \longrightarrow P[z]) .$$

(ii) *The torsion group of* $A[z, z^{-1}]$ *is such that*

$$K_1(A[z, z^{-1}]) = K_1(A) \oplus K_0(A) \oplus \widetilde{\mathrm{Nil}}_0(A) \oplus \widetilde{\mathrm{Nil}}_0(A) ,$$

with an isomorphism

$$K_1(A) \oplus K_0(A) \oplus \widetilde{\mathrm{Nil}}_0(A) \oplus \widetilde{\mathrm{Nil}}_0(A) \longrightarrow K_1(A[z, z^{-1}]) ;$$
$$\begin{aligned}(\tau, [P], [P^+, \nu^+], [P^-, \nu^-]) \longrightarrow\ & i_!\tau + \tau(-z : P[z, z^{-1}] \longrightarrow P[z, z^{-1}]) \\ & + \tau(1 - z\nu^+ : P^+[z, z^{-1}] \longrightarrow P^+[z, z^{-1}]) \\ & + \tau(1 - z^{-1}\nu^- : P^-[z, z^{-1}] \longrightarrow P^-[z, z^{-1}]) .\end{aligned}$$

(iii) *The Whitehead group of* $A[z, z^{-1}]$ *is such that*

$$Wh(A[z, z^{-1}]) = Wh(A) \oplus \widetilde{K}_0(A) \oplus \widetilde{\mathrm{Nil}}_0(A) \oplus \widetilde{\mathrm{Nil}}_0(A) . \qquad \square$$

Definition 21.13 A *Mayer–Vietoris presentation* (C^+, C^-) of a finite based f.g. free $A[z, z^{-1}]$-module chain complex C is a based f.g. free $A[z]$-module subcomplex $C^+ \subset C$ together with a based f.g. free $A[z^{-1}]$-module subcomplex $C^- \subset C$ such that $C^+ \cap C^- \subset C$ is a based f.g. free A-module subcomplex, with

$$C = A[z, z^{-1}] \otimes_{A[z]} C^+ = A[z, z^{-1}] \otimes_{A[z^{-1}]} C^- ,$$

and such that the basis elements of $C^+, C^-, C^+ \cap C^-$ are each of the type $z^N b$ for some basis element $b \in C$ and $N \in \mathbb{Z}$. $\square$

Proposition 21.14 (Ranicki [124]) (i) *Every finite based f.g. free* $A[z, z^{-1}]$*-module chain complex* C *admits a Mayer–Vietoris presentation* (C^+, C^-).

(ii) *Given a finite based f.g. free* $A[z, z^{-1}]$*-module chain complex* C *and a Mayer–Vietoris presentation* (C^+, C^-) *define the finite based f.g. free* A*-module chain complexes*

$$D = C^+ \cap C^- , \quad E = C^+ \cap \zeta C^- .$$

The injections

$$f^+ : D \longrightarrow E ; x \longrightarrow x , \quad f^- : D \longrightarrow E ; y \longrightarrow \zeta y$$

are such that there is defined a short exact sequence of finite based f.g. free $A[z,z^{-1}]$-module chain complexes

$$0 \longrightarrow D[z,z^{-1}] \xrightarrow{f^+ - z^{-1}f^-} E[z,z^{-1}] \longrightarrow C \longrightarrow 0$$

with torsion $\tau = 0 \in Wh(A[z,z^{-1}])$. In particular, C is simple chain equivalent to $\mathcal{W}(f^+,f^-)$.

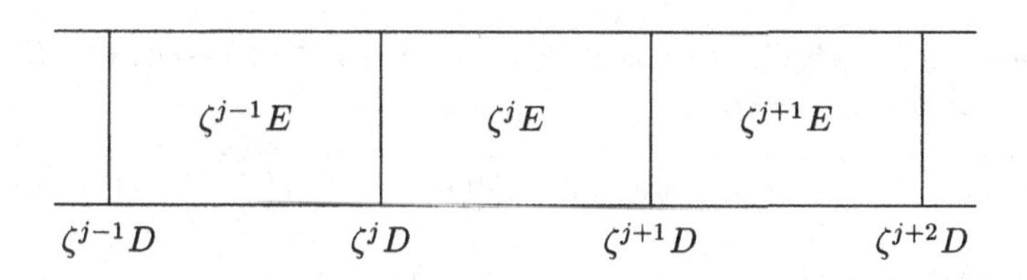

□

Example 21.15 Let W be a connected finite CW complex with fundamental group $\pi_1(W) = \pi \times \mathbb{Z}$. Let $\widetilde{W}$ be the universal cover of W, so that $\overline{W} = \widetilde{W}/\pi$ is an infinite cyclic cover of W with $\pi_1(\overline{W}) = \pi$, and a generating covering translation $\zeta : \overline{W} \longrightarrow \overline{W}$ induces $\zeta_* = 1 : \pi \longrightarrow \pi$. Assume that $\overline{W}$ has a CW π_1-fundamental domain $(V;U,\zeta U)$, so that $\pi_1(U) = \pi_1(V) = \pi$ and $W = \mathcal{W}(f^+,f^-)$ with

$$f^+ : U \longrightarrow V \; ; \; x \longrightarrow x \; , \quad f^- : U \longrightarrow V \; ; \; x \longrightarrow \zeta x \; .$$

Let

$$\overline{W}^+ = \bigcup_{j=0}^{\infty} \zeta^j V \; , \quad \overline{W}^- = \bigcup_{j=-\infty}^{-1} \zeta^j V \subset \overline{W} = \bigcup_{j=-\infty}^{\infty} \zeta^j V \; ,$$

and let $\widetilde{W}^+, \widetilde{W}^- \subset \widetilde{W}$ be the lifts of $\overline{W}^+, \overline{W}^- \subset \overline{W}$. The cellular chain complex

$$C(\widetilde{W}) = \mathcal{W}(\tilde{f}^+,\tilde{f}^-) = \mathcal{C}(\tilde{f}^+ - z^{-1}\tilde{f}^- : C(\widetilde{U})[z,z^{-1}] \longrightarrow C(\widetilde{V})[z,z^{-1}])$$

is a finite complex of based f.g. free $\mathbb{Z}[\pi][z,z^{-1}]$-modules with a Mayer–Vietoris presentation $(C(\widetilde{W}^+), C(\widetilde{W}^-))$ such that

$$\begin{aligned}
C(\widetilde{W}^+) &= \text{coker}(\tilde{f}^+ - z^{-1}\tilde{f}^- : zC(\widetilde{U})[z] \longrightarrow C(\widetilde{V})[z]) \; , \\
C(\widetilde{W}^-) &= \text{coker}(\tilde{f}^+ - z^{-1}\tilde{f}^- : z^{-1}C(\widetilde{U})[z^{-1}] \longrightarrow z^{-1}C(\widetilde{V})[z^{-1}]) \; , \\
C(\widetilde{W}^+) \cap C(\widetilde{W}^-) &= C(\widetilde{U}) \; , \quad C(\widetilde{W}^+) \cap \zeta C(\widetilde{W}^-) = C(\widetilde{V}) \; .
\end{aligned}$$

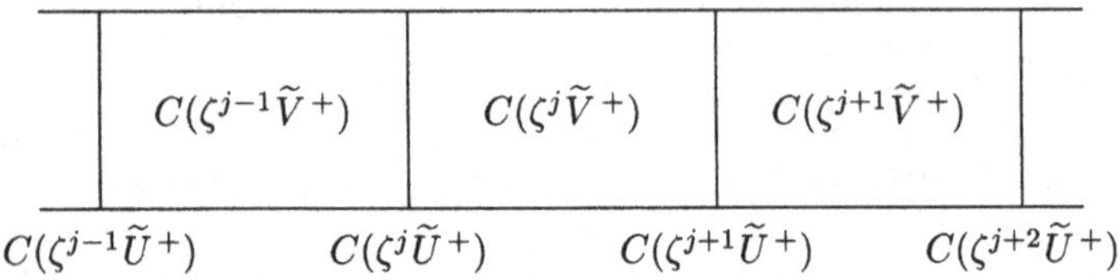

$C(\zeta^{j-1}\widetilde{V}^+)$	$C(\zeta^{j}\widetilde{V}^+)$	$C(\zeta^{j+1}\widetilde{V}^+)$

$C(\zeta^{j-1}\widetilde{U}^+)$ $C(\zeta^{j}\widetilde{U}^+)$ $C(\zeta^{j+1}\widetilde{U}^+)$ $C(\zeta^{j+2}\widetilde{U}^+)$ □

By analogy with 13.18:

Proposition 21.16 *Given A-module chain maps $i^+ : C \longrightarrow E$, $i^- : C \longrightarrow F$, $j^+ : D \longrightarrow F$, $j^- : D \longrightarrow E$ let*

$$E \cup_C F = \mathcal{C}(\begin{pmatrix} i^+ \\ i^- \end{pmatrix} : C \longrightarrow E \oplus F) \ , \ E \cup_D F = \mathcal{C}(\begin{pmatrix} j^+ \\ j^- \end{pmatrix} : D \longrightarrow E \oplus F) \ .$$

The algebraic mapping coequalizers of the chain maps

$$f(i,j)^+ : C \xrightarrow{i^+} E \longrightarrow E \cup_D F \ , \ f(i,j)^- : C \xrightarrow{i^-} F \longrightarrow E \cup_D F \ ,$$

$$g(i,j)^+ : D \xrightarrow{j^+} F \longrightarrow F \cup_C E \ , \ g(i,j)^- : D \xrightarrow{j^-} E \longrightarrow F \cup_C E$$

are related by a canonical $A[z,z^{-1}]$-module chain equivalence

$$\mathcal{W}(f(i,j)^+, f(i,j)^-) \simeq \mathcal{W}(g(i,j)^+, g(i,j)^-) \ ,$$

which is simple if C, D, E, F are finite and based f.g. free.

Proof The $A[z,z^{-1}]$-module chain complex

$$B = \mathcal{C}(\begin{pmatrix} i^+ & j^+ \\ z^{-1}i^- & j^- \end{pmatrix} : (C \oplus D)[z,z^{-1}] \longrightarrow (E \oplus F)[z,z^{-1}])$$

can be cut open along either C or D, so that both $\mathcal{W}(f(i,j)^+, f(i,j)^-)$ and $\mathcal{W}(g(i,j)^+, g(i,j)^-)$ are chain equivalent to B. □

22

Algebraic bands

An 'algebraic band' is the chain complex analogue of a CW band. We shall now recall from Ranicki [124, Chapter 20] the algebraic band version of the Whitehead torsion obstruction of Farrell [47] and Siebenmann [145] for fibring a manifold band over S^1. In Chapters 23–25 we shall develop the chain complex analogues of forward and reverse tameness, relaxation, and ribbons, which will then be applied in Chapter 26 to obtain an algebraic version of the homotopy theoretic twist glueing of Chapter 19.

Definition 22.1 A *chain complex band* is a finite based f.g. free $A[z,z^{-1}]$-module chain complex C which is A-finitely dominated, so that the projective class $[C] \in K_0(A)$ is defined. □

Proposition 22.2 (i) *If C is a finite based f.g. free A-module chain complex and $h : C \longrightarrow C$ is a chain equivalence the algebraic mapping torus $T(h)$ is an $A[z,z^{-1}]$-module chain complex band.*

(ii) *If C is a finitely dominated A-module chain complex and $h : C \longrightarrow C$ is a chain equivalence then any finite based f.g. free A-module chain complex E chain equivalent to $T(h)$ is an $A[z,z^{-1}]$-module chain complex band. If $(D, f : C \longrightarrow D, g : D \longrightarrow C, gf \simeq 1 : C \longrightarrow C)$ is a finite domination of C then $E = T(fhg : D \longrightarrow D)$ is such a chain complex band in the canonical simple chain homotopy type of $T(h)$.* □

For any f.g. free $A[z,z^{-1}]$-module chain complex C there are defined exact sequences of $A[z,z^{-1}]$-module chain complexes

$$0 \longrightarrow C[z,z^{-1}] \xrightarrow{1 - z\zeta^{-1}} C[z,z^{-1}] \longrightarrow C \longrightarrow 0 \, ,$$

$$0 \longrightarrow C[z,z^{-1}] \xrightarrow{1 - z^{-1}\zeta} C[z,z^{-1}] \longrightarrow C \longrightarrow 0$$

with

$$\zeta : C \longrightarrow C ; x \longrightarrow zx ,$$

$$C[z,z^{-1}] \longrightarrow C ; \sum_{j=-\infty}^{\infty} x_j z^j \longrightarrow \sum_{j=-\infty}^{\infty} \zeta^j(x_j) .$$

If C is a band the induced $A[z,z^{-1}]$-module chain maps

$$\begin{aligned} T'(\zeta^{-1}) &= \mathcal{C}(1 - z\zeta^{-1} : C[z,z^{-1}] \longrightarrow C[z,z^{-1}]) \longrightarrow C , \\ T(\zeta) &= \mathcal{C}(1 - z^{-1}\zeta : C[z,z^{-1}] \longrightarrow C[z,z^{-1}]) \longrightarrow C \end{aligned}$$

are $A[z,z^{-1}]$-module chain equivalences.

Definition 22.3 The *fibring obstructions* of an $A[z,z^{-1}]$-module chain complex band C are

$$\begin{aligned} \Phi^+(C) &= \tau(T'(\zeta^{-1}) \longrightarrow C) , \\ \Phi^-(C) &= \tau(T(\zeta) \longrightarrow C) \in Wh(A[z,z^{-1}]) . \end{aligned}$$

□

Proposition 22.4 *Let W be a connected finite CW complex with fundamental group $\pi_1(W) = \pi \times \mathbb{Z}$ and universal cover $\widetilde{W}$, so that $\overline{W} = \widetilde{W}/\pi$ is an infinite cyclic cover of W with $\pi_1(\overline{W}) = \pi$, and*

$$\mathbb{Z}[\pi_1(W)] = \mathbb{Z}[\pi \times \mathbb{Z}] = \mathbb{Z}[\pi][z,z^{-1}] .$$

Let $c : W \longrightarrow S^1$ be a map inducing

$$c_* = \text{projection} : \pi_1(W) = \pi \times \mathbb{Z} \longrightarrow \pi_1(S^1) = \mathbb{Z} .$$

Then (W,c) is an (untwisted) CW band if and only if the cellular $\mathbb{Z}[\pi][z,z^{-1}]$-module chain complex $C(\widetilde{W})$ is a chain complex band, in which case the fibring obstructions of (W,c) are the fibring obstructions of $C(\widetilde{W})$:

$$\Phi^{\pm}(W,c) = \Phi^{\pm}(C(\widetilde{W})) \in Wh(\pi_1(\overline{W}) \times \mathbb{Z}) .$$

□

For any finite based f.g. free $A[z,z^{-1}]$-module chain complex C and any Mayer–Vietoris presentation (C^+, C^-) let

$$\begin{aligned} f^+ &: D = C^+ \cap C^- \longrightarrow E = C^+ \cap \zeta C^- ; x \longrightarrow x , \\ f^- &: D = C^+ \cap C^- \longrightarrow E = C^+ \cap \zeta C^- ; y \longrightarrow \zeta y , \end{aligned}$$

so that as in 21.14 there is defined an exact sequence

$$0 \longrightarrow D[z,z^{-1}] \xrightarrow{f^+ - z^{-1}f^-} E[z,z^{-1}] \longrightarrow C \longrightarrow 0$$

with $\tau = 0 \in Wh(A[z,z^{-1}])$. The A-module chain maps

$$g^+ : E \longrightarrow C^+ \ ; \ x \longrightarrow x \ , \quad g^- : E \longrightarrow C^- \ ; \ x \longrightarrow \zeta^{-1}x \ ,$$
$$\zeta^+ : C^+ \longrightarrow C^+ \ ; \ x \longrightarrow \zeta x \ , \quad \zeta^- : C^- \longrightarrow C^- \ ; \ x \longrightarrow \zeta^{-1}x$$

are such that there are defined commutative squares

$$\begin{array}{ccc} D & \xrightarrow{f^-} & E \\ {\scriptstyle g^+f^+}\downarrow & & \downarrow{\scriptstyle g^+} \\ C^+ & \xrightarrow{\zeta^+} & C^+ \end{array} \qquad \begin{array}{ccc} D & \xrightarrow{f^+} & E \\ {\scriptstyle g^-f^-}\downarrow & & \downarrow{\scriptstyle g^-} \\ C^- & \xrightarrow{\zeta^-} & C^- \end{array}$$

giving rise to exact sequences

$$0 \longrightarrow D \xrightarrow{\begin{pmatrix} f^+ \\ g^-f^- \end{pmatrix}} E \oplus C^- \xrightarrow{(-g^- \ \zeta^-)} C^- \longrightarrow 0 \ ,$$
$$0 \longrightarrow D \xrightarrow{\begin{pmatrix} f^- \\ g^+f^+ \end{pmatrix}} E \oplus C^+ \xrightarrow{(-g^+ \ \zeta^+)} C^+ \longrightarrow 0$$

and chain equivalences

$$(g^-f^-, g^-) : \mathcal{C}(f^+ : D \longrightarrow E) \xrightarrow{\simeq} \mathcal{C}(\zeta^- : C^- \longrightarrow C^-) \ ,$$
$$(g^+f^+, g^+) : \mathcal{C}(f^- : D \longrightarrow E) \xrightarrow{\simeq} \mathcal{C}(\zeta^+ : C^+ \longrightarrow C^+) \ .$$

Proposition 22.5 (Ranicki [124, Chapter 20 ; 126]) (i) *The fibring obstructions of an $A[z,z^{-1}]$-module chain complex band C are simple chain homotopy invariants, such that $\Phi^+(C) = \Phi^-(C) = 0 \in Wh(A[z,z^{-1}])$ if and only if C is simple chain equivalent to the algebraic mapping torus $T(h)$ of a simple self chain equivalence $h : F \longrightarrow F$ of a finite based f.g. free A-module chain complex F.*

(ii) *A finite based f.g. free $A[z,z^{-1}]$-module chain complex C is a band if and only if for any Mayer–Vietoris presentation (C^+, C^-) the A-module chain complexes C^+, C^- are finitely dominated. If C is a band then C/C^+, C/C^- are also finitely dominated, and the fibring obstructions of C are given by*

$$\begin{aligned} \Phi^+(C) &= (\phi^+, -[C^-], [C/C^+, \zeta], [C/C^-, \zeta^{-1}]) \ , \\ \Phi^-(C) &= (\phi^-, [C^+], [C/C^+, \zeta], [C/C^-, \zeta^{-1}]) \\ &\in Wh(A[z,z^{-1}]) \ = \ Wh(A) \oplus \widetilde{K}_0(A) \oplus \widetilde{\mathrm{Nil}}_0(A) \oplus \widetilde{\mathrm{Nil}}_0(A) \end{aligned}$$

with $[C/C^+, \zeta]$, $[C/C^-, \zeta^{-1}]$ the classes of the chain homotopy nilpotent

A-module chain maps

$$\zeta : C/C^+ \longrightarrow C/C^+ \ ; \ x \longrightarrow zx \ ,$$
$$\zeta^{-1} : C/C^- \longrightarrow C/C^- \ ; \ x \longrightarrow z^{-1}x \ .$$

The torsions

$$\begin{aligned}\phi^+ &= -\tau((g^-f^-,g^-) : \mathcal{C}(f^+ : D \longrightarrow E) \longrightarrow \mathcal{C}(\zeta^- : C^- \longrightarrow C^-)) \ , \\ \phi^- &= -\tau((g^+f^+,g^+) : \mathcal{C}(f^- : D \longrightarrow E) \longrightarrow \mathcal{C}(\zeta^+ : C^+ \longrightarrow C^+)) \in Wh(A)\end{aligned}$$

are defined using the canonical simple chain homotopy types on $\mathcal{C}(\zeta^+ : C^+ \longrightarrow C^+)$ and $\mathcal{C}(\zeta^- : C^- \longrightarrow C^-)$ given by 6.3, and are such that

$$\phi^+ - \phi^- = \tau(\zeta : C \longrightarrow C) \in Wh(A) \ .$$

(iii) *The fibring obstructions are simple chain homotopy invariants of an $A[z,z^{-1}]$-module chain complex band C. The difference*

$$\begin{aligned}\Phi^+(C) - \Phi^-(C) &= (\phi^+ - \phi^-, -[C^+] - [C^-], 0, 0) \\ &= (\tau(\zeta : C \longrightarrow C), -[C], 0, 0) \\ &= \tau(-z^{-1}\zeta : C[z,z^{-1}] \longrightarrow C[z,z^{-1}]) \\ \in Wh(A[z,z^{-1}]) &= Wh(A) \oplus \widetilde{K}_0(A) \oplus \widetilde{\mathrm{Nil}}_0(A) \oplus \widetilde{\mathrm{Nil}}_0(A)\end{aligned}$$

is a chain homotopy invariant of C.

(iv) *For any chain equivalence $h : C' \longrightarrow C$ of $A[z,z^{-1}]$-module chain complex bands*

$$\begin{aligned}\tau(h) &= \Phi^+(C) - \Phi^+(C') \\ &= \Phi^-(C) - \Phi^-(C') \in Wh(A[z,z^{-1}]) \ .\end{aligned}$$

□

Example 22.6 Given finite based f.g. free A-module chain complexes D, E and chain equivalences $f^+, f^- : D \longrightarrow E$ define an $A[z,z^{-1}]$-module chain complex band

$$C = \mathcal{C}(f^+ - z^{-1}f^- : D[z,z^{-1}] \longrightarrow E[z,z^{-1}]) \ .$$

The fibring obstructions of C are given by

$$\begin{aligned}\Phi^+(C) &= (\tau(f^-), 0, 0, 0) \ , \\ \Phi^-(C) &= (\tau(f^+), 0, 0, 0) \\ &\in Wh(A[z,z^{-1}]) = Wh(A) \oplus \widetilde{K}_0(A) \oplus \widetilde{\mathrm{Nil}}_0(A) \oplus \widetilde{\mathrm{Nil}}_0(A) \ .\end{aligned}$$

□

Example 22.7 If C is a finitely dominated A-module chain complex and $h : C \longrightarrow C$ is a chain equivalence the algebraic mapping torus $T(h)$ has the canonical simple chain homotopy type of a chain complex band (22.2 (ii)), with respect to which

$$\begin{aligned} \Phi^+(T(h)) &= (\tau(h), -[C], 0, 0) , \\ \Phi^-(T(h)) &= (0, 0, 0, 0) \\ \in Wh(A[z, z^{-1}]) &= Wh(A) \oplus \widetilde{K}_0(A) \oplus \widetilde{\mathrm{Nil}}_0(A) \oplus \widetilde{\mathrm{Nil}}_0(A) . \end{aligned}$$

□

23

Algebraic tameness

We shall now develop algebraic analogues of tameness for $A[z]$- and $A[z, z^{-1}]$-module chain complexes for any ring A, corresponding to the geometric tameness properties of the ends of infinite cyclic covers of finite CW complexes. The algebraic theory of tameness will be applied in 23.22 to prove that an end $\overline{W}^+$ of an infinite cyclic cover $\overline{W}$ of a finite CW complex W with $\pi_1(W) = \pi_1(\overline{W}) \times \mathbb{Z}$ is forward (resp. reverse) tame if and only if the cellular $\mathbb{Z}[\pi_1(\overline{W})]$-module chain complex $C(\widetilde{W}^+)$ is forward (resp. reverse) tame.

Definition 23.1 (i) The *formal power series extension* of A is the ring

$$A[[z]] = \{\sum_{j=0}^{\infty} a_j z^j \mid a_j \in A\} ,$$

without any finiteness conditions on the coefficients a_j. Similarly for $A[[z^{-1}]]$, which is isomorphic to $A[[z]]$.

(ii) The *Novikov polynomial extension* of A is the ring

$$A((z)) = A[[z]][z^{-1}] = \{ \sum_{j=-\infty}^{\infty} a_j z^j \mid a_j \in A, \{j \leq 0 \mid a_j \neq 0\} \text{ finite}\}$$

obtained from $A[[z]]$ by inverting z. Similarly for $A((z^{-1}))$, which is isomorphic to $A((z))$.

(iii) The *formal Laurent polynomial extension* $A[[z, z^{-1}]]$ is the $A[z, z^{-1}]$-bimodule consisting of all the formal power series

$$A[[z, z^{-1}]] = \{ \sum_{j=-\infty}^{\infty} a_j z^j \mid a_j \in A\} . \qquad \square$$

Note that

$$\begin{aligned}
&A[[z]] \subset A((z)) \subset A[[z,z^{-1}]] ,\\
&A[[z^{-1}]] \subset A((z^{-1})) \subset A[[z,z^{-1}]] ,\\
&A[[z]] \cap A[[z^{-1}]] \ = \ A \ , \quad A((z)) \cap A((z^{-1})) \ = \ A[z,z^{-1}] .
\end{aligned}$$

Remark 23.2 A compact manifold M fibres over S^1 if and only if it admits a Morse function $c : M \longrightarrow S^1$ without critical points. Novikov [106] applied the rings $A((z)), A((z^{-1}))$ to the Morse theory of S^1-valued functions on compact manifolds M with $\pi_1(M) = \mathbb{Z}$, $A = \mathbb{Z}$. Farber [44, 45] proved that in this case there exists a Morse function $c : M \longrightarrow S^1$ with the minimum number of critical points given by $\mathbb{Z}((z))$-coefficient homology, recovering the fibring theorem of Browder and Levine [13]. Pazhitnov [108] applied the Novikov rings with $A = \mathbb{Z}[\pi]$ to the Morse theory of S^1-valued functions on compact manifolds M with $\pi_1(M) = \pi \times \mathbb{Z}$ for any π, recovering the fibring obstruction of Farrell [46, 47] and Siebenmann [145] as an $A((z))$-coefficient Reidemeister torsion. The main result of [108] gives a direct proof that a manifold band (M, c) with $\dim(M) \geq 6$ fibres over S^1 if and only if the $A((z))$-coefficient Reidemeister torsion is 0. See also Ranicki [126, 128]. □

Definition 23.3 Let C^+ be a finite f.g. free $A[z]$-module chain complex.

(i) The *locally finite* chain complex of C^+ is the induced $A[[z]]$-module chain complex

$$C^{+,lf} \ = \ A[[z]] \otimes_{A[z]} C^+ \ .$$

(ii) The *end complex* of C^+ is the $A[z]$-module chain complex

$$e(C^+) \ = \ \mathcal{C}(i : C^+ \longrightarrow C^{+,lf})_{*+1} \ ,$$

with $i : C^+ \longrightarrow C^{+,lf}$ the inclusion.

(iii) C^+ is *reverse tame* if it is A-finitely dominated, in which case the projective class $[C^+] \in K_0(A)$ is defined.

(iv) C^+ is *reverse collared* if it is chain homotopy A-finite, i.e. A-module chain equivalent to a finite f.g. free A-module chain complex. □

Similarly for an $A[z^{-1}]$-module chain complex C^-, with

$$C^{-,lf} \ = \ A[[z^{-1}]] \otimes_{A[z^{-1}]} C^- \ , \ e(C^-) \ = \ \mathcal{C}(i : C^- \longrightarrow C^{-,lf})_{*+1} \ .$$

Example 23.4 As in 21.15 let W be a connected finite CW complex with $\pi_1(W) = \pi \times \mathbb{Z}$, such that the infinite cyclic cover $\overline{W}$ has a CW π_1-fundamental domain. The corresponding Mayer–Vietoris presentation $(C(\widetilde{W}^+), C(\widetilde{W}^-))$ of $C(\widetilde{W})$ is such that

$$\begin{aligned} C(\widetilde{W}) &= \mathbb{Z}[\pi][z,z^{-1}] \otimes_{\mathbb{Z}[\pi][z]} C(\widetilde{W}^+) = \mathbb{Z}[\pi][z,z^{-1}] \otimes_{\mathbb{Z}[\pi][z^{-1}]} C(\widetilde{W}^-)\ , \\ C^{lf,\pi}(\widetilde{W}^+) &= \mathbb{Z}[\pi][[z]] \otimes_{\mathbb{Z}[\pi][z]} C(\widetilde{W}^+)\ , \\ C^{lf,\pi}(\widetilde{W}^-) &= \mathbb{Z}[\pi][[z^{-1}]] \otimes_{\mathbb{Z}[\pi][z^{-1}]} C(\widetilde{W}^-)\ . \end{aligned}$$

It is clear that if $\overline{W}^+$ is reverse tame (resp. collared) then the $\mathbb{Z}[\pi][z]$-module chain complex $C(\widetilde{W})^+$ is reverse tame (resp. collared) – see 23.22 below for the converse. □

Proposition 23.5 *A finite f.g. free $A[z]$-module chain complex C^+ is reverse collared if and only if C^+ is reverse tame and $[C^+] = 0 \in \widetilde{K}_0(A)$.*

Proof The reduced projective class $[D] \in \widetilde{K}_0(A)$ of any finitely dominated A-module chain complex D is such that $[D] = 0$ if and only if D is chain homotopy A-finite. □

Definition 23.6 A commutative square of rings and morphisms

$$\begin{array}{ccc} A & \xrightarrow{f} & B \\ {\scriptstyle f'}\big\downarrow & & \big\downarrow{\scriptstyle g} \\ B' & \xrightarrow{g'} & A' \end{array}$$

is *cartesian* if the sequence of additive groups

$$0 \longrightarrow A \xrightarrow{\begin{pmatrix} f \\ f' \end{pmatrix}} B \oplus B' \xrightarrow{(g\ \ -g')} A' \longrightarrow 0$$

is exact. □

Proposition 23.7 (Ranicki [124, 126]) (i) *The various polynomial extensions of a ring A fit into cartesian squares of rings*

$$\begin{array}{ccc} A[z] & \longrightarrow & A[z,z^{-1}] \\ \big\downarrow & & \big\downarrow \\ A[[z]] & \longrightarrow & A((z)) \end{array} \qquad \begin{array}{ccc} A[z^{-1}] & \longrightarrow & A[z,z^{-1}] \\ \big\downarrow & & \big\downarrow \\ A[[z^{-1}]] & \longrightarrow & A((z^{-1})) \end{array}$$

(ii) *A Mayer–Vietoris presentation* (C^+,C^-) *of a finite based f.g. free* $A[z,z^{-1}]$*-module chain complex* C *determines exact sequences*

$$0 \longrightarrow C^+ \longrightarrow C^{+,lf} \oplus C \longrightarrow A((z)) \otimes_{A[z]} C^+ \longrightarrow 0 ,$$
$$0 \longrightarrow C^+ \cap C^- \longrightarrow C^{+,lf} \oplus C^- \longrightarrow A((z)) \otimes_{A[z]} C^+ \longrightarrow 0 ,$$
$$0 \longrightarrow C^+ \cap C^- \longrightarrow C^+ \oplus C^{-,lf} \longrightarrow A((z^{-1})) \otimes_{A[z]} C^+ \longrightarrow 0 ,$$

with

$$A((z)) \otimes_{A[z]} C^+ = A((z)) \otimes_{A[z^{-1}]} C^- = A((z)) \otimes_{A[z,z^{-1}]} C ,$$
$$A((z^{-1})) \otimes_{A[z]} C^+ = A((z^{-1})) \otimes_{A[z^{-1}]} C^- = A((z^{-1})) \otimes_{A[z,z^{-1}]} C .$$

(iii) *A finite f.g. free* $A[z]$*-module chain complex* C^+ *is reverse tame if and only if* $H_*(A((z^{-1})) \otimes_{A[z]} C^+) = 0$.

(iii)′ *A finite f.g. free* $A[z^{-1}]$*-module chain complex* C^- *is reverse tame if and only if* $H_*(A((z)) \otimes_{A[z^{-1}]} C^-) = 0$.

(iv) *A finite f.g. free* $A[z,z^{-1}]$*-module chain complex* C *is A-finitely dominated if and only if*

$$H_*(A((z)) \otimes_{A[z,z^{-1}]} C) = H_*(A((z^{-1})) \otimes_{A[z,z^{-1}]} C) = 0 .$$

(v) *A finite based f.g. free* $A[z,z^{-1}]$*-module chain complex* C *is a band if and only if for any Mayer–Vietoris presentation* (C^+,C^-) *of* C *the A-module chain maps*

$$C^+ \longrightarrow C^{+,lf} \oplus C \; ; \; x \longrightarrow (x,x) ,$$
$$C^- \longrightarrow C^{-,lf} \oplus C \; ; \; x \longrightarrow (x,x)$$

are homology equivalences, in which case they are chain equivalences. □

Since $A[[z,z^{-1}]]$ is not a ring, the notion of '$A[[z,z^{-1}]]$-module' does not quite make sense. We shall only use it in the following context: if M is an $A[z,z^{-1}]$-module the 'induced $A[[z,z^{-1}]]$-module' is the induced $A[z,z^{-1}]$-module

$$M^{lf} = A[[z,z^{-1}]] \otimes_{A[z,z^{-1}]} M ,$$

constructed using the $A[z,z^{-1}]$-bimodule structure of $A[[z,z^{-1}]]$.

Definition 23.8 Let C be a finite f.g. free $A[z,z^{-1}]$-module chain complex. The *locally finite* chain complex of C is the induced $A[[z,z^{-1}]]$-module chain complex

$$C^{lf} = A[[z,z^{-1}]] \otimes_{A[z,z^{-1}]} C .$$ □

Proposition 23.9 *Let C be a finite f.g. free $A[z,z^{-1}]$-module chain complex.*

(i) *The homology of the locally finite chain complex C^{lf} fits into a long exact sequence of $A[z,z^{-1}]$-modules*

$$\begin{aligned}\dots \longrightarrow H_{r+1}(C^{lf}) &\xrightarrow{\partial} H_r(C)\\ &\longrightarrow H_r(A((z))\otimes_{A[z,z^{-1}]} C)\oplus H_r(A((z^{-1}))\otimes_{A[z,z^{-1}]} C)\\ &\qquad\qquad\longrightarrow H_r(C^{lf}) \longrightarrow \dots\ .\end{aligned}$$

(ii) *C is A-finitely dominated if and only if the connecting morphisms in (i) are isomorphisms $\partial : H_{*+1}(C^{lf}) \cong H_*(C)$.*

Proof (i) This is the homology exact sequence induced by the short exact sequence of A-modules

$$0 \longrightarrow C \longrightarrow (A((z))\otimes_{A[z,z^{-1}]} C)\oplus(A((z^{-1}))\otimes_{A[z,z^{-1}]} C) \longrightarrow C^{lf} \longrightarrow 0\ .$$

(ii) Immediate from (i) and 23.7 (iv). □

Example 23.10 Let W be a connected finite CW complex with a map $c : W \longrightarrow S^1$ such that

$$c_* \;=\; \text{projection} \;:\; \pi_1(W) \;=\; \pi\times\mathbb{Z} \longrightarrow \pi_1(S^1) \;=\; \mathbb{Z}\ ,$$

and let $\widetilde{W}$ be the universal cover of W. The infinite cyclic cover $\overline{W} = c^*\mathbb{R} = \widetilde{W}/\pi$ of W is finitely dominated (i.e. (W,c) is a band) if and only if the finite f.g. free $\mathbb{Z}[\pi][z,z^{-1}]$-module chain complex $C(\widetilde{W})$ is $\mathbb{Z}[\pi]$-finitely dominated (6.8 (i)). Thus by 23.9 (W,c) is a band if and only if the $\mathbb{Z}[\pi][z,z^{-1}]$-module morphisms

$$\partial \;:\; H^{lf,\pi}_{*+1}(\widetilde{W}) \;=\; H_{*+1}(C(\widetilde{W})^{lf}) \longrightarrow H_*(\widetilde{W}) \;=\; H_*(C(\widetilde{W}))$$

are isomorphisms. In particular, this gives an algebraic proof of the isomorphism $H^{lf,\pi}_{*+1}(\widetilde{W}) \cong H_*(\widetilde{W})$ for a band (W,c) obtained geometrically in 15.7 (i). □

Let C be an $A[z,z^{-1}]$-module chain complex band with a Mayer–Vietoris presentation (C^+,C^-), and let

$$D \;=\; C^+\cap C^-\ ,\ E \;=\; C^+\cap\zeta C^-\ .$$

Write the inclusions as

$$f^+ \;:\; D \longrightarrow E\ ;\ x \longrightarrow x\ \ ,\ \ f^- \;:\; D \longrightarrow E\ ;\ y \longrightarrow \zeta y\ ,$$

so that there are defined exact sequences

$$0 \longrightarrow D[z,z^{-1}] \xrightarrow{f^+ - z^{-1}f^-} E[z,z^{-1}] \longrightarrow C \longrightarrow 0 ,$$
$$0 \longrightarrow zD[z] \xrightarrow{f^+ - z^{-1}f^-} E[z] \longrightarrow C^+ \longrightarrow 0 ,$$
$$0 \longrightarrow z^{-1}D[z^{-1}] \xrightarrow{f^+ - z^{-1}f^-} z^{-1}E[z^{-1}] \longrightarrow C^- \longrightarrow 0 .$$

Proposition 23.11 (i) *The A-module chain maps*

$$u : D \longrightarrow C \oplus C^{+,lf} \oplus C^{-,lf} ; \ x \longrightarrow (x,x,x) ,$$
$$v : E \longrightarrow C \oplus C^{+,lf} \oplus C^{-,lf} ; \ y \longrightarrow (y,y,\zeta^{-1}y)$$

are chain equivalences, such that there are defined commutative squares

$$\begin{array}{ccc} D & \xrightarrow{\quad f^+ \quad} & E \\ u \big\downarrow \simeq & & \simeq \big\downarrow v \\ C \oplus C^{+,lf} \oplus C^{-,lf} & \xrightarrow{1 \oplus 1 \oplus \zeta^{-,lf}} & C \oplus C^{+,lf} \oplus C^{-,lf} \end{array}$$

$$\begin{array}{ccc} D & \xrightarrow{\quad f^- \quad} & E \\ u \big\downarrow \simeq & & \simeq \big\downarrow v \\ C \oplus C^{+,lf} \oplus C^{-,lf} & \xrightarrow{\zeta \oplus \zeta^{+,lf} \oplus 1} & C \oplus C^{+,lf} \oplus C^{-,lf} \end{array}$$

(ii) *The fibring obstructions of C are such that*

$$\begin{aligned} \Phi^+(C) &= (\phi^+, -[C^-], -[C^{+,lf},\zeta^{+,lf}], -[C^{-,lf},\zeta^{-,lf}]) , \\ \Phi^-(C) &= (\phi^-, [C^+], -[C^{+,lf},\zeta^{+,lf}], -[C^{-,lf},\zeta^{-,lf}]) \\ &\in Wh(A[z,z^{-1}]) \ = \ Wh(A) \oplus \widetilde{K}_0(A) \oplus \widetilde{\mathrm{Nil}}_0(A) \oplus \widetilde{\mathrm{Nil}}_0(A) \end{aligned}$$

with

$$\phi^+ = \tau(v^{-1}(\zeta \oplus 1 \oplus 1)u : D \longrightarrow E) \ , \ \phi^- = \tau(v^{-1}u : D \longrightarrow E) \in Wh(A)$$

such that $\phi^+ - \phi^- = \tau(\zeta : C \longrightarrow C)$.

Proof (i) The A-module chain maps u, v are chain equivalences since the natural A-module chain maps $C^+ \longrightarrow C^{+,lf} \oplus C$, $C^- \longrightarrow C^{-,lf} \oplus C$ are chain equivalences (by 23.7 (v)) and there are defined chain homotopy squares

$$\begin{array}{ccc} D & \longrightarrow & C^+ \\ \big\downarrow & & \big\downarrow j^+ \\ C^- & \xrightarrow{\ j^-\ } & C \end{array} \qquad \begin{array}{ccc} E & \longrightarrow & C^+ \\ \big\downarrow & & \big\downarrow j^+ \\ C^- & \xrightarrow{\ \zeta j^-\ } & C \end{array}$$

with $j^+ : C^+ \longrightarrow C$, $j^- : C^- \longrightarrow C$ the inclusions.

(ii) There are defined chain equivalences

$$(C/C^+, \zeta) \simeq S(C^{+,lf}, \zeta^{+,lf}) , \ (C/C^-, \zeta^{-1}) \simeq S(C^{-,lf}, \zeta^{-,lf}) ,$$

so that

$$[C/C^+, \zeta] = -[C^{+,lf}, \zeta^{+,lf}] , \ [C/C^-, \zeta^{-1}] = -[C^{-,lf}, \zeta^{-,lf}] \in \widetilde{\text{Nil}}_0(A) .$$

Combining this with (i) and 22.5 gives the expressions for the fibring obstructions $\Phi^+(C), \Phi^-(C) \in Wh(A[z, z^{-1}])$. □

Definition 23.12 Let C^+ be a finite f.g. free $A[z]$-module chain complex.

(i) C^+ is *forward tame* if the A-module chain map

$$\zeta^{+,lf} : C^{+,lf} \longrightarrow C^{+,lf} ; x \longrightarrow zx$$

is chain homotopy nilpotent, that is $(\zeta^{+,lf})^k \simeq 0$ for some $k \geq 0$.

(ii) C^+ is *forward collared* if it is forward tame and $C^{+,lf}$ is chain homotopy A-finite, i.e. A-module chain equivalent to a finite f.g. free A-module chain complex. □

Similarly for an $A[z^{-1}]$-module chain complex C^-, with

$$\zeta^{-,lf} : C^{-,lf} \longrightarrow C^{-,lf} ; x \longrightarrow z^{-1}x .$$

Proposition 23.13 (Ranicki [130]) (i) *A finite f.g. free $A[z]$-module chain complex C^+ is such that*

$$H_*(A[z, z^{-1}] \otimes_{A[z]} C^+) = 0$$

if and only if C^+ is A-finitely dominated and the A-module chain map $\zeta^+ : C^+ \longrightarrow C^+; x \longrightarrow zx$ is chain homotopy nilpotent.

(ii) *A finite f.g. free $A[[z]]$-module chain complex B is such that*

$$H_*(A((z)) \otimes_{A[[z]]} B) = 0$$

if and only if B is A-finitely dominated and the A-module chain map $\zeta : B \longrightarrow B; x \longrightarrow zx$ is chain homotopy nilpotent. □

Definition 23.14 Let C^+ be a finite f.g. free $A[z]$-module chain complex. A *cofinite neighbourhood (of infinity)* $D \subseteq C^+$ is a f.g. free A-module subcomplex such that $z^k C^+ \subseteq D$ for some $k \geq 0$. □

Proposition 23.15 *Let C^+ be a finite f.g. free $A[z]$-module chain complex.*
(i) *The following conditions are equivalent:*

(a) *C^+ is forward tame,*
(b) $H_*(A((z)) \otimes_{A[z]} C^+) = 0$,
(c) *the inclusion $j : D \longrightarrow C^+$ of a cofinite neighbourhood $D \subseteq C^+$ is such that $j^{lf} \simeq 0 : D^{lf} \longrightarrow C^{+,lf}$,*
(d) *the inclusions*

$$\begin{aligned} i^+ &: C^+ \longrightarrow C^{+,lf} = A[[z]] \otimes_{A[z]} C^+ \ ; \ x \longrightarrow 1 \otimes x \ , \\ q^+ &: C^+ \longrightarrow C = A[z,z^{-1}] \otimes_{A[z]} C^+ \ ; \ x \longrightarrow 1 \otimes x \end{aligned}$$

are the components of a homology equivalence

$$\begin{pmatrix} i^+ \\ q^+ \end{pmatrix} : C^+ \xrightarrow{\simeq} C^{+,lf} \oplus C \ ,$$

(e) *the composite A-module chain map*

$$e(C^+) = \mathcal{C}(i^+ : C^+ \longrightarrow C^{+,lf})_{*+1} \longrightarrow C^+ \xrightarrow{q^+} C$$

is a homology equivalence.

(ii) *If C^+ is forward tame then $C^{+,lf}$ is A-finitely dominated, and the nilpotent class of $(C^{+,lf}, \zeta^{+,lf})$ is such that*

$$[C^{+,lf}, \zeta^{+,lf}] = -[C/C^+, \zeta] = -[C^-/(C^+ \cap C^-), \zeta] \in \mathrm{Nil}_0(A) \ .$$

(iii) *C^+ is forward collared if and only if $H_*(A((z)) \otimes_{A[z]} C^+) = 0$ and*

$$[C^{+,lf}] = 0 \in \widetilde{K}_0(A) \ .$$

(iv) *If there exists a cofinite neighbourhood $D \subseteq C^+$ such that $D^{lf} \simeq 0$ then C^+ is forward collared.*

Proof (i) (a) $\Longrightarrow$ (b) For any $k \geq 0$ there is defined an exact sequence of A-module chain complexes

$$0 \longrightarrow C^{+,lf} \xrightarrow{(\zeta^{+,lf})^k} C^{+,lf} \longrightarrow C^+/z^k C^+ \longrightarrow 0$$

which is split in each degree, so that there is defined an A-module chain equivalence

$$C^+/z^k C^+ \simeq \mathcal{C}((\zeta^{+,lf})^k : C^{+,lf} \longrightarrow C^{+,lf}) \ .$$

Moreover, C^+/z^kC^+ is a finite f.g. free A-module chain complex. If $k \geq 0$ is so large that $(\zeta^{+,lf})^k \simeq 0 : C^{+,lf} \longrightarrow C^{+,lf}$ there is defined an A-module chain equivalence

$$C^+/z^kC^+ \simeq C^{+,lf} \oplus SC^{+,lf} .$$

Thus $C^{+,lf}$ is A-finitely dominated, and by 23.13 (ii)

$$H_*(A((z)) \otimes_{A[z]} C^+) = H_*(A((z)) \otimes_{A[[z]]} C^{+,lf}) = 0 .$$

(b) $\Longrightarrow$ (a) Immediate from 23.13 (ii).

(a) $\Longrightarrow$ (c) If $(\zeta^{+,lf})^k \simeq 0 : C^{+,lf} \longrightarrow C^{+,lf}$ the inclusion $j : D \longrightarrow C^+$ of the cofinite neighbourhood $D = z^kC^+ \subseteq C^+$ is such that $j^{lf} \simeq 0 : D^{lf} \longrightarrow C^{+,lf}$.

(c) $\Longrightarrow$ (a) For any cofinite neighbourhood $D \subseteq C^+$ let $k \geq 0$ be so large that $z^kC^+ \subseteq D$, in which case there are factorizations

$$(\zeta^+)^k : C^+ \longrightarrow D \xrightarrow{j} C^+ ,$$

$$(\zeta^{+,lf})^k : C^{+,lf} \longrightarrow D^{lf} \xrightarrow{j^{lf}} C^{+,lf} .$$

If $j^{lf} \simeq 0$ then $(\zeta^{+,lf})^k \simeq 0 : C^{+,lf} \longrightarrow C^{+,lf}$.

(b) $\Longleftrightarrow$ (d) Immediate from 23.7 (ii).

(d) $\Longleftrightarrow$ (e) Trivial.

(ii) From (i) and 23.13 (ii) with $B = C^{+,lf}$, noting that the proof of 23.11 (ii) and excision give A-module chain equivalences

$$S(C^{+,lf}, \zeta^{+,lf}) \simeq (C/C^+, \zeta) \simeq (C^-/(C^+ \cap C^-), \zeta) .$$

(iii) Immediate from (i).

(iv) Apply (i) and (iii), noting that $j^{lf} \simeq 0 : D^{lf} \simeq 0 \longrightarrow C^{+,lf}$ and $[C^{+,lf}] = [D^{lf}] = 0 \in \widetilde{K}_0(A)$. □

Similarly for an $A[z^{-1}]$-module chain complex:

Proposition 23.15′ *Let C^- be a finite f.g. free $A[z^{-1}]$-module chain complex.*

(i) *The following conditions are equivalent:*

(a) *C^- is forward tame,*

(b) $H_*(A((z^{-1})) \otimes_{A[z^{-1}]} C^-) = 0$,

(c) *the inclusion $j : D \longrightarrow C^-$ of a cofinite neighbourhood $D \subseteq C^-$ is such that $j^{lf} \simeq 0 : D^{lf} \longrightarrow C^{-,lf}$,*

(d) *the inclusions*

$$i^- : C^- \longrightarrow C^{-,lf} = A[[z^{-1}]] \otimes_{A[z^{-1}]} C^- ; \ x \longrightarrow 1 \otimes x ,$$

$$q^- : C^- \longrightarrow C = A[z, z^{-1}] \otimes_{A[z^{-1}]} C^- ; \ x \longrightarrow 1 \otimes x$$

are the components of a homology equivalence

$$\begin{pmatrix} i^- \\ q^- \end{pmatrix} : C^- \xrightarrow{\simeq} C^{-,lf} \oplus C ,$$

(e) *the composite A-module chain map*

$$e(C^-) = \mathcal{C}(i^- : C^- \longrightarrow C^{-,lf})_{*+1} \longrightarrow C^- \xrightarrow{q^-} C$$

is a homology equivalence.

(ii) *If C^- is forward tame then $C^{-,lf}$ is A-finitely dominated, and the nilpotent class of $(C^{-,lf},\zeta^{-,lf})$ is such that*

$$[C^{-,lf},\zeta^{-,lf}] = -[C/C^-,\zeta^{-1}] = -[C^+/(C^+\cap C^-),\zeta^{-1}] \in \mathrm{Nil}_0(A) .$$

(iii) *C^- is forward collared if and only if $H_*(A((z^{-1})\otimes_{A[z^{-1}]}C^-) = 0$ and*

$$[C^{-,lf}] = 0 \in \widetilde{K}_0(A) .$$

(iv) *If there exists a cofinite neighbourhood $D \subseteq C^-$ such that $D^{lf} \simeq 0$ then C^- is forward collared.* □

In the following three propositions it is assumed that C is a finite based f.g. free $A[z,z^{-1}]$-module chain complex with a Mayer–Vietoris presentation (C^+,C^-).

Proposition 23.16 *The following conditions are equivalent:*

(i) *C^+ is forward tame,*
(ii) *C^- is reverse tame,*
(iii) *$H_*(A((z))\otimes_{A[z,z^{-1}]}C) = 0$,*
(iv) *the natural A-module chain map $e(C^+)\longrightarrow C$ is a homology equivalence,*
(v) *the A-module chain map*

$$C^+ \longrightarrow C\oplus C^{+,lf} \; ; \; x \longrightarrow (x,x)$$

is a homology equivalence.

If these conditions are satisfied

$$[C^{+,lf}] = [C^+]-[C] = [C^+\cap C^-]-[C^-] \in K_0(A) ,$$
$$[C^{+,lf},\zeta^{+,lf}] = -[C/C^+,\zeta] = -[C^-/(C^+\cap C^-),\zeta] \in \mathrm{Nil}_0(A) .$$

Proof Combine 23.7 and 23.15. □

Reversing the role of C^+ and C^- gives:

Proposition 23.16′ *The following conditions are equivalent:*

(i) C^- *is forward tame,*
(ii) C^+ *is reverse tame,*
(iii) $H_*(A((z^{-1}))\otimes_{A[z,z^{-1}]}C)=0$,
(iv) *the natural A-module chain map* $e(C^-)\longrightarrow C$ *is a homology equivalence,*
(v) *the A-module chain map*

$$C^- \longrightarrow C\oplus C^{-,lf} \ ;\ x\longrightarrow (x,x)$$

is a homology equivalence.

If these conditions are satisfied

$$[C^{-,lf}] = [C^+]-[C] = [C^+\cap C^-]-[C^-]\in K_0(A)\ ,$$
$$[C^{-,lf},\zeta^{-,lf}] = -[C/C^-,\zeta^{-1}] = -[C^+/(C^+\cap C^-),\zeta^{-1}]\in \mathrm{Nil}_0(A)\ .\ \square$$

Together, 23.16 and 23.16′ give:

Proposition 23.17 *The following conditions are equivalent:*

(i) C *is a chain complex band,*
(ii) C^+ *is forward and reverse tame,*
(iii) $H_*(A((z))\otimes_{A[z]}C^+)=H_*(A((z^{-1}))\otimes_{A[z]}C^+)=0$,
(iv) *the natural A-module chain maps* $e(C^+)\longrightarrow C$, $e(C^-)\longrightarrow C$ *are both homology equivalences.*

If these conditions are satisfied

$$\begin{aligned}[C] &= [C^+]+[C^-]-[C^+\cap C^-]\\ &= [C^+]-[C^{+,lf}] = [C^-]-[C^{-,lf}]\in K_0(A)\end{aligned}$$

and

$$[C^{+,lf},\zeta^{+,lf}] = -[C/C^+,\zeta] = -[C^-/(C^+\cap C^-),\zeta]\ ,$$
$$[C^{-,lf},\zeta^{-,lf}] = -[C/C^-,\zeta^{-1}] = -[C^+/(C^+\cap C^-),\zeta^{-1}]\in \mathrm{Nil}_0(A)\ .$$

$\square$

Example 23.18 Let

$$C = T(h) = \mathcal{C}(1-zh : D[z,z^{-1}]\longrightarrow D[z,z^{-1}])$$

be the algebraic mapping torus of a chain map $h : D \longrightarrow D$ for a finite based f.g. free A-module chain complex D. The finite f.g. free $A[z]$-module chain complex

$$C^+ = \mathcal{C}(1 - zh : D[z] \longrightarrow D[z])$$

is forward collared, and (equivalently) the finite f.g. free $A[z^{-1}]$-module chain complex

$$C^- = \mathcal{C}(1 - zh : z^{-1}D[z^{-1}] \longrightarrow D[z^{-1}])$$

is reverse collared, with A-module chain equivalences

$$C^{+,lf} \simeq 0 \ , \ C^- \simeq D \ .$$

In general, C^+ is not reverse tame and (equivalently) C^- is not forward tame. See 23.25 for an explicit example of such non-tameness. If $h : D \longrightarrow D$ is a chain equivalence then C is a chain complex band, with C^+ reverse collared, C^- forward collared and $C \simeq D$ A-module chain homotopy finite. □

Example 23.19 Let A be an integral domain, and let

$$p(z) = \sum_{j=m}^{n} a_j z^j \in A[z] \ \ (a_j \in A)$$

be a polynomial over A, with $a_m, a_n \neq 0 \in A$.

(i) The polynomial $p(z)$ is a unit in $A((z))$ (resp. $A((z^{-1}))$) if and only if a_m (resp. a_n) is a unit in A.

(ii) The 1-dimensional f.g. free $A[z]$-module chain complex

$$C^+ = \mathcal{C}(p(z) : A[z] \longrightarrow A[z])$$

is forward (resp. reverse) tame if and only if a_m (resp. a_n) is a unit in A.

(iii) The 1-dimensional based f.g. free $A[z, z^{-1}]$-module chain complex

$$C = \mathcal{C}(p(z) : A[z, z^{-1}] \longrightarrow A[z, z^{-1}])$$

is a band if and only if a_m, a_n are both units in A. □

Example 23.20 For any ring A and central non-zero divisor $s \in A$ the localization of A inverting s and the s-adic completion of A (2.28) are such that

$$\begin{aligned} A[1/s] &= A[z]/(1 - zs) \ , \\ \widehat{A}_s &= \varprojlim_k (A/s^k A) = A[[z]]/(z - s) \ . \end{aligned}$$

In fact, $A[1/s] = A[z]/(1 - zs)$ is just a restatement of the identification of 2.28 (i)

$$A[1/s] = \varinjlim(A \xrightarrow{s} A \xrightarrow{s} A \xrightarrow{s} A \longrightarrow \ldots) \ .$$

The localization of the completion $\widehat{A}_s[1/s]$ fits into the cartesian square of rings

$$\begin{array}{ccc} A & \longrightarrow & A[1/s] \\ \big\downarrow & & \big\downarrow \\ \widehat{A}_s & \longrightarrow & \widehat{A}_s[1/s] \end{array}$$

(See Ranicki [119] for the algebraic K- and L-theory of such squares). The 1-dimensional chain complexes

$$\begin{aligned} C^+ &= \mathcal{C}(1 - zs : A[z] \longrightarrow A[z]) \ , \\ C^- &= \mathcal{C}(1 - zs : z^{-1}A[z^{-1}] \longrightarrow A[z^{-1}]) \\ &\simeq \mathcal{C}(z^{-1} - s : A[z^{-1}] \longrightarrow A[z^{-1}]) \ , \\ C &= \mathcal{C}(1 - zs : A[z, z^{-1}] \longrightarrow A[z, z^{-1}]) \\ &= A[z, z^{-1}] \otimes_{A[z]} C^+ \ = \ A[z, z^{-1}] \otimes_{A[z^{-1}]} C^- \end{aligned}$$

define a Mayer–Vietoris presentation (C^+, C^-) of C such that

$$\begin{aligned} &H_0(C^+) \ = \ H_0(C) \ = \ A[1/s] \ , \\ &H_0(C^{+,lf}) \ = \ H_0(A[[z]] \otimes_{A[z]} C^+) \ = \ 0 \ , \\ &H_0(C^-) \ = \ A \ , \\ &H_0(C^{-,lf}) \ = \ H_0(A[[z^{-1}]] \otimes_{A[z^{-1}]} C^-) \ = \ \widehat{A}_s \ , \\ &H_0(A((z)) \otimes_{A[z]} C^+) \ = \ H_0(A((z)) \otimes_{A[z^{-1}]} C^-) \ = \ 0 \ , \\ &H_0(A((z^{-1})) \otimes_{A[z]} C^+) \ = \ H_0(A((z^{-1})) \otimes_{A[z^{-1}]} C^-) \ = \ \widehat{A}_s[1/s] \ . \end{aligned}$$

Thus C^+ is forward tame and C^- is reverse tame. The following conditions are equivalent:

(i) C^+ is reverse tame,
(ii) C^- is forward tame,
(iii) C is a band,
(iv) $s \in A$ is a unit,
(v) $\widehat{A}_s = 0$. □

Remark 23.21 Note that for $s = z \in A = B[z]$ the localization–completion square in 23.20 is just the localization–completion square in 23.7 (i)

$$\begin{array}{ccc} B[z] & \longrightarrow & B[z,z^{-1}] \\ \downarrow & & \downarrow \\ B[[z]] & \longrightarrow & B((z)) \end{array}$$

□

Given a connected CW complex W and a $\mathbb{Z}[\pi_1(W)]$-module Λ the Λ-*coefficient homology* of W is defined as usual by

$$H_*(W;\Lambda) \;=\; H_*(\Lambda \otimes_{\mathbb{Z}[\pi_1(W)]} C(\widetilde{W}))\ ,$$

with $C(\widetilde{W})$ the cellular $\mathbb{Z}[\pi_1(W)]$-module chain complex of the universal cover $\widetilde{W}$.

Proposition 23.22 *Let W be a connected finite CW complex with $\pi_1(W) = \pi \times \mathbb{Z}$, so that $\overline{W} = \widetilde{W}/\pi$ is an infinite cyclic cover of W with $\zeta_* = 1 : \pi_1(\overline{W}) = \pi \longrightarrow \pi$. Given a fundamental domain $(V;U,\zeta U)$ for $\overline{W}$ let*

$$\overline{W}^+ \;=\; \bigcup_{j=0}^{\infty} \zeta^j V\ ,\quad \overline{W}^- \;=\; \bigcup_{j=-\infty}^{-1} \zeta^j V \subseteq \overline{W}\ .$$

(i) *The following conditions are equivalent :*

(a) *the CW complex $\overline{W}^+$ is forward tame,*
(b) *the CW complex $\overline{W}^-$ is reverse tame,*
(c) *the natural map $e(\overline{W}^+) \longrightarrow \overline{W}$ is a homotopy equivalence,*
(d) *the finite f.g. free $\mathbb{Z}[\pi][z]$-module chain complex $C(\widetilde{W}^+)$ is forward tame,*
(e) *the finite f.g. free $\mathbb{Z}[\pi][z^{-1}]$-module chain complex $C(\widetilde{W}^-)$ is reverse tame,*
(f) $H_*(W;\mathbb{Z}[\pi]((z))) = 0$.

If these conditions are satisfied

$$[\overline{W}^+]^{lf} \;=\; -[\overline{W}^-] \in \widetilde{K}_0(\mathbb{Z}[\pi_1(\overline{W})])\ .$$

Similarly with the role of $\overline{W}^+, \overline{W}^-$ reversed.

(ii) *The following conditions are equivalent :*

(a) *$\overline{W}$ is infinite simple homotopy equivalent to an infinite cyclic cover $\overline{X}$ of a finite CW complex X with $\overline{X}^+$ forward collared,*

(b) $\overline{W}$ *is infinite simple homotopy equivalent to an infinite cyclic cover* $\overline{X}$ *of a finite CW complex* X *with* $\overline{X}^-$ *reverse collared,*

(c) *the finite f.g. free* $\mathbb{Z}[\pi][z]$*-module chain complex* $C(\widetilde{W}^+)$ *is forward collared,*

(d) *the finite f.g. free* $\mathbb{Z}[\pi][z^{-1}]$*-module chain complex* $C(\widetilde{W}^-)$ *is reverse collared,*

(e) $H_*(W;\mathbb{Z}[\pi]((z))) = 0$ *and* $[\overline{W}^+]^{lf} = 0 \in \widetilde{K}_0(\mathbb{Z}[\pi_1(\overline{W})])$,

(f) $H_*(W;\mathbb{Z}[\pi]((z))) = 0$ *and* $[\overline{W}^-] = 0 \in \widetilde{K}_0(\mathbb{Z}[\pi_1(\overline{W})])$.

Similarly with the role of $\overline{W}^+, \overline{W}^-$ *reversed.*

(iii) *The following conditions are equivalent:*

(a) *the CW complex* $\overline{W}$ *is finitely dominated,*

(b) $C(\widetilde{W})$ *is a chain complex band,*

(c) $\overline{W}^+$ *is both forward and reverse tame,*

(d) $\overline{W}^-$ *is both forward and reverse tame,*

(e) $H_*(W;\mathbb{Z}[\pi]((z))) = H_*(W;\mathbb{Z}[\pi]((z^{-1}))) = 0$.

Proof Combine 13.15 and 23.15, 23.16, 23.17. □

Example 23.23 Let $W = T(h)$ be the mapping torus of a self map $h : K \longrightarrow K$ of a connected finite CW complex K. Let $\overline{W} = \overline{T}(h)$ be the canonical infinite cyclic cover of W, with classifying map $c : W \longrightarrow S^1$, and define $\overline{W}^+ = \overline{T}^+(h)$, $\overline{W}^- = \overline{T}^-(h)$ as in 14.6 (vii). Then $\overline{W}^+$ is forward collared and $\overline{W}^-$ is reverse collared, but in general $\overline{W}^+$ is not reverse tame and $\overline{W}^-$ is not forward tame (cf. 23.25 below). If $h : K \longrightarrow K$ is a homotopy equivalence then (W, c) is a CW band, with $\overline{W}^+$ reverse collared, $\overline{W}^-$ forward collared and $\overline{W} \simeq K$ homotopy finite. □

In general, $\overline{W}^+$ and $\overline{W}^-$ need be neither forward nor reverse tame, and $\overline{W}$ need not be finitely dominated:

Example 23.24 Let $W = S^1 \vee S^1$, the figure 8 space:

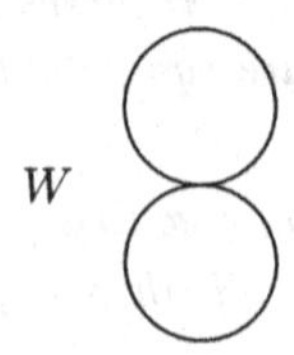

and let $\overline{W} = c^*\mathbb{R}$ be the infinite cyclic cover of W classified by a projection $c : W \longrightarrow S^1$ collapsing one of the circles to the base point:

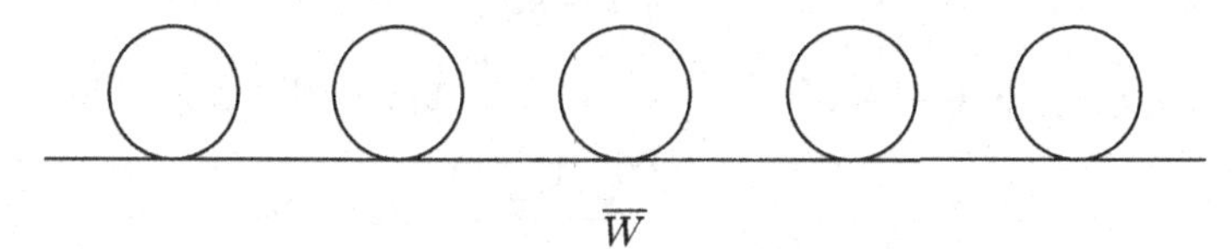

$\overline{W}$

The cellular $\mathbb{Z}[z, z^{-1}]$-module chain complex of $\overline{W}$ is given by

$$C(\overline{W}) \ : \ \dots \longrightarrow 0 \longrightarrow \mathbb{Z}[z,z^{-1}] \oplus \mathbb{Z}[z,z^{-1}] \xrightarrow{(1-z \ \ 0)} \mathbb{Z}[z,z^{-1}] \ .$$

The homology groups

$$H_1(W;\mathbb{Z}((z))) \ = \ \mathbb{Z}((z)) \ , \ \ H_1(W;\mathbb{Z}((z^{-1}))) \ = \ \mathbb{Z}((z^{-1}))$$

are non-zero, so that $\overline{W}^+$, $\overline{W}^-$ are neither forward nor reverse tame, and $\overline{W}$ is not finitely dominated. Actually, for the non-reverse-tameness and non-finite-domination in this case it is enough to note that the homology groups

$$H_1(\overline{W}^+) \ = \ \mathbb{Z}[z] \ , \ \ H_1(\overline{W}^-) \ = \ \mathbb{Z}[z^{-1}] \ , \ \ H_1(\overline{W}) \ = \ \mathbb{Z}[z,z^{-1}]$$

are not finitely generated (cf. 8.6). □

Example 23.25 Fix an integer $s \geq 2$. The mapping torus $T(s : S^1 \longrightarrow S^1)$ has fundamental group

$$\begin{aligned}\pi_1(T(s)) \ &= \ \mathbb{Z} *_s \mathbb{Z} \ = \ \{x, z \,|\, zxz^{-1} = x^s\} \\ &= \ \mathbb{Z}[1/s] \times_s \mathbb{Z} \ ,\end{aligned}$$

an extension of $\mathbb{Z}[1/s]$ by $\mathbb{Z}$. The canonical infinite cyclic cover $\overline{T}(s)$ is classified by the projection

$$\pi_1(T(s)) \longrightarrow \mathbb{Z} \ ; \ x \longrightarrow 0 \ , \ z \longrightarrow 1 \ ,$$

with

$$\pi_1(\overline{T}(s)) \ = \ \langle x \rangle \ = \ \ker(\pi_1(T(s)) \longrightarrow \mathbb{Z})$$

such that there is defined an isomorphism

$$\mathbb{Z}[1/s] \xrightarrow{\simeq} \pi_1(\overline{T}(s)) \ ; \ \sum_{k=0}^{\infty}(n_k/s^k) \longrightarrow \prod_{k=0}^{\infty} z^{-k}x^{n_k}z^k \ .$$

Note that $\pi_1(\overline{T}(s)) = \mathbb{Z}[1/s]$ is not finitely generated. The cellular $\mathbb{Z}[z, z^{-1}]$-

module chain complex of the infinite cyclic cover $\overline{T}(s)$ is

$$C(\overline{T}(s)) \ : \ 0 \longrightarrow \mathbb{Z}[z,z^{-1}] \xrightarrow{\begin{pmatrix} 1-sz \\ 0 \end{pmatrix}} \mathbb{Z}[z,z^{-1}] \oplus \mathbb{Z}[z,z^{-1}] \xrightarrow{(0 \ \ 1-z)} \mathbb{Z}[z,z^{-1}] \ ,$$

and

$$\begin{aligned} H_1(T(s)) &= \mathbb{Z}_{s-1} \oplus \mathbb{Z} \ , \ H_1(\overline{T}(s)) = \mathbb{Z}[1/s] \ , \\ H_1(T(s);\mathbb{Z}((z))) &= 0 \ , \ H_1(T(s);\mathbb{Z}((z^{-1}))) = \widehat{\mathbb{Z}}_s[1/s] = \widehat{\mathbb{Q}}_s \ , \\ H_1^{lf}(\overline{T}(s)) &= \widehat{\mathbb{Z}}_s \oplus \mathbb{Z} \ . \end{aligned}$$

The space $\overline{T}(s)$ is not finitely dominated, and is neither forward nor reverse tame. The positive half

$$\overline{T}^+(s) = \mathrm{Tel}(s) \simeq \underrightarrow{\mathrm{hocolim}}\,(S^1 \xrightarrow{s} S^1 \xrightarrow{s} S^1 \longrightarrow \dots)$$

is forward collared but not finitely dominated (and hence not reverse tame), with

$$\begin{aligned} \overline{T}^+(s) &\simeq \overline{T}(s) \ , \ H_1(\overline{T}^+(s)) = \mathbb{Z}[1/s] \ , \\ H_1^{lf}(\overline{T}^+(s)) &= 0 \ , \ e(\overline{T}^+(s)) \simeq \overline{T}(s) \ . \end{aligned}$$

The negative half $\overline{T}^-(s)$ is the mapping cotelescope $\mathcal{W}(s)$ (2.16) – it is homotopy equivalent (but not proper homotopy equivalent) to S^1, and it is reverse collared but not forward tame, with

$$\begin{aligned} \overline{T}^-(s) &\simeq S^1 \ , \ H_1(\overline{T}^-(s)) = \mathbb{Z} \ , \ H_1^{lf}(\overline{T}^-(s)) = \widehat{\mathbb{Z}}_s \ , \\ e(\overline{T}^-(s)) &\simeq \underleftarrow{\mathrm{holim}}\,(S^1 \xleftarrow{s} S^1 \xleftarrow{s} S^1 \longleftarrow \dots) \\ &\simeq \underleftarrow{\lim}\,(S^1 \xleftarrow{s} S^1 \xleftarrow{s} S^1 \longleftarrow \dots) \\ &= \text{the } s\text{-adic solenoid } \widehat{S}^1_s \ (2.18) \ . \end{aligned}$$

See 23.28 for the homology of the s-adic solenoid. □

Example 23.26 For any finite f.g. free $A[z]$-module chain complex C^+ the subcomplexes $z^jC^+ \subseteq C^+$ ($j \geq 0$) define an inverse system of $A[z]$-module chain complexes

$$C^+ \supset zC^+ \supset z^2C^+ \supset \dots \supset z^jC^+ \supset z^{j+1}C^+ \supset \dots$$

with

$$\varprojlim_j z^j C^+ = \bigcap_{j=0}^{\infty} z^j C^+ = 0 \ , \quad \varprojlim_j{}^1 (C^+/z^j C^+) = 0 \ ,$$

$$\varprojlim_j (C^+/z^j C^+) = A[[z]] \otimes_{A[z]} C^+ = C^{+,lf} \ ,$$

$$\varprojlim_j{}^1 z^j C^+ = (A[[z]]/A[z]) \otimes_{A[z]} C^+ = C^{+,lf}/C^+ \simeq e(C^+)_{*-1} \ .$$

By 2.19 there are defined short exact sequences

$$0 \longrightarrow \varprojlim_j{}^1 H_{r+1}(C^+/z^j C^+) \longrightarrow H_r(C^{+,lf}) \longrightarrow \varprojlim_j H_r(C^+/z^j C^+) \longrightarrow 0 \ ,$$

$$0 \longrightarrow \varprojlim_j{}^1 H_{r+1}(z^j C^+) \longrightarrow H_r(e(C^+)) \longrightarrow \varprojlim_j H_r(z^j C^+) \longrightarrow 0 \ .$$

The A-modules

$$\begin{aligned} M_r &= \ker(1 - z^{-1}\zeta^+ : H_r(C^+)((z)) \longrightarrow H_r(C^+)((z))) \\ &= \mathrm{im}(H_{r+1}(A((z)) \otimes_{A[z]} C^+) \longrightarrow H_r(C^+)((z))) \ , \\ N_r &= \mathrm{coker}(1 - z^{-1}\zeta^+ : H_r(C^+)((z)) \longrightarrow H_r(C^+)((z))) \\ &= \mathrm{im}(H_r(C^+)((z)) \longrightarrow H_r(A((z)) \otimes_{A[z]} C^+)) \end{aligned}$$

are such that there are defined exact sequences of A-modules

$$0 \longrightarrow N_r \longrightarrow H_r(A((z)) \otimes_{A[z]} C^+) \longrightarrow M_{r-1} \longrightarrow 0 \ ,$$

$$0 \longrightarrow M_r \longrightarrow \varprojlim_j H_r(z^j C^+) \longrightarrow H_r(C) \longrightarrow N_r \longrightarrow \varprojlim_j{}^1 H_r(z^j C^+) \longrightarrow 0$$

with

$$C = A[z, z^{-1}] \otimes_{A[z]} C^+ = \mathrm{coker}(1 - z^{-1}\zeta^+ : C^+[z^{-1}] \longrightarrow C^+[z^{-1}]) \ ,$$

$$\varprojlim_j H_r(z^j C^+) = \ker(1 - z^{-1}\zeta^+ : H_r(C^+)[[z]] \longrightarrow H_r(C^+)[[z]]) \ ,$$

$$\varprojlim_j{}^1 H_r(z^j C^+) = \mathrm{coker}(1 - z^{-1}\zeta^+ : H_r(C^+)[[z]] \longrightarrow H_r(C^+)[[z]]) \ .$$

By 23.15 the following conditions are equivalent:

(i) C^+ is forward tame,
(ii) $H_*(A((z)) \otimes_{A[z]} C^+) = 0$,
(iii) $H_*(e(C^+)) = H_*(C)$,
(iv) $M_* = N_* = 0$,
(v) $H_*(C) = \varprojlim_j H_*(z^j C^+)$, $\varprojlim_j{}^1 H_*(z^j C^+) = 0$.

Condition (v) corresponds to

$$H_*(e(W)) = \varprojlim_j H_*(W_j) \ , \ \varprojlim_j{}^1 H_*(W_j) = 0$$

for a forward tame space W with a sequence of closed cocompact subspaces $W \supseteq W_0 \supseteq W_1 \supseteq \dots$ such that $\bigcap_j W_j = \emptyset$ (7.10 (ii)). □

Example 23.27 As in 23.20 let A be a ring with a central non-zero divisor $s \in A$. For the reverse collared 1-dimensional f.g. free $A[z]$-module chain complex C^+ defined by

$$d = z - s : C_1^+ = A[z] \longrightarrow C_0^+ = A[z]$$

the A-modules in 23.26 are

$$\begin{aligned}
&H_0(C^+) = A \ , \ H_0(C) = A[1/s] \ , \\
&H_0(C^{+,lf}) = \varprojlim_j H_0(C^+/z^jC^+) = \varprojlim_j (A/s^jA) = \widehat{A}_s \ , \\
&M_* = 0 \ , \ N_0 = \widehat{A}_s[1/s] \ , \ N_r = 0 \ (r \neq 0) \ , \\
&H_0(e(C^+)) = \varprojlim_j H_0(z^jC^+) = \varprojlim_j s^jA = \bigcap_{j=0}^{\infty} s^jA \ , \\
&H_{-1}(e(C^+)) = \varprojlim_j{}^1 H_0(z^jC^+) = \varprojlim_j{}^1 s^jA = \widehat{A}_s/A \ ,
\end{aligned}$$

so that C^+ is forward tame if and only if $s \in A$ is a unit. □

Example 23.28 For any integer $s \geq 2$ let $W = \overline{T}^-(s) = \mathcal{W}(s : S^1 \longrightarrow S^1)$ be the mapping cotelescope of 23.25. Let $p : W \longrightarrow [0, \infty)$ be the canonical proper map, so that the cofinite subcomplexes

$$W_j = p^{-1}[j, \infty) \subset W \ \ (j \geq 0)$$

define an inverse system

$$W = W_0 \supset W_1 \supset \dots \supset \bigcap_j W_j = \emptyset \ .$$

Each W_j is a copy of W, and the inclusions $W_j \longrightarrow W_{j-1}$ are given up to homotopy by

$$W_j \simeq S^1 \xrightarrow{s} W_{j-1} \simeq S^1 \ .$$

The cellular chain complex of W is given by

$$C(W) : \mathbb{Z}[z] \xrightarrow{\begin{pmatrix} z - s \\ 0 \end{pmatrix}} \mathbb{Z}[z] \oplus \mathbb{Z}[z] \xrightarrow{(0 \ \ z-1)} \mathbb{Z}[z]$$

with

$$C(W_j) = z^j C(W) \subseteq C(W) \quad (j \geq 0) .$$

The end space $e(W)$ is homotopy equivalent to the s-adic solenoid $\widehat{S}^1_s = \varprojlim(s : S^1 \longrightarrow S^1)$ (2.18). The exact sequences of 4.16

$$0 \longrightarrow \varprojlim_j{}^1 \pi_{r+1}(W_j) \longrightarrow \pi_r(e(W)) \longrightarrow \varprojlim_j \pi_r(W_j) \longrightarrow 0 ,$$

$$0 \longrightarrow \varprojlim_j{}^1 H_{r+1}(W_j) \longrightarrow H_r^\infty(W) \longrightarrow \varprojlim_j H_r(W_j) \longrightarrow 0$$

combined with 23.27 (with $A = \mathbb{Z}$) give the singular homology of the s-adic solenoid $\widehat{S}^1_s$ to be

$$H_r(\widehat{S}^1_s) = H_r(e(W)) = H_r^\infty(W) = \begin{cases} \mathbb{Z} \oplus (\widehat{\mathbb{Z}}_s/\mathbb{Z}) & \text{if } r = 0 , \\ 0 & \text{otherwise} . \end{cases}$$

($\widehat{S}^1_s$ is connected, but not path-connected.) Regard $\widehat{S}^1_s$ as the intersection

$$\widehat{S}^1_s = \bigcap_{j=0}^{\infty} T_j \subset \mathbb{R}^3$$

of solid tori

$$\mathbb{R}^3 \supset T_0 \supset T_1 \supset \ldots \supset T_{j-1} \supset T_j \supset \ldots \supset \widehat{S}^1_s ,$$

such that each inclusion $T_j \subset T_{j-1}$ induces $s : \pi_1(T_j) = \mathbb{Z} \longrightarrow \pi_1(T_{j-1}) = \mathbb{Z}$. Now $\widehat{S}^1_s$ is a compact metric space, so that the Steenrod homology groups $H^{st}_*(\widehat{S}^1_s)$ are defined – some of the properties of H^{st}_* were recalled at the end of Chapter 4. The short exact sequences of Milnor [95]

$$0 \longrightarrow \varprojlim_j{}^1 H^{st}_{*+1}(T_j) \longrightarrow H^{st}_*(\widehat{S}^1_s) \longrightarrow \varprojlim_j H^{st}_*(T_j) \longrightarrow 0$$

in the non-trivial dimensions $* = 0, 1$ involve the inverse and direct systems of the inverse systems of abelian groups

$$1 : H^{st}_0(T_j) = H_0(T_j) = \mathbb{Z} \longrightarrow H^{st}_0(T_{j-1}) = H_0(T_{j-1}) = \mathbb{Z} ,$$

$$s : H^{st}_1(T_j) = H_1(T_j) = \mathbb{Z} \longrightarrow H^{st}_1(T_{j-1}) = H_1(T_{j-1}) = \mathbb{Z}$$

so that the Steenrod homology of $\widehat{S}^1_s$ is the same as the singular homology

$$H^{st}_*(\widehat{S}^1_s) = H_*(\widehat{S}^1_s) . \qquad \square$$

24

Relaxation techniques

'Relaxation' is the name given by Siebenmann [145] to the idempotent map

$$\begin{aligned} Wh(A_\alpha[z,z^{-1}]) &= Wh(A,\alpha) \oplus \widetilde{\mathrm{Nil}}_0(A,\alpha) \oplus \widetilde{\mathrm{Nil}}_0(A,\alpha^{-1}) \\ &\longrightarrow Wh(A_\alpha[z,z^{-1}]) ; \\ w &= (x,y^+,y^-) \longrightarrow w' = (x,0,0) , \end{aligned}$$

which is defined for any ring A and automorphism $\alpha : A \longrightarrow A$. The geometric twist glueing construction of [145] associated to a manifold band (W,c) an h-cobordant 'relaxed' manifold band (W',c') with fibring obstruction

$$\Phi^+(W',c') = \Phi^+(W,c)' \in Wh(\pi_1(W)) ,$$

such that $((W')',(c')')$ is homeomorphic to (W',c'). We have already developed homotopy theoretic twist glueing in Chapter 19, defining the relaxation of a CW π_1-band (W,c) as the 1-twist glueing $(W',c') = (W[1],c[1])$. In Chapters 24–26 we shall develop the chain complex analogues of twist glueing and relaxation in the special case $\alpha = 1 : A \longrightarrow A$, with

$$Wh(A,1) = Wh(A) \oplus \widetilde{K}_0(A) \ , \ \widetilde{\mathrm{Nil}}_0(A,1) = \widetilde{\mathrm{Nil}}_0(A) .$$

Definition 24.1 A finite f.g. free $A[z]$-module chain complex C^+ is *relaxed* if it is forward tame and

$$[C^{+,lf},\zeta^{+,lf}] = 0 \in \widetilde{\mathrm{Nil}}_0(A) . \qquad \square$$

Proposition 24.2 *Let C^+ be a finite f.g. free $A[z]$-module chain complex.*
(i) *If*

$$\zeta^{+,lf} \simeq 0 : C^{+,lf} \longrightarrow C^{+,lf}$$

then C^+ is relaxed.

(ii) *If C^+ is forward tame the inclusion $q^+ : C^+ \longrightarrow C = A[z,z^{-1}] \otimes_{A[z]} C^+$ is a chain homotopy surjection, and if*

$$\zeta^+ \simeq p^+ \zeta q^+ : C^+ \longrightarrow C^+$$

for a chain homotopy injection $p^+ : C \longrightarrow C^+$ splitting q^+ then C^+ is relaxed.

Proof (i) It is clear that if $\zeta^{+,lf} \simeq 0 : C^{+,lf} \longrightarrow C^{+,lf}$ then C^+ is forward tame. Also

$$[C^{+,lf}, \zeta^{+,lf}] = [C^{+,lf}, 0] = 0 \in \widetilde{\mathrm{Nil}}_0(A) ,$$

so that C^+ is relaxed.

(ii) As C^+ is forward tame the inclusions

$$\begin{aligned} i^+ &: C^+ \longrightarrow C^{+,lf} = A[[z]] \otimes_{A[z]} C^+ , \\ q^+ &: C^+ \longrightarrow C = A[z,z^{-1}] \otimes_{A[z]} C^+ \end{aligned}$$

are the components of an A-module chain equivalence

$$\begin{pmatrix} i^+ \\ q^+ \end{pmatrix} : C^+ \xrightarrow{\simeq} C^{+,lf} \oplus C ,$$

by 23.15 (i). Let

$$j^+ : C^{+,lf} \longrightarrow C^+ , \quad p^+ : C \longrightarrow C^+$$

be the A-module chain maps which are the components of a chain homotopy inverse

$$\begin{pmatrix} i^+ \\ q^+ \end{pmatrix}^{-1} = (j^+ \ \ p^+) : C^{+,lf} \oplus C \xrightarrow{\simeq} C^+ ,$$

so that there is defined a map of A-module chain homotopy direct sum systems

$$\begin{array}{ccccc} C & \underset{q^+}{\overset{p^+}{\rightleftarrows}} & C^+ & \underset{j^+}{\overset{i^+}{\rightleftarrows}} & C^{+,lf} \\ \Big\downarrow \zeta & & \Big\downarrow \zeta^+ & & \Big\downarrow \zeta^{+,lf} \\ C & \underset{q^+}{\overset{p^+}{\rightleftarrows}} & C^+ & \underset{j^+}{\overset{i^+}{\rightleftarrows}} & C^{+,lf} \end{array}$$

If there exists an A-module chain homotopy $\zeta^+ \simeq p^+ \zeta q^+ : C^+ \longrightarrow C^+$ then $\zeta^{+,lf} \simeq 0 : C^{+,lf} \longrightarrow C^{+,lf}$, and C^+ is relaxed by (i). □

For any integer $q \geq 1$ define a ring morphism

$$q : A[z] \longrightarrow A[z] ; \ \sum_{j=0}^{\infty} a_j z^j \longrightarrow \sum_{j=0}^{\infty} a_j z^{jq} .$$

Definition 24.3 The *q-fold transfer*

$$q^! : \{A[z]\text{-modules}\} \longrightarrow \{A[z]\text{-modules}\} \; ; \; M \longrightarrow q^! M$$

is the induced functor: for any $A[z]$-module M the induced $A[z]$-module $q^! M$ has the same A-module structure as M but z acts by z^q. □

Example 24.4 If C^+ is a forward tame finite f.g. free $A[z]$-module chain complex and $q \geq 1$ is so large that $(\zeta^{+,lf})^q \simeq 0 : C^{+,lf} \longrightarrow C^{+,lf}$ then $q^! C^+$ is a finite f.g. free $A[z]$-module chain complex such that $\zeta^{+,lf} \simeq 0 : q^! C^{+,lf} \longrightarrow q^! C^{+,lf}$, so that $q^! C^+$ is relaxed by 24.2 (i). □

Definition 24.5 The *algebraic mapping telescope* of an A-module chain map $f : B \longrightarrow B$ is the $A[z]$-module chain complex

$$\mathrm{Tel}(f) \;=\; \mathcal{C}(1 - zf : B[z] \longrightarrow B[z]) \; .$$ □

Proposition 24.6 *If B is a finite f.g. free A-module chain complex and $f : B \longrightarrow B$ is a chain map, the algebraic mapping telescope* $\mathrm{Tel}(f)$ *is a relaxed finite f.g. free $A[z]$-module chain complex.*

Proof Immediate from

$$\mathrm{Tel}(f)^{lf} \;=\; \mathcal{C}(1 - zf : B[[z]] \longrightarrow B[[z]]) \;\simeq\; 0 \; .$$ □

Example 24.7 Let X be a connected finite CW complex with universal cover $\widetilde{X}$, and let $h : X \longrightarrow X$ be a cellular map such that $h_* = 1 : \pi_1(X) = \pi \longrightarrow \pi$. The mapping telescope of h

$$\mathrm{Tel}(h) \;=\; \Big(\coprod_{j=0}^{\infty} X \times I \times \{j\}\Big) \Big/ (x, 1, j) = (h(x), 0, j+1)$$

is an infinite CW complex with $\pi_1(\mathrm{Tel}(h)) = \pi$, such that the cellular chain complex of the universal cover $\widetilde{\mathrm{Tel}(h)}$ is the algebraic mapping telescope

$$C(\widetilde{\mathrm{Tel}(h)}) \;=\; \mathcal{C}(1 - z\widetilde{h} : C(\widetilde{X})[z] \longrightarrow C(\widetilde{X})[z]) \; ,$$

which is a relaxed finite f.g. free $\mathbb{Z}[\pi][z]$-module chain complex. □

Definition 24.1′ A finite f.g. free $A[z^{-1}]$-module chain complex C^- is *relaxed* if it is forward tame and

$$[C^{-,lf}, \zeta^{-,lf}] \;=\; 0 \in \widetilde{\mathrm{Nil}}_0(A) \; .$$

□

For any finite f.g. free $A[z]$-module chain complex C^+ there exists a finite f.g. free A-module subcomplex $E \subset C^+$ such that

$$\sum_{j=0}^{\infty} \zeta^j(E) = C^+$$

and

$$D = E \cap \zeta^{-1}(E) \subset C^+ \subset C = A[z,z^{-1}] \otimes_{A[z]} C^+$$

is a finite f.g. free A-module subcomplex, with the A-module chain maps

$$f^+ : D \longrightarrow E \; ; \; x \longrightarrow x \; ,$$
$$f^- : D \longrightarrow E \; ; \; y \longrightarrow \zeta y$$

such that there are defined exact sequences

$$0 \longrightarrow zD[z] \xrightarrow{f^+ - z^{-1}f^-} E[z] \longrightarrow C^+ \longrightarrow 0 \; ,$$
$$0 \longrightarrow D[z,z^{-1}] \xrightarrow{f^+ - z^{-1}f^-} E[z,z^{-1}] \longrightarrow C \longrightarrow 0 \; .$$

The $A[z^{-1}]$-module subcomplex

$$C^- = \sum_{j=\infty}^{-1} \zeta^j E \subset C$$

is such that (C^+, C^-) is a Mayer–Vietoris presentation of C.

Proposition 24.8 (i) *If C^+ is such that for some choice of $E \subset C^+$ the A-module chain map*

$$E/D \longrightarrow C^+/D$$

is a chain homotopy split surjection then C^+ is relaxed.

(ii) *If C^+ is such that for some choice of $E \subset C^+$ the A-module chain map*

$$\zeta^{-1}E/D \longrightarrow C^-/D$$

is a chain homotopy split surjection then C^- is relaxed.

Proof (i) We have that the inclusion

$$\zeta^{-1}C^+/C^+ = \zeta^{-1}E/D \longrightarrow C/C^+ = C^-/D$$

is a chain homotopy split surjection, so that C^- is reverse tame. By 23.16 C^+ is forward tame, and

$$[C^{+,lf}, \zeta^{+,lf}] = -[C/C^+, \zeta] \in \widetilde{\mathrm{Nil}}_0(A) \; .$$

It follows from the exact sequence

$$0 \longrightarrow \zeta^{-1}C^+/C^+ \longrightarrow C/C^+ \longrightarrow C/\zeta^{-1}C^+ \longrightarrow 0$$

that

$$\zeta \simeq 0 \;:\; C/C^+ \longrightarrow C/\zeta^{-1}C^+ \xrightarrow{\zeta} C/C^+ \;,$$

so that

$$[C/C^+,\zeta] \;=\; 0 \in \widetilde{\mathrm{Nil}}_0(A)$$

and C^+ is relaxed.

(ii) As for (i), using

$$[C^{-,lf},\zeta^{-,lf}] \;=\; -[C/C^+,\zeta] \in \widetilde{\mathrm{Nil}}_0(A) \;. \qquad \square$$

Example 24.9 Let W be a connected finite CW complex with a map $c : W \longrightarrow S^1$ such that

$$c_* \;=\; \text{projection} \;:\; \pi_1(W) \;=\; \pi \times \mathbb{Z} \longrightarrow \mathbb{Z} \;.$$

Let $(V;U,\zeta U)$ be a fundamental domain for the infinite cyclic cover $\overline{W} = c^*\mathbb{R}$ of W, and let $\widetilde{W}$ be the universal cover of W.

(i) If V dominates $\overline{W}^+ = \bigcup_{j=0}^{\infty} \zeta^j V$ rel U then $C(\widetilde{W}^+)$ is a relaxed finite f.g. free $\mathbb{Z}[\pi][z]$-module chain complex, by 24.8 (i). Thus if (W,c) is a positively relaxed CW band (15.14 (i)) then $C(\widetilde{W}^+)$ is relaxed.

(ii) If $\zeta^{-1}V$ dominates $\overline{W}^- = \bigcup_{j=-\infty}^{-1} \zeta^j V$ rel U then $C(\widetilde{W}^-)$ is a relaxed finite f.g. free $\mathbb{Z}[\pi][z^{-1}]$-module chain complex, by 24.8 (i)$'$. Thus if (W,c) is a negatively relaxed CW band (15.14 (ii)) then $C(\widetilde{W}^-)$ is relaxed. $\square$

Example 24.10 (i) The engulfing technique of Siebenmann [145] and Siebenmann, Guillou and Hähl [149] shows that for an n-dimensional manifold band (W,c) with $n \geq 6$ and $\pi_1(W) = \pi \times \mathbb{Z}$ the $\mathbb{Z}[\pi][z]$-module chain complex $C(\widetilde{W}^+)$ is relaxed if and only if there exists an isotopy

$$h_t^+ \;:\; \overline{W} \longrightarrow \overline{W} \;\;(0 \leq t \leq 1)$$

such that:

(a) $h_t^+(x) = x$ for $x \in \overline{W}^-$,
(b) $h_t^+(y) \in \overline{W}^+$ for $y \in \overline{W}^+$,
(c) $h_0^+ = \text{identity} : \overline{W} \longrightarrow \overline{W}$,
(d) $h_1^+(\overline{W}) \subseteq \zeta\overline{W}^-$.

If there exists such an isotopy $\{h_t^+\}$ the inclusion

$$g^+ \;:\; V \;=\; \overline{W}^+ \cap \zeta\overline{W}^- \longrightarrow \overline{W}^+$$

and the map

$$h_1^+|_{\overline{W}^+} \;:\; \overline{W}^+ \longrightarrow V$$

are such that there is defined a homotopy

$$h_t^+|_{\overline{W}^+} \ : \ \text{identity} \ \simeq \ g^+(h_1^+|_{\overline{W}^+}) \ : \ \overline{W}^+ \longrightarrow \overline{W}^+$$

rel $U = \overline{W}^+ \cap \overline{W}^-$, so that V dominates $\overline{W}^+$ rel U and $C(\widetilde{W}^+)$ is relaxed by 24.8 (i).

(i)′ As for (i), but reversing the role of the ends $\overline{W}^+, \overline{W}^-$. □

Proposition 24.11 *The following conditions on an $A[z,z^{-1}]$-module chain complex band C are equivalent:*

(i) *for any Mayer–Vietoris presentation (C^+, C^-) of C the $A[z]$-module chain complex C^+ and the $A[z^{-1}]$-module chain complex C^- are relaxed,*

(ii) *the $\widetilde{\text{Nil}}$-components of the fibring obstructions $\Phi^+(C), \Phi^-(C) \in Wh(A[z,z^{-1}])$ are 0, that is*

$$[C/C^+, \zeta] \ = \ [C/C^-, \zeta^{-1}] \ = \ 0 \in \widetilde{\text{Nil}}_0(A) \ . \quad \square$$

Definition 24.12 An $A[z,z^{-1}]$-module chain complex band C is *relaxed* if it satisfies the conditions of 24.11. □

Example 24.13 (i) If $f^+, f^- : D \longrightarrow E$ are chain equivalences of finite based f.g. free A-module chain complexes then

$$C \ = \ \mathcal{C}(f^+ - z^{-1}f^- : D[z,z^{-1}] \longrightarrow E[z,z^{-1}])$$

is a relaxed $A[z,z^{-1}]$-module chain complex band, with fibring obstructions

$$\begin{aligned} \Phi^+(C) \ &= \ (\tau(f^-), 0, 0, 0) \ , \ \ \Phi^-(C) \ = \ (\tau(f^+), 0, 0, 0) \\ &\in Wh(A[z,z^{-1}]) \ = \ Wh(A) \oplus \widetilde{K}_0(A) \oplus \widetilde{\text{Nil}}_0(A) \oplus \widetilde{\text{Nil}}_0(A) \end{aligned}$$

as in 22.7.

(ii) If C is a finitely dominated A-module chain complex and $h : C \longrightarrow C$ is a chain equivalence, then any based finite f.g. free $A[z,z^{-1}]$-module chain complex T in the canonical simple chain homotopy type of the algebraic mapping torus $T(h)$ is a relaxed $A[z,z^{-1}]$-module chain complex band with fibring obstructions

$$\begin{aligned} \Phi^+(T) \ &= \ (\tau(h), -[C], 0, 0) \ , \ \ \Phi^-(T) \ = \ (0, 0, 0, 0) \\ &\in Wh(A[z,z^{-1}]) \ = \ Wh(A) \oplus \widetilde{K}_0(A) \oplus \widetilde{\text{Nil}}_0(A) \oplus \widetilde{\text{Nil}}_0(A) \end{aligned}$$

as in 22.8. In particular, if $(D, f : C \longrightarrow D, g : D \longrightarrow C, gf \simeq 1 : C \longrightarrow C)$ is a finite domination of C then $T = T(fhg : D \longrightarrow D)$ is a relaxed $A[z,z^{-1}]$-module chain complex band in the canonical simple chain homotopy type of $T(h)$. □

Remark 24.14 The property of being relaxed is a chain homotopy invariant of a finite f.g. free $A[z]$-module chain complex C^+, and likewise for a finite f.g. free $A[z^{-1}]$-module chain complex C^-. The property of being relaxed is a simple chain homotopy invariant of an $A[z,z^{-1}]$-module chain complex band C, but it is not in general a chain homotopy invariant property. See Chapter 26 below for the construction of the 'relaxation' of an $A[z,z^{-1}]$-module chain complex band C, which is a relaxed $A[z,z^{-1}]$-module chain complex band C' in the chain homotopy type of C. □

Example 24.15 (i) The following conditions on an untwisted CW band (W,c) with $\pi_1(W)=\pi\times\mathbb{Z}$ are equivalent:

(a) W is simple homotopy equivalent to a CW π_1-band (also denoted by (W,c)) with a π_1-fundamental domain $(V;U,\zeta U)$ for $\overline{W}$ such that the inclusions

$$\begin{aligned} f^+ &: U \longrightarrow V\ ;\ x \longrightarrow x\ ,\\ f^- &: U \longrightarrow V\ ;\ y \longrightarrow \zeta y \end{aligned}$$

are homotopy equivalences,

(b) $\Phi^+(W)\ ,\ \Phi^-(W)\in \mathrm{im}(Wh(\pi)\longrightarrow Wh(\pi\times\mathbb{Z}))\ ,$

(c) W is relaxed and $\overline{W}^+,\overline{W}^-$ are homotopy finite.

If W satisfies these conditions the fibring obstructions are given by 24.13 (i) to be

$$\begin{aligned} \Phi^+(W) &= (\tau(f^-),0,0,0)\ ,\ \Phi^-(W)\ =\ (\tau(f^+),0,0,0)\\ &\in Wh(\pi\times\mathbb{Z})\ =\ Wh(\pi)\oplus\widetilde{K}_0(\mathbb{Z}[\pi])\oplus\widetilde{\mathrm{Nil}}_0(\mathbb{Z}[\pi])\oplus\widetilde{\mathrm{Nil}}_0(\mathbb{Z}[\pi])\ . \end{aligned}$$

(ii) A manifold band (W,c) *pseudo-fibres* over S^1 if $\overline{W}$ admits a (manifold) fundamental domain $(V;U,\zeta U)$ which is an h-cobordism. For $n\geq 6$ a manifold band (W,c) with $\pi_1(W)=\pi\times\mathbb{Z}$ pseudo-fibres if and only if the conditions in (i) are satisfied (cf. 24.16 below).

(iii) If $h:\overline{W}\longrightarrow\overline{W}$ is a homotopy equivalence such that $h_*=1:\pi\longrightarrow\pi$ then the fundamental domain $V(h)$ of the h-twist glueing $W(h)$ (19.8) is homotopy finite. Any choice of simple homotopy type on $V(h)$ determines a simple homotopy type on $W(h)$, with respect to which $(W(h),c(h))$ is a relaxed CW band with fibring obstructions of the type

$$\begin{aligned} \Phi^+(W(h)) &= (\phi^+(h),-[\overline{W}^-],0,0)\ ,\ \Phi^-(W(h))\ =\ (\phi^-(h),[\overline{W}^+],0,0)\\ &\in Wh(\pi\times\mathbb{Z})\ =\ Wh(\pi)\oplus\widetilde{K}_0(\mathbb{Z}[\pi])\oplus\widetilde{\mathrm{Nil}}_0(\mathbb{Z}[\pi])\oplus\widetilde{\mathrm{Nil}}_0(\mathbb{Z}[\pi]) \end{aligned}$$

for some $\phi^+(h),\phi^-(h)\in Wh(\pi)$ with $\phi^+(h)-\phi^-(h)=\tau(h)$. See Chapter 26 below for a more detailed account.

(iv) Let $h : X \longrightarrow X$ be a self homotopy equivalence of a finitely dominated CW complex X, such that $h_* = 1 : \pi_1(X) = \pi \longrightarrow \pi$. Any finite CW complex W in the canonical simple homotopy type of the mapping torus $T(h)$ determines a relaxed CW band (W, c) with

$$c_* = \text{projection} : \pi_1(W) = \pi_1(T(h)) = \pi \times \mathbb{Z} \longrightarrow \mathbb{Z} .$$

The fibring obstructions of such W are given by 24.13 (ii) to be

$$\begin{aligned}\Phi^+(W) &= (\tau(h), -[X], 0, 0) \ , \ \Phi^-(W) = (0,0,0,0) \\ &\in Wh(\pi \times \mathbb{Z}) = Wh(\pi) \oplus \widetilde{K}_0(\mathbb{Z}[\pi]) \oplus \widetilde{\mathrm{Nil}}_0(\mathbb{Z}[\pi]) \oplus \widetilde{\mathrm{Nil}}_0(\mathbb{Z}[\pi]) .\end{aligned}$$

If $\delta = (Y, f : X \longrightarrow Y, g : Y \longrightarrow X, gf \simeq 1 : X \longrightarrow X)$ is a finite domination of X then $(W = T(fhg : Y \longrightarrow Y), c)$ is such a relaxed CW band. In particular, this applies to $h = 1 : X \longrightarrow X$, showing that for any finitely dominated CW complex X every finite CW complex W in the canonical simple homotopy type of $X \times S^1$ (e.g. $W = T(fg : Y \longrightarrow Y)$ for any finite domination δ of X) determines a relaxed CW band (W, c) with the infinite cyclic cover $\overline{W}$ homotopy equivalent to X.

(v) If (W, c) is a CW band with a fundamental domain $(V; U, \zeta U)$ such that the inclusion

$$V \longrightarrow \overline{W}^+ = \bigcup_{j=0}^{\infty} \zeta^j V$$

is a homotopy surjection rel U, and the inclusion

$$V \longrightarrow \zeta \overline{W}^- = \bigcup_{j=-\infty}^{0} \zeta^j V$$

is a homotopy surjection rel ζU, then (W, c) is a relaxed CW band, by 24.13.

(vi) Let (W, c) be an untwisted n-dimensional manifold band with $n \geq 6$ and $\pi_1(W) = \pi \times \mathbb{Z}$, and let $C = C(\widetilde{W})$ be the corresponding $\mathbb{Z}[\pi][z, z^{-1}]$-module chain complex band. The geometric relaxation technique of Siebenmann [145] associates to (W, c) a relaxed manifold band (W', c') (= the ζ-twist glueing $(W[\zeta], c[\zeta])$ in the terminology of Chapter 19) such that $\overline{W} = \overline{W}'$, with a fundamental domain $(V'; U', \zeta' U')$ such that V' dominates $\overline{W}'^+$ rel U' and V' dominates $\zeta' \overline{W}'^-$ rel $\zeta' U'$ via isotopies $\{h_t'^+\}$, $\{h_t'^-\}$ of $\overline{W}$ as in 24.10. The manifold band (W, c) is a relaxed CW band if and only W is homeomorphic to the relaxation W'. The following conditions on an untwisted n-dimensional manifold band (W_1, c_1) are equivalent:

(a) there exists an h-cobordism $(M; W, W_1)$ with torsion one of the nilpotent class components of $\Phi^{\pm}(W,c) = \Phi^{\pm}(C)$,

$$\begin{aligned}\tau(M \longrightarrow W) &= (0,0,[C/C^+,\zeta],0) \\ \in Wh(\pi \times \mathbb{Z}) &= Wh(\pi) \oplus \widetilde{K}_0(\mathbb{Z}[\pi]) \oplus \widetilde{\mathrm{Nil}}_0(\mathbb{Z}[\pi]) \oplus \widetilde{\mathrm{Nil}}_0(\mathbb{Z}[\pi]) ,\end{aligned}$$

(b) (W_1, c_1) is homeomorphic to the relaxation (W', c'),

(c) (W_1, c_1) is h-cobordant to (W, c), and is a relaxed CW band. □

Example 24.16 Let $f : W \longrightarrow X \times S^1$ be a homotopy equivalence, with W a closed n-dimensional manifold and X a finite $(n-1)$-dimensional geometric Poincaré complex. Then (W, c) is an untwisted manifold band with

$$c : W \xrightarrow{f} X \times S^1 \xrightarrow{p} S^1 \quad (p = \text{proj.}) ,$$

and $(X \times S^1, p)$ is a geometric Poincaré band. Moreover, f induces a simple homotopy equivalence

$$T(\zeta : \overline{W} \longrightarrow \overline{W}) \simeq X \times S^1 ,$$

giving $T(\zeta)$ the canonical simple homotopy type. The fibring obstructions of W determine each other by Poincaré duality:

$$\Phi^+(W) = (-)^{n-1}\Phi^-(W)^* \in Wh(\pi \times \mathbb{Z}) .$$

Let $\pi_1(X) = \pi$, so that

$$\pi_1(W) = \pi_1(X \times S^1) = \pi \times \mathbb{Z}$$

and let $\widetilde{W}$ be the universal cover of W. The torsion of f agrees up to sign with one of the fibring obstructions of W:

$$\begin{aligned}\tau(f) &= -\tau(T(\zeta) \longrightarrow W) \\ &= -\Phi^+(W) = -\Phi^+(C) \in Wh(\pi \times \mathbb{Z}) ,\end{aligned}$$

where $C = C(\widetilde{W})$ is the $\mathbb{Z}[\pi][z, z^{-1}]$-module chain complex band. The image $[\tau(f)] \in \mathrm{coker}(Wh(\pi) \longrightarrow Wh(\pi \times \mathbb{Z}))$ is given by

$$\begin{aligned}[\tau(f)] &= -[\Phi^+(W)] = -[\Phi^+(C)] \\ &= ([C^-], -[C/C^+,\zeta], -[C/C^-,\zeta^{-1}]) \\ &\in \widetilde{K}_0(\mathbb{Z}[\pi]) \oplus \widetilde{\mathrm{Nil}}_0(\mathbb{Z}[\pi]) \oplus \widetilde{\mathrm{Nil}}_0(\mathbb{Z}[\pi]) ,\end{aligned}$$

and the $\widetilde{\mathrm{Nil}}_0$-components are $\pm$-dual to each other by Poincaré duality:

$$[C/C^+,\zeta] = (-)^n[C/C^-,\zeta^{-1}]^* \in \widetilde{\mathrm{Nil}}_0(\mathbb{Z}[\pi]) .$$

(i) Make f transverse at $X \times \{\text{pt.}\} \subset X \times S^1$, so that the restriction

$$g = f| : U = f^{-1}(X \times \{\text{pt.}\}) \longrightarrow X$$

is a degree 1 normal map. By definition, the homotopy equivalence f is *split* if it is homotopic to a map (also denoted by f) such that $g : U \longrightarrow X$ is a homotopy equivalence.

The splitting theorem of Farrell and Hsiang [49] shows that for $n \geq 6$ the following conditions are equivalent:

(a) f is split,
(b) (W, c) pseudo-fibres,
(c) $[\tau(f)] = 0$,
(d) $[C^-] = 0 \in \widetilde{K}_0(\mathbb{Z}[\pi])$ and $[C/C^+, \zeta] = 0 \in \widetilde{\mathrm{Nil}}_0(\mathbb{Z}[\pi])$,
(e) the fibring obstruction is such that

$$\Phi^+(W) = \Phi^+(C) \in \mathrm{im}(Wh(\pi) \longrightarrow Wh(\pi \times \mathbb{Z})) .$$

(See Ranicki [124, 10.9] for the analogous chain complex splitting results.)

(ii) The homotopy equivalence $f : W \longrightarrow X \times S^1$ is *split by a homotopy open strip* if it is homotopic to a map (also denoted by f) such that

$$U = f^{-1}(X \times \{\mathrm{pt.}\}) \subset W$$

has an open neighbourhood $Z \subset W$ with

$$f(Z) \subseteq X \times (-1, 1) \subset X \times S^1 ,$$

and such that the restriction $f| : Z \longrightarrow X \times (-1, 1)$ is a homotopy equivalence, in which case

$$\zeta \simeq 0 : C/C^+ \longrightarrow C/C^+ \ , \ \zeta^{-1} \simeq 0 : C/C^- \longrightarrow C/C^-$$

and C is a relaxed chain complex band. For $n \geq 6$ the following conditions are equivalent:

(a) f is split by a homotopy open strip,
(b) $[C/C^+, \zeta] = 0 \in \widetilde{\mathrm{Nil}}_0(\mathbb{Z}[\pi])$,
(c) C^+ is relaxed,
(d) C is relaxed,
(e) W is homeomorphic to the relaxation W' ,
(f) the map $c : W \simeq X \times S^1 \longrightarrow S^1$ is homotopic to a manifold approximate fibration,
(g) the fibring obstruction is relaxed, that is

$$\Phi^+(W) = \Phi^+(C) \in \mathrm{im}(Wh(\pi) \oplus \widetilde{K}_0(\mathbb{Z}[\pi]) \longrightarrow Wh(\pi \times \mathbb{Z})) .$$

The equivalence of (a) and (b) is due to Farrell and Hsiang [49, p. 835]. □

Remark 24.17 Let (W, c) be an ANR band, with universal cover $\widetilde{W}$ and infinite cyclic cover $\overline{W}$. For the usual sake of simplicity assume that (W, c) is untwisted, so that

$$\zeta_* = 1 : \pi_1(\overline{W}) = \pi \longrightarrow \pi \ , \ \pi_1(W) = \pi \times \mathbb{Z} \ .$$

The fibring obstructions $\Phi^+(W, c)$, $\Phi^-(W, c)$ (15.11) differ by

$$\Phi^-(W, c) - \Phi^+(W, c) = \tau(r) \in Wh(\pi \times \mathbb{Z}) \ ,$$

the torsion of the homeomorphism

$$r : T(\zeta^{-1}) \longrightarrow T(\zeta) \ ; \ (x, t) \longrightarrow (x, 1-t)$$

with respect to the canonical simple homotopy types (14.3). The torsion $\tau(r)$ is relaxed, with components (up to sign) the torsion of $\zeta : \overline{W} \longrightarrow \overline{W}$ and the finiteness obstruction of $\overline{W}$

$$\begin{aligned} \tau(r) &= \tau(-z\widetilde{\zeta}^{-1} : C(\widetilde{W})[z, z^{-1}] \longrightarrow C(\widetilde{W})[z, z^{-1}]) \\ &= (-\tau(\zeta), [\overline{W}], 0, 0) \\ &\in Wh(\pi \times \mathbb{Z}) = Wh(\pi) \oplus \widetilde{K}_0(\mathbb{Z}[\pi]) \oplus \widetilde{\mathrm{Nil}}_0(\mathbb{Z}[\pi]) \oplus \widetilde{\mathrm{Nil}}_0(\mathbb{Z}[\pi]) \end{aligned}$$

(Ranicki [124, p. 159]). As already noted in 15.12, if (W, c) is an n-dimensional manifold band the fibring obstructions are Poincaré dual to each other,

$$\Phi^+(W, c) = (-)^{n-1}\Phi^-(W, c)^* \in Wh(\pi \times \mathbb{Z}) \ ,$$

and for $n \geq 6$ $\Phi^+(W, c) = \Phi^-(W, c) = 0$ if and only if $c : W \longrightarrow S^1$ is homotopic to the projection of a fibre bundle. Chapman and Ferry [27, 28] investigated the fibring properties of an ANR band (W, c), assuming W to be a locally compact separable metric space and $c : W \longrightarrow S^1$ to be a (Hurewicz) fibration – in this generality, $\Phi^+(W, c)$ and $\Phi^-(W, c)$ need not be Poincaré dual. Their results have the following reformulation in terms of $\Phi^+(W, c)$ and $\Phi^-(W, c)$:

(i) [27, Thm. 3; 28, Thm. 4] c is fibre homotopy equivalent to a PL fibration (= PL map of compact polyhedra which is a fibration) if and only if c is fibre homotopy equivalent to the projection of a compact Hilbert cube manifold fibre bundle, if and only if

$$\tau(r) = 0 \in Wh(\pi \times \mathbb{Z}) \ .$$

(ii) [28, Thm. 6] If W is a compact Hilbert cube manifold then c is homotopic to the projection of a fibre bundle if and only if

$$\Phi^+(W, c) = \Phi^-(W, c) = 0 \in Wh(\pi \times \mathbb{Z}) \ .$$

The two conditions in (ii) are equivalent to $\tau(r) = 0$ (the first obstruction of [28]) and $\Phi^+(W,c) = 0$ (the second obstruction of [28]). In both [27] and [28] it was assumed that $\overline{W}$ is homotopy finite. If $(K, \phi : \overline{W} \simeq K)$ is a finite structure on $\overline{W}$ then K is a finite CW complex with a self homotopy equivalence $h = \phi\zeta\phi^{-1} : K \longrightarrow K$, and there is defined a homotopy equivalence $T(h) \longrightarrow W$, such that

$$\tau(r) \;=\; i_*\tau(h) \;\;,\;\; \Phi^+(W,c) \;=\; \tau(T(h) \longrightarrow W) \in Wh(\pi \times \mathbb{Z})$$

with $i_* : Wh(\pi) \longrightarrow Wh(\pi \times \mathbb{Z})$ the inclusion. However, it is not necessary to assume that $\overline{W}$ is homotopy finite in the reformulation. □

25

Algebraic ribbons

'Algebraic ribbons' are the chain complex analogues of the geometric ribbons of Chapter 15. We shall now develop the algebraic theory of ribbons, in the context of the bounded algebra of Pedersen and Weibel [110] and Ranicki [124]. A chain complex ribbon is a finite chain complex C in the category $\mathbb{C}_{\mathbb{R}}(A)$ of $\mathbb{R}$-bounded A-modules (for some ring A) with the end properties of a chain complex band. In Chapter 26 we shall use algebraic ribbons to develop the algebraic theory of twist glueing. An $A[z,z^{-1}]$-module chain complex band is an example of a chain complex ribbon; in 26.6 it will be shown that every chain complex ribbon C is simple chain equivalent to a chain complex band $\widehat{C}$, the 'wrapping up' of C. In Chapter 27 we shall describe the effects of wrapping up in algebraic K- and L-theory.

We refer to Ferry and Pedersen [58] and Ranicki [124] for accounts of bounded topology and algebra, only repeating the most essential definitions here.

Definition 25.1 Let A be a ring, and let B be a metric space. The *B-bounded A-module category* $\mathbb{C}_B(A)$ is the additive category with objects B-graded A-modules

$$M \;=\; \sum_{x\in B} M(x)$$

such that each $M(x)$ is based f.g. free, with $\{y\in B\,|\,d(x,y)<r, M(y)\neq 0\}$ finite for each $x\in B$, $r>0$. A morphism $f:M\longrightarrow N$ in $\mathbb{C}_B(A)$ is an A-module morphism such that there exists a number $b\geq 0$ for which the composite

$$f(y,x)\;:\;M(x)\xrightarrow{\text{incl.}} M\xrightarrow{f} N\xrightarrow{\text{proj.}} N(y)$$

is 0 whenever $d(x,y)>b$. □

Definition 25.2 The *B-bounded Whitehead group* of a ring A is

$$Wh_B(A) = \operatorname{coker}(K_1(\mathbb{C}_B(\mathbb{Z})) \longrightarrow K_1(\mathbb{C}_B(A))) . \qquad \square$$

A B-bounded CW complex (X, d) is a CW complex X with a proper cellular map $d : X \longrightarrow B$ such that the diameters of the images in B of the cells $e \subset X$ are uniformly bounded, that is there exists a bound $b \geq 0$ with diameter$(d(e)) < b$ for all $e \subset X$. If $\widetilde{X}$ is a regular cover of X with group of covering translations π the cellular $\mathbb{Z}[\pi]$-module chain complex $C(\widetilde{X})$ is defined in $\mathbb{C}_B(\mathbb{Z}[\pi])$. A B-bounded map $f : (X, d) \longrightarrow (Y, e)$ of B-bounded CW complexes is a proper cellular map $f : X \longrightarrow Y$ such that there exists a bound $b \geq 0$ with

$$d_B(e(f(x)), d(x)) \leq b \quad (x \in X) .$$

A B-bounded map f induces a chain map $\widetilde{f} : C(\widetilde{X}) \longrightarrow C(\widetilde{Y})$ in $\mathbb{C}_B(\mathbb{Z}[\pi])$ for any regular cover $\widetilde{Y}$ of Y with group of covering translations π, with $\widetilde{X} = f^*\widetilde{Y}$ the pullback cover of X. A B-bounded homotopy equivalence $f : (X, d) \longrightarrow (Y, e)$ of B-bounded CW complexes has a B-bounded torsion

$$\tau_B(f) = \tau_B(\widetilde{f} : C(\widetilde{X}) \longrightarrow C(\widetilde{Y})) \in Wh_B(\mathbb{Z}[\pi_1(X)]) .$$

A B-bounded h-cobordism has B-bounded torsion in $Wh_B(\mathbb{Z}[\pi_1])$, and there are bounded versions of the h- and s-cobordism theorems; in the bounded version of the Wall surgery theory a B-bounded normal map has a surgery obstruction in $L_*(\mathbb{C}_B(\mathbb{Z}[\pi_1]))$ (Ferry and Pedersen [58]). For bounded surgery theory A is a ring with involution (e.g. a group ring), and $\mathbb{C}_B(A)$ is an additive category with involution, so that the L-groups $L_*(\mathbb{C}_B(A))$ are defined as in Ranicki [124].

Definition 25.3 Let A be a ring.

(i) Let $\mathbb{M}(A)$ be the additive category of A-modules. Define the *sum* and *product* functors

$$\sum : \mathbb{C}_B(A) \longrightarrow \mathbb{M}(A) \; ; \; M \longrightarrow M = \sum_{x \in B} M(x) ,$$

$$\prod : \mathbb{C}_B(A) \longrightarrow \mathbb{M}(A) \; ; \; M \longrightarrow M^{lf} = \prod_{x \in B} M(x) .$$

The inclusion $i : M \longrightarrow M^{lf}$ defines a natural transformation from $\sum$ to $\prod$.

(ii) The *end complex* of a chain complex C in $\mathbb{C}_B(A)$ is the A-module chain complex

$$e(C) = \mathcal{C}(i : C \longrightarrow C^{lf})_{*+1} . \qquad \square$$

Example 25.4 Let (X,d) be a connected B-bounded CW complex, so that the cellular chain complex $C(\widetilde{X})$ of the universal cover $\widetilde{X}$ of X is a finite chain complex in $\mathbb{C}_B(\mathbb{Z}[\pi])$ with $\pi = \pi_1(X)$. The locally π-finite cellular chain complex of $\widetilde{X}$ (5.5 (i)) is

$$C^{lf,\pi}(\widetilde{X}) \; = \; C(\widetilde{X})^{lf} \; .$$

Let $\widetilde{e(X)} = p^*\widetilde{X}$ be the cover of the end complex $e(X)$ obtained from $\widetilde{X}$ by pullback along the projection $p : e(X)\longrightarrow X; \omega\longrightarrow\omega(0)$. If X is forward tame the $\mathbb{Z}[\pi]$-module chain complex at ∞ $C^{\infty,\pi}(\widetilde{X})$ is homology equivalent to the end complex $e(C(\widetilde{X}))$ of $C(\widetilde{X})$. □

Definition 25.5 A *subobject* $M' \subseteq M$ of an object M in $\mathbb{C}_B(A)$ is the object

$$M' \; = \; M(B') \; = \; \sum_{y\in B'} M(y)$$

determined by a subset $B' \subseteq B$, with $M(B')(x) = 0$ for $x \in B\backslash B'$. □

We shall be mainly concerned with the $\mathbb{R}$-bounded category $\mathbb{C}_{\mathbb{R}}(A)$ here.

Definition 25.6 A *covering translation* of an object M in $\mathbb{C}_{\mathbb{R}}(A)$ is an isomorphism $\zeta : M\longrightarrow M$ such that $\zeta(x+1,x) : M(x)\longrightarrow M(x+1)$ is a basis-preserving isomorphism for each $x \in \mathbb{R}$, and $\zeta(y,x) = 0 : M(x)\longrightarrow M(y)$ for $y \neq x+1$. □

Proposition 25.7 *The additive category of based f.g. free $A[z,z^{-1}]$-modules is equivalent to the category $\mathbb{C}_{\mathbb{R}}(A)^{\mathbb{Z}}$ of pairs (M,ζ) with M an object in $\mathbb{C}_{\mathbb{R}}(A)$ and $\zeta : M\longrightarrow M$ a covering translation, with morphisms $f : (M,\zeta)\longrightarrow (M',\zeta')$ defined by morphisms $f : M\longrightarrow M'$ in $\mathbb{C}_{\mathbb{R}}(A)$ such that $\zeta' f = f\zeta$.*
Proof If L is a based f.g. free A-module the induced based f.g. free $A[z,z^{-1}]$-module

$$M \; = \; L[z,z^{-1}] \; = \; \sum_{j=-\infty}^{\infty} z^j L$$

is an object in $\mathbb{C}_{\mathbb{R}}(A)$ with

$$M(x) \; = \; \begin{cases} z^j L & \text{if } x = j \in \mathbb{Z} \, , \\ 0 & \text{if } x \in \mathbb{R}\backslash\mathbb{Z} \end{cases}$$

and with a covering translation $\zeta : M\longrightarrow M$, defining an equivalence

$$\{\text{based f.g. free } A[z,z^{-1}]\text{-modules}\} \longrightarrow \mathbb{C}_{\mathbb{R}}(A)^{\mathbb{Z}} \; ; \; L[z,z^{-1}] \longrightarrow (M,\zeta) \; . \; \square$$

Similarly for the additive category of based f.g. free $A[z]$-modules. A finite based f.g. free $A[z]$-module chain complex C^+ can be regarded as a chain complex in $\mathbb{C}_{[0,\infty)}(A)$ together with an isomorphism $\zeta^+ : C^+ \longrightarrow C^+$ such that $\zeta^+(y,x) = 0$ for $y \neq x+1$, and the end complex $e(C^+)$ defined in 25.3 is just the end complex as defined in 23.3.

Definition 25.8 A *Mayer–Vietoris presentation* (C^+, C^-) of a finite chain complex C in $\mathbb{C}_{\mathbb{R}}(A)$ is the exact sequence

$$0 \longrightarrow C^+ \cap C^- \xrightarrow{\begin{pmatrix} j^+ \\ -j^- \end{pmatrix}} C^+ \oplus C^- \xrightarrow{(q^+ \;\; q^-)} C \longrightarrow 0$$

determined by subcomplexes $C^+, C^- \subseteq C$ such that

$$C_r^+ = C_r[-N_r^+, \infty) \ , \ C_r^- = C_r(-\infty, N_r^-]$$

for some real numbers $N_r^+, N_r^- \geq 0$, with

$$j^+ : C^+ \cap C^- \longrightarrow C^+ \ , \ j^- : C^+ \cap C^- \longrightarrow C^- \ ,$$
$$q^+ : C^+ \longrightarrow C \ , \ q^- : C^- \longrightarrow C$$

the inclusions.

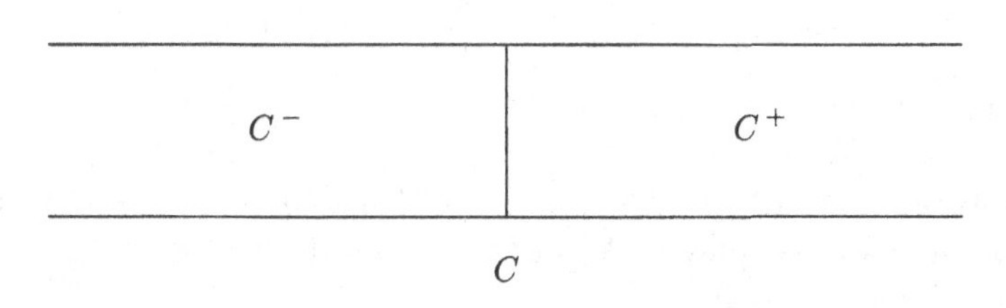

□

Remark 25.9 (i) Note that $C^+ \cap C^-$ is defined in $\mathbb{C}_{[-N,N]}(A)$ for some bound N (e.g. $\max\{N_r^+, N_r^-\}$), so that it is a finite chain complex of based f.g. free A-modules.

(ii) A Mayer–Vietoris presentation (C^+, C^-) of a based f.g. free $A[z,z^{-1}]$-module chain complex C (21.13) is a Mayer–Vietoris presentation of C regarded as a chain complex in $\mathbb{C}_{\mathbb{R}}(A)$ such that $\zeta(C^+) \subset C^+$, $\zeta^{-1}(C^-) \subset C^-$.

□

Proposition 25.10 (Ranicki [124, Chapters 6,7,8]) (i) *Every finite chain complex C in $\mathbb{C}_{\mathbb{R}}(A)$ admits Mayer–Vietoris presentations (C^+, C^-). If $h : C \longrightarrow C$ is a covering translation there exist Mayer–Vietoris presentations (C^+, C^-) such that $h(C^+) \subset C^+$, $h^{-1}(C^-) \subset C^-$.*

(ii) *The $\mathbb{R}$-bounded Whitehead group of A is isomorphic to the reduced projective class group of A :*

$$Wh(\mathbb{C}_{\mathbb{R}}(A)) \cong \widetilde{K}_0(A) \ .$$

(iii) *Every contractible finite chain complex C in $\mathbb{C}_{\mathbb{R}}(A)$ has a torsion $\tau_{\mathbb{R}}(C) \in Wh_{\mathbb{R}}(A)$ with image $[C^+] = -[C^-] \in \widetilde{K}_0(A)$, for any Mayer–Vietoris presentation (C^+, C^-) of C.* □

Remark 25.11 The $\mathbb{R}$-bounded Whitehead torsion of an $\mathbb{R}$-bounded homotopy equivalence $f : (X, d) \longrightarrow (Y, e)$ of $\mathbb{R}$-bounded *CW* complexes is

$$\tau_{\mathbb{R}}(f) \;=\; [\mathcal{C}(\widetilde{f}^+ : C(\widetilde{X}^+) \longrightarrow C(\widetilde{Y}^+))] \in Wh(\mathbb{C}_{\mathbb{R}}(\mathbb{Z}[\pi])) \;=\; \widetilde{K}_0(\mathbb{Z}[\pi])$$

with $\pi = \pi_1(X) = \pi_1(Y)$. This is also the infinite torsion (11.1):

$$\tau^{lf}(f) \in Wh^{lf}(X) \;=\; \widetilde{K}_0(\mathbb{Z}[\pi]) \,.$$ □

Definition 25.12 A *chain complex ribbon* is a finite chain complex C in $\mathbb{C}_{\mathbb{R}}(A)$ such that for some Mayer–Vietoris presentation (C^+, C^-) the A-module chain maps

$$C^+ \longrightarrow C \oplus C^{+,lf} \; ; \; x^+ \longrightarrow (x^+, x^+) \,,$$
$$C^- \longrightarrow C \oplus C^{-,lf} \; ; \; x^- \longrightarrow (x^-, x^-)$$

are chain equivalences, or equivalently such that the composites

$$e(C^+) \longrightarrow C^+ \xrightarrow{q^+} C \;\;,\;\; e(C^-) \longrightarrow C^- \xrightarrow{q^-} C$$

are chain equivalences. □

Example 25.13 If $(X, d : X \longrightarrow \mathbb{R})$ is a *CW* ribbon with $\pi_1(X) = \pi$ then the cellular chain complex $C(\widetilde{X})$ of the universal cover $\widetilde{X}$ of X is a chain complex ribbon in $\mathbb{C}_{\mathbb{R}}(\mathbb{Z}[\pi])$. □

Proposition 25.14 (i) *If C is a chain complex ribbon in $\mathbb{C}_{\mathbb{R}}(A)$ with a Mayer–Vietoris presentation (C^+, C^-) the natural A-module chain maps*

$$C^+ \longrightarrow C \oplus C^{+,lf} \;\;,\;\; C^- \longrightarrow C \oplus C^{-,lf}$$

are chain equivalences. Moreover, the A-module chain map

$$u \;:\; C^+ \cap C^- \longrightarrow C \oplus C^{+,lf} \oplus C^{-,lf} \; ; \; x \longrightarrow (x, x, x)$$

is a chain equivalence, so that each of $C, C^+, C^-, C^{lf}, C^{+,lf}, C^{-,lf}$ is A-finitely dominated.

(ii) *A finite chain complex C in $\mathbb{C}_{\mathbb{R}}(A)$ is a chain complex ribbon if and only if C^{lf} is chain equivalent to a free A-module chain complex and the connecting maps in the Mayer–Vietoris exact sequence*

$$\ldots \longrightarrow H_{r+1}(C^{lf}) \xrightarrow{\partial} H_r(C) \longrightarrow H_r(e(C^+) \longrightarrow C) \oplus H_r(e(C^-) \longrightarrow C)$$
$$\longrightarrow H_r(C^{lf}) \longrightarrow \ldots$$

are isomorphisms $\partial : H_{*+1}(C^{lf}) \cong H_*(C)$, *for any Mayer–Vietoris presentation* (C^+, C^-).

(iii) *A finite based f.g. free* $A[z,z^{-1}]$*-module chain complex* C *is a chain complex band (i.e.* C *is* A*-finitely dominated) if and only if* C *is a chain complex ribbon in* $\mathbb{C}_{\mathbb{R}}(A)$.

Proof (i) For any Mayer–Vietoris presentations (C^+, C^-), (C'^+, C'^-) of C there exists a Mayer–Vietoris presentation (C''^+, C''^-) such that $C^+ \subseteq C''^+$, $C'^+ \subseteq C''^+$, with C''/C, C''/C' finite f.g. free, so that

$$C^{+,lf}/C^+ \;=\; C'^{+,lf}/C'^+ \;=\; C''^{+,lf}/C''^+ \; .$$

The chain equivalence $C^+ \cap C^- \simeq C^{+,lf} \oplus C^{-,lf} \oplus C$ follows from the chain equivalences $C^+ \simeq C \oplus C^{+,lf}$, $C^- \simeq C \oplus C^{-,lf}$ and the short exact sequence

$$0 \longrightarrow C^+ \cap C^- \longrightarrow C^+ \oplus C^- \longrightarrow C \longrightarrow 0 \; .$$

(ii) The natural A-module chain map defines a chain equivalence

$$e(C^+) \oplus e(C^-) \;\simeq\; e(C) \; .$$

The Mayer–Vietoris exact sequence of the statement is induced by the short exact sequence of A-module chain complexes

$$0 \longrightarrow C \longrightarrow \mathcal{C}(e(C^+) \longrightarrow C) \oplus \mathcal{C}(e(C^-) \longrightarrow C) \longrightarrow D \longrightarrow 0$$

with $D = \mathcal{C}(e(C^+) \oplus e(C^-) \longrightarrow C) \simeq C^{lf}$. If C^{lf} is chain equivalent to a free A-module chain complex then so are $C^{+,lf}$, $C^{-,lf}$, and the chain maps $e(C^+) \longrightarrow C$, $e(C^-) \longrightarrow C$ are chain equivalences if and only if they are homology equivalences. The connecting maps ∂ are isomorphisms if and only if $H_*(e(C^+) \longrightarrow C) = H_*(e(C^-) \longrightarrow C) = 0$ (as in 23.9).

(iii) Immediate from 23.17. □

We shall use the following result in Chapter 26:

Proposition 25.15 *If* (C^+, C^-) *is a Mayer–Vietoris presentation of a finite chain complex* C *in* $\mathbb{C}_{\mathbb{R}}(A)$ *the commutative square of inclusions*

$$\begin{array}{ccc} D & \xrightarrow{\;j^+\;} & C^+ \\ {\scriptstyle j^-}\big\downarrow & & \big\downarrow{\scriptstyle q^+} \\ C^- & \xrightarrow{\;q^-\;} & C \end{array}$$

is chain homotopy cartesian, with $D = C^+ \cap C^-$.

Proof Let $P = \mathcal{C}((q^+\ q^-) : C^+ \oplus C^- \longrightarrow C)_{*+1}$, with $P_r = C^+_r \oplus C^-_r \oplus C_{r+1}$. The A-module chain map $D \longrightarrow P; x \longrightarrow (j^+(x), -j^-(x), 0)$ is a homology equivalence of free A-module chain complexes, so that it is a chain equivalence. □

26

Algebraic twist glueing

We shall now develop the algebraic theory of twist glueing for chain complex ribbons, constructing relaxed chain complex bands by a direct translation of the homotopy theoretic twist glueing of Chapter 19. Algebraic wrapping up and relaxation are the special cases of algebraic twist glueing by the identity and a covering translation.

Given a chain complex ribbon C in $\mathbb{C}_{\mathbb{R}}(A)$ and a Mayer–Vietoris presentation (C^+, C^-) write

$$\begin{aligned} p^+ &: C \simeq e(C^+) \longrightarrow C^+ , \\ p^- &: C \simeq e(C^-) \longrightarrow C^- . \end{aligned}$$

Use 25.15 to define the A-module chain maps

$$\begin{aligned} i^+ &= (1, q^-p^+) : C^+ \longrightarrow D , \\ i^- &= (q^+p^-, 1) : C^- \longrightarrow D , \end{aligned}$$

and to define chain homotopies

$$\begin{aligned} j^+i^+ &\simeq 1 : C^+ \longrightarrow C^+ \ , \ j^-i^- \simeq 1 : C^- \longrightarrow C^- , \\ q^+p^+ &\simeq 1 : C \longrightarrow C \ , \ q^-p^- \simeq 1 : C \longrightarrow C , \end{aligned}$$

with $j^\pm : D = C^+ \cap C^- \longrightarrow C^\pm$, $q^\pm : C^\pm \longrightarrow C$ the inclusions. Exactly as in Chapter 19 there are defined chain homotopy commutative diagrams

$$\begin{array}{ccc} C & \xrightarrow{p^+} & C^+ \\ {\scriptstyle p^-}\downarrow & & \downarrow{\scriptstyle i^+} \\ C^- & \xrightarrow{i^-} & D \end{array}$$

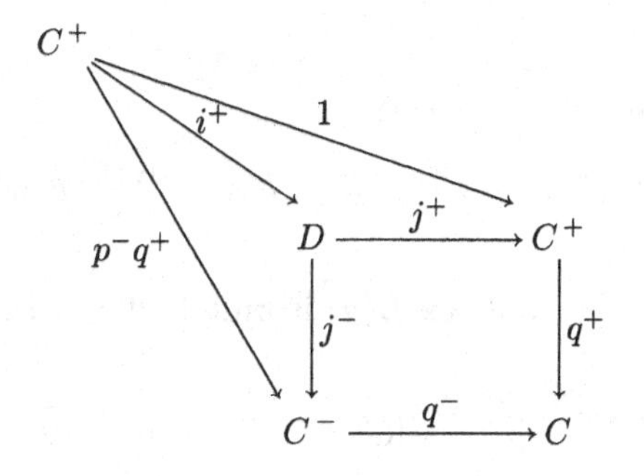

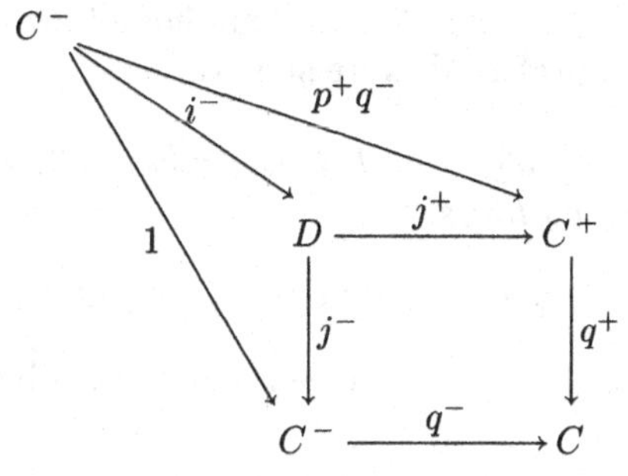

For an A-module chain equivalence $h : C \longrightarrow C$ the A-module chain complex defined by

$$E(h) = \mathcal{C}(\begin{pmatrix} p^+h \\ p^- \end{pmatrix} : C \longrightarrow C^+ \oplus C^-)$$

is such that there is defined a chain homotopy cartesian square

$$\begin{array}{ccc} C & \xrightarrow{p^+h} & C^+ \\ {\scriptstyle p^-}\Big\downarrow & & \Big\downarrow{\scriptstyle k^+(h)} \\ C^- & \xrightarrow{k^-(h)} & E(h) \end{array}$$

with

$$\begin{aligned} k^+(h) \ &: \ C^+ \longrightarrow E(h) \ ; \ x^+ \longrightarrow (x^+, 0, 0) \ , \\ k^-(h) \ &: \ C^- \longrightarrow E(h) \ ; \ x^- \longrightarrow (0, x^-, 0) \ . \end{aligned}$$

Define also the A-module chain maps

$$f^+(h) = k^+(h)j^+ \ , \ f^-(h) = k^-(h)j^- \ : D = C^+ \cap C^- \longrightarrow E(h) \ .$$

Definition 26.1 Given a chain complex ribbon C in $\mathbb{C}_{\mathbb{R}}(A)$, a Mayer–Vietoris presentation (C^+, C^-) and an A-module chain equivalence $h : C \longrightarrow C$ define the *algebraic h-twist glueing* to be the $A[z, z^{-1}]$-module chain complex

$$\begin{aligned} C(h) &= \mathcal{W}(f^+(h), f^-(h)) \\ &= \mathcal{C}(f^+(h) - z^{-1}f^-(h) : D[z, z^{-1}] \longrightarrow E(h)[z, z^{-1}]) . \end{aligned}$$ □

The A-module chain complex $E(h)$ is finitely dominated, with projective class

$$[E(h)] = [C^+] + [C^-] - [C] = [D] \in \text{im}(K_0(\mathbb{Z}) \longrightarrow K_0(A)) .$$

Thus $[E(h)] = 0 \in \widetilde{K}_0(A)$, and $E(h)$ is chain homotopy finite, but in general $E(h)$ does not have a preferred finite structure.

Proposition 26.2 *The algebraic h-twist glueing $C(h)$ is chain equivalent to the algebraic mapping torus*

$$T(h) = \mathcal{C}(1 - z^{-1}h : C[z, z^{-1}] \longrightarrow C[z, z^{-1}]) .$$

Proof Since $q^+p^+ \simeq 1 \simeq q^-p^- : C \longrightarrow C$ it is possible to choose a chain homotopy

$$e : q^+p^+ \simeq q^-p^- : C \longrightarrow C .$$

The A-module chain map $g(h) : E(h) \longrightarrow C$ defined by

$$g(h) = (q^+ \quad hq^- \quad he) : E(h)_r = C_r^+ \oplus C_r^- \oplus C_{r-1} \longrightarrow C_r$$

is such that there is defined a commutative diagram

$$\begin{array}{ccccc} D & \xrightarrow{f^+(h)} & E(h) & \xleftarrow{f^-(h)} & D \\ \Big\downarrow{\scriptstyle q^+j^+} & & \Big\downarrow{\scriptstyle g(h)} & & \Big\downarrow{\scriptstyle q^+j^+} \\ C & \xrightarrow{1} & C & \xleftarrow{h} & C \end{array}$$

The induced $A[z, z^{-1}]$-module chain map

$$\begin{pmatrix} g(h) & 0 \\ 0 & q^+j^+ \end{pmatrix} : C(h) \longrightarrow T(h)$$

is a chain equivalence by 21.16, since (as in the geometric case in 19.9) the chain maps

$$\begin{aligned} j^+ &: D \longrightarrow C^+ \ , \ j^- : D \longrightarrow C^- \ , \\ p^+h &: C \longrightarrow C^+ \ , \ p^- : C \longrightarrow C^- \end{aligned}$$

are such that up to chain homotopy

$$\begin{aligned} f^+(h) &: D \xrightarrow{j^+} C^+ \longrightarrow \mathcal{W}(p^+h, p^-) = E(h) , \\ f^-(h) &: D \xrightarrow{j^-} C^- \longrightarrow \mathcal{W}(p^+h, p^-) = E(h) , \\ h &: C \xrightarrow{p^+h} C^+ \xrightarrow{q^+} \mathcal{W}(j^+, j^-) \simeq C , \\ 1 &: C \xrightarrow{p^-} C^- \xrightarrow{q^-} \mathcal{W}(j^+, j^-) \simeq C . \end{aligned}$$

□

Proposition 26.3 *Let C be a chain complex ribbon in $\mathbb{C}_{\mathbb{R}}(A)$ and let $h : C \longrightarrow C$ be an isomorphism which is either a covering translation or the identity, so that there exists a Mayer–Vietoris presentation (C^+, C^-) with $h(C^+) \subseteq C^+$, $h^{-1}(C^-) \subseteq C^-$. The algebraic h-twist glueing $C(h)$ is $A[z, z^{-1}]$-module chain equivalent to a relaxed chain complex band $C[h]$ which is simple chain equivalent to C (in $\mathbb{C}_{\mathbb{R}}(A)$). The algebraic fibring obstructions are of the type*

$$\begin{aligned} \Phi^+(C[h]) &= (\phi[h]^+, -[C^-], 0, 0) \ , \quad \Phi^-(C[h]) = (\phi[h]^-, [C^+], 0, 0) \\ &\in Wh(A[z, z^{-1}]) = Wh(A) \oplus \widetilde{K}_0(A) \oplus \widetilde{\mathrm{Nil}}_0(A) \oplus \widetilde{\mathrm{Nil}}_0(A) , \end{aligned}$$

with

$$\phi[h]^+ - \phi[h]^- = \tau(h) \in Wh(A) .$$

Proof The subcomplex

$$E[h] = C^+ \cap h(C^-) \subset C$$

is a finite based f.g. free A-module chain complex. The A-module chain maps given by

$$\begin{aligned} f_h^+ &: D \longrightarrow E[h] \ ; \ x \longrightarrow x , \\ f_h^- &: D \longrightarrow E[h] \ ; \ x \longrightarrow h(x) , \\ g_h^+ &: E[h] \longrightarrow C^+ \ ; \ y \longrightarrow y , \\ g_h^- &: E[h] \longrightarrow C^- \ ; \ y \longrightarrow h^{-1}(y) , \\ k^+[h] &= f_h^+ i^+ : C^+ \longrightarrow D \longrightarrow E[h] , \\ k^-[h] &= f_h^- i^- : C^- \longrightarrow D \longrightarrow E[h] \end{aligned}$$

are such that there are defined chain homotopy cartesian squares

$$\begin{array}{ccc} E[h] & \xrightarrow{g_h^+} & C^+ \\ {\scriptstyle g_h^-}\big\downarrow & & \big\downarrow{\scriptstyle q^+} \\ C^- & \xrightarrow{hq^-} & C \end{array} \qquad \begin{array}{ccc} C & \xrightarrow{p^+h} & C^+ \\ {\scriptstyle p^-}\big\downarrow & & \big\downarrow{\scriptstyle k^+[h]} \\ C^- & \xrightarrow{k^-[h]} & E[h] \end{array}$$

Define the finite based f.g. free $A[z, z^{-1}]$-module chain complex

$$C[h] \;=\; \mathcal{W}(f^+[h], f^-[h])$$

with

$$\begin{aligned} f^+[h] &= k^+[h]j^+ = f_h^+ i^+ j^+ : D \longrightarrow E[h] , \\ f^-[h] &= k^-[h]j^- = f_h^- i^- j^- : D \longrightarrow E[h] . \end{aligned}$$

The A-module chain map

$$w[h] \;:\; E[h] \longrightarrow E(h) \;;\; x \longrightarrow (x, x, 0)$$

is a chain equivalence such that up to chain homotopy

$$\begin{aligned} f^+[h] &: D \xrightarrow{f^+(h)} E(h) \xrightarrow{w[h]^{-1}} E[h] , \\ f^-[h] &: D \xrightarrow{f^-(h)} E(h) \xrightarrow{w[h]^{-1}} E[h] . \end{aligned}$$

The induced $A[z, z^{-1}]$-module chain map

$$(1, w[h]) \;:\; C[h] \longrightarrow C(h)$$

is a chain equivalence, so that $C[h]$ is A-module chain equivalent to C and $C[h]$ is a chain complex band. The inclusions $C^+ \longrightarrow E[h]$, $C^- \longrightarrow z^{-1}E[h]$ determine A-module chain equivalences

$$\begin{aligned} C^+ \longrightarrow C[h]^+ &= \mathcal{C}(f^+[h] - z^{-1}f^-[h] : \sum_{k=1}^{\infty} z^k D \longrightarrow \sum_{k=0}^{\infty} z^k E[h]) , \\ C^- \longrightarrow C[h]^- &= \mathcal{C}(f^+[h] - z^{-1}f^-[h] : \sum_{k=-\infty}^{-1} z^k D \longrightarrow \sum_{k=-\infty}^{-1} z^k E[h]) \end{aligned}$$

which in turn determine chain equivalences

$$\begin{aligned} (C/C^+, 0) &\simeq (C[h]/C[h]^+, 0) \simeq S^{-1}(C[h]^{+,lf}, \zeta^{+,lf}) , \\ (C/C^-, 0) &\simeq (C[h]/C[h]^-, 0) \simeq S^{-1}(C[h]^{-,lf}, \zeta^{-,lf}) . \end{aligned}$$

Thus $C[h]$ is a relaxed chain complex band, with

$$(C[h]^{+,lf}, \zeta^{+,lf}) \;=\; (C[h]^{-,lf}, \zeta^{-,lf}) \;=\; 0 \in \widetilde{\mathrm{Nil}}_0(A) .$$

The composite of $(1, w[h])$ and the A-module chain equivalence

$$C(h) \longrightarrow C\ ;\ \sum_{j=-\infty}^{\infty} z^j(x_j, y_j) \longrightarrow \sum_{j=-\infty}^{\infty} h^j x_j$$

is an A-module chain equivalence which is chain homotopic to a chain equivalence $F[h] : C[h] \longrightarrow C$ in $\mathbb{C}_{\mathbb{R}}(A)$ (the algebraic analogue of the proper homotopy equivalence $F[h] : X[h] \longrightarrow X$ in 19.12 (ii)), with torsion

$$\tau_{\mathbb{R}}(F[h]) = [C^+] - [C[h]^+] = 0 \in Wh(\mathbb{C}_{\mathbb{R}}(A)) = \widetilde{K}_0(A)\ .$$

In order to compute the fibring obstructions $\Phi^+(C[h])$, $\Phi^-(C[h])$ consider the A-module chain equivalences given by 23.11

$$\begin{aligned} u[h] &: D \longrightarrow C[h] \oplus C[h]^{+,lf} \oplus C[h]^{-,lf}\ ;\ x \longrightarrow (x, x, x)\ , \\ v[h] &: E[h] \longrightarrow C[h] \oplus C[h]^{+,lf} \oplus C[h]^{-,lf}\ ;\ y \longrightarrow (y, y, h^{-1}y)\ , \end{aligned}$$

so that

$$\begin{aligned} \Phi^+(C[h]) &= (\phi[h]^+, -[C[h]^-], 0, 0)\ ,\quad \Phi^-(C[h]) = (\phi[h]^-, [C[h]^+], 0, 0) \\ &\in Wh(A[z, z^{-1}]) = Wh(A) \oplus \widetilde{K}_0(A) \oplus \widetilde{\mathrm{Nil}}_0(A) \oplus \widetilde{\mathrm{Nil}}_0(A) \end{aligned}$$

with

$$\begin{aligned} \phi[h]^+ &= \tau(v[h]^{-1}(h \oplus 1 \oplus 1)u[h] : D \longrightarrow E[h])\ , \\ \phi[h]^- &= \tau(v[h]^{-1}u[h] : D \longrightarrow E[h]) \in Wh(A) \end{aligned}$$

such that $\phi[h]^+ - \phi[h]^- = \tau(h)$. □

The algebraic twist glueing of 26.1 is the algebraic analogue of the *CW* twist glueing of Chapter 19:

Proposition 26.4 *Let (X, d) be a CW ribbon, and let $h : X \longrightarrow X$ be a homotopy equivalence, so that the CW h-twist glueing $(X(h), d(h))$ is defined as in 19.8. Assume $h_* = 1 : \pi_1(X) = \pi \longrightarrow \pi$, so that $\pi_1(X(h)) = \pi \times \mathbb{Z}$.*

(i) *The cellular $\mathbb{Z}[\pi][z, z^{-1}]$-module chain complex of the universal cover $\widetilde{X}(h)$ of $X(h)$ is given up to chain equivalence by the algebraic $\widetilde{h}$-twist glueing of the chain complex ribbon $C(\widetilde{X})$ in $\mathbb{C}_{\mathbb{R}}(\mathbb{Z}[\pi])$*

$$C(\widetilde{X}(h)) = C[\widetilde{h} : C(\widetilde{X}) \longrightarrow C(\widetilde{X})]\ .$$

(ii) *If h is either a covering translation or the identity (as in 19.12) then the cellular $\mathbb{Z}[\pi][z, z^{-1}]$-module chain complex of the universal cover $\widetilde{X}[h]$ of the CW band $(X[h], d[h])$ is given up to simple chain equivalence by*

$$C(\widetilde{X}[h]) = C[\widetilde{h} : C(\widetilde{X}) \longrightarrow C(\widetilde{X})]\ ,$$

with the fibring obstructions given by 26.3:

$$\begin{aligned}\Phi^{\pm}(X[h],d[h]) &= \Phi^{\pm}(C[\widetilde{h}])) = (\phi^{\pm}[\widetilde{h}],\mp[\overline{X}^{\pm}],0,0)\\ \in Wh(\pi\times\mathbb{Z}) &= Wh(\pi)\oplus\widetilde{K}_0(\mathbb{Z}[\pi])\oplus\widetilde{\mathrm{Nil}}_0(\mathbb{Z}[\pi])\oplus\widetilde{\mathrm{Nil}}_0(\mathbb{Z}[\pi])\ .\end{aligned}$$

In particular, the homotopy equivalence $T(h)\simeq X[h]$ *has torsion*

$$\tau(T(h)\simeq X[h]) = \Phi^-(X[h],d[h])\in Wh(\pi\times\mathbb{Z})\ .$$

Proof Immediate from 26.3. □

Definition 26.5 The *wrapping up* of a chain complex ribbon C in $\mathbb{C}_{\mathbb{R}}(A)$ is the relaxed $A[z,z^{-1}]$-module chain complex band

$$\widehat{C} = C[1]\ .$$ □

Proposition 26.6 (i) *The wrapping up of a chain complex ribbon* C *in* $\mathbb{C}_{\mathbb{R}}(A)$ *is an* $A[z,z^{-1}]$*-module chain complex band*

$$\widehat{C} = C[1] = \mathcal{C}(i^+j^+ - z^{-1}i^-j^- : D[z,z^{-1}]\longrightarrow D[z,z^{-1}])$$

which is simple chain equivalent to C, *with*

$$\begin{aligned}\Phi^+(\widehat{C}) &= (0,-[C^-],0,0)\ ,\ \Phi^-(\widehat{C}) = (0,[C^+],0,0)\\ &\in Wh(A[z,z^{-1}]) = Wh(A)\oplus\widetilde{K}_0(A)\oplus\widetilde{\mathrm{Nil}}_0(A)\oplus\widetilde{\mathrm{Nil}}_0(A)\ .\end{aligned}$$

(ii) *Let* C *be an* $A[z,z^{-1}]$*-module chain complex, let* $C^!$ *denote* C *regarded as a finitely dominated* A*-module chain complex, and let* $\zeta^!:C^!\longrightarrow C^!$ *denote the covering translation* $\zeta:C\longrightarrow C$ *regarded as an* A*-module isomorphism. Let* z' *denote an invertible indeterminate which commutes with* z, *and let*

$$\begin{aligned}\Sigma &= \mathcal{C}(1-z:\mathbb{Z}[z,z^{-1}]\longrightarrow\mathbb{Z}[z,z^{-1}])\ ,\\ \Sigma' &= \mathcal{C}(1-z':\mathbb{Z}[z',z'^{-1}]\longrightarrow\mathbb{Z}[z',z'^{-1}])\ .\end{aligned}$$

The fibring obstructions of the $A[z,z^{-1},z',z'^{-1}]$*-module chain complex band* $C\otimes\Sigma'$ *vanish,*

$$\Phi^+(C\otimes\Sigma') = \Phi^-(C\otimes\Sigma') = 0\in Wh(A[z,z^{-1},z',z'^{-1}])\ ,$$

and there is defined a simple $A[z,z^{-1},z',z'^{-1}]$*-module chain equivalence*

$$C\otimes\Sigma' \simeq T(\widehat{\zeta}':\widehat{C}'\longrightarrow\widehat{C}')$$

with $\widehat{C}'$ *a copy of* $\widehat{C}$ *defined over* $A[z',z'^{-1}]$ *and*

$$\widehat{\zeta}'\simeq\zeta^!\otimes 1\ :\ \widehat{C}' \simeq C^!\otimes\Sigma' \longrightarrow \widehat{C}' \simeq C^!\otimes\Sigma'$$

a simple $A[z',z'^{-1}]$*-module chain equivalence.*

Proof (i) Immediate from 26.3, noting that for $h = 1 : C \longrightarrow C$

$$f_1^+ = f_1^- = 1 : D \longrightarrow E[1] = D ,$$
$$u[1] = v[1] : D \longrightarrow C[1] \oplus C[1]^{+,lf} \oplus C[1]^{-,lf} .$$

(ii) This is the algebraic analogue of 20.8. □

Example 26.7 The wrapping up of a CW ribbon (X,d) is a relaxed CW π_1-band

$$(W,c) = (\widehat{X},\widehat{d})$$

such that the relaxed $\mathbb{Z}[\pi][z,z^{-1}]$-module chain complex band $C(\widetilde{W}) = \widehat{C}$ is the algebraic wrapping up of the chain complex ribbon $C = C(\widetilde{X})$, by the special case $h = 1 : X \longrightarrow X$ of 26.4, with $\pi = \pi_1(X)$. The application of 26.6 (i) gives the fibring obstructions of $\widehat{X}$ to be

$$\Phi^+(W,c) = -[X^-] \ , \ \Phi^-(W,c) = [X^+] \in \widetilde{K}_0(\mathbb{Z}[\pi]) \subseteq Wh(\pi \times \mathbb{Z}) .$$

Moreover, 26.6 (ii) shows that the infinite simple homotopy equivalence $G : X \times S^1 \simeq W \times \mathbb{R}$ of 20.3 (iii) is not (in general) simple as a homotopy equivalence with respect to the canonical simple homotopy types, since it has torsion

$$\tau(G) = \Phi^-(W,c) = [X^+] \in \widetilde{K}_0(\mathbb{Z}[\pi]) \subseteq Wh(\pi \times \mathbb{Z}) .$$

□

Definition 26.8 The *relaxation* of an $A[z,z^{-1}]$-module chain complex band C is the ζ-twist glueing

$$C' = C[\zeta : C \longrightarrow C] .$$

□

Proposition 26.9 *Let C be an $A[z,z^{-1}]$-module chain complex band. Given a Mayer–Vietoris presentation (C^+,C^-) let*

$$f^+ : D = C^+ \cap C^- \longrightarrow E = C^+ \cap \zeta C^- \ ; \ x \longrightarrow x ,$$
$$f^- : D = C^+ \cap C^- \longrightarrow E = C^+ \cap \zeta C^- \ ; \ y \longrightarrow \zeta y$$

so that there is defined an exact sequence

$$0 \longrightarrow D[z,z^{-1}] \xrightarrow{f^+ - z^{-1}f^-} E[z,z^{-1}] \longrightarrow C \longrightarrow 0 .$$

The relaxation of C is given up to simple chain equivalence by

$$C' = \mathcal{C}(f'^+ - z^{-1}f'^- : D[z,z^{-1}] \longrightarrow E[z,z^{-1}])$$

with

$$f'^+ = f^+[\zeta] \ , \ f'^- = f^-[\zeta] : D \longrightarrow E$$

such that there are defined chain homotopy commutative squares

$$\begin{array}{ccc}
D & \xrightarrow{\quad f'^{+} \quad} & E \\
{\scriptstyle u}\downarrow\simeq & & \simeq\downarrow{\scriptstyle v} \\
C \oplus C^{+,lf} \oplus C^{-,lf} & \xrightarrow{1 \oplus 1 \oplus 0} & C \oplus C^{+,lf} \oplus C^{-,lf}
\end{array}$$

$$\begin{array}{ccc}
D & \xrightarrow{\quad f'^{-} \quad} & E \\
{\scriptstyle u}\downarrow\simeq & & \simeq\downarrow{\scriptstyle v} \\
C \oplus C^{+,lf} \oplus C^{-,lf} & \xrightarrow{\zeta \oplus 0 \oplus 1} & C \oplus C^{+,lf} \oplus C^{-,lf}
\end{array}$$

with u, v as in 23.11. □

As in Chapter 24 define the Whitehead group relaxation map

$$\begin{aligned}
Wh(A[z,z^{-1}]) &= Wh(A) \oplus \widetilde{K}_0(A) \oplus \widetilde{\mathrm{Nil}}_0(A) \oplus \widetilde{\mathrm{Nil}}_0(A) \\
&\longrightarrow Wh(A[z,z^{-1}]) \; ; \\
x &= (a,b,c,d) \longrightarrow x' = (a,b,0,0) \; .
\end{aligned}$$

Proposition 26.10 *Let C be an $A[z,z^{-1}]$-module chain complex band.*

(i) *The relaxation C' is a relaxed chain complex band with fibring obstructions the relaxations of the fibring obstructions of C:*

$$\Phi^{\pm}(C') = \Phi^{\pm}(C)' \in Wh(A[z,z^{-1}]) \; .$$

In the notation of 23.11

$$\begin{aligned}
\Phi^{+}(C') &= (\phi^{+}, -[C^{-}], 0, 0) \;\; , \;\; \Phi^{-}(C') = (\phi^{-}, [C^{+}], 0, 0) \\
&\in Wh(A[z,z^{-1}]) = Wh(A) \oplus \widetilde{K}_0(A) \oplus \widetilde{\mathrm{Nil}}_0(A) \oplus \widetilde{\mathrm{Nil}}_0(A)
\end{aligned}$$

with

$$\begin{aligned}
\Phi^{+}(C) &= (\phi^{+}, -[C^{-}], -[C^{+,lf}, \zeta^{+,lf}], -[C^{-,lf}, \zeta^{-,lf}]) \; , \\
\Phi^{-}(C) &= (\phi^{-}, [C^{+}], -[C^{+,lf}, \zeta^{+,lf}], -[C^{-,lf}, \zeta^{-,lf}]) \\
&\in Wh(A[z,z^{-1}]) = Wh(A) \oplus \widetilde{K}_0(A) \oplus \widetilde{\mathrm{Nil}}_0(A) \oplus \widetilde{\mathrm{Nil}}_0(A) \; .
\end{aligned}$$

(ii) *The relaxation C' is chain equivalent to C, with torsion*

$$\begin{aligned}\tau(C \simeq C') &= \Phi^+(C') - \Phi^+(C) = \Phi^-(C') - \Phi^-(C) \\ &= (0, 0, [C^{+,lf}, \zeta^{+,lf}], [C^{-,lf}, \zeta^{-,lf}]) \\ &\in Wh(A[z, z^{-1}]) = Wh(A) \oplus \widetilde{K}_0(A) \oplus \widetilde{\mathrm{Nil}}_0(A) \oplus \widetilde{\mathrm{Nil}}_0(A)\end{aligned}$$

the $\widetilde{\mathrm{Nil}}$-components of the fibring obstructions $\Phi^{\pm}(C) \in Wh(A[z, z^{-1}])$.

(iii) *C is relaxed if and only if $\tau(C \simeq C') = 0 \in Wh(A[z, z^{-1}])$.*

(iv) $\tau(C' \simeq (C')') = 0 \in Wh(A[z, z^{-1}])$.

Proof Apply 26.3 with $h = \zeta : C \longrightarrow C$. □

Example 26.11 If (W, c) is an untwisted CW π_1-band with relaxation (W', c') (20.7) and $\pi_1(W) = \pi \times \mathbb{Z}$ then $C(\widetilde{W})$ is a $\mathbb{Z}[\pi][z, z^{-1}]$-module chain complex band with relaxation

$$C(\widetilde{W})' = C(\widetilde{W}')$$

(up to simple chain equivalence) and

$$\begin{aligned}\Phi^{\pm}(W, c) &= \Phi^{\pm}(C(\widetilde{W})) \\ &= (\phi^{\pm}, \mp[\overline{W}^{\pm}], -[C^{lf,\pi}(\widetilde{W}^+), \zeta^{+,lf}], -[C^{lf,\pi}(\widetilde{W}^-), \zeta^{-,lf}]), \\ \Phi^{\pm}(W', c') &= \Phi^{\pm}(C(\widetilde{W})') = \Phi^{\pm}(C(\widetilde{W}))' = (\phi^{\pm}, \mp[\overline{W}^{\pm}], 0, 0) \\ &\in Wh(\pi \times \mathbb{Z}) = Wh(\pi) \oplus \widetilde{K}_0(\mathbb{Z}[\pi]) \oplus \widetilde{\mathrm{Nil}}_0(\mathbb{Z}[\pi]) \oplus \widetilde{\mathrm{Nil}}_0(\mathbb{Z}[\pi]) .\end{aligned}$$

□

Proposition 26.12 *An untwisted CW π_1-band (W, c) with $\pi_1(W) = \pi \times \mathbb{Z}$ is simple homotopy equivalent to a relaxed CW π_1-band if and only if the $\mathbb{Z}[\pi][z, z^{-1}]$-module chain complex band $C(\widetilde{W})$ is relaxed.*

Proof If (W, c) is an untwisted CW π_1-band which is simple homotopy equivalent to a relaxed CW π_1-band then $C(\widetilde{W})$ is a relaxed chain complex band over $\mathbb{Z}[\pi][z, z^{-1}]$, by 24.9 and the simple chain homotopy invariance of the nilpotent classes of a chain complex band C

$$[C^{+,lf}, \zeta^{+,lf}] , [C^{-,lf}, \zeta^{-,lf}] \in \widetilde{\mathrm{Nil}}_0(A) .$$

Conversely, if (W, c) is an untwisted CW π_1-band such that $C(\widetilde{W})$ is a relaxed chain complex band then (W, c) is simple homotopy equivalent to the relaxation (W', c'), which is a relaxed CW π_1-band. □

By analogy with 20.12:

Proposition 26.13 *An $A[z,z^{-1}]$-module chain complex band C and its relaxation C' are related by the* relaxation algebraic h-cobordism $(B;C,C')$ *with B an $A[z,z^{-1}]$-module chain complex band and $C\longrightarrow B$, $C'\longrightarrow B$ chain equivalences such that*

$$\begin{aligned}
\Phi^+(B) &= (\phi^+,-[C^-],-[C^{+,lf},\zeta^{+,lf}],0)\ ,\\
\Phi^-(B) &= (\phi^-,[C^+],0,-[C^{-,lf},\zeta^{-,lf}])\ ,\\
\tau(C\longrightarrow B) &= (0,0,0,[C^{-,lf},\zeta^{-,lf}])\ ,\\
\tau(C'\longrightarrow B) &= (0,0,[C^{+,lf},\zeta^{+,lf}],0)\\
\in Wh(A[z,z^{-1}]) &= Wh(A)\oplus\widetilde{K}_0(A)\oplus\widetilde{\text{Nil}}_0(A)\oplus\widetilde{\text{Nil}}_0(A)\ .
\end{aligned}$$

Proof Any finite based f.g. free $A[z,z^{-1}]$-module chain complex B with a chain equivalence $C\longrightarrow B$ such that $\tau(C\longrightarrow B)=(0,0,0,[C^{-,lf},\zeta^{-,lf}])$ will satisfy the other conditions. The wrapping up $\widehat{C}$ can be used to give an explicit construction of B, by analogy with the construction of the relaxation CW h-cobordism $(Z;W,W')$ in 20.12. Define

$$B = \mathcal{C}(g^+ - z^{-1}g^- : E[z,z^{-1}]\longrightarrow F[z,z^{-1}])$$

with F the algebraic mapping cylinder of a simple A-module chain equivalence

$$\mathcal{C}(\begin{pmatrix}\widehat{f}^-\\ f'^+\end{pmatrix} : D\longrightarrow E\oplus E) \xrightarrow{\simeq} \mathcal{C}(\begin{pmatrix}f^-\\ \widehat{f}^+\end{pmatrix} : D\longrightarrow E\oplus E)$$

and $g^+,g^- : E\longrightarrow F$ the chain maps appearing in the commutative diagram

$$\begin{array}{ccccc}
D & \xrightarrow{\widehat{f}^+} & E & \xleftarrow{\widehat{f}^-} & D\\
{\scriptstyle f^+}\big\downarrow & & \big\downarrow{\scriptstyle g^+} & & \big\downarrow{\scriptstyle f'^+}\\
E & \longrightarrow & F & \longleftarrow & E\\
{\scriptstyle f^-}\big\uparrow & & \big\uparrow{\scriptstyle g^-} & & \big\uparrow{\scriptstyle f'^-}\\
D & \xrightarrow{\widehat{f}^+} & E & \xleftarrow{\widehat{f}^-} & D
\end{array}$$

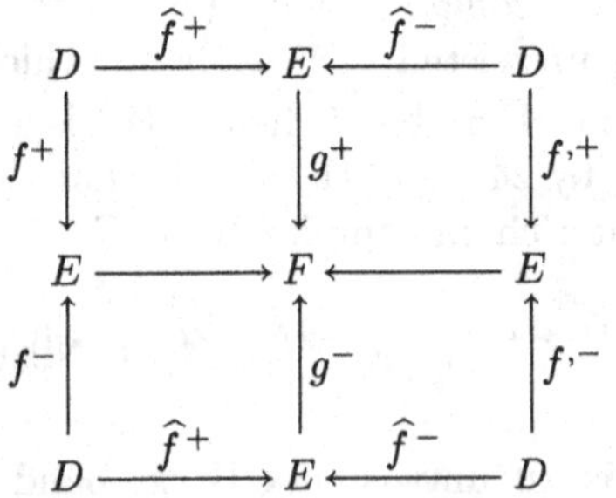

which is chain equivalent (via u,v) to the commutative diagram

$$\begin{array}{ccccc}
C \oplus C^{+,lf} \oplus C^{-,lf} & \xrightarrow{1\oplus 1\oplus 0} & C \oplus C^{+,lf} \oplus C^{-,lf} & \xleftarrow{1\oplus 0\oplus 1} & C \oplus C^{+,lf} \oplus C^{-,lf} \\
\downarrow{\scriptstyle 1\oplus 1\oplus \zeta^{-,lf}} & & \downarrow{\scriptstyle 1\oplus 1\oplus 0} & & \downarrow{\scriptstyle 1\oplus 1\oplus 0} \\
C \oplus C^{+,lf} \oplus C^{-,lf} & \xrightarrow{1\oplus 1\oplus 0} & C \oplus C^{+,lf} \oplus C^{-,lf} & \xleftarrow{1\oplus 0\oplus 1} & C \oplus C^{+,lf} \oplus C^{-,lf} \\
\uparrow{\scriptstyle \zeta\oplus \zeta^{+,lf}\oplus 1} & & \uparrow{\scriptstyle \zeta\oplus \zeta^{+,lf}\oplus 1} & & \uparrow{\scriptstyle \zeta\oplus 0\oplus 1} \\
C \oplus C^{+,lf} \oplus C^{-,lf} & \xrightarrow{1\oplus 1\oplus 0} & C \oplus C^{+,lf} \oplus C^{-,lf} & \xleftarrow{1\oplus 0\oplus 1} & C \oplus C^{+,lf} \oplus C^{-,lf}
\end{array}$$

□

Remark 26.14 Let $(W, \partial W)$ be an open n-dimensional manifold with compact boundary and one end. In Chapter 17 we used geometric wrapping up to show that if $n \geq 5$ and W is both forward and reverse tame then the end has an open neighbourhood $V \subset W$ which is the infinite cyclic cover $(V, d) = (\overline{U}, \overline{c})$ of a relaxed manifold band (U, c). Moreover, it was shown that $(U, c) = (\widehat{V}, \widehat{d})$ is the wrapping up of the ribbon (V, d), with a homotopy equivalence $e(W) \simeq \overline{U}$ and a homeomorphism $\overline{U} \times S^1 \cong U \times \mathbb{R}$. If $(W, \partial W) = (\overline{M}^+, \overline{M}^+ \cap \overline{M}^-)$ is an end of the infinite cyclic cover $\overline{M} = c^*\mathbb{R}$ of a manifold band (M, c) then U, V are such that $V = \overline{U} \cong \overline{M}$, $M \times S^1 \cong T(\widehat{\zeta})$ for a homeomorphism $\widehat{\zeta} : U \longrightarrow U$ in the homotopy class of $\zeta \times 1 : U \simeq \overline{M} \times S^1 \longrightarrow \overline{M} \times S^1 \simeq U$ (17.10). The underlying (simple) homotopy types were obtained in Chapter 19 using homotopy theoretic twist glueing. In the terminology of Chapter 19 the wrapping up $\widehat{W} = M[1]$ is a manifold in the canonical simple homotopy type of the 1-twist glueing $M(1)$ of M. By 26.4

$$C(\widetilde{U}) \;=\; C(\widetilde{M(1)}) \;=\; C[1 : C(\widetilde{M}) \longrightarrow C(\widetilde{M})] \;=\; \widehat{C(\widetilde{M})}\ .$$

The relaxation CW h-cobordism $(Z; W, W')$ of 20.12 between an untwisted CW π_1-band (W, c) and the relaxation (W', c') induces the chain complex h-cobordism $(B; C, C')$ of 26.13, with

$$B \;=\; C(\widetilde{Z})\ ,\ C \;=\; C(\widetilde{W})\ ,\ C' \;=\; C(\widetilde{W}')\ .$$

The relaxation of an untwisted manifold band (W, c) is an untwisted manifold band (W', c'), and there exists a manifold h-cobordism $(Z; W, W')$ realizing the chain complex h-cobordism $(B; C, C')$, verifying the correspondence of the manifold and CW relaxation h-cobordisms (20.13 (i)). □

27

Wrapping up in algebraic K- and L-theory

The geometric wrapping up construction of Chapter 17 and the algebraic wrapping up construction of Chapter 26 are now related to the splitting theorems for the algebraic K- and L-groups of a Laurent polynomial extension $A[z,z^{-1}]$ (with involution on A and $\overline{z}=z^{-1}$ in L-theory)

$$\begin{aligned} K_1(A[z,z^{-1}]) &= K_1(A)\oplus K_0(A)\oplus \widetilde{\mathrm{Nil}}_0(A)\oplus \widetilde{\mathrm{Nil}}_0(A)\ , \\ L_h^n(A[z,z^{-1}]) &= L_h^n(A)\oplus L_p^{n-1}(A)\ , \\ L_n^h(A[z,z^{-1}]) &= L_n^h(A)\oplus L_{n-1}^p(A) \end{aligned}$$

and the corresponding results for the $\mathbb{R}$-bounded category $\mathbb{C}_{\mathbb{R}}(A)$

$$\begin{aligned} K_1(\mathbb{C}_{\mathbb{R}}(A)) &= K_1(A[z,z^{-1}])^{INV} &= K_0(A)\ , \\ L^n(\mathbb{C}_{\mathbb{R}}(A)) &= L_h^n(A[z,z^{-1}])^{INV} &= L_p^{n-1}(A)\ , \\ L_n(\mathbb{C}_{\mathbb{R}}(A)) &= L_n^h(A[z,z^{-1}])^{INV} &= L_{n-1}^p(A) \end{aligned}$$

with INV denoting the subgroup of the elements invariant under the transfers induced from all the finite covers $q: S^1 \longrightarrow S^1$ $(q\geq 1)$ of S^1. We refer to Ranicki [124] for the chain complex treatment of these splitting theorems.

Regard S^1 as the unit circle $\{z\in\mathbb{C}\,|\,|z|=1\}$ in the complex plane. For each integer $q\geq 1$ complex multiplication defines a q-fold covering of S^1 by itself

$$q: S^1 \longrightarrow S^1\ ;\ z\longrightarrow z^q\ ,$$

corresponding to injections of the rings $A[z]$, $A[z,z^{-1}]$ into themselves by

$$\begin{aligned} q &: A[z]\longrightarrow A[z]\ ;\ z\longrightarrow z^q\ , \\ q &: A[z,z^{-1}]\longrightarrow A[z,z^{-1}]\ ;\ z\longrightarrow z^q\ . \end{aligned}$$

In Chapter 24 we considered the q-fold transfer

$$q^!: \{A[z]\text{-modules}\}\longrightarrow \{A[z]\text{-modules}\}\ ;\ M\longrightarrow M^!\ .$$

Definition 27.1 The *q-fold transfer* functor

$$q^! \; : \; \{A[z,z^{-1}]\text{-modules}\} \longrightarrow \{A[z,z^{-1}]\text{-modules}\} \; ; \; M \longrightarrow M^!$$

sends an $A[z,z^{-1}]$-module M to the $A[z,z^{-1}]$-module $M^!$ with the same additive group and

$$A[z,z^{-1}] \times M^! \longrightarrow M^! \; ; \; (z,x) \longrightarrow z^q x \; . \qquad \square$$

Proposition 27.2 (Ranicki [124, Chapters 12,18]) (i) *The q-fold transfer map induced in algebraic K-theory is such that*

$$\begin{aligned} q^! \; = \; & q \oplus 1 \oplus q^! \oplus q^! \; : \\ & K_1(A[z,z^{-1}]) \; = \; K_1(A) \oplus K_0(A) \oplus \widetilde{\mathrm{Nil}}_0(A) \oplus \widetilde{\mathrm{Nil}}_0(A) \\ & \quad \longrightarrow K_1(A[z,z^{-1}]) \; = \; K_1(A) \oplus K_0(A) \oplus \widetilde{\mathrm{Nil}}_0(A) \oplus \widetilde{\mathrm{Nil}}_0(A) \end{aligned}$$

with

$$q^! \; : \; \widetilde{\mathrm{Nil}}_0(A) \longrightarrow \widetilde{\mathrm{Nil}}_0(A) \; ; \; (P,\nu) \longrightarrow (P,\nu^q) \; .$$

The transfer invariant subgroup of $K_1(A[z,z^{-1}])$

$$K_1(A[z,z^{-1}])^{INV} \; = \; \{x \in K_1(A[z,z^{-1}]) \,|\, q^!(x) = x \text{ for } q \geq 2\}$$

is such that there is defined an isomorphism

$$K_0(A) \xrightarrow{\simeq} K_1(A[z,z^{-1}])^{INV} \; ; \; [P] \longrightarrow \tau(-z : P[z,z^{-1}] \longrightarrow P[z,z^{-1}]) \; .$$

(ii) *The q-fold transfer maps induced in symmetric L-theory are such that*

$$\begin{aligned} q^! \; = \; q \oplus 1 : \; & L^n_h(A[z,z^{-1}]) \; = \; L^n_h(A) \oplus L^{n-1}_p(A) \\ & \longrightarrow L^n_h(A[z,z^{-1}]) \; = \; L^n_h(A) \oplus L^{n-1}_p(A) \; . \end{aligned}$$

The transfer invariant subgroup of $L^n_h(A[z,z^{-1}])$

$$L^n_h(A[z,z^{-1}])^{INV} \; = \; \{x \in L^n_h(A[z,z^{-1}]) \,|\, q^!(x) = x \text{ for } q \geq 2\}$$

is such that there is defined an isomorphism

$$L^{n-1}_p(A) \xrightarrow{\simeq} L^n_h(A[z,z^{-1}])^{INV} \; ; \; (C,\phi) \longrightarrow (C,\phi) \otimes \sigma^*(S^1)$$

with (C,ϕ) *any finitely dominated* $(n-1)$*-dimensional symmetric Poincaré complex over* A*, and* $\sigma^*(S^1)$ *the* 1*-dimensional symmetric Poincaré complex over* $\mathbb{Z}[z,z^{-1}]$ *of* S^1*. Similarly for quadratic L-theory* L_*.

(iii) *The forgetful functor*

$$\{\text{based f.g. free } A[z,z^{-1}]\text{-modules}\} \longrightarrow \mathbb{C}_{\mathbb{R}}(A)$$

induces isomorphisms

$$K_1(A[z,z^{-1}])^{INV} \cong K_1(\mathbb{C}_{\mathbb{R}}(A)) ,$$
$$L^n_h(A[z,z^{-1}])^{INV} \cong L^n(\mathbb{C}_{\mathbb{R}}(A)) .$$

Similarly for the Whitehead group Wh and quadratic L-theory L_*. □

Let W be a space with a map $c : W \longrightarrow S^1$. The pullback along c of the q-fold cover $q : S^1 \longrightarrow S^1$ of S^1 is the q-fold cover $(W^!, c^!)$ of (W,c) given by

$$W^! = \{(x,y) \in W \times S^1 \,|\, c(x) = y^q\} \ , \ c^!(x,y) = y \in S^1$$

with

$$\zeta^! = \zeta^q : \overline{W}^! = \overline{W} \longrightarrow \overline{W}^! = \overline{W} \ , \ W^! = \overline{W}^!/\zeta^! = \overline{W}/\zeta^q .$$

$$\begin{array}{ccc} W^! & \xrightarrow{c^!} & S^1 \\ \downarrow & & \downarrow q \\ W & \xrightarrow{c} & S^1 \end{array}$$

If $(V;U,\zeta U)$ is a fundamental domain for the infinite cyclic cover $\overline{W}$ of W then

$$\bigcup_{j=0}^{q-1}(\zeta^j V;\zeta^j U,\zeta^{j+1}U) = (\bigcup_{j=0}^{q-1}\zeta^j V;U,\zeta^q U)$$

is a fundamental domain for the infinite cyclic cover $\overline{W}^!$ of $W^!$, with identifications

$$W = V/(U=\zeta U) ,$$
$$W^! = (\bigcup_{j=0}^{q-1}\zeta^j V)/(U=\zeta^q U) ,$$
$$\overline{W}^! = \overline{W} = \bigcup_{j=-\infty}^{\infty}\zeta^j V .$$

If W is connected then so is $W^!$, and the morphism of fundamental groups induced by the covering projection $W^! \longrightarrow W$ fits into an exact sequence

$$\{1\} \longrightarrow \pi_1(W^!) \longrightarrow \pi_1(W) \longrightarrow \mathbb{Z}_q \longrightarrow \{1\} .$$

Proposition 27.3 *Let (W,c) be an untwisted CW band, so that $\zeta_* = 1 : \pi_1(\overline{W}) = \pi \longrightarrow \pi$ and*

$$\pi_1(W) = \pi \times \mathbb{Z} \ , \ \mathbb{Z}[\pi_1(W)] = \mathbb{Z}[\pi][z,z^{-1}] .$$

(i) *The q-fold cover* $(W^!, c^!)$ *of* (W, c) *induced from the q-fold cover* $q : S^1 \longrightarrow S^1$ *is an untwisted CW band with fibring obstructions*

$$\Phi^{\pm}(W^!, c^!) = q^!\Phi^{\pm}(W, c) \in Wh(\pi \times \mathbb{Z}) .$$

(ii) *If* (W, c) *is an n-dimensional geometric Poincaré band then so is* $(W^!, c^!)$, *with symmetric signature*

$$\sigma^*(W^!) = q^!\sigma^*(W) \in L^n_h(\mathbb{Z}[\pi][z, z^{-1}]) .$$

Proof If W is a CW complex with a map $c : W \longrightarrow S^1$ inducing

$$c_* = \text{projection} : \pi_1(W) = \pi \times \mathbb{Z} \longrightarrow \pi_1(S^1) = \mathbb{Z}$$

then the covering projection $W^! \longrightarrow W$ of the induced q-fold cover $W^!$ of W induces the injection of fundamental groups

$$\pi_1(W^!) = \pi \times \mathbb{Z} \longrightarrow \pi_1(W) = \pi \times \mathbb{Z} ; (g, z) \longrightarrow (g, z^q) .$$

The universal cover $\widetilde{W}^!$ of $W^!$ is the universal cover $\widetilde{W}$ of W with the q-fold $\pi \times \mathbb{Z}$-action, so that as a $\mathbb{Z}[\pi][z, z^{-1}]$-module chain complex

$$C(\widetilde{W}^!) = q^!C(\widetilde{W}) .$$

Similarly for the finiteness obstruction, Whitehead torsion, the fibring obstructions and the symmetric signature. □

The algebraic K-theory effect of wrapping up is given by:

Proposition 27.4 (i) *The wrapping up of a CW ribbon* (X, d) *is a relaxed CW band* $(\widehat{X}, \widehat{d})$ *which is transfer invariant: for every finite cover* $q : S^1 \longrightarrow S^1$ *of* S^1 *the pullback q-fold cover* $(\widehat{X}^!, \widehat{d}^!)$ *is simple homotopy equivalent to* $(\widehat{X}, \widehat{d})$. *The fibring obstructions of* $(\widehat{X}, \widehat{d})$ *are given by*

$$\begin{aligned} \Phi^+(\widehat{X}, \widehat{d}) &= -[X^-] \ , \ \Phi^-(\widehat{X}, \widehat{d}) = [X^+] \\ &\in Wh(\pi \times \mathbb{Z})^{INV} = \widetilde{K}_0(\mathbb{Z}[\pi]) . \end{aligned}$$

(ii) *A CW band* (W, c) *is simple homotopy equivalent to the wrapping up* $(\widehat{X}, \widehat{d})$ *of a CW ribbon* (X, d) *if and only if it is relaxed and there exists a homotopy* $\zeta \simeq 1 : \overline{W} \longrightarrow \overline{W}$.

(iii) *Let* $f : (X, d) \longrightarrow (Y, e)$ *be an* $\mathbb{R}$*-bounded homotopy equivalence of CW ribbons, with* $\pi_1(X) = \pi_1(Y) = \pi$. *The isomorphisms*

$$Wh(\mathbb{C}_{\mathbb{R}}(\mathbb{Z}[\pi])) \cong Wh(\pi \times \mathbb{Z})^{INV} \cong \widetilde{K}_0(\mathbb{Z}[\pi])$$

send the torsion $\tau_{\mathbb{R}}(f) \in Wh(\mathbb{C}_{\mathbb{R}}(\mathbb{Z}[\pi]))$ *to the transfer invariant torsion* $\tau(\widehat{f}) \in Wh(\pi \times \mathbb{Z})^{INV}$ *of the induced homotopy equivalence of the wrapping up relaxed CW bands*

$$\widehat{f} : \widehat{X} \simeq X \times S^1 \xrightarrow{f \times 1} Y \times S^1 \simeq \widehat{Y} ,$$

and to $[Y^+]-[X^+]\in\widetilde{K}_0(\mathbb{Z}[\pi])$.

Proof (i) The fundamental domain $(V;U,\zeta U)$ for $\overline{W}$ given by 19.12 is such that the inclusions are idempotents

$$f^+ = i^+j^+ \ , \ f^- = i^-j^- \ : U \longrightarrow V = U \ .$$

There exists a rel ∂ simple homotopy equivalence

$$(V;U,\zeta U)\cup(\zeta V;\zeta U,\zeta^2 U) \ \simeq \ (V;U,\zeta U)\ ,$$

so that for any $q\geq 2$ there exists a rel ∂ simple homotopy equivalence

$$\bigcup_{j=0}^{q-1}(\zeta^j V;\zeta^j U,\zeta^{j+1}U) \ \simeq \ (V;U,\zeta U)$$

between the fundamental domains of $\overline{W}^!$ and $\overline{W}$. The fibring obstructions have already been computed in 26.6 for the corresponding algebraic wrapping up.

(ii) If (W,c) is simple homotopy equivalent to a wrapping up then it is relaxed and $\zeta\simeq 1:\overline{W}\longrightarrow\overline{W}$ by 19.9. Conversely, if (W,c) is relaxed and $\zeta\simeq 1$ then

$$\begin{aligned}\Phi^+(W,c) \ &= \ -[\overline{W}^-] \ , \ \Phi^-(W,c) \ = \ [\overline{W}^+] \\ &\in Wh(\pi\times\mathbb{Z})^{INV} \ = \ \widetilde{K}_0(\mathbb{Z}[\pi])\ ,\end{aligned}$$

so that (W,c) is simple homotopy equivalent to the wrapping up $(\widehat{X},\widehat{d})$ of the CW ribbon $(X,d)=(\overline{W},\overline{c})$

(iii) Immediate from 27.3 and (i). □

The algebraic L-theory effect of wrapping up is given by:

Proposition 27.5 *An n-dimensional geometric Poincaré ribbon (X,d) is a finitely dominated $(n-1)$-dimensional geometric Poincaré complex. The wrapping up of (X,d) is a transfer invariant relaxed n-dimensional geometric Poincaré band $(\widehat{X},\widehat{d})$ with infinite cyclic cover infinite simple homotopy equivalent to (X,d). If X is connected with universal cover $\widetilde{X}$ and $\pi_1(X)=\pi$ the isomorphisms*

$$Wh(\mathbb{C}_{\mathbb{R}}(\mathbb{Z}[\pi])) \ \cong \ Wh(\pi\times\mathbb{Z})^{INV} \ \cong \ \widetilde{K}_0(\mathbb{Z}[\pi])$$

send the $\mathbb{R}$-bounded torsion

$$\tau_{\mathbb{R}}(X) \ = \ \tau([X]\cap - : C(\widetilde{X})^{n-*}\longrightarrow C(\widetilde{X}))\in Wh(\mathbb{C}_{\mathbb{R}}(\mathbb{Z}[\pi]))$$

to the transfer invariant torsion $\tau(\widehat{X})\in Wh(\pi\times\mathbb{Z})^{INV}$ and to

$$[X^+]+(-)^{n-1}[X^-]^*\in\widetilde{K}_0(\mathbb{Z}[\pi])\ .$$

The isomorphisms

$$L^n(\mathbb{C}_{\mathbb{R}}(\mathbb{Z}[\pi])) \cong L^n_h(\mathbb{Z}[\pi][z,z^{-1}])^{INV} \cong L^{n-1}_p(\mathbb{Z}[\pi])$$

send the $\mathbb{R}$*-bounded symmetric signature* $\sigma^*(X) \in L^n(\mathbb{C}_{\mathbb{R}}(\mathbb{Z}[\pi]))$ *to the transfer invariant symmetric signature* $\sigma^*(\widehat{X}) \in L^n_h(\mathbb{Z}[\pi][z,z^{-1}])^{INV}$ *and to the projective symmetric signature* $\sigma^*(X) \in L^{n-1}_p(\mathbb{Z}[\pi])$. *Similarly for quadratic* L*-theory and the surgery obstructions of normal maps.* □

Remark 27.6 The theory of chain complex ribbons developed in Chapter 25 has an evident generalization to algebraic Poincaré ribbons, i.e. chain complex ribbons with abstract Poincaré duality. The algebraic K- and L-theory interpretations of wrapping up (27.4, 27.5) apply also to chain complex ribbons and their algebraic Poincaré analogues. □

Part Four: Appendices

Appendix A. Locally finite homology with local coefficients

A.1. Regular covers and singular homology and locally finite singular homology with local coefficients

In this section W denotes a path-connected space.

Let $p : \widetilde{W} \longrightarrow W$ be a regular cover with group of translations π. It is well-known that the ordinary singular chain complex $S(\widetilde{W})$ is a chain complex of $\mathbb{Z}[\pi]$-modules and has an interpretation in terms of local coefficients. Namely, $S(\widetilde{W})$ is isomorphic to $S_*(W;\Gamma)$ where Γ is a local system (described below) with $\Gamma_x \cong \mathbb{Z}[\pi]$ for each $x \in W$ and $S_*(W;\Gamma)$ is the singular chain complex of W with coefficients in the local system Γ (see Whitehead [169, p. 278] and Spanier [150, p. 179]).

The goal here is to generalize this to the locally finite case. However, when $S_*(W;\Gamma)$ is replaced by the locally finite $S_*^{lf}(W;\Gamma)$ we get a chain complex isomorphic to $S^{lf,\pi}(\widetilde{W})$, the locally π-finite singular chain complex, rather than $S^{lf}(\widetilde{W})$. This should help convince the reader that $S^{lf,\pi}(\widetilde{W})$ is a more natural object than $S^{lf}(\widetilde{W})$.

We begin by setting up the notation needed to describe the local $\mathbb{Z}[\pi]$ coefficient system on W.

Assume that W and $\widetilde{W}$ are path-connected. For each $x \in W$ fix $\tilde{x} \in p^{-1}(x)$ and let

$$\pi_x \;=\; \pi_1(W,x)/p_*\pi_1(\widetilde{W},\tilde{x}) \;.$$

Of course, $\pi_x \cong \pi$ for each $x \in W$, but we need to have an explicit isomorphism γ_x on hand. To this end, for each $x \in W$ define a bijection

$$\alpha_x \;:\; \pi \xrightarrow{\simeq} p^{-1}(x) \;;\; g \longrightarrow g(\tilde{x})$$

and also a bijection

$$\beta_x \ : \ p^{-1}(x) \xrightarrow{\simeq} \pi_x \ ; \ y \longrightarrow [p\omega]$$

where ω is a path in $\widetilde{W}$ from $\tilde{x}$ to y. Let

$$\gamma_x \ = \ \beta_x \circ \alpha_x \ : \ \pi \longrightarrow \pi_x \ .$$

Then γ_x is a group isomorphism.

Given a path $\omega : I \longrightarrow W$ define an isomorphism

$$\pi_\omega \ : \ \pi_{\omega(0)} \xrightarrow{\simeq} \pi_{\omega(1)} \ ; \ [\lambda] \longrightarrow [\omega^{-1} * \lambda * \omega] \ .$$

Let $\Gamma_x = \mathbb{Z}[\pi_x]$ for $x \in W$. The isomorphism $\gamma_x : \pi \xrightarrow{\simeq} \pi_x$ induces a ring isomorphism, also denoted by γ_x,

$$\gamma_x \ : \ \mathbb{Z}[\pi] \xrightarrow{\simeq} \Gamma_x \ ,$$

which thereby gives Γ_x the structure of a $\mathbb{Z}[\pi]$-module.

For each path $\omega : I \longrightarrow W$ π_ω extends to an isomorphism

$$\Gamma_\omega \ : \ \Gamma_{\omega(0)} \xrightarrow{\simeq} \Gamma_{\omega(1)} \ ,$$

so that there is defined a local system Γ of $\mathbb{Z}[\pi]$-modules on W [150, p. 179].

Using the local system Γ we can construct the usual singular chain complex of W with coefficients in Γ as well as the locally finite singular chain complex of W with coefficients in Γ.

Let $S_*(W;\Gamma) = \{S_q(W;\Gamma), \partial\}$ be the singular chain complex of W with local $\mathbb{Z}[\pi]$-coefficients as defined in [150, p. 179]. For $\sigma \in W^{\Delta^q}$, $\Gamma(\sigma)$ is the $\mathbb{Z}[\pi]$-module of Γ sections of σ. An element of $\Gamma(\sigma)$ is a function

$$s \ : \ \Delta^q \longrightarrow \coprod_{y \in \Delta^q} \Gamma_{\sigma(y)}$$

such that $s(y) \in \Gamma_{\sigma(y)}$ for each $y \in \Delta^q$ and for every path $\omega : I \longrightarrow \Delta^q$

$$\Gamma_{\sigma \circ \omega}(s(\omega(0))) \ = \ s(\omega(1)) \ .$$

Then

$$\begin{aligned} S_q(W;\Gamma) \ = \ \{\text{functions } c : W^{\Delta^q} \longrightarrow \coprod_{\sigma \in W^{\Delta^q}} \Gamma(\sigma) \, | \\ c \text{ is finitely non-zero and } c(\sigma) \in \Gamma(\sigma) \text{ for each } \ \sigma \in W^{\Delta^q}\} \ . \end{aligned}$$

The boundary operator ∂ is defined by restriction [150, p. 179].

The locally finite singular chain complex $S_*^{lf}(W;\Gamma) = \{S_q^{lf}(W;\Gamma), \partial\}$ with coefficients in Γ is defined as follows.

$$\begin{aligned} S_q^{lf}(W;\Gamma) = \{&\text{functions } c : W^{\Delta^q} \longrightarrow \coprod_{\sigma \in W^{\Delta^q}} \Gamma(\sigma) \mid \\ &c(\sigma) \in \Gamma(\sigma) \text{ for each } \sigma \in W^{\Delta^q} \text{ and } \{\sigma(\Delta^q) \mid c(\sigma) \neq 0\} \\ &\text{is a locally finite family in } W\} . \end{aligned}$$

The boundary operator ∂ is induced from the one above [150, p. 182].

On the other hand, the singular chain complex $S_*(\widetilde{W})$, the locally finite singular chain complex $S^{lf}(\widetilde{W})$ and the locally π-finite singular chain complex $S^{lf,\pi}(\widetilde{W})$ are also chain complexes of $\mathbb{Z}[\pi]$-modules.

Let 0 denote the zeroth vertex of Δ^q. Then $S_q(\widetilde{W})$ is a free $\mathbb{Z}[\pi]$-module with basis consisting of those singular q-simplexes $\sigma : \Delta^q \longrightarrow \widetilde{W}$ such that $\sigma(0) = \tilde{x}$ for some $x \in W$. Thus

$$S_q(\widetilde{W}) \cong \bigoplus \mathbb{Z}[\pi]$$

where the direct sum is over the set of singular q-simplexes in W. On the other hand, $S_q^{lf,\pi}(\widetilde{W})$ is not a free $\mathbb{Z}[\pi]$-module, and it is not even clear when it is a non-trivial direct product.

Proposition A.1 *There are isomorphisms of $\mathbb{Z}[\pi]$-module chain complexes*

$$A : S_*(\widetilde{W}) \xrightarrow{\simeq} S_*(W;\Gamma) \ , \quad A^{lf,\pi} : S_*^{lf,\pi}(\widetilde{W}) \xrightarrow{\simeq} S_*^{lf}(W;\Gamma) \ .$$

Proof For a basis element $\sigma \in S^q(\widetilde{W})$ define $A(\sigma) \in S_q(W;\Gamma)$ by

$$A(\sigma)(\tau) = \begin{cases} 0 & \text{if } \tau \neq p \circ \sigma \ , \\ 1 & \text{if } \tau = p \circ \sigma \ . \end{cases}$$

Since $A(\sigma)(\tau) \in \Gamma(\tau)$, we have

$$A(\sigma)(\tau) : \Delta^q \longrightarrow \coprod_{y \in \Delta^q} \Gamma_{\tau(y)} \ .$$

To say that $A(\sigma)(\tau) = 1$ when $\tau = p \circ \sigma$, we mean that

$$A(\sigma)(\tau)(y) = 1 \in \Gamma_{\tau(y)} = \mathbb{Z}[\pi_{\tau(y)}] \text{ for each } y \in \Delta^q \ .$$

To investigate the effect of A on an arbitrary element of $S^q(\widetilde{W})$ we proceed as follows.

For each $\sigma \in W^{\Delta^q}$, let $\phi_0 : \Gamma(\sigma) \longrightarrow \Gamma_{\sigma(0)}$ be the isomorphism of

[150, p. 179], $\phi_0(s) = s(\sigma(0))$. Then let $\delta_\sigma : \mathbb{Z}[\pi] \longrightarrow \Gamma(\sigma)$ be the isomorphism which is the composition

$$\delta_\sigma : \mathbb{Z}[\pi] \xrightarrow{\gamma_{\sigma(0)}} \Gamma_{\sigma(0)} \xrightarrow{\phi_0^{-1}} \Gamma(\sigma) .$$

An arbitrary element of $S_q(\widetilde{W})$ can be uniquely written as

$$\sum_{\tau \in W^{\Delta^q}} r_\tau \tilde{\tau}$$

where $r_\tau \in \mathbb{Z}[\pi]$, the set $\{r_\tau \,|\, r_\tau \neq 0\}$ is finite, and $\tilde{\tau} : \Delta^q \longrightarrow \widetilde{W}$ is the unique lift of τ with $\tilde{\tau}(0) = \widetilde{\tau(0)}$. Then

$$A(\sum_{\tau \in W^{\Delta^q}} r_\tau \tilde{\tau})(\sigma) = \delta_\sigma(r_\sigma) \in \Gamma(\sigma) .$$

Clearly, $\sigma \longrightarrow \delta_\sigma(r_\sigma)$ is finitely non-zero.

We now describe the inverse B for A. Let

$$c : W^{\Delta^q} \longrightarrow \coprod_{\sigma \in W^{\Delta^q}} \Gamma(\sigma)$$

be a finitely non-zero function such that $c(\sigma) \in \Gamma(\sigma)$ for each $\sigma \in W^{\Delta^q}$. Then

$$Bc = \sum_{\tau \in W^{\Delta^q}} \delta_\tau^{-1}(c\tau)\tilde{\tau}$$

where as above $\tilde{\tau}$ is the unique lift of τ with $\tilde{\tau}(0) = \widetilde{\tau(0)}$.

These formulas generalize easily to the locally π-finite case as follows. An element of $S_q^{lf,\pi}(\widetilde{W})$ can be uniquely written as

$$\prod_{\tau \in W^{\Delta^q}} r_\tau \tilde{\tau}$$

where $r_\tau \in \mathbb{Z}[\pi]$, the set $\{\tau(\Delta^q) \,|\, r_\tau \neq 0\}$ is locally finite, and $\tilde{\tau}$ is the unique lift of τ with $\tilde{\tau}(0) = \widetilde{\tau(0)}$. Then define

$$A^{lf,\pi} : S_q^{lf,\pi}(\widetilde{W}) \longrightarrow S_q^{lf}(W;\Gamma)$$

by

$$A^{lf,\pi}(\prod_{\tau \in W^{\Delta^q}} r_\tau \tilde{\tau})(\sigma) = \delta_\sigma(r_\sigma) \in \Gamma(\sigma) .$$

Clearly, the function $\sigma \longrightarrow \delta_\sigma(r_\sigma)$ has the correct local finiteness property and an inverse $B^{lf,\pi}$ for $A^{lf,\pi}$ can be defined similarly to B. □

Remarks A.2 (i) It might be worthwhile to note how the local system Γ_x depends on the choice of the basepoints $\tilde{x} \in p^{-1}(x)$. First note that the group π_x is independent of $\tilde{x}$, but γ_x is not. For if $x' \in p^{-1}(x)$ is another choice inducing the isomorphism $\gamma'_x : \pi \longrightarrow \pi_x$, let λ_x be a path in $\widetilde{W}$ from $\tilde{x}$ to x'. If $\hat{\lambda}_x : \pi_x \longrightarrow \pi_x$ is the inner automorphism given by

$$\hat{\lambda}_x(a) = [p\lambda_x] * a * [p\lambda_x]^{-1} \text{ for each } a \in \pi_x,$$

then $\gamma_x = \hat{\lambda}_x \circ \gamma'_x$. Since the isomorphism $\gamma_x : \mathbb{Z}[\pi] \longrightarrow \Gamma_x$ depends on the choice of $\tilde{x}$, so does the $\mathbb{Z}[\pi]$-module structure on Γ_x. In fact a different choice of basepoints, say $x' \in p^{-1}(x)$, induces a different local system Γ' which need not be isomorphic to Γ (in the sense of Steenrod [157]). However, the inner automorphisms $\hat{\lambda}_x : \pi_x \longrightarrow \pi_x$ induce isomorphisms $\hat{\lambda}_x : \Gamma_x \longrightarrow \Gamma'_x$ which in turn induce isomorphisms of chain complexes $\hat{\lambda}_x : S_*(W;\Gamma) \cong S_*(W;\Gamma')$.

(ii) In order to introduce another local system Λ on W consider the $\mathbb{Z}[\pi]$-module $\mathbb{Z}[[\pi]]$. It has elements written as formal products $\prod_{g\in\pi} n_g g$ with $n_g \in \mathbb{Z}$. Addition is defined termwise and $\mathbb{Z}[\pi]$ acts on $\mathbb{Z}[[\pi]]$ via

$$\sum_{h\in\pi} m_h h \cdot \prod_{g\in\pi} n_g g = \prod_{k\in\pi} l_k k \text{ where } l_k = \sum_{hg\,=\,k} m_h n_g \,.$$

Likewise the $\mathbb{Z}[\pi]$-module structure on $\mathbb{Z}[\pi_x]$ induces a $\mathbb{Z}[\pi]$-module structure on

$$\Lambda_x = \mathbb{Z}[[\pi_x]] \text{ for } x \in W$$

and the morphism $\Gamma_\omega : \Gamma_{\omega(0)} \longrightarrow \Gamma_{\omega(1)}$ induced by a path $\omega : I \longrightarrow W$ induces a $\mathbb{Z}[\pi]$-module morphism $\Lambda_\omega : \Lambda_{\omega(0)} \longrightarrow \Lambda_{\omega(1)}$. Thus, we have a local system Λ of $\mathbb{Z}[\pi]$-modules on W. It follows that we can consider the $\mathbb{Z}[\pi]$-module chain complexes $S_*(W;\Lambda)$ and $S_*^{lf}(W;\Lambda)$. It may be conjectured that $S_*^{lf}(\widetilde{W})$ and $S_*^{lf}(W;\Lambda)$ have isomorphic homology groups if W is locally contractible and paracompact. □

A.2. Cellular homology with local coefficients

Let W be a CW complex, let R be a commutative ring, and let Γ be an arbitrary local system of R-modules on W, where we use the notation and terminology of [150].

Define the nth *cellular chain module of* W *with local coefficients* Γ by

$$C_n(W;\Gamma) = H_n(W^n, W^{n-1};\Gamma) = H_n(S_*(W^n, W^{n-1};\Gamma))$$

with

$$S_p(W^n, W^{n-1};\Gamma) = \frac{S_p(W^n;\Gamma)}{S_p(W^{n-1};\Gamma)}$$

(as in [150, p. 181]). The cellular chain complex $C_*(W;\Gamma) = \{C_n(W;\Gamma), \partial\}$ of W with local coefficients has boundary operator

$$\begin{aligned} \partial \; : \; C_n(W;\Gamma) \; &= H_n(W^n, W^{n-1};\Gamma) \\ &\longrightarrow C_{n-1}(W;\Gamma) \; = \; H_{n-1}(W^{n-1}, W^{n-2};\Gamma) \end{aligned}$$

given by the connecting morphism for the triple (W^n, W^{n-1}, W^{n-2}).

This is spelled out in Whitehead [169, p. 283], where it is asserted that

$$H_n(C_*(W;\Gamma)) \;\cong\; H_n(W;\Gamma)$$

generalizing the classical case of global coefficients. We shall also outline a proof in section A.4 below.

Another straightforward generalization of the classical case is also given in Whitehead [169, p. 282], namely the direct sum representation of $C_n(W;\Gamma)$ described as follows. Let I_n be an indexing set for the set of n-cells of W and let $h_\alpha : \Delta^n \longrightarrow W$ $(\alpha \in I_n)$ be a characteristic map for the αth n-cell of W. If $\Gamma(h_\alpha)$ denotes the R-module of Γ sections, there is a natural morphism $\Gamma(h_\alpha) \longrightarrow S_n(W;\Gamma)$ [150, p. 180] which represents $S_n(W;\Gamma)$ as a direct sum

$$C_n(W;\Gamma) \;\cong\; \sum_{\alpha \in I_n} \Gamma(h_\alpha)\;.$$

A.3. Locally finite cellular homology with local coefficients

Locally finite cellular homology with global coefficients is defined by Geoghegan [63] where it is called *infinite cellular homology.* We need to generalize [63] to the case of local coefficients.

Let W and Γ be as in section A.2 above, but now assume that W is locally finite.

Define the nth *locally finite cellular chain module of W with local coefficients* Γ by

$$C_n^{lf}(W;\Gamma) \;=\; \prod_{\alpha \in I_n} \Gamma(h_\alpha)$$

where h_α is the characteristic map given above.

The *locally finite cellular chain complex of W with local coefficients* Γ

$$C_*^{lf}(W;\Gamma) \;=\; \{C_n^{lf}(W;\Gamma), \partial\}$$

has boundary operator induced by the boundary operator for $C_*(W;\Gamma)$ (this requires that W be locally finite) which in turn is the connecting morphism for the triple (W^n, W^{n-1}, W^{n-2}).

In Proposition A.6 below we shall prove that

$$H_n(C_*^{lf}(W;\Gamma)) \cong H_n^{lf}(W;\Gamma)$$

when W is a strongly locally finite CW complex (the module on the right is the locally finite singular homology module of W with local coefficients Γ, namely $H_n(S_*^{lf}(W;\Gamma))$).

A.4. Local $\mathbb{Z}[\pi]$-coefficients and cellular homology

Let Γ be the local system of $\mathbb{Z}[\pi]$ coefficients on the connected locally finite CW complex W as described in section A.1 above.

Note that there is a commutative diagram

$$\begin{array}{ccc} S_*(\widetilde{W}^{n-1}) & \xrightarrow{A} & S_*(W^{n-1};\Gamma) \\ \downarrow & & \downarrow \\ S_*(\widetilde{W}^n) & \xrightarrow{A} & S_*(W^n;\Gamma) \end{array}$$

where the horizontal maps are the isomorphisms of $\mathbb{Z}[\pi]$-module chain complexes given by Proposition A.1 above. We therefore have an induced isomorphism $S_*(\widetilde{W}^n, \widetilde{W}^{n-1}) \cong S_*(W^n, W^{n-1};\Gamma)$.

Proposition A.3 (i) *The cellular chain complex $C_*(\widetilde{W})$ of $\widetilde{W}$ is isomorphic as a chain complex of $\mathbb{Z}[\pi]$-modules to the cellular chain complex $C_*(W;\Gamma)$ of W with local coefficients Γ*

$$C_*(\widetilde{W}) = C_*(W;\Gamma) .$$

(ii) *The locally π-finite cellular chain complex $C_*^{lf,\pi}(\widetilde{W})$ of $\widetilde{W}$ is isomorphic as a chain complex of $\mathbb{Z}[\pi]$-modules to the locally finite cellular chain complex $C_*^{lf}(W;\Gamma)$ of W with local coefficients Γ*

$$C_*^{lf,\pi}(\widetilde{W}) = C_*^{lf}(W;\Gamma) .$$

Proof (i) Identify

$$\begin{aligned} C_n(\widetilde{W}) &= H_n(S_*(\widetilde{W}^n, \widetilde{W}^{n-1})) \\ &\cong H_n(S_*(W^n, W^{n-1};\Gamma)) = C_n(W;\Gamma) . \end{aligned}$$

Of course, one needs to check that the isomorphisms induced by A commute with the boundary operators.

(ii) These complexes are each obtained from the ordinary ones in (i) by replacing direct sums of copies of $\mathbb{Z}[\pi]$ by direct products of the same copies of $\mathbb{Z}[\pi]$. The isomorphism of the direct sums is such that it induces an isomorphism of direct products. □

A.5. The local coefficient version of Wall's intermediate chain complex

Let W be a CW complex and Γ a local system of R-modules on W. The filtration of W by its p-skeleta W^p induces a filtration of $S_*(W;\Gamma)$ by letting $F^pS_*(W;\Gamma)$ be the image of $S_*(W^p;\Gamma) \longrightarrow S_*(W;\Gamma)$ under the inclusion induced chain map. Following Wall [164, p. 130] define

$$D_p(W;\Gamma) = \ker(\partial : F^pS_p(W;\Gamma) \longrightarrow F^pS_{p-1}(W;\Gamma)/F^{p-1}S_{p-1}(W;\Gamma)) .$$

Then $D_*(W;\Gamma)$ is a subcomplex of $S_*(W;\Gamma)$ and $C_*(W;\Gamma)$ is a quotient complex of $D_*(W;\Gamma)$. The following proposition follows from the argument of Lemma 1 of [164].

Proposition A.4 *The natural chain maps*

$$S_*(W;\Gamma) \hookleftarrow D_*(W;\Gamma) \longrightarrow C_*(W;\Gamma)$$

are chain equivalences of R-module chain complexes. In particular, $S_*(W;\Gamma)$ and $C_*(W;\Gamma)$ are chain equivalent. □

A.6. The weak equivalence of locally finite singular and cellular homology

Let W be a countable, strongly locally finite CW complex and let Γ be a local system of R-modules on W. There exist a sequence of cofinite subcomplexes $W_1 \supseteq W_2 \supseteq \cdots$ such that $\bigcap\limits_i W_i = \emptyset$.

Proposition A.5 *There exist isomorphisms of R-module chain complexes*

$$\text{(i)}\ S_*^{lf}(W;\Gamma) \cong \varprojlim_i S_*(W,W_i;\Gamma) ,$$

$$\text{(ii)}\ C_*^{lf}(W;\Gamma) \cong \varprojlim_i C_*(W,W_i;\Gamma) .$$

Proof (i) is just the local coefficient version of 3.16. It is also derived by Spanier [150, Theorem 9.12].

(ii) is clear from the direct sum and direct product descriptions above. In the constant coefficient case it was also observed by Geoghegan [63]. □

Observe that for each integer n the bonding maps in each of the following inverse sequences are epimorphisms:

$$S_n(W, W_1; \Gamma) \longleftarrow S_n(W, W_2; \Gamma) \longleftarrow S_n(W, W_3; \Gamma) \longleftarrow \cdots ,$$
$$D_n(W, W_1; \Gamma) \longleftarrow D_n(W, W_2; \Gamma) \longleftarrow D_n(W, W_3; \Gamma) \longleftarrow \cdots ,$$
$$C_n(W, W_1; \Gamma) \longleftarrow C_n(W, W_2; \Gamma) \longleftarrow C_n(W, W_3; \Gamma) \longleftarrow \cdots .$$

It follows that

$$\varprojlim{}^1 S_n(W, W_i; \Gamma) = \varprojlim{}^1 D_n(W, W_i; \Gamma) = \varprojlim{}^1 C_n(W, W_i; \Gamma) = 0 ,$$

and there is a commutative diagram with exact rows

$$\begin{array}{ccccccccc}
0 \to \varprojlim{}^1 H_{n+1}(S_*(W, W_i; \Gamma)) & \to & H_n(\varprojlim S_*(W, W_i; \Gamma)) & \to & \varprojlim H_n(S_*(W, W_i; \Gamma)) \to 0 \\
\uparrow & & \uparrow & & \uparrow \\
0 \to \varprojlim{}^1 H_{n+1}(D_*(W, W_i; \Gamma)) & \to & H_n(\varprojlim D_*(W, W_i; \Gamma)) & \to & \varprojlim H_n(D_*(W, W_i; \Gamma)) \to 0 \\
\downarrow & & \downarrow & & \downarrow \\
0 \to \varprojlim{}^1 H_{n+1}(C_*(W, W_i; \Gamma)) & \to & H_n(\varprojlim C_*(W, W_i; \Gamma)) & \to & \varprojlim H_n(C_*(W, W_i; \Gamma)) \to 0
\end{array}$$

(cf. Massey [89, Appendix], Geoghegan [63], Spanier [150, p. 186]). The vertical maps on the sides are isomorphisms by Proposition A.4 above. The 5-lemma implies that the vertical maps in the middle are also isomorphisms. Define a subcomplex $D^{lf}_*(W; \Gamma)$ of the singular chain complex $S^{lf}_*(W; \Gamma)$ by

$$D^{lf}_p(W; \Gamma) = \ker(\partial : F^p S^{lf}_p(W; \Gamma) \longrightarrow F^p S^{lf}_{p-1}(W; \Gamma)/F^{p-1} S^{lf}_{p-1}(W; \Gamma)) ,$$

such that $C^{lf}_*(W; \Gamma)$ is a quotient complex of $D^{lf}_*(W; \Gamma)$. Applying Proposition A.5, we have the following analogue of 4.7:

Proposition A.6 *If W is a countable, strongly locally finite CW complex and Γ is any local system on W, then the natural chain maps*

$$S^{lf}_*(W; \Gamma) \hookleftarrow D^{lf}_*(W; \Gamma) \longrightarrow C^{lf}_*(W; \Gamma)$$

are homology equivalences of $\mathbb{Z}$-module chain complexes. In particular, $S^{lf}_*(W; \Gamma)$ and $C^{lf}_*(W; \Gamma)$ are homology equivalent and

$$H^{lf}_*(W; \Gamma) = H_*(S^{lf}_*(W; \Gamma)) = H_*(C^{lf}_*(W; \Gamma)) . \qquad \square$$

Question *Are $S^{lf}_*(W; \Gamma)$ and $C^{lf}_*(W; \Gamma)$ chain equivalent?*

Note that this is the case if the $\mathbb{Z}$-module chain complexes $S^{lf}_*(W; \Gamma)$,

$C_*^{lf}(W;\Gamma)$ are chain equivalent to free $\mathbb{Z}$-module chain complexes, e.g. if they are finitely dominated.

Define a subcomplex $D_*^{lf,\pi}(\widetilde{W})$ of the singular chain complex $S_*^{lf,\pi}(\widetilde{W})$ by

$$D_p^{lf,\pi}(\widetilde{W}) \;=\; \ker(\partial : F^p S_p^{lf,\pi}(\widetilde{W}) \longrightarrow F^p S_{p-1}^{lf,\pi}(\widetilde{W})/F^{p-1}S_{p-1}^{lf,\pi}(\widetilde{W})) \;,$$

such that $C_*^{lf,\pi}(\widetilde{W})$ is a quotient complex of $D_*^{lf,\pi}(\widetilde{W})$.

Proposition A.7 *If W is a countable, strongly locally finite CW complex and $\widetilde{W}$ is a regular cover of W with group of covering translations π, then the natural chain maps*

$$S_*^{lf,\pi}(\widetilde{W}) \;\hookleftarrow\; D_*^{lf,\pi}(\widetilde{W}) \longrightarrow C_*^{lf,\pi}(\widetilde{W})$$

are homology equivalences of $\mathbb{Z}[\pi]$-module chain complexes. In particular, $S_*^{lf,\pi}(\widetilde{W})$ and $C_*^{lf,\pi}(\widetilde{W})$ are homology equivalent and

$$H_*^{lf,\pi}(\widetilde{W}) \;=\; H_*(S_*^{lf,\pi}(\widetilde{W})) \;=\; H_*(C_*^{lf,\pi}(\widetilde{W})) \;.$$

Proof This is a special case of Proposition A.6, using Propositions A.1, A.3 to identify $S_*^{lf,\pi}(\widetilde{W}) = S_*^{lf}(W;\Gamma)$ etc. □

Appendix B. A brief history of end spaces

Path spaces are widely used in topology, notably in homotopy theory and Morse theory. They have also been long used to describe the local topological behaviour of spaces. In 1955 John Nash [102] used path spaces to give a homotopy theoretic model of the tangent space of a smooth manifold. Given a smooth manifold M and $x \in M$, Nash considered the path spaces

$$\begin{aligned} T(M,x) &= \{\omega : I \longrightarrow M \mid \omega(0) = x\,,\ \omega(t) \neq x \text{ for all } t > 0\}\ , \\ T(M) &= \bigcup_{x \in M} T(M,x) \subseteq M^I\ . \end{aligned}$$

If $\dim(M) = m$ then each $T(M,x)$ is homotopy equivalent to S^{m-1}, and the map

$$T(M) \longrightarrow M\ ;\ \omega \longrightarrow \omega(0)$$

is a fibration with fibre S^{m-1}. Nash observed that this fibration is fibre homotopy equivalent to the tangent sphere bundle of M. Consequently, the fibre homotopy type of the tangent sphere bundle of a smooth manifold depends only on the underlying topological type of the manifold. This observation constituted Nash's proof of Thom's theorem on the topological invariance of the Stiefel–Whitney classes.

Hu [69] studied the path space $T(X,x)$ where X is an arbitrary space. He used the algebraic topology of $T(X,x)$ to define the local algebraic topology of X at x. Later, Hu [70, 71] considered a related construction to model the normal, as opposed to the tangential, direction. Given a subspace $Y \subseteq X$ and a point $y \in Y$, let

$$\begin{aligned} N(X,Y,y) &= \{\omega : I \longrightarrow X \mid \omega(0) = y,\ \omega(t) \notin Y \text{ for all } t > 0\}\ , \\ N(X,Y) &= \bigcup_{y \in Y} N(X,Y,y) \subseteq X^I\ . \end{aligned}$$

In certain very special cases Hu showed that evaluation at 0 defines a fibration $N(X,Y) \longrightarrow Y$.

Fadell [43] used Hu's generalization to give a homotopy theoretic model of the normal bundle of a submanifold $P \subset M$ without assuming a smooth structure on M. In [43] it is proved that for a p-dimensional locally flat topological submanifold P of an m-dimensional topological manifold M, the evaluation at 0 defines a fibration $N(M,P) \longrightarrow P$ with fibre S^{m-p-1}. If M is a smooth manifold and P is a smooth submanifold, then $N(M,P) \longrightarrow P$ is fibre homotopy equivalent to the normal sphere bundle of P in M. Consequently, the fibre homotopy type of the normal sphere bundle depends only on the underlying topological type. In fact, every finite CW complex P can be embedded in $M = S^m$ (m large) with a regular neighbourhood $(W, \partial W)$. If $m - p \geq 3$ then P is a p-dimensional Poincaré complex if and only if the homotopy fibre of the inclusion $\partial W \longrightarrow W$ is S^{m-p-1}, in which case

$$S^{m-p-1} \longrightarrow N(S^m, P) \simeq \partial W \longrightarrow P \simeq W$$

is the normal fibration of Spivak [152], which depends only on the homotopy type of P.

The path space constructions of Nash, Hu, and Fadell gave homotopy theoretic models for tangent and normal sphere bundles for topological manifolds, even when those manifolds have no smooth structure. Later, Milnor [98] invented microbundles as a better substitute for the tangent bundle of a topological manifold. Then Kister [85] and Mazur used that theory to construct a well-defined tangent bundle (rather than just a microbundle) of a topological manifold. On the other hand, for a locally flat topological submanifold P of a topological manifold M, Rourke and Sanderson [138] showed that P need not have a normal bundle neighbourhood in M. Thus, Fadell's construction has stood as the best model for a normal bundle in the general topological setting. For more recent results in this direction, see Hughes, Taylor and Williams [78].

Quinn [116] used path space models in his work on stratified spaces. In order for a pair (X, Y) to be a homotopically stratified set in the sense of [116], it is necessary that the evaluation map $N(X,Y) \longrightarrow Y$ be a fibration, although in this generality the fibre need not be a sphere. $N(X,Y)$ is called the *homotopy link* of Y in X in [116] (see 12.11 for the definition).

The homotopy model for the behaviour at infinity of a space W is the Nash (spherical) tangent space $T(W^\infty, \infty)$ of the one-point compactification W^∞ at ∞. Of course, $T(W^\infty, \infty)$ is the same space as the Hu–Fadell normal space $N(W^\infty, \{\infty\})$. Quinn would consider W^∞ to be a stratified set with two strata, W and $\{\infty\}$. Then $T(W^\infty, \infty)$ is the homotopy link of $\{\infty\}$ in W^∞. All of these points of view yield the same object, namely the end space $e(W)$. Freedman and Quinn [60, p. 214] call $e(W)$ a *homotopy collar* of W.

A different point of view which often arises in studying the end theory of W is to focus attention on the inverse system of complements of compact subsets of W. When W can be written as an ascending union

$$K_1 \subseteq K_2 \subseteq K_3 \subseteq \ldots \subseteq \bigcup_{i=1}^{\infty} K_i = W$$

of compact subspaces, one considers the inverse sequence of inclusions

$$W \backslash K_1 \longleftarrow W \backslash K_2 \longleftarrow W \backslash K_3 \longleftarrow \ldots .$$

(This is the point of view of Porter [111], for example.) The homotopy inverse limit of this sequence is homotopy equivalent to the end space $e(W)$. This model for $e(W)$ has been exploited by Edwards and Geoghegan [39], and by Edwards and Hastings [40] for problems related to proper homotopy theory, shape theory, and end theory.

Appendix C. A brief history of wrapping up

In this appendix we use wrapping up to denote the geometric compactification procedure which passes from a non-compact space X with a proper map $X \longrightarrow \mathbb{R}^n$ to a compact space $\widehat{X}$ with a map $\widehat{X} \longrightarrow T^n$. The wrapping up of Chapter 17 is the special case $n = 1$. Wrapping up sometimes goes under the name 'belt buckle trick', 'furling' and it is a special case of the 'torus trick'. In this appendix we shall give a brief history of the development and applications of wrapping up. In the applications, it is also frequently useful to consider the passage in the reverse direction.

In 1964 M. Brown (unpublished) proved that if X, Y are compact Hausdorff spaces which are related by a homeomorphism $f : X \times \mathbb{R} \longrightarrow Y \times \mathbb{R}$ then $X \times S^1, Y \times S^1$ are related by a homeomorphism $\widehat{f} : X \times S^1 \longrightarrow Y \times S^1$. Three proofs are available in the literature: Siebenmann [142, 145] and Edwards and Kirby [41]. For manifold X, Y the result was obtained by h-cobordism theory (17.3).

'Novikov first exploited a torus *furling* idea in 1965 to prove the topological invariance of rational Pontrjagin classes. And this led to Sullivan's partial proof of the Hauptvermutung. Kirby's *unfurling* of the torus was a fresh idea that proved revolutionary.' (Siebenmann [146, footnote p. 135]).

Edwards and Kirby [41] used wrapping up to prove the local contractibility of the homeomorphism group of a compact manifold. Siebenmann [145] developed a general twist glueing construction, with wrapping up as a special case, and used it to analyse the the obstruction to fibring manifolds over S^1 (cf. Chapter 17). Chapman [25, 26] used this form of wrapping up to obtain approximation results for manifolds, such as the sucking principle (16.13). Hughes [72] developed a parametrized wrapping up, which led to the classification of manifold approximate fibrations by Hughes, Taylor and Williams [77].

Various forms and applications of geometric wrapping up have also appeared in the works of Anderson and Hsiang [2], Bryant and Pacheco [17], Burghelea, Lashof and Rothenberg [18], Ferry [53], Freedman and Quinn [60], Madsen and Rothenberg [88], Prassidis [112], Rosenberg and Weinberger [137], Steinberger and West [158], Weinberger [166], Weiss and Williams [167], ...

References

[1] E. Akin, "Manifold phenomena in the theory of polyhedra", Trans. A.M.S. **143**, 413–473 (1969)

[2] D. R. Anderson and W. C. Hsiang, "The functors K_{-i} and pseudoisotopies of polyhedra", Ann. of Maths. **105**, 201–223 (1977)

[3] W. Ballmann, M. Gromov and V. Schroeder, *Manifolds of Nonpositive Curvature*, Progress in Mathematics, Vol. **61**, Birkhäuser (1985)

[4] H. Bass, *Algebraic K-theory*, Benjamin (1968)

[5] ––, A. Heller and R. Swan, "The Whitehead group of a polynomial extension", Publ. Math. I. H. E. S. **22**, 61–80 (1964)

[6] R. Benedetti and C. Petronio, *Lectures on Hyperbolic Geometry*, Universitext, Springer (1992)

[7] K. Borsuk, *Theory of Shape*, PWN, Polish Scientific Publishers (1975)

[8] N. Bourbaki, *General Topology*, Part 1, Addison–Wesley (1966)

[9] A. K. Bousfield and D. M. Kan, *Homotopy limits, completions and localizations*, Lecture Notes in Mathematics **304**, Springer (1972)

[10] M. G. Brin and T. L. Thickstun, "On the proper Steenrod homotopy groups, and proper embeddings of planes into 3-manifolds", Trans. A.M.S. **289**, 737–755 (1985)

[11] –– and ––, *3-manifolds which are end 1-movable*, Mem. A.M.S. **411** (1989)

[12] W. Browder, "Structures on $M \times \mathbb{R}$", Proc. Camb. Phil. Soc. **61**, 337–345 (1965)

[13] –– and J. Levine, "Fibering manifolds over the circle", Comm. Math. Helv. **40**, 153–160 (1966)

[14] ––, –– and G. R. Livesay, "Finding a boundary for an open manifold", Amer. J. Math. **87**, 1017–1028 (1965)

[15] –– and F. Quinn, "A surgery theory for G-manifolds and stratified sets", Manifolds–Tokyo 1973, 27–36, Univ. of Tokyo Press (1975)

[16] M. Brown, "Locally flat embeddings of topological manifolds", Proc. 1961 Georgia Conf. on the Topology of 3-manifolds and related topics, 83–91, Prentice–Hall (1962)

[17] J. L. Bryant and P. S. Pacheco, "K_{-i} obstructions to factoring an open manifold", Topology and its Applications **29**, 107–139 (1988)

[18] D. Burghelea, R. Lashof and M. Rothenberg, *Groups of Automorphisms of Manifolds*, Lecture Notes in Mathematics **473**, Springer (1975)

[19] S. Cappell, "A splitting theorem for manifolds", Inv. Math. **33**, 69–170 (1976)

[20] -- and J. Shaneson, "The codimension two placement problem, and homology equivalent manifolds", Ann. of Maths. **99**, 277–348 (1974)

[21] G. Carlsson and E. K. Pedersen, "Controlled algebra and the Novikov conjectures for K- and L-theory", Topology **34**, 731–758 (1995)

[22] T. A. Chapman, "Cell-like mappings of Hilbert cube manifolds: applications to simple homotopy theory", Bull. A.M.S. **79**, 1286–1291 (1973)

[23] --, *Lectures on Hilbert Cube Manifolds*, C.B.M.S. Regional Conf. Series in Mathematics **28**, A.M.S. (1976)

[24] --, "Simple homotopy theory for ANR's", General Topology and its Applications **7**, 165–174 (1977)

[25] --, "Approximation results in Hilbert cube manifolds", Trans. A.M.S. **262**, 303–334 (1980)

[26] --, *Approximation results in topological manifolds*, Mem. A.M.S. **251** (1981)

[27] -- and S. Ferry, "Hurewicz fiber maps with ANR fibers", Topology **16**, 131–143 (1977)

[28] -- and --, "Fibering Hilbert cube manifolds over ANRs", Comp. Math. **36**, 7–35 (1978)

[29] -- and --, "Approximating homotopy equivalences by homeomorphisms", Amer. J. Math. **101**, 583–607 (1979)

[30] M. Cohen, *A Course in Simple Homotopy Theory*, Graduate Texts in Mathematics, Vol. **10**, Springer (1973)

[31] R. Conelly, "A new proof of Brown's collaring theorem", Proc. A.M.S. **27**, 180–182 (1971)

[32] E. H. Connell, "A topological h-cobordism theorem", Ill. J. Math. **11**, 300–309 (1967)

[33] A. Connes, *Noncommutative geometry*, Academic Press (1994)

[34] F. X. Connolly and B. Vajiac, "An end theorem for stratified spaces" (preprint)

[35] D. S. Coram, "Approximate fibrations – a geometric perspective", Shape Theory and Geometric Topology, Proc. 1980 Dubrovnik Conf., Lecture Notes in Mathematics **870**, 37–47, Springer (1981)

[36] -- and P. F. Duvall, "Approximate fibrations", Rocky Mountain J. Math. **7**, 275–288 (1977)

[37] R. Diestel, "The end structure of a graph: recent results and open problems", Discrete Math. **100**, 313–327 (1992)

[38] J. Dugundji, *Topology*, Allyn and Bacon (1966)

[39] D. A. Edwards and R. Geoghegan, "Shapes of complexes, ends of manifolds, homotopy limits and the Wall obstruction", Ann. of Maths. **101**, 521–535 (1975)

[40] -- and H. M. Hastings, "Čech and Steenrod Homotopy Theories with Applications to Geometric Topology", Lecture Notes in Mathematics **542**, Springer (1976)

[41] R. D. Edwards and R. C. Kirby, "Deformations of spaces of imbeddings", Ann. of Maths. **93**, 63–88 (1971)

[42] D. B. A. Epstein, "Ends", Proc. 1961 Georgia Conf. on the Topology of 3-manifolds and related topics, 110–117, Prentice–Hall (1962)

[43] E. Fadell, "Generalized normal bundles for locally-flat embeddings", Trans. A.M.S. **114**, 488–513 (1965)

[44] M. Sh. Farber, "Mappings into a circle with minimal number of critical points, and multidimensional knots", Dokl. Akad. Nauk S. S. S. R. **279**, 43–46 (1984)

[45] --, "Exactness of the Novikov inequalities", Functional Analysis and its Applications **19**, 49–59 (1985)

[46] F. T. Farrell, "The obstruction to fibering a manifold over a circle", Proc. 1970 I.C.M. Nice, Vol. 2, 69–72 (1971)

[47] --, "The obstruction to fibering a manifold over a circle", Indiana Univ. J. **21**, 315–346 (1971)

[48] -- and W. C. Hsiang, "A formula for $K_1(R_\alpha[T])$", Proc. Symp. A.M.S. **17**, 192–218 (1970)

[49] -- and --, "Manifolds with $\pi_1 = G \times_\alpha T$", Amer. J. Math. **95**, 813–845 (1973)

[50] -- and L. Jones, *Classical aspherical manifolds*, C.B.M.S. Regional Conf. Series in Mathematics **75**, A.M.S. (1990)

[51] --, L. Taylor and J. Wagoner, "The Whitehead theorem in the proper category", Compositio Math. **27**, 1–23 (1973)

[52] -- and J. Wagoner, "Algebraic torsion for infinite simple homotopy types", Comm. Math. Helv. **47**, 502–513 (1972)

[53] S. Ferry, "Approximate fibrations with nonfinite fibres", Proc. A.M.S. **64**, 335–345 (1977)

[54] --, "Homotoping ϵ-maps to homeomorphisms", Amer. J. Math. **101**, 567–582 (1979)

[55] --, "A simple homotopy approach to finiteness obstruction theory", Shape Theory and Geometric Topology, Proc. 1980 Dubrovnik Conf., Lecture Notes in Mathematics **870**, 73–81, Springer (1981)

[56] --, "Remarks on Steenrod homology", Proc. 1993 Oberwolfach Conf. on the Novikov Conjectures, and Index Theorems and Rigidity, Vol. 2, L.M.S. Lecture Notes **227**, 148–166, Cambridge (1995)

[57] -- and E. K. Pedersen, "Some mildly wild circles in S^n arising from algebraic K-theory", K-theory **4**, 479–499 (1991)

[58] –– and ––, "Epsilon surgery theory", Proc. 1993 Oberwolfach Conf. on the Novikov Conjectures, Index Theorems and Rigidity, Vol. 2, L.M.S. Lecture Notes **227**, 167–226, Cambridge (1995)

[59] ––, A. A. Ranicki and J. Rosenberg, "A history and survey of the Novikov conjecture", Proc. 1993 Oberwolfach Conf. on the Novikov Conjectures, Index Theorems and Rigidity, Vol. 1, L.M.S. Lecture Notes **226**, 7–66, Cambridge (1995)

[60] M. Freedman and F. Quinn, *The topology of 4-manifolds*, Princeton (1990)

[61] H. Freudenthal, "Über die Enden topologischer Räume und Gruppen", Math. Zeit. **33**, 692–713 (1931)

[62] R. Geoghegan, "A note on the vanishing of $\lim^1$", J. Pure App. Alg. **17**, 113–116 (1980)

[63] ––, *Topological Methods in Group Theory*, (to appear)

[64] V. L. Golo, "On an invariant of open manifolds", Izv. Akad. Nauk SSSR, ser. mat. **31**, 1091–1104 (1967), English translation: Math. USSR - Izvestija **1**, 1041–1054 (1967)

[65] M. Gromov, "Positive curvature, macroscopic dimension, spectral gaps and higher signatures", Functional Analysis on the Eve of the 21st Century: In Honor of I. M. Gelfand's 80th Birthday, Vol. 2, Progress in Mathematics **132**, 1–213, Birkhäuser (1996)

[66] J. G. Hocking and G. S. Young, *Topology*, Addison–Wesley (1961), Dover (1988)

[67] L. Hofer, *Wall's endelighedsobstruktion for endeligt dominerede CW-komplekser*, Odense Master's thesis (1983)

[68] H. Hopf, "Enden offener Räume und unendliche diskontinuierliche Gruppen", Comm. Math. Helv. **16**, 81–100 (1944)

[69] S. T. Hu, "Algebraic local invariants of topological spaces", Compositio Math. **13**, 173–218 (1958)

[70] ––, "Isotopy invariants of topological spaces", Proc. Roy. Soc. Lond. A **255**, 331–366 (1960)

[71] ––, "Fibrings of enveloping spaces", Proc. L.M.S. (3) **11**, 691–707 (1961)

[72] B. Hughes, "Approximate fibrations on topological manifolds", Michigan Math. J. **32**, 167–183 (1985)

[73] ––, "Controlled homotopy topological structures", Pacific J. Math. **133**, 69–97 (1988)

[74] ––, "Geometric topology of stratified spaces", (to appear)

[75] –– and S. Prassidis, "Control and relaxation over the circle", (to appear)

[76] ––, L. Taylor, S. Weinberger and B. Williams, "Neighborhoods in stratified spaces with two strata", (to appear)

[77] ––, –– and B. Williams, "Bundle theories for topological manifolds", Trans. A.M.S. **319**, 1–65 (1990)

[78] ––, –– and ––, "Manifold approximate fibrations are approximately bundles", Forum Math. **3**, 309–325 (1991)

[79] ––, –– and ––, "Bounded homeomorphisms over Hadamard manifolds", Math. Scand. **73**, 161–176 (1993)

[80] D. S. Kahn, J. Kaminker and C. Schochet, "Generalized homology theories on compact metric spaces", Michigan Math. J. **24**, 203–224 (1977)

[81] J. Kaminker and C. Schochet, "K-theory and Steenrod homology: applications to the Brown–Douglas–Fillmore theory of operator algebras", Trans. A.M.S. **227**, 63–107 (1977)

[82] I. Kaplansky, *Infinite Abelian Groups*, Michigan (1954)

[83] M. Kervaire, "Le théorème de Barden–Mazur–Stallings", Comm. Math. Helv. **40**, 31–42 (1965)

[84] R. Kirby and L. C. Siebenmann, *Foundational essays on topological manifolds, smoothings, and triangulations*, Ann. of Maths. Stud. **88**, Princeton (1977)

[85] J. Kister, "Microbundles are fibre bundles", Ann. of Maths. **80**, 190–199 (1964)

[86] E. Laitinen, "End homology and duality", Forum Math. **8**, 121–133 (1996)

[87] W. Lück and A. A. Ranicki, "Chain homotopy projections", J. of Algebra **120**, 361–391 (1989)

[88] I. Madsen and M. Rothenberg, "On the classification of G-spheres III.", (to appear)

[89] W. S. Massey, *Homology and cohomology: an approach based on Alexander–Spanier cochains*, Marcel Dekker (1978)

[90] J. Mather, *Notes on Topological Stability*, Harvard Univ. (1970)

[91] M. Mather, "Counting homotopy types of manifolds", Topology **4**, 93–94 (1965)

[92] S. Maumary, "Proper surgery groups and Wall–Novikov groups", Proc. 1972 Battelle Seattle Conf. on Algebraic K-theory, Vol. III, Lecture Notes in Mathematics **343**, 526–539, Springer (1973)

[93] M. Mihalik, "Semistability at the end of a group extension", Trans. A.M.S. **277**, 307–321 (1983)

[94] –– and S. Tschantz, *Semistability of amalgamated free products and HNN-extensions*, Mem. A.M.S. **471** (1992)

[95] J. Milnor, "On the Steenrod homology theory", (originally issued as a preprint, 1961) Proc. 1993 Oberwolfach Conf. on the Novikov Conjectures, Index Theorems and Rigidity, Vol. 1, L.M.S. Lecture Notes **226**, 79–96, Cambridge (1995)

[96] ––, "Two complexes which are homeomorphic but combinatorially distinct", Ann. of Maths. **74**, 575–590 (1961)

[97] ––, "On axiomatic homology theory", Pacific J. Math. **12**, 337–341 (1962)

[98] ––, "Microbundles, Part I", Topology **3** (Suppl. **1**), 53–88 (1964)
[99] ––, "Whitehead torsion", Bull. A.M.S. **72**, 358–426 (1966)
[100] ––, "Infinite cyclic covers", Proc. 1967 Conf. on the Topology of Manifolds, 115–133, Prindle, Weber and Schmidt (1968)
[101] J. Munkres, *Elements of Algebraic Topology*, Benjamin/Cummings (1974)
[102] J. Nash, "A path space and the Stiefel–Whitney classes", Proc. N.A.S. **41**, 320–321 (1955)
[103] S. P. Novikov, "Rational Pontrjagin classes. Homeomorphism and homotopy type of closed manifolds I.", Izv. Akad. Nauk SSSR, ser. mat. **29**, 1373–1388 (1965), English translation: A.M.S. Transl. (2) **66**, 214–230 (1968)
[104] ––, "Manifolds with free abelian fundamental group and applications (Pontrjagin classes, smoothings, high dimensional knots)", Izv. Akad. Nauk SSSR, ser. mat. **30**, 207–246 (1966), English translation: A.M.S. Transl. (2) **67**, 1–42 (1969)
[105] ––, "The algebraic construction and properties of hermitian analogues of K-theory for rings with involution, from the point of view of the hamiltonian formalism. Some applications to differential topology and the theory of characteristic classes", Izv. Akad. Nauk SSSR, ser. mat. **34**, 253–288, 478–500 (1970), English translation: Math. USSR–Izv. **4**, 257–292, 479–505 (1970)
[106] ––, "The hamiltonian formalism and a multivalued analogue of Morse theory", Uspekhi Mat. Nauk **37**, 3–49 (1982), English translation: Russ. Math. Surv. **37**, 1–56 (1982)
[107] C. D. Papakyriakopoulos, "On Dehn's lemma and the asphericity of knots", Ann. of Maths. **66**, 1–26 (1957)
[108] A. V. Pazhitnov, "Surgery on the Novikov complex", K-theory (to appear)
[109] E. K. Pedersen and A. A. Ranicki, "Projective surgery theory", Topology **19**, 239–254 (1980)
[110] –– and C. Weibel, "A nonconnective delooping of algebraic K-theory", Algebraic and Geometric Topology, Proc. 1983 Rutgers Conf., Lecture Notes in Mathematics **1126**, 166–181, Springer (1985)
[111] T. Porter, "Proper homotopy theory", Handbook of Algebraic Topology, 127–167, Elsevier Science (1995)
[112] S. Prassidis, "The Bass–Heller–Swan formula for the equivariant topological Whitehead group", K-theory **5**, 395–448 (1992)
[113] F. Quinn, "Open book decompositions, and the bordism of automorphisms", Topology **18**, 55–73 (1979)
[114] ––, "Ends of maps I.", Ann. of Maths. **110**, 275–331 (1979)
[115] ––, "Ends of maps II.", Inv. Math. **68**, 353–424 (1982)
[116] ––, "Homotopically stratified sets", J. A.M.S. **1**, 441–499 (1988)

[117] A. A. Ranicki, "Algebraic L-theory I. Foundations", Proc. L.M.S. (3) **27**, 101–125 (1973)

[118] ––, "Algebraic L-theory II. Laurent extensions", Proc. L.M.S. (3) **27**, 126–158 (1973)

[119] ––, *Exact sequences in the algebraic theory of surgery*, Mathematical Notes **26**, Princeton (1981)

[120] ––, "The algebraic theory of finiteness obstruction", Math. Scand. **57**, 105–126 (1985)

[121] ––, "The algebraic theory of torsion I. Foundations", Algebraic and Geometric Topology, Proc. 1983 Rutgers Conf., Lecture Notes in Mathematics **1126**, 199–237, Springer (1985)

[122] ––, "Algebraic and geometric splittings of the K- and L-groups of polynomial extensions", Transformation Groups, Proc. 1985 Poznań Conf., Lecture Notes in Mathematics **1217**, 321–363, Springer (1986)

[123] ––, "The algebraic theory of torsion II. Products", K-theory **1**, 115–170 (1987)

[124] ––, *Lower K- and L-theory*, L.M.S. Lecture Notes **178**, Cambridge (1992)

[125] ––, *Algebraic L-theory and topological manifolds*, Tracts in Mathematics **102**, Cambridge (1992)

[126] ––, "Finite domination and Novikov rings", Topology **34**, 619–632 (1995)

[127] ––, "On the Novikov conjecture", Proc. 1993 Oberwolfach Conf. on the Novikov Conjectures, Index Theorems and Rigidity, Vol. 1, L.M.S. Lecture Notes **226**, 272–337, Cambridge (1995)

[128] ––, "The bordism of automorphisms of manifolds from the algebraic L-theory point of view", Prospects in Topology; proceedings of a conference in honor of William Browder, Ann. of Maths. Studies **138**, 314–327, Princeton (1995)

[129] ––, "An explicit projection", available electronically on WWW from http://www.maths.ed.ac.uk/people/aar (preprint)

[130] ––, *The algebraic theory of bands*, (to appear)

[131] –– (ed.), *The Hauptvermutung book*, A collection of papers on the topology of manifolds by A. J. Casson, D. P. Sullivan, M. A. Armstrong, C. P. Rourke, G. E. Cooke and A. A. Ranicki, K-monographs in Mathematics **1**, Kluwer (1996)

[132] –– and M. Yamasaki, "Controlled K-theory", Topology and its Applications **61**, 1–59 (1995)

[133] J. G. Ratcliffe, *Foundations of Hyperbolic Manifolds*, Graduate Texts in Mathematics, Vol. **149**, Springer (1994)

[134] F. Raymond, "The end point compactification of manifolds", Pacific J. Math. **10**, 947–963 (1960)

[135] J. Roe, *Index theory, coarse geometry and topology of manifolds*, C.B.M.S. Regional Conf. Series in Mathematics, A.M.S. (to appear)

[136] J. Rosenberg, *Algebraic K-theory and its Applications*, Graduate Texts in Mathematics, Vol. **147**, Springer (1994)

[137] -- and S. Weinberger, "Higher G-signatures for Lipschitz Manifolds", K-theory **7**, 101–132 (1993)

[138] C. P. Rourke and B. J. Sanderson, "On topological regular neighbourhoods", Compositio Math. **22**, 387–425 (1970)

[139] -- and --, *Introduction to piecewise-linear topology*, Ergebnisse der Mathematik und ihrer Grenzgebiete, Vol. **69**, Springer (1972)

[140] L. Siebenmann, *The obstruction to finding the boundary of an open manifold of dimension greater than five*, Princeton Ph.D. thesis (1965)

[141] --, "The structure of tame ends", Notices A.M.S. **66T–G7**, 861, (1966)

[142] --, "Pseudo-annuli and invertible cobordisms", Arch. Math. **19**, 528–535 (1968)

[143] --, "A torsion invariant for bands", Notices A.M.S. **68T–G7**, 811 (1968)

[144] --, "Infinite simple homotopy theory", Indag. Math. **32**, 479–495 (1970)

[145] --, "A total Whitehead torsion obstruction to fibering over the circle", Comm. Math. Helv. **45**, 1–48 (1970)

[146] --, "Topological manifolds", Proc. 1970 Nice I. C. M., Vol. 2, 133–163, Gauthier-Villars (1971) (reprinted in [84], 307–337)

[147] --, "Deformations of homeomorphisms on stratified sets", Comm. Math. Helv. **47**, 123–165 (1971)

[148] --, "Regular (or canonical) open neighbourhoods", General Topology and its Applications **3**, 51–61 (1973)

[149] --, L. Guillou and H. Hähl, "Les voisinages ouverts réguliers : critéres homotopiques d'existence", Ann. scient. Éc. norm. Sup. (4) **7**, 431–462 (1974)

[150] E. Spanier, "Singular homology and cohomology with local coefficients, and duality for manifolds", Pacific J. Math. **160**, 165–200 (1993)

[151] E. Specker, "Die erste Cohomologiegruppe von Überlagerungen und Homotopie-Eigenschaften drei dimensionaler Mannigfaltigkeiten", Comm. Math. Helv. **23**, 303–333 (1949)

[152] M. Spivak, "Spaces satisfying Poincaré duality", Topology **6**, 77–101 (1967)

[153] J. Stallings, "On fibering certain 3-manifolds", Proc. 1961 Georgia Conf. on the Topology of 3-manifolds and related topics, 95–100, Prentice–Hall (1962)

[154] --, "On the piecewise linear structure of euclidean space", Proc. Camb. Phil. Soc. **58**, 481–488 (1962)

[155] – –, "On infinite processes leading to differentiability in the complement of a point", Differential and Combinatorial Topology, 245–254, Princeton (1965)

[156] N. E. Steenrod, "Regular cycles of compact metric spaces", Ann. of Maths. **41**, 833–851 (1940)

[157] – –, "Homology with local coefficients", Ann. of Maths. **44**, 610–627 (1943)

[158] M. Steinberger and J. West, "Approximation by equivariant homeomorphisms. I", Trans. A.M.S. **302**, 297–317 (1987)

[159] D. Stone, *Stratified polyhedra*, Lecture Notes in Mathematics **252**, Springer (1972)

[160] L. Taylor, *Surgery on paracompact manifolds*, Berkeley Ph.D. thesis (1972)

[161] R. Thom, "Ensembles et morphismes stratifiés", Bull. A.M.S. **75**, 240–284 (1969)

[162] G. S. Ungar, "Conditions for a mapping to have the slicing structure property", Pacific J. Math. **30**, 549–553 (1969)

[163] C. T. C. Wall, "Finiteness conditions for CW complexes", Ann. of Maths. **81**, 56–69 (1965)

[164] – –, "Finiteness conditions for CW complexes II.", Proc. Roy. Soc. London Ser. A **295**, 129–139 (1965)

[165] – –, *Surgery on compact manifolds*, Academic Press (1970)

[166] S. Weinberger, *The topological classification of stratified spaces*, Chicago (1994)

[167] M. Weiss and B. Williams, "Automorphisms of manifolds and algebraic K-theory: I", K-theory **1**, 575–626 (1988)

[168] J. E. West, "Mapping Hilbert cube manifolds to ANR's", Ann. of Maths. **106**, 1–18 (1977)

[169] G. W. Whitehead, *Elements of Homotopy Theory*, Graduate Texts in Mathematics, Vol. **61**, Springer (1978)

[170] H. Whitney, "Elementary structure of real algebraic varieties", Ann. of Maths. **66**, 545–566 (1957)

Index